ESSAYS IN STRUCTURAL CHEMISTRY

Dr L. A. Woodward

Essays in Structural Chemistry

Edited by

A. J. Downs—*Inorganic Chemistry Laboratory, University of Oxford*
D. A. Long—*School of Chemistry, University of Bradford*
L. A. K. Staveley—*Inorganic Chemistry Laboratory, University of Oxford*

PLENUM PRESS
NEW YORK

First published 1971

U.S. edition published by Plenum Press, a division of Plenum Publishing Corporation, 227 West 17th Street, New York, N.Y. 10011.

SBN 306 30525-9

Library of Congress Catalog Card Number 76-144136

Printed in Great Britain

These essays are presented to

L. A. WOODWARD

on the occasion of his retirement

by some of his colleagues and former students

Contents

Table for conversion to S.I. Units

Physical quantity	Unit used in this book	S.I. unit	Conversion factor (multiply by this factor to obtain value in S.I. unit)
wavelength	ångström	nanometer (nm)	10^{-1}
frequency	reciprocal second	hertz (Hz)	1
wave number	reciprocal centimetre	reciprocal metre (m^{-1})	10^2
force constant	millidyne per ångström	newton per metre ($N\ m^{-1}$)	10^2
energy	kilocalorie	joule (J)	$4{\cdot}184 \times 10^3$
	electron volt	joule per molecule ($J\ mole^{-1}$)	$1{\cdot}602 \times 10^{-19}$
	electron volt per mole	joule per mole ($J\ mol^{-1}$)	$9{\cdot}648 \times 10^4$
electric field strength	e.s.u. cm^{-2}	volt per metre ($V\ m^{-1}$)	$2{\cdot}998 \times 10^4$
pressure	atmosphere	newton per square metre ($N\ m^{-2}$)	$1{\cdot}013 \times 10^5$
magnetic flux density	gauss	tesla (T)	10^{-4}
permittivity (ϵ_0)	$4\pi\epsilon_0 = 1$	farad per metre ($F\ m^{-1}$)	$1{\cdot}113 \times 10^{-10}$
molecular density	amagat	moles per cubic metre ($mole\ m^{-3}$)	44·617

Editorial Foreword

This book deals with selected aspects of structural chemistry, concentrating particularly on molecular and Raman spectroscopy. The authors of the various chapters were chosen from friends, colleagues and past students of Len Woodward. It is our hope that the book will prove useful both to honours students and to research workers.

We would like to thank all our contributors for their willing cooperation in this endeavour. We are also grateful to all those who have given permission for the reproduction of copyright material from other publications; specific acknowledgments are made in each chapter. We are particularly indebted to the Principal and Fellows of Jesus College, Oxford, and the artist, H. A. Freeth, R.A., for permission to reproduce the portrait of Len Woodward which forms the frontispiece. Our thanks are also due to Mrs. J. Stevenson, who undertook a great deal of the secretarial work associated with the organization of this volume, and to Mr. P. Espé who photographed the portrait.

The royalties from the sale of this book will, in the first instance, go to Jesus College, Oxford, and will be used for the establishment of a prize to be associated with Len Woodward's name.

Finally, the vexed matter of units. Many physical scientists would plead utility as much as mere conservatism for the retention of such internationally recognized units as the angstrom, the reciprocal centimetre and the calorie. We have preferred, on utilitarian grounds, the familiar to the novel, however rational, and have not adopted S.I. units. However, a conversion table is appended.

A. J. Downs
D. A. Long
L. A. K. Staveley

Preface

This volume to mark the retirement of Leonard Ary Woodward is a measure of the affection and respect his colleagues and former pupils hold for him.

Len Woodward came up from Maidenhead School to Lincoln College, Oxford, in 1922 having won an Open Scholarship in Chemistry. Here he took Mathematical Moderations, an examination unusual for chemists at that time, distinguished himself by winning the Gibbs' Scholarship in Chemistry, and finally graduated with First Class Honours in the Chemistry School. This was followed by his appointment as a junior demonstrator in the laboratory. In 1928 he was awarded a D.S.I.R. Senior Research Grant and chose to work at Leipzig under the fertile direction of Professor P. Debye. Here was a stimulating atmosphere with newly-fermenting ideas on molecular polarities, polarizabilities and oscillations. Following Robert Herrick's vision

> 'Now, when I cast mine eyes and see
> That brave vibration each way free;
> Oh, how that glittering taketh me.'

he was attracted by the recently discovered Raman effect, and in 1931, in the main subject of Physics with subsidiaries of Chemistry and the Theory of Functions, together with a thesis on 'Raman spectra of electrolyte solutions', he was awarded his doctorate with the distinction *summa cum laude*.

His first published work, a paper on the Raman effect of nitric acid in solution, appeared in *Nature*[1] in 1930; some of his results for this system were also contained in a paper by Meyer concerning the Raman spectra of aqueous media.[2] The exploitation of the Raman effect for the study of strong electrolytes and of electrolytic dissociation was the dominant theme of subsequent publications[3–8] in the period 1930–34; one of the most striking of these described the now classical experiment demonstrating the binuclear nature of the mercurous ion.[8] In 1930 two papers appeared in the *Proceedings of the Royal Society* under the names of Woodward and Sidgwick describing spectrometric investigations of indicators[9] and of the effect of a neutral salt on the dissociation of acetic acid.[10] Details of technical advances relevant to the practice of Raman spectroscopy were published in 1934–35.[11, 12] His complete mastery of the principles and practice of Raman spectroscopy is shown by his lucid article on this subject in the *Annual Reports of the Chemical Society* for 1934,[13] an article of the greatest value to chemists at that time and still useful for students. Two short excerpts from this account give some impression of the clarity and aptness of language which have always been characteristic features of his scientific writing. Thus in referring to the individuality of Raman scattering as a molecular property, he says, 'Each different scattering molecular species gives its own characteristic Raman spectrum—writes (as it were) its autograph on the photographic plate.' Later on this same imagery is deftly taken up again with the sentence, 'In the above applications, the "molecular autograph" is used merely for identification purposes, and no attempt is made to deduce character from the handwriting.'

His return to England, now married to Jane McCall Donaldson, coincided with the period of economic depression, and posts were hard to find. For two years he was Massey Research Fellow at Nottingham, moving then to be assistant lecturer in the Physics Department of Manchester University, followed by appointment as Scientific Officer at the Fuel Research Station of D.S.I.R. Even during this difficult period up to 1939, with H. H. Barber he carried out studies of blood-serum lipoids,[14] while with F. S. Sinnatt,[15] C. M. Cawley[16, 17] and E. F. G. Herington[18] he published accounts of the work of the Fuel Research Station, which included an exploration of hydrogenation reactions.

Only in 1939 was he able to return to his chosen field of Raman spectroscopy, which, with H. J. V. Tyrrell, he applied to aqueous solutions of potassium hydrogen fluoride and hydrofluoric acid.[19] This was made possible by his election to an Official Fellowship and Tutorship at Jesus College, Oxford. Here, just across the road from his old college, he found congenial surroundings as colleague of D. L. Chapman. The war, however, brought Jesus College Laboratory into the national effort, and a team of chemists, directed by Chapman, Woodward and G. O. Jones, undertook secret work for the Tube Alloy Programme involving chemical and physical studies of diffusion membranes and their resistance to corrosive vapours.

On Chapman's retirement in 1944 Woodward took charge of the Leoline Jenkins' Laboratories, where some research on Raman spectroscopy had been undertaken in the early days of the war. This activity was resumed after the war, and continued in the College laboratories until they were closed in 1947, when the work was transferred to the University laboratories. The facilities available for spectroscopy in the College laboratories were indeed primitive by modern standards. Those who worked with Woodward at that time recall that the construction of a quite modest light source depended on the resourcefulness of the College carpenter, the collaboration of the Morris radiator factory and the help of the glassblower from the Clarendon laboratory.

In 1947 Woodward was appointed Demonstrator in the Inorganic Chemistry Laboratory of the University, and here he was now able to build up adequate facilities for spectroscopic research. In 1966 his services in the Laboratory were marked by his election to the Aldrichian Praelectorship in Chemistry.

Apart from some early studies of the kinetics of the base-catalysed bromination of ethyl cyclopentanone-2-carboxylate[20] and of molybdenum oxide catalysts for the reduction of phenol to benzene,[21] the development and application of the Raman spectrometer became the abiding research interest. Thus, 1949 saw the publication of accounts of an effective recording spectrometer for Raman spectroscopy[22] and of some of the first reliable evaluations of mean derived polarizabilities based on observed relative intensities in the Raman spectra of some Group IV tetrahalides;[23] the analytical applications of the Raman effect were also briefly reviewed.[24] In the following year a paper[25] drew attention to the interesting and still unexploited possibility of inducing vibrational and rotational absorption in a molecule by the application of an electrostatic field. The influence of refraction on the relative intensities of Raman lines[26] and on the total light flux received by a Raman spectrograph[27] was investigated with J. H. B. George. At about this time there appeared two papers on the Raman and infrared spectra of germanium tetrafluoride,[28,29] first of a notable series of papers on the vibrational spectra of simple compounds; in the period up to 1956 the following compounds were examined in this way: selenium tetrafluoride,[30] selenium oxyfluoride,[31] trifluorochloroethylene,[32] nitryl fluoride,[33] osmium tetroxide,[34] sulphur tetrafluoride,[35] fused gallium dichloride[36] and phosphonium iodide.[37] The presentation of the spectroscopic evidence that the sulphur tetrafluoride molecule has a structure with C_{2v} symmetry[35] remains to this day a model of how such delicate scientific arguments should proceed. Studies on the intensities of Raman lines were extended to the symmetrical neopentane molecule.[38,39] Another important advance started at about this time was the use of the Raman effect to explore the phenomena of ion-pairing and complex-formation in solution. Beginning with another classical paper about the vibrational Raman effect of an electrostatically bound ion-pair, of relevance to the nature of the undissociated part of thallium(I) hydroxide in solution,[40] a series of papers described the development of this research area to include studies of the complexity of the silver(I) ion in solution,[41] of complex halide anions of cadmium(II),[42] mercury(II),[42] gallium(III)[43, 44] and indium(III),[45] of the BH_4^- ion in liquid ammonia,[46] and of the PF_6^- ion in aqueous solution.[47] The growing significance of the Raman effect as a chemical technique was made plain in a most convincing review article that appeared in 1956.[48] At about this time the Oxford laboratory had become under Woodward's direction internationally renowned for its work on Raman spectroscopy.

Subsequent research proceeded along the following main lines:

(i) Technical improvements were made, as in the design of new sources for the excitation of Raman spectra.[49]

(ii) Various compounds were subjected to spectroscopic examination as a guide principally to their molecular constitution and stereochemistry: such compounds were trifluoroacetaldehyde,[50] disulphur and ditellurium decafluorides,[51, 52] tetrakis(trifluorophosphine) derivatives of nickel,[53, 54] palladium[55] and platinum,[55] fused gallium dibromide,[56] trimethylboron,[57] -gallium[58] and -indium,[58] dimethylmercury,[59] methylmercuric derivatives,[60] pentafluorosulphur chloride,[61] thionyl tetrafluoride,[62] vanadium tetrachloride[63] and phosphorus pentachloride.[64] A fruitful collaboration with members of the Inorganic Chemistry Department at Cambridge, notably E. A. V. Ebsworth and N. Sheppard, led to an extensive study of the vibrational spectra of volatile silicon–hydrogen compounds: in this way some of the first evidence of a planar Si_3N skeleton in trisilylamine was deduced.[65] A most elegant experiment, described in 1959,[66] demonstrated unequivocally by the substitution of ^{18}O for ^{16}O that the Si—O—Si skeleton of disilyl ether is bent, notwithstanding the earlier confusion presented by the vibrational spectra of this compound. Other silicon– and germanium–hydrogen compounds investigated at Oxford were disilyl sulphide and selenide,[67] tetrasilylhydrazine,[68] monomethyl-,[69] dimethyl-[70] and trimethylsilane,[70] silyl isothiocyanate,[71] trisilylphosphine,[72, 73] trisilyl-arsine and -stibine,[74] methyldisilylamines,[75] trigermylphosphine,[76] and germyl isothiocyanate.[77]

(iii) The measured intensity of Raman scattering was interpreted for tetrahalides[78, 79] and tetramethyls[80, 81] of the Group IV elements. Some problems touching on vibrational amplitudes[79] and molecular force fields (see, for example, references 52, 79 and 82) of these and similar systems were also explored; more recently some important general features of the machine-computation of force constants have been delineated.[83]

(iv) The characterization by Raman spectroscopy of complex species in solution grew apace to include $[SnBr_6]^{2-}$,[84] $[SnCl_3]^-$ and $[SnBr_3]^-$,[85] $[PbCl_6]^{2-}$,[86] $[GaI_4]^-$ and $[InI_4]^-$,[87] $[InCl_4]^-$ and $[FeCl_4]^-$,[88] transient chloride and bromide complexes believed to contain gallium(I),[89] $Hg(CN)_2$ and $[Hg(CN)_4]^{2-}$,[90] $[CH_3Hg\cdot OH_2]^+$ and CH_3HgONO_2,[91, 92] $[HgX]^+$ (X = Cl, Br or I) as well as a species thought to be $[Hg—I—Hg]^{3+}$ formed at very high mercury/iodide ratios,[93] $[(CH_3Hg)_3O]^+$ and $[(CH_3Hg)_3S]^+$,[94] $[(CH_3)_2Tl]^+$,[95] $[O_3OsN]^-$,[96] $[PdCl_6]^{2-}$,[97] $[PtX_6]^{2-}$ (X = F, Cl or Br),[97, 98] $[ReX_6]^{2-}$ (X = Cl or Br) and $[OsCl_6]^{2-}$,[99] $[ThCl_6]^{2-}$ and $[UCl_6]^{2-}$,[100] $[MnX_4]^{2-}$,[101, 102] $[CoX_4]^{2-}$,[102] $[NiX_4]^{2-}$ (X = Cl, Br or I),[102] and $[M(CN)_8]^{3-}$ and $[M(CN)_8]^{4-}$ (M = Mo or W).[103] An anomalous intensity pattern found for the Raman spectra of $[PtCl_6]^{2-}$ and $[PtBr_6]^{2-}$ but not of $[PtF_6]^{2-}$ represents[97, 98] just one of the intriguing features thus brought to light. In the interpretation of such results subtleties such as π-bonding or Jahn-Teller effects[99] were invariably nicely judged to sustain and not to overstretch the staple of the argument. More recently measurements focussed on the intensities of Raman lines due to solute species have been applied to the quantitative study of labile equilibria in aqueous solution. In addition to a review of some aspects of this work, which appeared in 1965,[104] papers have described the results of such investigations for the dissociation of methanesulphonic acid[105] and of methylmercuric salts[92, 106] in aqueous solution.

Since 1956 there have appeared four more articles relating to more general features of molecular spectroscopy. These comprise a short review,[107] an account of new advances in Raman spectroscopy up to 1959,[108] a short review with data relevant to the Raman spectra of some inorganic species,[109] and the introductory chapter of a recent book about Raman spectroscopy,[110] which has all the qualities of clarity and elegance distinguishing the *Quarterly Review* article of 1956.

Altogether, Len Woodward has published over a hundred papers, mostly on the vibrational spectroscopy of comparatively simple aggregates, with special emphasis on Raman spectra, and has maintained his position as one of the leading experts in that field. His work has always been characterized by a high degree of self-criticism and meticulous accuracy combined with clarity. He has always avoided self-advertisement and has tended to shrink from the rough-and-tumble of scientific congresses. His merits, however, have been recognized by his invitations abroad, in 1958 to New Hampshire, in 1962 to the Mellon Research Institute, Pittsburgh, as Visiting Research Fellow, in 1966 to New Zealand and Australia as Commonwealth Prestige Fellow, and in 1968 to the University of Tennessee, Knoxville, as American National Science Foundation Senior Foreign Scientist.

Woodward has fitted perfectly into the dual system at Oxford. His intense interests in research in the University Laboratory have been matched by his equally devoted duties in his College. Here he has been an extremely successful tutor and has served the College in many active ways, as, for example, Dean for eleven years, Vice-Principal from 1960 to 1963, and Steward of the Senior Common Room and Manager of the College Cellar from 1962 up to his retirement.

With his friends and acquaintances his relationships have been easy, helpful, sincere and entirely free from self-importance. With himself he appears to be much more severe, for his relaxations have never been of a trivial character. Until he had virtually exhausted its possibilities he was a serious collector of lepidoptera, and once had the unique experience of being 'dive-bombed' by a specimen of the rare Purple Emperor in an Oxfordshire wood. He is a member of the Alpine Garden Society and has long pursued the study and cultivation of alpines. In an even more intellectual vein he was for many years a composer of 2-move chess problems, and as a member of the British Chess Problem Society published over fifty of them. The standards he has continually set himself have been the admiration and a source of inspiration to all students who have felt his influence. This may be judged from the spontaneous comment of one former student, Dr. J. R. Hall, who concluded his contribution to the present volume with the following: 'I cannot close without paying personal tribute to my mentor, a shrewd chemist, the familiar figure on the motorized cycle, the unraveller of problems on walks through the Parks, who displayed a penchant for the subtle joke and the limerick.'

E.J.B.
A.J.D.

References

1. Woodward, L. A., *Nature*, 1930, **126**, 58.
2. Meyer, E. H. L., *Phys. Z.*, 1930, **31**, 699.
3. Woodward, L. A., *Phys. Z.*, 1930, **31**, 792.
4. Woodward, L. A., *Phys. Z.*, 1931, **32**, 212.
5. Woodward, L. A., *Phys. Z.*, 1931, **32**, 261.
6. Woodward, L. A., *Phys. Z.*, 1931, **32**, 777.
7. Woodward, L. A., and Horner, R. G., *Proc. Roy. Soc.*, 1934, *A*, **144**, 129.
8. Woodward, L. A., *Phil. Mag.*, 1934, [7] **18**, 823.
9. Sidgwick, N. V., Worboys, W. J., and Woodward, L. A., *Proc. Roy. Soc.*, 1930, *A*, **129**, 537.
10. Sidgwick, N. V., and Woodward, L. A., *Proc. Roy. Soc.*, 1930, *A*, **130**, 1.
11. Woodward, L. A., *Proc. Roy. Soc.*, 1934, *A*, **144**, 118.

12. Woodward, L. A., and Horner, R. G., *J. Sci. Instrum.*, 1935, **12**, 17.
13. Woodward, L. A., *Ann. Reports Chem. Soc.*, 1934, **31**, 21.
14. Barber, H. H., and Woodward, L. A., *Biochem. J.*, 1936, **30**, 1463, 1467.
15. Sinnatt, F. S., and Woodward, L. A., *Fuel Econ. Rev.*, 1937, 39.
16. Cawley, C. M., and Woodward, L. A., *Ann. Rev. Petroleum Technology*, 1938, **3**, 373.
17. Cawley, C. M., and Woodward, L. A., *Ann. Rev. Petroleum Technology*, 1939, **4**, 381.
18. Herington, E. F. G., and Woodward, L. A., *Trans. Faraday Soc.*, 1939, **35**, 958.
19. Woodward, L. A., and Tyrrell, H. J. V., *Trans. Faraday Soc.*, 1942, **38**, 513.
20. Bell, R. P., Smith, R. D., and Woodward, L. A., *Proc. Roy. Soc.*, 1948, *A*, **192**, 479.
21. Woodward, L. A., and Glover, A. T., *Trans. Faraday. Soc.*, 1948, **44**, 608.
22. Miller, C. H., Long, D. A., Woodward, L. A., and Thompson, H. W., *Proc. Phys. Soc.*, 1949, *A*, **62**, 401.
23. Woodward, L. A., and Long, D. A., *Trans. Faraday Soc.*, 1949, **45**, 1131.
24. Woodward, L. A., *Ann. Reports Chem. Soc.*, 1949, **46**, 276.
25. Woodward, L. A., *Nature*, 1950, **165**, 198.
26. Woodward, L. A., and George, J. H. B., *Nature*, 1951, **167**, 193.
27. Woodward, L. A., and George, J. H. B., *Proc. Phys. Soc.*, 1951, *B*, **64**, 780.
28. Caunt, A. D., Short, L. N., and Woodward, L. A., *Nature*, 1951, **168**, 557.
29. Caunt, A. D., Short, L. N., and Woodward, L. A., *Trans. Faraday Soc.*, 1952, **48**, 873.
30. Rolfe, J. A., Woodward, L. A., and Long, D. A., *Trans. Faraday Soc.*, 1953, **49**, 1388.
31. Rolfe, J. A., and Woodward, L. A., *Trans. Faraday Soc.*, 1955, **51**, 778.
32. Rolfe, J. A., and Woodward, L. A., *Trans. Faraday Soc.*, 1954, **50**, 1030.
33. Dodd, R. E., Rolfe, J. A., and Woodward, L. A., *Trans. Faraday Soc.*, 1956, **52**, 145.
34. Woodward, L. A., and Roberts, H. L., *Trans. Faraday Soc.*, 1956, **52**, 615.
35. Dodd, R. E., Woodward, L. A., and Roberts, H. L., *Trans. Faraday Soc.*, 1956, **52**, 1052.
36. Woodward, L. A., Garton, G., and Roberts, H. L., *J. Chem. Soc.*, 1956, 3723.
37. Woodward, L. A., and Roberts, H. L., *Trans. Faraday Soc.*, 1956, **52**, 1458.

38. Long, D. A., Matterson, A. H. S., and Woodward, L. A., *Proc. Roy. Soc.*, 1954, *A*, **224**, 33.
39. Matterson, A. H. S., and Woodward, L. A., *Proc. Roy. Soc.*, 1955, *A*, **231**, 514.
40. George, J. H. B., Rolfe, J. A., and Woodward, L. A., *Trans. Faraday Soc.*, 1953, **49**, 375.
41. Waters, D. N., and Woodward, L. A., *J. Chem. Soc.*, 1954, 3250.
42. Rolfe, J. A., Sheppard, D. E., and Woodward, L. A., *Trans. Faraday Soc.*, 1954, **50**, 1275.
43. Woodward, L. A., and Nord, A. A., *J. Chem. Soc.*, 1955, 2655.
44. Woodward, L. A., and Nord, A. A., *J. Chem. Soc.*, 1956, 3721.
45. Woodward, L. A., and Bill, P. T., *J. Chem. Soc.*, 1955, 1699.
46. Woodward, L. A., and Roberts, H. L., *J. Chem. Soc.*, 1956, 1170.
47. Woodward, L. A., and Anderson, L. E., *J. Inorg. Nuclear Chem.*, 1956, **3**, 326.
48. Woodward, L. A., *Quart. Rev.*, 1956, **10**, 185.
49. Woodward, L. A., and Waters, D. N., *J. Sci. Instrum.*, 1957, **34**, 222.
50. Dodd, R. E., Roberts, H. L., and Woodward, L. A., *J. Chem. Soc.*, 1957, 2783.
51. Dodd, R. E., Woodward, L. A., and Roberts, H. L., *Trans. Faraday Soc.*, 1957, **53**, 1545.
52. Woodward, L. A., and Roberts, H. L., *Trans. Faraday Soc.*, 1957, **53**, 1557.
53. Woodward, L. A., and Hall, J. R., *Nature*, 1958, **181**, 831.
54. Woodward, L. A., and Hall, J. R., *Spectrochim. Acta*, 1960, **16**, 654.
55. Edwards, H. G. M., and Woodward, L. A., *Spectrochim. Acta*, 1970, **26A**, 897.
56. Woodward, L. A., Greenwood, N. N., Hall, J. R., and Worrall, I. J., *J. Chem. Soc.*, 1958, 1505.
57. Woodward, L. A., Hall, J. R., Dixon, R. N., and Sheppard, N., *Spectrochim. Acta*, 1959, **15**, 249.
58. Hall, J. R., Woodward, L. A., and Ebsworth, E. A. V., *Spectrochim. Acta*, 1964, **20**, 1249.
59. Woodward, L. A., *Spectrochim. Acta*, 1963, **19**, 1963.
60. Goggin, P. L., and Woodward, L. A., *Trans. Faraday Soc.*, 1966, **62**, 1423.
61. Cross, L. H., Roberts, H. L., Goggin, P. L., and Woodward, L. A., *Trans. Faraday Soc.*, 1960, **56**, 945.
62. Goggin, P. L., Roberts, H. L., and Woodward, L. A., *Trans. Faraday Soc.*, 1961, **57**, 1877.
63. Creighton, J. A., Woodward, L. A., and Dove, M. F. A., *Spectrochim. Acta*, 1962, **18**, 267.

64. Taylor, M. J., and Woodward, L. A., *J. Chem. Soc.*, 1963, 4670.
65. Ebsworth, E. A. V., Hall, J. R., Mackillop, M. J., McKean, D. C., Sheppard, N., and Woodward, L. A., *Spectrochim. Acta*, 1958, **13**, 202.
66. McKean, D. C., Taylor, R., and Woodward, L. A., *Proc. Chem. Soc.*, 1959, 321.
67. Ebsworth, E. A. V., Taylor, R., and Woodward, L. A., *Trans. Faraday Soc.*, 1959, **55**, 211.
68. Aylett, B. J., Hall, J. R., McKean, D. C., Taylor, R., and Woodward, L. A., *Spectrochim. Acta*, 1960, **16**, 747.
69. Ball, D. F., Carter, T., McKean, D. C., and Woodward, L. A., *Spectrochim. Acta*, 1964, **20**, 1721.
70. Ball, D. F., Goggin, P. L., McKean, D. C., and Woodward, L. A., *Spectrochim. Acta*, 1960, **16**, 1358.
71. Ebsworth, E. A. V., Mould, R., Taylor, R., Wilkinson, G. R., and Woodward, L. A., *Trans. Faraday Soc.*, 1962, **58**, 1069.
72. Davidson, G., Ebsworth, E. A. V., Sheldrick, G. M., and Woodward, L. A., *Chem. Comm.*, 1965, 122.
73. Davidson, G., Ebsworth, E. A. V., Sheldrick, G. M., and Woodward, L. A., *Spectrochim. Acta*, 1966, **22**, 67.
74. Davidson, G., Woodward, L. A., Ebsworth, E. A. V., and Sheldrick, G. M., *Spectrochim. Acta*, 1967, **23A**, 2609.
75. Buttler, M. J., McKean, D. C., Taylor, R., and Woodward, L. A., *Spectrochim. Acta*, 1965, **21**, 1379.
76. Cradock, S., Davidson, G., Ebsworth, E. A. V., and Woodward, L. A., *Chem. Comm.*, 1965, 515.
77. Davidson, G., Woodward, L. A., Mackay, K. M., and Robinson, P., *Spectrochim. Acta*, 1967, **23A**, 2383.
78. Long, D. A., Spencer, T. V., Waters, D. N., and Woodward, L. A., *Proc. Roy. Soc.*, 1957, *A*, **240**, 499.
79. Chantry, G. W., and Woodward, L. A., *Trans. Faraday Soc.*, 1960, **56**, 1110.
80. Waters, D. N., and Woodward, L. A., *Proc. Roy. Soc.*, 1958, *A*, **246**, 119.
81. Taylor, K. A., and Woodward, L. A., *Proc. Roy. Soc.*, 1961, *A*, **264**, 558.
82. Woodward, L. A., *Trans. Faraday Soc.*, 1958, **54**, 1271.
83. Bruton, M. J., and Woodward, L. A., *Spectrochim. Acta*, 1967, **23A**, 175.
84. Woodward, L. A., and Anderson, L. E., *J. Chem. Soc.*, 1957, 1284.
85. Woodward, L. A., and Taylor, M. J., *J. Chem. Soc.*, 1962, 407.
86. Creighton, J. A., and Woodward, L. A., *Trans. Faraday Soc.*, 1962, **58**, 1077.
87. Woodward, L. A., and Singer, G. H., *J. Chem. Soc.*, 1958, 716.

88. Woodward, L. A., and Taylor, M. J., *J. Chem. Soc.*, 1960, 4473.
89. Woodward, L. A., and Taylor, M. J., *J. Inorg. Nuclear Chem.*, 1965, **27**, 737.
90. Woodward, L. A., and Owen, H. F., *J. Chem. Soc.*, 1959, 1055.
91. Goggin, P. L., and Woodward, L. A., *Trans. Faraday Soc.*, 1962, **58**, 1495.
92. Clarke, J. H. R., and Woodward, L. A., *Trans. Faraday Soc.*, 1966, **62**, 3022.
93. Clarke, J. H. R., and Woodward, L. A., *Trans. Faraday Soc.*, 1965, **61**, 207.
94. Clarke, J. H. R., and Woodward, L. A., *Spectrochim. Acta*, 1967, **23A**, 2077.
95. Goggin, P. L., and Woodward, L. A., *Trans. Faraday Soc.*, 1960, **56**, 1591.
96. Woodward, L. A., Creighton, J. A., and Taylor, K. A., *Trans. Faraday Soc.*, 1960, **56**, 1267.
97. Woodward, L. A., and Creighton, J. A., *Spectrochim. Acta*, 1961, **17**, 594.
98. Woodward, L. A., and Ware, M. J., *Spectrochim. Acta*, 1963, **19**, 775.
99. Woodward, L. A., and Ware, M. J., *Spectrochim. Acta*, 1964, **20**, 711.
100. Woodward, L. A., and Ware, M. J., *Spectrochim. Acta*, 1968, **24A**, 921.
101. Edwards, H. G. M., Ware, M. J., and Woodward, L. A., *Chem. Comm.*, 1968, 540.
102. Edwards, H. G. M., Woodward, L. A., Gall, M. J., and Ware, M. J., *Spectrochim. Acta*, 1970, **26A**, 287.
103. Parish, R. V., Simms, P. G., Wells, M. A., and Woodward, L. A., *J. Chem. Soc. A*, 1968, 2882.
104. Woodward, L. A., *Pure and Appl. Chem.*, 1965, **11**, 473.
105. Clarke, J. H. R., and Woodward, L. A., *Trans. Faraday Soc.*, 1966, **62**, 2226.
106. Clarke, J. H. R., and Woodward, L. A., *Trans. Faraday Soc.*, 1968, **64**, 1041.
107. Woodward, L. A., *Contemporary Phys.*, 1960, **1**, 319.
108. Woodward, L. A., *Ann. Reports Chem. Soc.*, 1959, **56**, 67.
109. Woodward, L. A., 'Handbook of Analytical Chemistry', ed. L. Meites, McGraw-Hill, New York, 1963, p. 6–171.
110. Woodward, L. A., 'Raman Spectroscopy: Theory and Practice', ed. H. A. Szymanski, Plenum Press, New York, 1967, p. 1.

CHAPTER 1

Molecular Force Fields and Valency

J. W. LINNETT

A. Introduction

The study of intramolecular force fields for small nuclear displacements provides a means of extending our knowledge and understanding of valency. For the chemist, the most useful expression for the force field is in terms of bond-length changes and bond-angle changes. The quadratic form for small displacements is then

$$2V = \sum_i k_i \Delta r_i^2 + \sum_{ij} k_{ij} \Delta r_i \Delta r_j + \sum_p k_p \Delta \alpha_p^2 + \sum_{pq} k_{pq} \Delta \alpha_p \Delta \alpha_q + \sum_{ip} k_{ip} \Delta r_i \Delta \alpha_p \qquad (1.1)$$

where the summation is over all bonds and all interbond angles; $\Delta r_i = r_i - r_{ei}$, where r_{ei} is the equilibrium length of the *i*th bond, and $\Delta \alpha_p = \alpha_p - \alpha_{ep}$, where α_{ep} is the equilibrium value of the *p*th angle. There are two types of constants in equation (1.1): the main constants k_i (bond) and k_p (angle), and the interaction constants k_{ij} (bond–bond), k_{pq} (angle–angle) and k_{ip} (bond–angle).

The main constants k_i measure the energy change occurring when the bond *i* only is altered in length, the rest of the molecule being left unchanged (if that is possible). The same holds for the k_p in relation to angles. The interaction constants measure the coupling between different bonds, different angles, and bonds and angles. They have a rather special value in that they provide data about the interaction between different parts of the same molecule which are of great interest and which help us to understand the nature of valence forces. This information cannot be obtained from other measurements. A bond force constant k_i is linked to the 'strength of the bond' and hence is related to the equilibrium length and the bond dissociation energy which provide an alternative source of information. The interaction constants are not linked so directly to any other properties, although later their relation to bond-length and interbond-angle changes will be discussed. The bending force constants also have a particular interest because they provide us with knowledge of the rigidity of the molecule to angular distortion. This again is not really obtainable in any other way.

Other force fields which also have a physical basis have been proposed. For instance, Urey and Bradley[1] suggested that terms allowing for the interaction of non-bonded atoms should be included specifically in the potential energy function. This is clearly a reasonable suggestion, for we know from other sources that the interaction between non-bonded atoms can have significant effects. Heath and Linnett[2] examined this for a number of tetrahedral molecules such as carbon tetrachloride (and other species) and concluded that, in certain ways, there were advantages in the Urey–Bradley type of force field. Simanouti[3] came to a similar conclusion. However, the magnitude of the terms needed sometimes produced other problems. Also, it inevitably becomes difficult to separate the potential energy function into a part that should be assigned to non-bonded interaction and a part that should be assigned to 'directed valency', because undoubtedly interbond angles are decided mainly by electronic effects which cannot conceivably be classed under the heading of 'interaction between non-bonded atoms'.

Also, an orbital-valency force field[4] was proposed which had a separate constant for each bond and it seemed at the time to be more closely related to the then current ideas of directed valency than the simple-valency force field. The latter includes inter-bond angle terms in the potential energy function (cf. equation (1.1)). In the former each atom is supposed to possess a set of directed orbitals (e.g. an octahedral set) and angular distortion is treated as the departure of the attached atoms from these particular directions. For example, the four carbon orbitals were directed to the apices of a regular tetrahedron. Distortion of the basic set could also occur by changes in hybridization and the concept of orbitals 'following' a particular distortion was added. This was able to account in a logical way for the relative magnitude of two bending frequencies of CH_4. The orbitals can follow the triply degenerate bending motion but cannot follow the doubly degenerate bending motion.

Certainly this approach is not an unreasonable one. However, the growth in our understanding of the importance of correlation effects suggests that perhaps a function which uses interbond angles may be preferable to one that considers the direction of each bond separately despite the obvious advantages of the latter in separating out the effects bond by bond for subsequent analysis and interpretation.

Much more use has been made of the simple bond-stretching constants than of angle-bending constants and interaction constants. Perhaps the reason is that much of our thinking about molecules is more easily conducted in terms of bonds so that the constant most directly related to them has been used most.

The number of constants in the general P.E. function of a molecule is equal to $\sum_i \frac{1}{2}n_i(n_i + 1)$ where n_i is the number of vibrations in a given symmetry class and the summation is over all symmetry classes. Some particular examples are: CO_2, 3; H_2O, 4; HCN, 4; ClNO, 6; CH_2O, 10; FCHO, 16; CH_3F, 12; CH_2F_2, 17; CH_2FCl, 25; C_2H_4, 18; $CH_2{=}CHF$, 51. It is apparent that even in simple molecules the number can be quite large and that the number increases with decreasing symmetry. Among these examples, the number of frequencies is equal to the number of constants for CO_2 only.

B. Determination of force constants

The main source of information about force constants is molecular vibration frequencies which are most frequently obtained from infrared and Raman spectra.[5] However, the number of known vibration frequencies is usually less than the number of constants in the most general potential energy function. Several procedures can therefore be adopted. First some interaction constants which are likely to be small can be set equal to zero. This is unsatisfactory because it involves prejudging the bonding in the molecule and this clearly reduces its power as a method of investigating bonding. The usefulness of the study is reduced because the results will depend on the arbitrary simplification that has been made. Alternatively, some force constants may be transferred from other molecules in which the bonding situation appears to be similar. Again, the result is to that extent prejudged and the transfer may later prove to be unjustifiable. Finally, an attempt may be made to obtain other data so that the number of arbitrary assumptions needed can be reduced.

The most obvious way of adding to the data is to determine the vibration frequencies of an isotopically substituted molecule. For example, for HCN three vibration frequencies are available but the general force field requires four constants. If the molecule DCN is then examined as well as HCN, sufficient data are immediately available. However, with other molecules the situation is not so satisfactory. Methyl halides, CH_3X, require twelve potential constants but have only six vibration frequencies. Addition of the data for the molecules CD_3X produces six more frequencies but there are two Teller–Redlich[6] relationships between the frequencies of the isotopic molecules which are independent of the force field. Therefore, only four pieces of data have been added. Therefore the spectrum of another isotopically substituted molecule must be obtained.

There are two further difficulties. Strictly, only fundamental frequencies corrected for anharmonicity should be used, but the anharmonic constants are difficult to obtain in all but the most simple molecules. Consequently, the corrections cannot usually be made and uncorrected 01-transition frequencies have to be used. Secondly, some parts of the force field are often rather insensitive to isotopic substitution. This is illustrated by the Glockler–Tung[7] graphs obtained by Smith and Linnett[8] for H_2O and D_2O. For each molecule separately three of the force constants are plotted as a function of the fourth, there being three frequencies and four constants in the general P.E. function. The lines and ellipses are almost coincident for H_2O and D_2O which shows that in this case isotopic substitution does not really provide a good method for fixing the precise value of the four constants.

Fortunately, other quantities that are dependent on the force field can be determined experimentally from molecular spectra. These are centrifugal stretching constants and Coriolis coupling coefficients. They do supplement the vibration frequency data much better. Because more data of this kind have become available in the last few years, the extent of our knowledge of general P.E. functions has increased.

C. The valence force field

The *simple* valence force field (S.V.F.F.) is

$$2V = \sum_i k_i \Delta r_i^2 + \sum_p k_p \Delta \alpha_p^2 \qquad (1.2)$$

That is, it consists only of the first two sets of terms in equation (1.1), all interaction constants being put equal to zero. It can be regarded as a possible first approximation, and it might be expected to be reasonably successful for the simplest molecules in which non-bonded interactions are small. It is, in fact, quite satisfactory for H_2O and CH_4, though even for NH_3 a considerable angle–angle interaction constant is necessary. This is because of the lone pair and the fact that there are two very different bending vibrations.

Regarding the S.V.F.F. as the starting point, what causes cross-terms to appear? How are the intramolecular interactions to be envisaged? Are we to suppose, for bond–bond interactions, that the alteration of one bond i modifies the equilibrium length of the other bond j according to an equation such as

$$r'_{ej} = r_{ej} + \beta \Delta r_i \tag{1.3}$$

or the force constant of the other bond according to

$$k'_j = k_j + \beta' \Delta r_i \tag{1.4}$$

In these, r_{ej} is the true equilibrium length and r'_{ej} the modified one; k_j is the true force constant and k'_j the modified one. The P.E. associated with this bond j when bond i is distorted by Δr_i will be:

$$\begin{aligned} \tfrac{1}{2}(k_j + \beta' \Delta r_i)(r_j - r'_{ej})^2 &= \tfrac{1}{2}(k_j + \beta' \Delta r_i)(\Delta r_j - \beta \Delta r_i)^2 \\ &= \tfrac{1}{2}k_j(\Delta r_j^2 + \beta^2 \Delta r_i^2 - 2\beta \Delta r_i \Delta r_j) \\ &\quad + \tfrac{1}{2}\beta' \Delta r_i(\Delta r_j - \beta \Delta r_i)^2 \end{aligned} \tag{1.5}$$

From this it is clear that the effect on the force constant does not contribute to the quadratic terms. Therefore, bond–bond cross-terms can be discussed solely in terms of the effect of the alteration of one bond on the equilibrium length of the other. The bond–angle and angle–angle cross-terms can be treated in a similar manner.

From equations (1.3) and (1.5), it is apparent that the bond–bond cross-term constant will be negative if the lengthening of one bond causes the equilibrium length of the other bond to increase, and positive if it causes the equilibrium length to decrease. The bond–bond cross-term constant in CO_2 is positive. Removal of one oxygen atom from CO_2 will be expected to leave a CO molecule (the atom must, of course, leave in an excited state). The equilibrium bond length in CO_2 is 1·16 Å and in CO, 1·13 Å. The behaviour on dissociation, therefore, corresponds correctly to the sign of the interaction constant. The same is true for the water molecule in which the bond–bond interaction constant is negative. In this case, the OH bond-length in the radical is larger (0·97 Å) than in the molecule H_2O (0·96 Å), so that again there is a correspondence between the sign of the constant and the behaviour on dissociation. In this case, both the cross-term constant and the change in length are much smaller than for CO_2. Hoare and Linnett[9] drew attention to other examples of this kind, but pointed out that such a correspondence might not always exist. Other examples could now be added to their list.

Thompson and Linnett[10] accounted for the sign of the cross-term in CO_2 in 1937 in terms of the Pauling resonance forms, which are

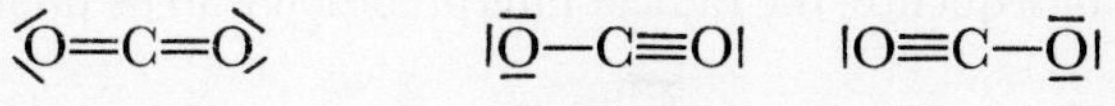

The increasing importance of one or other of the asymmetric forms during an asymmetric distortion accounts for the fact that such a distortion is 'easier' than a symmetric distortion and the sign of the constant is explained. The electronic structure can follow (i.e. accommodate itself to) such a distortion. I would now be inclined to consider the important structures to be

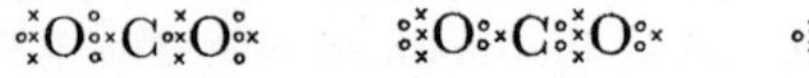

However, in terms of these, the explanation is precisely similar. The effect in H_2O is, as has been said, much smaller and is probably because the effective electronegativity of the oxygen atom is less in the radical than in the molecule. Or some might prefer an explanation based on changes in the hybridization of the orbitals of the oxygen atom. The two would be equivalent to one another.

So far the discussion of cross-term constants has been qualitative. For a quantitative discussion, it is useful to transform the observed interaction constant into the change in the equilibrium length of bond j produced by unit change in the length of bond i. This is the constant β in equation (1.3). From equation (1.5) it is apparent that β is equal to the calculated interaction constant divided by the value of k_j (though here it is necessary to warn the reader to examine the exact form of the P.E. function used in any given publication, as in some cases the numerical value of the interaction constant actually used should be divided by twice the value of the main constant rather than its value). This type of analysis has been used by Jones, Ryan and Asprey[12] and is certainly useful. Some figures they give for ClNO will be considered later.

Bond–angle and angle–angle interaction constants can be discussed in a similar way. In methyl compounds such as the halides CH_3X the interaction constant between the CX bond length and the HCH angle is significant. Its sign shows that as the carbon–halogen bond-length is increased (or the CC bond-length in ethane) the methyl group tends to become flatter. This can be related to the fact that, on dissociation, a methyl radical will be left which is probably planar, but certainly flatter than the methyl group in the molecules. Of course, it is possible that, near the equilibrium configuration, the elongation of the CX bond should be regarded as the movement of an X^- away from a CH_3^+. The same qualitative conclusion regarding the sign of the bond–angle interaction constant would follow. However, the system certainly would not dissociate into ions. Also, the dipole moment is not large enough to demand such an interpretation. Consequently the radical interpretation is to be preferred to the ionic one.

D. Examples of the use of force constants

In the remainder of this chapter, a number of molecules in which force constants are useful in relation to our consideration of valency and electronic structures will be discussed. These examples are chosen at random but they illustrate various ways in which force constants can be used.

(a) *Chlorine fluorides*

Christe and Sawodny[13] have calculated the ClF bond-stretching force constant in ClF, ClF_2^+ and ClF_2^-. The value for the diatomic molecule is 4·4 mdyne/Å. The constant in ClF_2^+ is a little greater: 4·8 mdyne/Å. The bond in ClF must presumably be a single (electron pair) bond. The force constant shows that the bonds in ClF_2^+ are similar. The formula must be

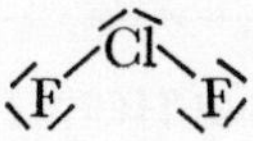

The positive charge is formally located on the chlorine rather than the fluorine atoms; this is reasonable. The small increase in the force constant can be ascribed to the presence of the positive charge which will strengthen the bond a little. On the other hand, the force constant in ClF_2^- is only half as big: 2·35. Is the electronic structure described best by

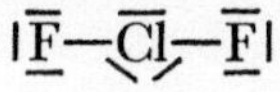

or by

In the former, the valence shell of the chlorine atom is expanded to ten electrons and the negative charge of the ion is formally carried by the chlorine atom. In the latter, the valence shells on all three atoms are octets, and the charge on the ion is formally shared between the two fluorine atoms. This is more reasonable in view of the relative electronegativity of fluorine and chlorine. Moreover, the low value of the force constant would appear to favour the second structure. The bond–bond interaction constant is +0·17. This has the sign that would be expected since the removal of one fluorine negative ion will leave ClF in which the bond has a higher force constant than in ClF_2^- and hence is almost certainly shorter. The interaction constant might have been expected to have been bigger. However, Christe *et al.*[14] have examined the cross-terms in $BrCl_2^-$ and ICl_2^- as well. Their results for these three ions are shown in Table 1.1

TABLE 1.1. *Bond force constants and bond–bond interaction constants in mdyne/Å for some linear triatomic ions*

Species	k_r triatomic	k_r diatomic	k_{rr}	$\frac{k_r \text{ (diatomic)}}{k_r \text{ (triatomic)}}$	k_{rr}/k_r
ClF_2^-	2·35	4·36	0·17	1·8	0·07
ICl_2^-	1·00	2·38	0·36	2·4	0·36
$BrCl_2^-$	1·02	2·83	0·47	2·8	0·46
HF_2^-	2·31	9·67	1·72	4·2	0·74
HCl_2^-	0·80	5·15	0·61	6·4	0·76

together with figures for HF_2^- and HCl_2^-. It can be seen that, as the ratio of the cross-term constant to the main constant (k_{rr}/k_r) increases, the ratio of the main constant for the diatomic molecule to that for the triatomic $(k_r \text{ (diatomic)}/k_r \text{ (triatomic)})$ also increases. Because bonds with higher force constants are shorter the correlation of data presented in Table 1.1 is understandable.

In the case of the interhalogen ions XY_2^- listed in Table 1.1 the electronegativity of the central ion is in all cases less than that of the outer atom Y. Hence the structure

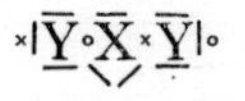

is, for all cases, reasonable. Likewise in the ions HX_2^-, the structures

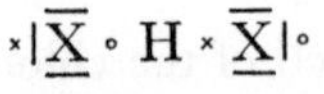

share the negative charge between the two X atoms, which is reasonable on electronegativity grounds. In all cases the force constant data support the hypothesis that the two one-electron bonds are present. An alternative statement of this is that the three atoms are held together by two electrons occupying a three-centre bonding orbital. The inclusion of end-to-end correlation in such a bonding orbital will lead to a situation which is best described in terms of two one-electron bonds. This is, therefore, equivalent to the formulation of Pimentel and of Rundle.[15]

(*b*) *Oxides of chlorine and fluorine*

In this section the molecules OX, OX_2, O_2X and O_2X_2 will be considered where X = H, F or Cl. The force-constant and bond-length results that must be considered are listed in Table 1.2. (These are taken, almost

TABLE 1.2. *Force constants and equilibrium bond-lengths*

	k (OX) mdyne/Å	r_e (OX) Å	k (OO) mdyne/Å	r_e (OO) Å
OH	7·1	0·97	—	—
OH_2	7·7	0·96	—	—
O_2H	6·5	—	6·2	—
O_2H_2	7·4	0·95	4·6	1·48
OF	5·4	—	—	—
OF_2	4·0	1·41	—	—
O_2F	1·3	—	10·5	—
O_2F_2	1·4	1·58	10·3	1·22
OCl	6·1	—	—	—
OCl_2	2·8	1·7	—	—
O_2Cl	1·3	—	9·7	—
O_2Cl_2	(6)	—	—	—
O_2	—	—	11·4	1·21

entirely, from a table prepared by J. J. Turner.[16]) The first feature of this table that must be stressed is that more force constants are available than bond lengths. Of course, the values of the force constants may not be exact because of uncertainties in the force field. Consequently, only the broad pattern must be used and small changes should be treated with caution. However, this set of molecules can only be discussed as a complete set if force constants are used.

The OH molecule must contain a single bond but OCl and OF could be described by

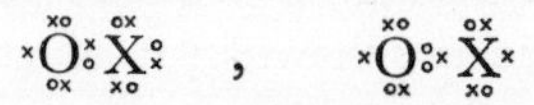

or by a resonance hybrid of the two. The former places zero formal charge on each atom but the bond is only a single one, whereas the latter places a formal charge of $-\frac{1}{2}$ on the oxygen and $+\frac{1}{2}$ on the halogen atom. Presumably, all the OX_2 molecules can be described accurately by

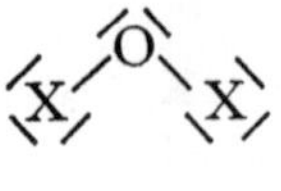

the bonds being single.

Clearly, the bond lengths and force constants in OH and OH_2 support the view that the bond in the hydroxyl radical is a single one. The force constants in OCl and OCl_2 suggest that the OCl bond in the diatomic radical is stronger than a two-electron one. This is understandable, because the lower electronegativity of the chlorine atom allows it to carry a formal positive charge of $+\frac{1}{2}$ in this species. Of course

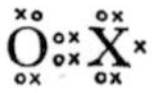 or 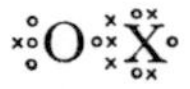

is also possible, the valence shell of the halogen carrying nine electrons; in these structures the formal charge on each atom is zero.

With OF, the expansion of the octet at both atoms is not allowed. Also, the high electronegativity of fluorine relative to oxygen makes it unlikely that the fluorine atom will carry a significant positive charge. Hence, the structure involving an electron pair bond between the two atoms in OF is likely to be the most important. The force-constant data support this: k (diatomic)/k (triatomic) is 1·35 for OF and OF_2 but 2·2 for OCl and OCl_2, indicating that the structure F—O is favoured relative to $\text{F}\overset{\bullet}{-}\text{O}$, while $\text{Cl}\overset{\bullet}{-}\text{O}$ is favoured relative to Cl—O.

From H_2O_2 the force-constant and bond-length data show that the structure is described satisfactorily by

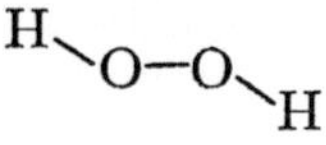

But clearly F_2O_2 is not to be described by such a formula. The force constant and length show that the OO bond is similar to that in O_2 and the OF bond is clearly weak as is shown by its length and the force constant. Formulae that account for this are

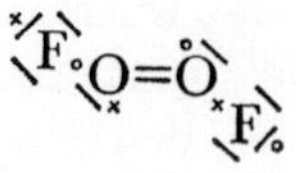 and 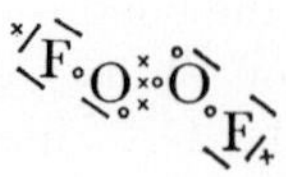

The reason for the difference between F_2O_2 and H_2O_2 is surely the high electronegativity of fluorine. In the above structure the fluorine atoms carry a formal charge of $-\frac{1}{2}$ and they do this by forming one-electron bonds to the oxygen atoms. As a consequence, the OO bond is of order two. There is the added advantage in these structures that the inter-electron repulsion energy is reduced.[17]

The OF and OO bonds in FO_2 are very similar to those in F_2O_2 and a probable structure is

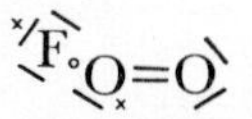

The formula of HO_2 will not have this form because the electronegativity of hydrogen is lower. The formula which places a zero formal charge on hydrogen is

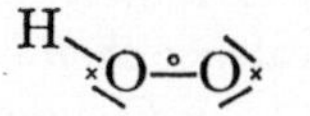

This, like all the other formulae proposed for the tri- and tetra-atomic species so far in this section assigns to each atom in the first short period eight electrons. The OO force constant (6·2) is lower than in O_2 (11·4), FO_2 (10·5) and F_2O_2 (10·3) but higher than in H_2O_2 (4·6). The above formula accounts for this and for the fact that the OH force constant is approximately the same as in OH, OH_2 and O_2H_2.

The ClO bond in $(ClO)_2$ is approximately the same as that in diatomic ClO according to Pimentel *et al.*[18] They infer that the bonding between two ClO groups is weak. The situation is certainly quite different from that in F_2O_2 and H_2O_2. Harcourt[19] would use his concept of 'increased valency' to account for a loose bonding between the two ClO groups which does not weaken the two ClO bonds (cf. NO_2)

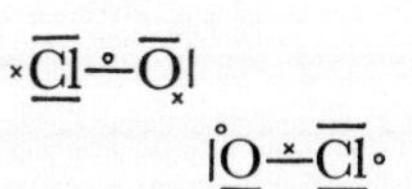

The bonding electron in the upper ClO group has a certain probability of being on the chlorine atom (and also a certain probability of being on the oxygen atom). Because there is a contribution involving its location on the chlorine, the electron of the same spin may be accepted from the

oxygen on the other ClO so that it is shared between the two oxygen atoms; in the wave function the octet rule will not be broken though, at first glance, this would appear to happen. Of course

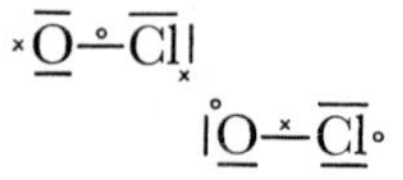

and

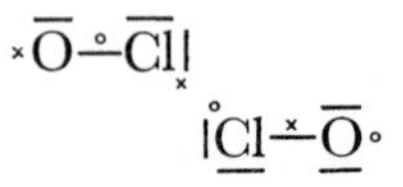

would also appear to be possible and they have been suggested by Pimentel *et al.*[18] The other possibility is that a loose association occurs by the expansion of the octet of the valence shell of the chlorine atom.

The contrast between H_2O_2, F_2O_2 and Cl_2O_2 is fascinating. Force-constant measurements show that they do not resemble one another. H_2O_2 has a classical structure. F_2O_2 is different because of the greater electronegativity of fluorine and the readiness with which its atoms assume a formal negative charge. On the other hand, because chlorine can assume a formal positive charge when attached to oxygen with chlorine diatomic ClO forms such a strong bond that, if a conventional dimer were formed, the loss of energy in the ClO bonds would not be counterbalanced by the formation of the weak OO single bond. Hence, only a weak force between two strongly bound ClO species occurs.

There remains ClO_2. Is this to be a chlorine atom loosely bound to a strong OO bond

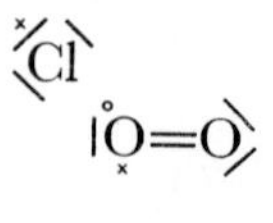

or an oxygen atom loosely bound to a strong ClO bond

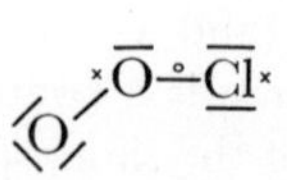

The latter involves formal charges of -1, $+\frac{1}{2}$, $+\frac{1}{2}$; the former $-\frac{1}{2}$, $+\frac{1}{2}$, 0. The number of bonding electrons is the same in each structure. Because the formal charges are lower for the structure which is similar to that of FO_2, and because the chlorine atom can assume a negative charge (cf. ClNO), this structure is preferred. Hence, ClO_2 resembles FO_2 even though Cl_2O_2 does not resemble F_2O_2.

The whole of the extraordinarily varied behaviour of the three sets of four molecules listed in Table 1.2 is satisfactorily rationalized by considering the formal charges on the various centres, providing that structures which allow the electrons of one spin set to adopt a different pattern from those of the other spin set are allowed and bonds containing an odd number of electrons are included. Except in the diatomic species the octet is maintained throughout (for OH there are less than eight electrons altogether outside the K-shell).

(*c*) *Nitrosyl halides*

Jones, Asprey and Ryan[12] have obtained the six constants for the general force field of FNO and ClNO. The FN force constant in the former molecule is 2·31 as against 4·4 in NF_3. The NO force constant is 15·9 which is approximately the same as that in diatomic NO, and is certainly larger than would be expected for a double bond. A formula

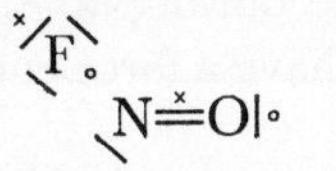

similar to that of F_2O_2 is probable. It accounts for the observed force constants (and also for the observed bond lengths[20]). The structure of ClNO is similar; the observed force constant of the NCl bond is low (1·3). ClNO resembles FNO for the same reason that ClOO resembles FOO. This was discussed in the last section.

The bond–bond cross term in FNO is positive and large (+2·5). The decrease in the FN equilibrium bond length per unit increase in the NO bond length is 1·1. This means that as the NO bond length is increased the structure

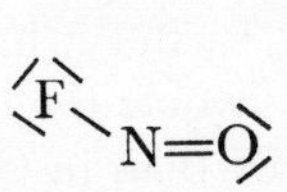

is favoured and the NF bond shortens considerably. On the other hand, the decrease in the NO equilibrium bond length per unit increase in the NF bond length is only 0·16.

The bond–bond interaction constant in ClNO is also quite large (1·53). In both molecules the interaction constant is greater than the NX bond force constant. The bond–angle interaction constants in both molecules are small. This might have been expected because there is no obvious effect which would cause them to be considerable.

(*d*) *Ketene and diazomethane*

The bond force constants in ketene[21] indicate that a reasonable structure is

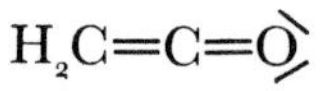

But, in the diazo compound, in order to reduce the high formal negative charge on the terminal nitrogen atom, some contribution may also be made by

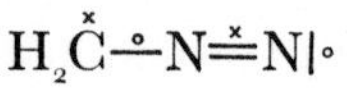

Most of the bending force constants have magnitudes that could reasonably be anticipated. However, the out-of-plane bending of the CH_2 group is found, in both molecules, to have a force constant which is only one-tenth of the other constants.

The above structure for ketene favours a planar arrangement of the nuclei. When an out-of-plane motion of the methylene occurs, the importance of the structure

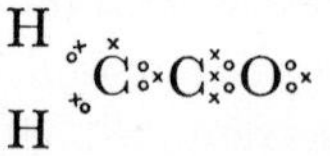

will increase because the electrons of one-spin set (represented by crosses) favour the molecule being non-planar. Because of this, the appropriate distortion will take place easily and the force constant will be low. For diazomethane the constant is half what it is for ketene ($0{\cdot}045 \times 10^{-11}$ as against $0{\cdot}086 \times 10^{-11}$ ergs rad^{-1}). But it has already been suggested that the electronic structure which favours the out-of-plane distortion is probably more important in CH_2NN than in CH_2CO because of the need to transfer electrons away from the terminal nitrogen atom so that the formal charge there shall not be too great. Therefore, the abnormally low bending force constants for this type of distortion in ketene and diazomethane (and their relative values) can be satisfactorily explained.

(e) *Thiazyl fluoride*

The molecule FSN is interesting structurally because the question arises as to whether the SN bond is best regarded as a triple bond or not. The structure

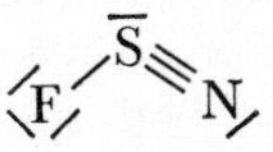

appears reasonable except that the triple bond S≡N is unexpected. The formula

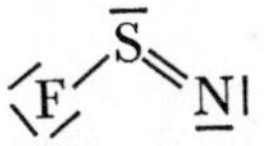

is unsatisfactory because the nitrogen atom carries a significant negative charge, and the sulphur a formal positive charge. On the other hand, a triple bond between sulphur and nitrogen is avoided.

The bond lengths have been determined by Kirchhoff and Wilson[22]: SF, 1·646 Å; SN, 1·446 Å. The interbond angle is 117°. The SF bond is longer than that in SF_6 (1·58 Å).

The general force field has been determined by Mirri and Guarnieri[23] using infrared and microwave data. The six force constants are k_{SN}, 10·71 mdyne/Å; k_{SF}, 2·87 mdyne/Å; k_{α}, 0·98 mdyne/Å; $k_{rr'}$, 0·10 mdyne/Å; $k_{SN\alpha}$, −0·06 mdyne/Å and $k_{SF\alpha}$, 0·03 mdyne/Å. Perhaps the most interesting feature is that the simple valency force field is so satisfactory and that the interaction constants are so small. This shows that distortion of one part of the molecule has little effect on the electron distribution in another part of the molecule.

The SF force constant is quite small. That in SF_6 is certainly greater.[24] So this confirms the observed change in bond length. The SN force constant is high and the bond is short. Hence, the SN bond is certainly a multiple one though it is difficult to be sure of the order. It is shorter than the bond in NS (1·496 Å) and, because that diatomic molecule is isoelectronic with NO, the bond order there might be expected to be $2\frac{1}{2}$.

The most important structure would seem to be that given at the beginning of this section, though it is probable that some contribution from

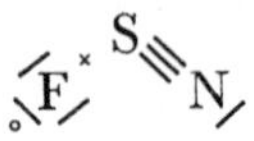

ought to be included, because of the high electronegativity of the fluorine. This would account for the greater length and low force constant of the SF bond.

A combination of the wave functions corresponding to these two structures would not lead one to expect a large bond–bond interaction constant, because, in both, the SN bond is a triple one. The situation is, therefore, quite different from CO_2. Also, both structures will favour a non-linear molecule with approximately the same interbond angle. Hence, the low values of the bond–angle interaction constants are understandable.

References

1. Urey, H. C., and Bradley, C. A., *Phys. Rev.*, 1931, **38**, 1969.
2. Heath, D. F., and Linnett, J. W., *Trans. Faraday Soc.*, 1948, **44**, 561.
3. Simanouti, T., *J. Chem. Phys.*, 1949, **17**, 245, 734, 848.
4. Heath, D. F., and Linnett, J. W., *Trans. Faraday Soc.*, 1948, **44**, 878, 884.
5. Herzberg, G., 'Infra-Red and Raman Spectra of Polyatomic Molecules,' van Nostrand, New York, 1945.
6. Redlich, O., *Z. Physik. Chem. B.*, 1935, **28**, 371.
7. Glockler, G., and Tung, J. Y., *J. Chem. Phys.*, 1945, **13**, 388.
8. Smith, S., and Linnett, J. W., *Trans. Faraday Soc.*, 1956, **52**, 891.
9. Hoare, M. F., and Linnett, J. W., *Trans. Faraday Soc.*, 1949, **45**, 844.
10. Thompson, H. W., and Linnett, J. W., *J. Chem. Soc.*, 1937, 1384.
11. Linnett, J. W., 'Electronic Structure of Molecules,' Methuen, London, 1964, p. 65.
12. Jones, L. H., Ryan, R. R., and Asprey, L. B., *J. Chem. Phys.*, 1967, **47**, 3371; 1968, **49**, 581.
13. Christe, K. O., and Sawodny, W., *Inorg. Chem.*, 1967, **6**, 313.
14. Christe, K. O., Sawodny, W., and Guertin, J. P., *Inorg. Chem.*, 1967, **6**, 1159.
15. Pimentel, G. C., *J. Chem. Phys.*, 1951, **19**, 446.
16. Turner, J. J., private communication.
17. Linnett, J. W., 'Electronic Structure of Molecules,' Methuen, London, 1964, p. 83.

18. Rochkind, M. M., and Pimentel, G. C., *J. Chem. Phys.*, 1967, **46**, 4481; Alcock, W. G., and Pimentel, G. C., *J. Chem. Phys.*, 1968, **48**, 2373; Arkell, A., and Schwager, I., *J. Amer. Chem. Soc.*, 1967, **89**, 5999.
19. Harcourt, R. D., *J. Chem. Education*, 1968, **45**, 779; *Australian J. Chem.*, 1969, **22**, 279.
20. Linnett, J. W., 'Electronic Structure of Molecules,' Methuen, London, 1964, p. 78.
21. Moore, C. B., and Pimentel, G. C., *J. Chem. Phys.*, 1963, **38**, 2816; 1964, **40**, 329, 342.
22. Kirchhoff, W. H., and Wilson, E. B., *J. Amer. Chem. Soc.*, 1963, **85**, 1726.
23. Mirri, A. M., and Guarnieri, A., *Spectrochim. Acta*, 1967, **23A**, 2159.
24. Simpson, C. J. S. M., and Linnett, J. W., *Trans. Faraday Soc.*, 1959, **55**, 857.

CHAPTER 2

The Hyper Raman Effect

D. A. LONG

A. Introduction

My first research problem in Raman spectroscopy was concerned with the measurement of vibrational Raman intensities. This involved a detailed analysis, based on the Placzek polarizability theory,[1] of the factors controlling Raman intensities. This revealed the considerable experimental problems which arise in making meaningful intensity measurements when the scattering sample is illuminated with an extensive non-directional light source like a mercury arc. With laser excitation of Raman spectra, these problems are much simplified because of the unique directional properties of laser radiation. This is but one of the many advantages that lasers possess as sources for Raman spectroscopy. Many of the other advantages are dealt with specifically in subsequent chapters: these include the study of rotational and vibration-rotation Raman spectra under high resolution (Chapter 6), the Raman spectra of crystals (Chapters 5 and 8), and the investigation of vibrational Raman spectra in the vapour phase (Chapter 5). The general ease with which good quality Raman spectra can be obtained from very small quantities of imperfect samples using laser excitation will also be apparent from the relative abundance of information now available from Raman spectroscopy and discussed in the several chapters dealing with vibrational assignments (particularly Chapters 3, 13, 16 and 17).

It is, perhaps, less widely appreciated that not only have lasers facilitated the exploitation of Raman spectroscopy, they have revealed new spectroscopic phenomena. In this chapter I will give an account of one such novel phenomenon, the hyper Raman effect. This effect may be given a satisfactory theoretical treatment by extending the Placzek theory to take into account higher-order terms. This extension of my first joint paper[2] with L. A. Woodward seemed an appropriate choice of subject for my contribution to this volume.

B. The origin of the hyper Raman effect

The dipole moment induced in a molecule by an incident electric field is related to the field strength by the series expansion†

$$\mu_i = 4\pi\epsilon_0(\alpha_{ij}E_j + \tfrac{1}{2}\beta_{ijk}E_jE_k + \tfrac{1}{6}\gamma_{ijkl}E_jE_kE_l\ldots) \qquad (2.1)$$

where the subscripts i, j, k, l denote Cartesian vectors x, y, z. The first order polarizability α is a second order tensor with components α_{ij}. It determines the contribution to the induced moment from the first power of the field E. The second and third order polarizabilities or first and second hyperpolarizabilities β and γ are respectively third and fourth order tensors with components β_{ijk} and γ_{ijkl}. They determine the contributions to the induced moment from E^2 and E^3. It should be noted that whereas α is *always* non-zero for a molecule, symmetry conditions require β to be zero for all molecules with a centre of symmetry. The existence of higher-order polarizabilities has been appreciated for some time and molecular hyperpolarizabilities have been the subject of a recent review by Buckingham and Orr.[3]

For normal field-strengths only the first polarizability α makes any significant contribution to the induced moment. This is a consequence of the smallness of β and γ compared with α. However for very large electric field-strengths the contribution to the induced moment from the first hyperpolarizability begins to be significant. Field-strengths of the required magnitude are readily attainable with certain types of lasers, as for example a ruby laser operating in the giant pulse mode.

It is well known that the components α_{ij} of α control Rayleigh scattering and Raman scattering arising from rotational transitions and that the derivatives of α_{ij} with respect to normal coordinates of vibration control vibrational Raman scattering. Light scattering phenomena, broadly described as non-linear in origin, also arise from the higher order polarizabilities. For example, with sufficiently intense incident radiation at frequency ν the first hyperpolarizability β gives rise to light scattering at frequency 2ν and $2\nu \pm \nu_M$ where ν_M is the frequency associated with a transition between two levels of the scattering molecule. These new frequencies may be described as hyper Rayleigh and hyper Raman

† The reasons for the choice of this definition involving ϵ_0 the permittivity of free space are discussed in Appendix I. For convenience the factor $4\pi\epsilon_0$ is omitted in the ensuing derivations.

scattering respectively since they are controlled by the hyperpolarizability. Hyper Rayleigh scattering is often referred to as optical frequency doubling and is a special case of optical harmonic generation. Higher harmonics will arise from the higher order polarizabilities, the second hyperpolarizability γ giving rise to third harmonic generation or frequency trebling and so on.

The possibility of a hyper Raman effect seems to have been suggested first by Decius and Rauch.[4] Some aspects of the theory of non-linear light scattering were developed by Kielich[5] and by Li.[6] The first experimental observation of hyper Raman spectra was reported by Terhune, Maker and Savage[7] who obtained spectra for water, fused quartz and carbon tetrachloride. Since the hyper Raman effect is many orders of magnitude weaker than the normal Raman effect (typically 10^{-6} of Raman intensities) their work represented a considerable experimental triumph. Late in 1965 Cyvin, Rauch and Decius[8] published an extensive treatment of the selection rules for the hyper Raman effect for molecular point groups. Peticolas *et al.*[9,10] have recently extended this treatment to include helical polymers. Long and Stanton[11] have given an account of the theory of the hyper Raman effect which includes a treatment of resonance phenomena.

A number of generalizations emerge from a comparison of the symmetry properties of the β_{ijk} with those of α_{ij} and μ_i, but undoubtedly the two most significant are (i) many modes *inactive* in both Raman and infrared are *hyper Raman active* (ii) *all* infrared active modes are hyper Raman active. Hyper Raman spectroscopy therefore offers a direct method for the investigation of molecular frequencies previously regarded as spectroscopically inaccessible. Furthermore since hyper Raman spectra necessarily yield all the frequency information obtainable from infrared spectroscopy and the experimental techniques for hyper Raman spectroscopy could be readily adapted for normal Raman spectroscopy, it becomes possible to think of obtaining Raman, hyper Raman and infrared spectra with essentially one piece of equipment.

The very low intensity of the hyper Raman effect calls for the development of highly sensitive spectrometers for light scattering spectroscopy. Terhune, Maker and co-workers have continued experimental work in this field and by gradual refinement of their techniques of detection have obtained a number of interesting results.[12] Further advances in experimental techniques are however needed before the hyper Raman effect can become a widely used tool.

It is intriguing to reflect on the general situation in which we now find ourselves. More complete spectroscopic information can be obtained from a higher-order effect like hyper Raman spectroscopy since the selection rules are less restrictive. However by their very nature such effects are weak and more difficult to observe.

It is timely to consider the theoretical foundations of the hyper Raman effect and this is the main purpose of this chapter. Following the theoretical treatment in sections C and D the selection rules are considered in section E. Experimental techniques and results are briefly treated in section F.

C. Classical treatment

The frequency dependence of the scattered radiation and an insight into the selection rules for the vibrational hyper Raman effect can be obtained from an extension of the classical treatment of light scattering to include the hyperpolarizability. In the following treatment the appropriate subscripts denoting the vector components of μ and E and the tensor components of α and β have been omitted.

The time dependence of the electric field E of the incident radiation of frequency ω_0 is given by

$$E = E_0 \cos \omega_0 t \tag{2.2}$$

where E_0 is the amplitude, and $\omega_0 = 2\pi\nu_0$.

The polarizability α and the hyperpolarizability β will in general be functions of the normal coordinate of vibration Q associated with some fundamental vibrational frequency ω_M of the scattering molecules. The appropriate relationships can be obtained by expanding α and β in power series in Q as follows

$$\alpha = \alpha_0 + \left(\frac{\partial \alpha}{\partial Q}\right)_0 Q + \text{(higher order terms)} \tag{2.3}$$

$$\beta = \beta_0 + \left(\frac{\partial \beta}{\partial Q}\right)_0 Q + \text{(higher order terms)} \tag{2.4}$$

where the subscript '0' on the derivatives indicates that these are to be taken at the equilibrium value of Q. The higher-order terms involve second and higher derivatives of α and β and second and higher powers of Q; they will be neglected for the time being.

The time dependence of the normal coordinate of vibration Q is given in the simple harmonic approximation by

$$Q = Q_0 \cos \omega_M t \tag{2.5}$$

where Q_0 is the amplitude.

Insertion of (2.5) in (2.3) and (2.4) gives the time dependence of α and

β resulting from the vibration of frequency ω_M.

$$\alpha = \alpha_0 + \left(\frac{\partial \alpha}{\partial Q}\right)_0 Q_0 \cos \omega_M t \tag{2.6}$$

$$\beta = \beta_0 + \left(\frac{\partial \beta}{\partial Q}\right)_0 Q_0 \cos \omega_M t \tag{2.7}$$

The dipole μ induced by the incident field E is given to the second order by

$$\mu = \alpha E + \tfrac{1}{2}\beta E^2 \tag{2.8}$$

Insertion in (2.8) of the time dependence of α, β and E given by equations (2.6), (2.7) and (2.2) respectively gives the following expression for the time dependence of μ.

$$\mu = \left[\alpha_0 + \left(\frac{\partial \alpha}{\partial Q}\right)_0\right] E_0\, Q_0 \cos \omega_0 t \cos \omega_M t + \tfrac{1}{2}\left[\beta_0 + \left(\frac{\partial \beta}{\partial Q}\right)_0\right] \times E_0^2\, Q_0 \cos^2 \omega_0 t \cos \omega_M t \tag{2.9}$$

With a little rearrangement and the use of well-known trigonometric relationships it can be shown that there are eight distinct frequency components in the total induced dipole. These are

(i) $\omega = 0$: a d.c. field term

$$\mu = \tfrac{1}{4}\beta_0 E_0^2 \tag{2.10}$$

(ii) $\omega = \omega_0$: Rayleigh scattering term

$$\mu = \alpha_0 E_0 \cos \omega_0 t \tag{2.11}$$

(iii) $\omega = \omega_0 \pm \omega_M$: Raman Stokes and anti-Stokes terms

$$\mu = \tfrac{1}{2}E_0\, Q_0 \left(\frac{\partial a}{\partial Q}\right)_0 \{\cos(\omega_0 + \omega_M)\, t + \cos(\omega_0 - \omega_M)\, t\} \tag{2.12}$$

(iv) $\omega = 2\omega_0$: hyper Rayleigh scattering term

$$\mu = \tfrac{1}{4}\beta_0 E_0^2 \cos 2\omega_0 t \tag{2.13}$$

(v) $\omega = 2\omega_0 \pm \omega_M$: hyper Raman Stokes and anti-Stokes terms

$$\mu = \tfrac{1}{8}\left(\frac{\partial \beta}{\partial Q}\right)_0 Q_0 E_0^2 \{\cos(2\omega_0 + \omega_M)\, t + \cos(2\omega_0 - \omega_M)\, t\} \tag{2.14}$$

(vi) $\omega = \omega_M$: molecular frequency term

$$\mu = \tfrac{1}{4}\left(\frac{\partial \beta}{\partial Q_0}\right) Q_0 E_0^2 \cos \omega_M t \tag{2.15}$$

In the above treatment, to avoid cumbersome formulae, it has been tacitly assumed that there is no phase difference between the electric field and the molecular vibration. In general this is not so, and the treatment can be made perfectly general by rewriting (2.5) as

$$Q = Q_0 \cos(\omega_M t + \delta) \qquad (2.16)$$

where δ is an arbitrary phase factor. The formulae (2.12), (2.14) and (2.15) must then be modified by replacing $\omega_M t$ by $(\omega_M t + \delta)$.

Three of these terms involving ω_0, and $\omega_0 \pm \omega_M$ corresponding to Rayleigh and Stokes and anti-Stokes Raman scattering, are well known and will not be considered further. The remaining five terms all involve the hyperpolarizability β and hence necessarily E_0^2. Two of these, the d.c. field term and the molecular frequency term, will not be considered further here. The other three terms with frequency dependence $2\omega_0$, and $2\omega_0 \pm \omega_M$ correspond to hyper Rayleigh and Stokes and anti-Stokes hyper Raman scattering. It can be seen that the hyper Rayleigh effect requires the scattering molecule to have a non-zero hyperpolarizability β; the hyper Raman effect associated with a molecular vibration frequency ω_M requires the first derivative of the hyperpolarizability $(\partial\beta/\partial Q)_0$ to be non-zero for this mode of vibration.

This treatment can be readily extended to take into account terms in E^3 which involve the second hyperpolarizability γ. The frequency components of the induced dipole then include terms in $3\omega_0$, ω_0, $3\omega_0 \pm \omega_M$ and $\omega_0 \pm \omega_M$. Thus, in addition to frequency trebling and a second hyper Raman effect, there are contributions to normal Rayleigh and Raman scattering from γ.

If higher order terms in the power series (2.3) and (2.4) for α and β are considered, overtones and combination bands become permitted in both the normal Raman and the hyper Raman effects. Some discussion of these factors in relation to the hyper Raman effect will be given later. If there is mechanical anharmonicity equation (2.5) no longer holds and overtones and combination bands can also arise in this way.

D. Density matrix treatment

(a) *Calculation of the second-order transition moment*

The theoretical treatment of the Raman effect that has proved most useful is that due to Placzek.[1] In this treatment Placzek found expressions for the wave functions of a molecule perturbed by the electric field associated with electromagnetic radiation of a given frequency and used them to calculate the transition moment between two states, k and n of the perturbed system. Rayleigh and Raman scattering may be accounted for, by considering only those terms in the transition moment which are linearly dependent on the incident electric field. If, however, terms in the transition moment which involve the square of the electric field strength are included, hyper Rayleigh and hyper Raman scattering may be treated. This procedure is somewhat tedious since it requires the calculation of the second order contributions to the perturbed wave functions. For the higher order approximations necessary for the treatment of non-linear phenomena it is more convenient to use the density matrix method. The difference between the two approaches may be illustrated as follows.

According to Placzek, the transition moment from a state k to a state n under the influence of a perturbing electric field E associated with electromagnetic radiation of a given frequency defined† by

$$E = A \exp(-i\omega t) + A^* \exp(i\omega t) \tag{2.17}$$

is given by the matrix element

$$P_{nk} = \langle \Psi_n' | M | \Psi_k' \rangle \tag{2.18}$$

where M is the dipole moment operator.

Ψ_n' is the time dependent wave function of the perturbed system in the state n and to a second order this is given by

$$\Psi_n' = \Psi_n + \sum_r (a_{rn}^{(1)} + a_{rn}^{(2)}) \Psi_r \tag{2.19}$$

where Ψ_n is the unperturbed time dependent wave function of the state n and the summation is over all states r of the system.

The first order coefficients $a_{rn}^{(1)}$ are linear in A and the second order coefficients $a_{rn}^{(2)}$ are quadratic in A. The transition moments for the hyper Rayleigh ($k = n$) and hyper Raman ($k \neq n$) effects may be obtained from the terms in (2.18) which involve A^2.

Stanton[13] has shown that a transition moment such as (2.18) can be expressed in an alternative form involving only the unperturbed wave functions of the system by making use of the density matrix. In its most general form the transition moment from state k to state n when the system is subject to a perturbation is given by

$$P_{nk} = \sum_{\mu,\nu} \langle \Psi_n | \rho^{(\mu)} M \rho^{(\nu)} | \Psi_k \rangle \tag{2.20}$$

where μ and ν are orders of approximation of the density operator.

A matrix element ρ_{ij} of the density operator ρ is defined by

$$\langle \psi_i | \rho | \psi_j \rangle = \overline{c_i c_j^*} \tag{2.21}$$

where the bar denotes an ensemble average and the c_i are expansion coefficients in the expansion of a perturbed wave function Ψ' in terms of

† This is a more general form than equation (2.2). It includes all polarization cases. For example for plane polarized light (say $A_x \neq 0$) equation (2.17) becomes $E = 2A_x \cos \omega t$ whence we see $2A_x = E_0$.

a complete set of orthonormal *time-independent* functions:

$$\Psi' = \sum_i c_i \psi_i \tag{2.22}$$

The matrix elements of ρ may be found in principle from the Liouville equation

$$\dot{\rho} = \mathrm{i}\hbar^{-1}[\rho, H] \tag{2.23}$$

but in practice the desired solution is obtained by expanding the density operator in a power series

$$\rho = \rho^{(0)} + \lambda\rho^{(1)} + \lambda^2 \rho^{(2)} \tag{2.24}$$

where the $\rho^{(\mu)}$ are various orders of approximation of the density operator. If the operator for the perturbed system is written in the form

$$H = H^0 + \lambda H' \tag{2.25}$$

where H^0 is the operator for the unperturbed system, then on substituting (2.24) and (2.25) into (2.23) and equating powers of the parameter λ, a hierarchy of equations is obtained the nth of which is given by

$$\mathrm{i}\hbar\dot{\rho}^{(n)} = [H^0, \rho^{(n)}] + [H', \rho^{(n-1)}] \tag{2.26}$$

Radiation damping may be introduced phenomenonologically by adding to the right hand side of equation (2.26) a term $\mathrm{i}\hbar\dot{\rho}$ (damping) which is defined as

$$\dot{\rho}\ (\text{damping})^{(n)} = -\Gamma\rho^{(n)} \tag{2.27}$$

The complete nth order Liouville equation then becomes

$$\mathrm{i}\hbar\dot{\rho}^{(n)} = [H^0, \rho^{(n)}] - \mathrm{i}\hbar\Gamma\rho^{(n)} + [H', \rho^{(n-1)}] \tag{2.28}$$

with corresponding matrix elements defined by

$$\mathrm{i}\hbar\dot{\rho}_{lm}^{(n)} = [H^0, \rho^{(n)}]_{lm} - \mathrm{i}\hbar\Gamma\rho_{lm}^{(n)} + [H', \rho^{(n-1)}]_{lm} \tag{2.29}$$

The total order of dependence of the transition moment (2.18) on the electric field strength is given by

$$\gamma = \mu + \nu \tag{2.30}$$

Thus terms in the transition moment which are second order in the field arise when $\gamma = 2$ (i.e. $\mu = 1$, $\nu = 1$; $\mu = 2$, $\nu = 0$ and $\mu = 0$, $\nu = 2$). Consequently we may write for the second order transition moment

$$P_{nk}^{(2)} = \langle \Psi_n | \rho^{(0)} M \rho^{(2)} + \rho^{(2)} M \rho^{(0)} + \rho^{(1)} M \rho^{(1)} | \Psi_k \rangle \tag{2.31}$$

This may be evaluated[11] by expanding the Ψ in complete sets of functions and introducing the solutions for the density matrix elements of the various orders.

Expansion of (2.31) in complete sets of functions gives

$$P_{nk}^{(2)} = \{\sum_{r,s} (\rho_{nr}^{(0)} M_{rs} \rho_{sk}^{(2)} + \rho_{nr}^{(2)} M_{rs} \rho_{sk}^{(0)} + \rho_{nr}^{(1)} M_{rs} \rho_{sk}^{(1)})\} \exp(-\mathrm{i}\omega_{kn} t) \quad (2.32)$$

Remembering that

$$\langle \Psi_s | \rho^0 | \Psi_r \rangle = \delta_{sr} \quad (2.33)$$

equation (2.32) becomes

$$P_{nk}^{(2)} = \sum_{r,s} \{(M_{ns} \rho_{sk}^{(2)} + \rho_{ns}^{(2)} M_{sk} + \rho_{nr}^{(1)} M_{rs} \rho_{sk}^{(1)})\} \exp(-\mathrm{i}\omega_{kn} t) \quad (2.34)$$

The solutions for the first and second order density matrix elements required in (2.34) are found to be

$$\rho_{lm}^{(1)} = \frac{\hbar^{-1} H'_{lm} (\rho_{mm}^{(0)} - \rho_{ll}^{(0)})}{(\omega_{ml} \pm \omega - \mathrm{i}\Gamma_{ml})} \quad (2.35)$$

$$\rho_{lm}^{(2)} = \frac{\hbar^{-1} [H, \rho^{(1)}]_{lm}}{(\omega_{ml} + f(\omega) - \mathrm{i}\Gamma_{ml})} \quad (2.36)$$

The + sign in the denominator of (2.35) occurs when the perturbation H' involves the first term in (2.17) and the − sign for the second term in (2.17).

The situation is a little more complicated in the case of (2.36) which effectively involves H' twice. Since H' involves E and this latter involves two field amplitudes A and A^* there are four possible products of two field terms. The quantity in (2.36) is to be interpreted in the following way

$$f(\omega) = 2\omega \text{ if } H' \text{ contains } AA$$

$$f(\omega) = -2\omega \text{ if } H' \text{ contains } A^* A^*$$

$$f(\omega) = 0 \text{ if } H' \text{ contains } AA^* \text{ or } A^* A$$

Using equations (2.35) and (2.36) and the above definitions for $f(\omega)$ substitution in (2.34) gives

$$P_{nk}^{(2)} = \hbar^{-2} \left\{ \sum_{r,s} \left[\frac{(M_{nr} \cdot A)(M_{rs})(M_{sk} \cdot A)}{(\omega_{sk} - \omega - \mathrm{i}\Gamma_{sk})(\omega_{rn} + \omega - \mathrm{i}\Gamma_{rn})} \right. \right.$$

$$+ \frac{(M_{nr} \cdot A)(M_{rs} \cdot A)(M_{sk})}{(\omega_{sn} + 2\omega - \mathrm{i}\Gamma_{sn})(\omega_{rn} + \omega \mathrm{i}\Gamma_{rn})}$$

$$\left. + \frac{(M_{ns})(M_{sr} \cdot A)(M_{rk} \cdot A)}{(\omega_{sk} - 2\omega - \mathrm{i}\Gamma_{sk})(\omega_{rk} - \omega - \mathrm{i}\Gamma_{rk})} \right] \exp(-\mathrm{i}(\omega_{kn} + 2\omega) t)$$

$$+ \sum_{r,s} \left[\frac{(M_{nr} \cdot A^*)(M_{rs})(M_{sk} \cdot A^*)}{(\omega_{sk} + \omega - \mathrm{i}\Gamma_{sk})(\omega_{rn} - \omega - \mathrm{i}\Gamma_{rn})} \right.$$

$$+ \frac{(M_{nr} \cdot A^*)(M_{rs} \cdot A^*)(M_{sk})}{(\omega_{sn} - 2\omega - \mathrm{i}\Gamma_{sn})(\omega_{rn} - \omega - \mathrm{i}\Gamma_{rn})}$$

$$+\frac{(M_{ns})\,(M_{sr}\cdot A^*)\,(M_{rk}\cdot A^*)}{(\omega_{sk}+2\omega-\mathrm{i}\Gamma_{sk})\,(\omega_{rk}+\omega-\mathrm{i}\Gamma_{rk})}\Bigg]\exp\left(-(\omega_{kn}-2\omega)\,t\right)$$

$$+\sum_{r,s}\Bigg[\frac{(M_{nr}\cdot A)\,(M_{rs})\,(M_{sk}\cdot A^*)}{(\omega_{sk}+\omega-\mathrm{i}\Gamma_{sk})\,(\omega_{rn}+\omega-\mathrm{i}\Gamma_{rn})}$$

$$+\frac{(M_{nr}\cdot A^*)\,(M_{rs})\,(M_{sk}\cdot A)}{(\omega_{sk}-\omega-\mathrm{i}\Gamma_{sk})\,(\omega_{rn}-\omega-\mathrm{i}\Gamma_{rn})}$$

$$+\frac{(M_{nr}\cdot A^*)\,(M_{rs}\cdot A)\,(M_{sk})}{(\omega_{sn}-\mathrm{i}\Gamma_{sn})\,(\omega_{rn}-\omega-\mathrm{i}\Gamma_{rn})}$$

$$+\frac{(M_{nr}\cdot A)\,(M_{rs}\cdot A^*)\,(M_{sk})}{(\omega_{sk}-\mathrm{i}\Gamma_{sk})\,(\omega_{rn}+\omega-\mathrm{i}\Gamma_{rn})}+\frac{(M_{ns})\,(M_{sr}\cdot A^*)\,(M_{rk}\cdot A)}{(\omega_{sk}-\mathrm{i}\Gamma_{sk})\,(\omega_{rk}-\omega-\mathrm{i}\Gamma_{rk})}$$

$$+\frac{(M_{ns})\,(M_{sr}\cdot A)\,(M_{rk}\cdot A^*)}{(\omega_{sk}-\mathrm{i}\Gamma_{sk})\,(\omega_{rk}+\omega-\mathrm{i}\Gamma_{rk})}\Bigg]\exp\left(-\mathrm{i}\omega_{kn}t\right)\Bigg\} \tag{2.37}$$

Equation (2.37) may be written in a condensed notation as follows

$$P_{nk}^{(2)}=P_{nk}^{(2)}(\omega_{kn}+2\omega)+P_{nk}^{(2)}(\omega_{kn}-2\omega)+P_{nk}^{(2)}(\omega_{kn})\,. \tag{2.38}$$

(b) The frequency dependence of the scattered radiation

The moment defined by (2.37) is complex. The corresponding real dipole may be found by application of Klein's condition which requires that a term of the form $M_{ji}\exp(-\mathrm{i}\omega_{ij}t)$ leads to scattering of frequency ω_{ij} if

$$\omega_{ij}>0 \tag{2.39}$$

The radiation then corresponds to the classically radiated intensity associated with the quantity

$$M_{ji}\exp(-\mathrm{i}\omega_{ij}t)+M_{ji}^*\exp(\mathrm{i}\omega_{ij}t) \tag{2.40}$$

It follows that the first term of the second order transition moment (2.37) leads to scattering if

$$\omega_{kn}+2\omega>0 \tag{2.41}$$

Under most circumstances likely to be encountered in practice:

$$|2\omega|>|\omega_{kn}| \tag{2.42}$$

Thus the sign of ω_{kn} in (2.41) is irrelevant and so

$$\omega_{kn}\gtrless 0 \tag{2.43}$$

In the case where $\omega_{kn}>0$ the initial state k is higher in energy than the final state n. Thus scattering through a downward transition is involved and this corresponds to hyper Raman anti-Stokes emission at $2\omega + |\omega_{kn}|$. Similarly when $\omega_{kn} < 0$, the initial state lies lower in energy than the final state and this corresponds to Stokes hyper Raman emission at $2\omega - |\omega_{kn}|$. The special case of hyper Rayleigh scattering at 2ω (or second harmonic generation) occurs when ω_{kn} is identically zero.

The second time dependence in (2.37) has the form $\exp -i(\omega_{kn} - 2\omega)t$ and this may lead to scattering when

$$\omega_{kn} - 2\omega > 0 \tag{2.44}$$

This implies that

$$\omega_{kn} > 2\omega \tag{2.45}$$

so that the transition ω_{kn} has a greater frequency than that of the incident radiation. Since ω must necessarily be positive, then ω_{kn} must also be positive and thus the initial state k must be higher in energy than state n. Again this corresponds to an induced emission from an excited state and the emission frequency $\omega_{kn} - 2\omega$ must be accompanied by another at 2ω to preserve energy balance.

The third, and final, time dependent portion of (2.37) involves a frequency ω_{kn}. Klein's condition requires that

$$\omega_{kn} > 0 \tag{2.46}$$

and so state k lies above state n. This is therefore an emission from an excited state. Under typical circumstances this would be a quantum of vibrational energy and would appear in the infrared region. This process is not a spontaneous relaxation however, for it involves a hyperpolarizability which interacts quadratically with the incident radiation field to induce the emission.

In the special case when ω_{kn} is zero the last term in (2.37) describes a static polarization of the system. This is sometimes termed optical rectification and has been observed in practice by Bass and co-workers.[14]

We shall only need to concern ourselves further with the first term in (2.37), that is

$$P_{kn}^{(2)}(\omega_{kn} + 2\omega) \tag{2.47}$$

(c) *The role of the intermediate states*

The first term in $P_{nk}^{(2)}$ which controls hyper Raman scattering (and also hyper Rayleigh if $k = n$) is relatively complicated. Summations over two sets of intermediate states r and s are involved whereas the corresponding first order term $P_{nk}^{(1)}$ for the Raman effect involves summation only over one set of states r. The intermediate states may lie above, below, or between the initial and final states for it is only the *sums* over unperturbed states which have meaning in the perturbed system.

In the sums over matrix elements of the dipole operator which occur in (2.37), only those intermediate states r and s are important for which there are finite transition moments connecting them both to each other and to the initial and final state. That is to say only those terms occur for which

$$M_{rk} \neq 0; \; M_{sr} \neq 0; \; M_{ns} \neq 0 \tag{2.48}$$

The first and third types of inequalities occur in the Raman effect but the second term does not. This fact is connected with the relaxation of selection rules in the hyper Raman effect, relative to those for the normal Raman effect.

Since there is a total of twenty-seven components in a third rank tensor, as opposed to nine in a second rank tensor, the hyperpolarizability will naturally have more entries to distribute among the various symmetry species of a given point group than the ordinary polarizability. More flexible selection rules are therefore to be expected. Physically, this is really a direct result of the existence of non-zero matrix elements which permit a transition between intermediate states, where the symmetry requirements are appropriate. This transition may allow a connection to be made between k and n via the intermediate states, where otherwise such a connection would not be possible.

(*d*) *Resonance conditions*

When the denominator in a polarizability expression tends to zero, the term is said to be resonant. The polarizability then assumes a very large value with consequent intense scattering. This state of affairs is known to occur in the ordinary Raman effect when there is an emission or absorption band essentially coincident with the incident frequency. In the ordinary Raman effect, there are two possible resonances which have been described previously by Placzek. However in the hyper Raman effect nine resonance processes may be distinguished by considering the various ways in which the frequency denominators in the hyper Raman polarizability (the terms in (2.37) with frequency dependence $(\omega_{kn} + 2\omega)$) can tend to zero.

(i) Resonance condition

$$\omega = \omega_{sk}; \; \omega_{rn} + \omega \neq 0 \tag{2.49}$$

Since of necessity $\omega > 0$ then $\omega_{sk} > 0$. Now by Klein's condition

$$2\omega + \omega_{kn} > 0 \tag{2.50}$$

for hyper Raman emission, thus

$$2\omega_s - 2\omega_k + \omega_k - \omega_n > 0 \tag{2.51}$$

Rearrangement of (2.51) yields

$$\omega_s - \omega_k > \omega_n - \omega_s \tag{2.52}$$

Since ω_{sk} is positive then from (2.52)

$$\omega_s \gtrless \omega_n \tag{2.53}$$

Thus from (2.52) and (2.53) it may be concluded that the state s must lie higher in energy than state k. It may lie between k and n, or above n, but in either event it must lie nearer n than to k. Thus this resonance may be summarized: if there is an absorption band of k terminating closer to n than to k then there will be resonance scattering.

(ii) Resonance condition

$$\omega = -\omega_{rn};\ \omega_{sk} - \omega \neq 0 \tag{2.54}$$

Since $\omega > 0$ then

$$\omega_n > \omega_r \tag{2.55}$$

Therefore state r lies lower in energy than state n.

Now by Klein's condition

$$2\omega + \omega_{kn} > 0$$

and thus

$$2(\omega_n - \omega_r) + (\omega_k - \omega_n) > 0 \tag{2.56}$$

Rearrangement of equation (2.56) gives

$$(\omega_n - \omega_r) + (\omega_k - \omega_r) > 0; \tag{2.57}$$

but then the term $(\omega_n - \omega_r)$ has been shown (2.55) to be positive and thus using this information and (2.57)

$$\omega_r \gtrless \omega_k \tag{2.58}$$

Note however from (2.57) that

$$\omega_n - \omega_r > \omega_r - \omega_k \tag{2.59}$$

It is thus concluded that r is an emission state of n and that it lies closer to k than to n. There is no restriction on the position of the state s.

(iii) Resonance condition

$$\omega = -\omega_{rn} \quad \text{and} \quad \omega = \omega_{sk} \tag{2.60}$$

This is a different case from the two discussed above because it is the first example of a double resonance. From the conditions (2.60) it is seen that

$$\omega_n > \omega_r \quad \text{and} \quad \omega_s > \omega_k \tag{2.61}$$

therefore r lies lower in energy than n but s lies above state k. Since this is so it is interesting to find the relative energies of r and s. Recalling the relation (2.50) and using (2.60), the following inequality is obtained

$$\omega_n - \omega_r + \omega_s - \omega_k + \omega_k - \omega_n > 0 \tag{2.62}$$

This immediately gives

$$\omega_s > \omega_r \tag{2.63}$$

Therefore for the given conditions of this double resonance the order of the intermediate states is unique. We have resonance if *r* is an emission termination of *n* and provided that *s* is an absorption termination of *k*.

(iv) Resonance condition

$$\omega = -\omega_{rn};\ 2\omega + \omega_{sn} \neq 0 \tag{2.64}$$

The result of this resonance is the same as that of case (b).

(v) Resonance condition

$$2\omega = -\omega_{sn};\ \omega + \omega_{rn} \neq 0 \tag{2.65}$$

Similar arguments to those used previously yield

$$\omega_n > \omega_s \tag{2.66}$$

Using (2.50) and the first relation of (2.65) leads to

$$\omega_n - \omega_s + \omega_k - \omega_n > 0 \tag{2.67}$$

Thus

$$\omega_k > \omega_s \tag{2.68}$$

The state *s* is therefore an emission termination of the final state. The position of state *r* is undetermined.

(vi) Resonance condition

$$2\omega = -\omega_{sn} \quad \text{and} \quad \omega = -\omega_{rn} \tag{2.69}$$

This is another case of double resonance.

These relations immediately yield the inequalities,

$$\omega_n > \omega_s;\ \omega_n > \omega_r \tag{2.70}$$

Using relation (2.50) it emerges that

$$\omega_k \gtrless \omega_r \tag{2.71}$$

and furthermore that

$$\omega_n - \omega_r > \omega_r - \omega_k \tag{2.72}$$

Hence *r* and *s* are emission terminations of *n*. In addition it is required that *r* lies nearer *k* than to *n*. This particular resonance is unusual in that both intermediate states *r* and *s* combine resonantly with the final state.

(vii) Resonance condition

$$2\omega = \omega_{sk};\ \omega_{rk} - \omega \neq 0 \tag{2.73}$$

The result of this case is similar to that of (i) even though the left hand condition in (2.73) is not identical to that in (i).

(viii) Resonance condition

$$\omega_{rk} = \omega; \ \omega_{sk} \neq 0 \tag{2.74}$$

This leads to the same result as in (i).

(ix) Resonance condition

$$\omega = \omega_{rk} \quad \text{and} \quad 2\omega = \omega_{sk} \tag{2.75}$$

This is a double resonance case

$$\omega_r > \omega_k; \ \omega_s > \omega_k \tag{2.76}$$

Analysing in the usual way, the following inequalities result

$$\omega_s > \omega_n; \ \omega_n \gtrless \omega_r; \ \omega_r - \omega_k > \omega_n - \omega_r \tag{2.77}$$

Thus there is resonance in this situation if there is an absorption band at the incident frequency and at its first harmonic. This is a second case wherein the intermediate levels combine resonantly with the same state (this time state k). It is necessary in addition that the energy of state s is higher than that of state r.

Some interesting comparisons may be drawn between the resonance Raman effect (R.R.E.) and the resonance hyper Raman effect (R.H.R.E.). The R.R.E. occurs when there is an absorption band of the initial state or an emission band of the final state, having the same frequency as that of the incident radiation. The R.H.R.E. may occur when there are absorption or emission bands at the incident frequency, its optical harmonic or simultaneously at both the incident frequency and its harmonic. In the cases of resonant emission states the intermediate state is likely to lie lower in energy than the initial state k.

In the R.R.E. it is necessary for an absorption (say) to occur at the frequency of the exciting radiation. Since the scattered radiation lies close to the exciting frequency, it is often absorbed by the Raman medium and hence observation of the effect is rendered difficult. However, in the R.H.R.E. it is possible that observation may be facilitated, in certain molecules, by their possessing absorption bands at the incident frequency ω but not at 2ω which is the region in which the scattering is observed. This being so the resonance hyper Raman scattering would suffer but little attenuation before detection.

In the case of resonance hyper Raman emission due to an absorption band at 2ω there is, of course, the question of absorption of the scattering. This does not preclude its observation but all the difficulties found with the normal R.R.E. would be encountered. In the case of a double resonance, the same remarks may be made but a very large gain will occur, thus facilitating observation.

Long and Stanton[11] have estimated that under favourable conditions a single resonance can enhance the intensity of hyper Raman scattering by a factor of about 10^6 and a double resonance by 10^{12}. It may be that with tuneable lasers it will be possible to bring many molecules into resonance so facilitating the observation of hyper Raman scattering with all its advantages of less restrictive selection rules.

(*e*) *The third order scattering tensor*

Provided Klein's condition that $\omega_{kn} + 2\omega$ should be positive is satisfied, the real dipole associated with hyper Raman scattering is given by

$$B_{nk} \exp\{-\mathrm{i}(\omega_{kn} + 2\omega)\, t\} + B^*_{nk} \exp\{\mathrm{i}(\omega_{kn} + 2\omega)\, t\} \tag{2.78}$$

where

$$B_{nk} = \frac{1}{\hbar^2} \sum_{r,s} \left[\frac{(M_{nr} \cdot A)\,(M_{rs})\,(M_{sk} \cdot A)}{(\omega_{sk} - \omega - \mathrm{i}\Gamma_{sk})\,(\omega_{rn} + \omega - \mathrm{i}\Gamma_{rn})} + \frac{(M_{nr} \cdot A)\,(M_{rs} \cdot A)\,(M_{sk})}{(\omega_{sn} + 2\omega - \mathrm{i}\Gamma_{sn})\,(\omega_{rn} + \omega - \mathrm{i}\Gamma_{rn})} + \frac{(M_{ns})\,(M_{sr} \cdot A)\,(M_{rk} \cdot A)}{(\omega_{sk} - 2\omega - \mathrm{i}\Gamma_{sk})\,(\omega_{rk} - \omega - \mathrm{i}\Gamma_{rk})} \right] \tag{2.79}$$

If $k = n$ the dipole (2.78) corresponds to Rayleigh scattering.

So far, spatial indices have been omitted but when these are introduced it follows from (2.79) that the relationship between a component of the vector B and components of the field A can be written as

$$[B_i]_{nk} = \tfrac{1}{2} \sum_{jk} [b_{ijk}]_{nk} A_j A_k \tag{2.80}$$

where

$$b_{ijk} = \frac{2}{\hbar^2} \sum_{r,s} \left[\frac{[M_j]_{nr}\,[M_i]_{rs}\,[M_k]_{sk}}{(\omega_{sk} - \omega - \mathrm{i}\Gamma_{sk})\,(\omega_{rn} + \omega - \mathrm{i}\Gamma_{rn})} + \frac{[M_j]_{nr}\,[M_k]_{rs}\,[M_i]_{sk}}{(\omega_{sn} + 2\omega - \mathrm{i}\Gamma_{sn})\,(\omega_{rn} + \omega - \mathrm{i}\Gamma_{rn})} + \frac{[M_i]_{ns}\,[M_j]_{sr}\,[M_k]_{rk}}{(\omega_{sk} - 2\omega - \mathrm{i}\Gamma_{sk})\,(\omega_{rk} - \omega - \mathrm{i}\Gamma_{rk})} \right] \tag{2.81}$$

and the factor $\frac{1}{2}$ is introduced for consistency with equation 2.1.

$[b_{ijk}]_{nk}$ is a matrix element of the *ijk* component of the hyper Raman tensor. Similarly $[b_{ijk}]_{kk}$ is a matrix element of the *ijk* component of the hyper Rayleigh tensor.

For a transition from a state k to a state n to give rise to hyper Raman scattering it is necessary that the matrix element of at least one of the components of the hyper Raman tensor be non-zero. It can be shown that $[b_{ijk}]_{nk}$ transforms as

$$\langle n|b_{ijk}|k\rangle \tag{2.82}$$

where b_{ijk} transforms as the product of Cartesian vectors *ijk*. This property enables us to define the general condition necessary for $[b_{ijk}]_{nk}$ to be non-zero: the triple direct product of the species of the wave functions for the two states n and k and b_{ijk} must contain the totally symmetric species.

The hyper Raman and Rayleigh scattering tensors are third order tensors. A non-symmetric third order tensor can have 27 distinct components but permutation symmetry of the spatial suffices of the quadratic field terms $(A_i A_j = A_j A_i)$ reduces the number of distinct components to 18. The hyper Rayleigh scattering is normally symmetric provided the components are real and the frequency ω lies in a region of low dispersion. There is then complete permutation symmetry of the spatial suffices *ijk* and so the number of distinct components is reduced to 10. The hyper Raman tensor also becomes symmetric in certain circumstances. The conditions to be satisfied are just those under which the hyperpolarizability theory, discussed in the next section, applies.

The *symmetric* hyper Raman scattering tensor will be given the special name of hyperpolarizability tensor and for the transition $k \rightarrow n$, the appropriate matrix element of the *ijk* tensor component will be written as $[\beta_{ijk}]_{nk}$. The corresponding term for the symmetric Rayleigh scattering tensor is, of course, $[\beta_{ijk}]_{kk}$. We shall only concern ourselves further in this chapter with the symmetric scattering tensors.

(*f*) *The hyperpolarizability theory*

The expression (2.37) for $P_{nk}^{(2)}$, despite its relatively complicated nature, has yielded, fairly easily, information on the frequency dependence of the second-order scattering, the role of the intermediate states and the effect of near resonance. It has also permitted the development of the concept of the third-order tensor and has enabled a perfectly general statement to be made about the symmetry conditions necessary for such a tensor to have non-zero components. However it cannot readily be developed further in this form.

We see from (2.81) that a matrix element of the third order scattering tensor involves sums over all the states r of the system where the complete definition of *each* state involves *all* the quantum numbers, electronic and nuclear of this state. This evaluation of (2.81) requires a knowledge of molecular levels which generally speaking we do not possess. Assuming the applicability of the Born–Oppenheimer theory, then, provided (i) the incident frequency ω is much lower than ω_{el} the lowest electronic transition frequency of the system, (ii) the ground electronic state is non-degenerate and (iii) ω, ω_{el} and any vibrational frequency ω_v satisfy the conditions $\omega \gg \omega_v$ and $\omega_{el} - \omega \gg \omega_v$, we may make a sweeping simplification and replace (2.81) by

$$[\beta_{ijk}]_{nk} = 2\langle n|\beta_{ijk}|k\rangle \tag{2.83}$$

where k and n are now only the nuclear quantum numbers of the initial and final states and β_{ijk} is the ijk component of the symmetric hyperpolarizability tensor, which is a function of the nuclear coordinates only.

The matrix elements of the form (2.83), unlike those of the form (2.81), may be evaluated readily. So far the indices ijk have been used quite generally to denote any Cartesian axis system. It is now necessary to define two sets of Cartesian axis systems: a space-fixed system XYZ and a system xyz fixed in the scattering molecule. The transformation of the components of β under a rotation is given by

$$\beta_{XYZ} = \sum_{xyz} \cos(X,x)\cos(Y,y)\cos(Z,z)\,\beta_{xyz} \tag{2.84}$$

where β_{xyz} is a function of the vibrational coordinates of the molecule and is independent of the rotational state and the orientation of the molecule. The terms $\cos(X,x)$, etc. are direction cosines between the X and x axes, etc. and are functions of the Eulerian angles.

If the initial state k is defined by a set of rotational quantum numbers R'' and a set of vibrational quantum numbers v'' and the final state n by a set of rotational and vibrational quantum numbers R' and v', then we may write

$$\begin{aligned}[\beta_{XYZ}]_{nk} &= [\beta_{XYZ}]R'v', R''v'' \\ &= \sum_{xyz} \langle R'|\cos(X,x)\cos(Y,y)\cos(Z,z)|R''\rangle\langle v'|\beta_{xyz}|v''\rangle \\ &= \sum_{xyz} [\cos(X,x)\cos(Y,y)\cos(Z,z)]_{R'R''}[\beta_{xyz}]_{v'v''}\end{aligned} \tag{2.85}$$

These rotational and vibrational matrix elements are of great importance for the further development of a theory of the hyper Raman effect. The nature of the restrictions on changes in the vibrational and rotational quantum numbers follows from a development of the matrix elements defined in (2.85). These restrictions and also the symmetry requirements for non-zero matrix elements of components of the hyper Raman tensor based on the transformation properties of (2.82) in general and (2.83) and (2.85) in particular, are dealt with in the next section.

E. Selection rules, symmetry and activity

This discussion will be restricted to vibrational transitions.

(a) Permitted changes in vibrational quantum numbers

The vibrational matrix element $[\beta_{xyz}]_{v'v''}$ may be expanded by expressing β_{xyz} as a Taylor series in the normal coordinates of vibration Q_p as follows:

$$[\beta_{xyz}]_{v'v''} = \beta_0 \langle v'|v''\rangle + \sum_p \left(\frac{\partial \beta_{xyz}}{\partial Q_p}\right)_0 \langle v'|Q_p|v''\rangle + \text{higher terms} \quad (2.86)$$

This expansion, the validity of which was tacitly assessed in the classical treatment in section C, is justified provided the conditions for the applicability of the hyperpolarizability theory are satisfied.

The resulting integrals are exactly similar to those which arise in the Raman effect when α is expanded as a power series in Q. Consequently the restrictions on changes in vibrational quantum numbers in the hyper Raman effect are the same as those for the normal Raman effect.

In the harmonic oscillator approximation, the first term in (2.86) is zero unless $v' = v''$ (that is $\Delta v = 0$) when

$$\beta_0 \langle v'|v''\rangle = \beta_0 \quad (2.87)$$

This term therefore gives rise to hyper Rayleigh scattering which is controlled by β_0, the equilibrium value of the hyperpolarizability.

The second term in (2.86) can only be different from zero if, for only one normal vibration, say the *p*th, the vibrational quantum number changes by unity (that is $\Delta v_p = \pm 1$). If the set of vibrational quantum numbers defining the initial state is v_1'', v_2'', ... v_p'' and the set for the final state v_1', v_2' ... v_p' ..., then the selection rule is $v_1'' = v_1'$, $v_2'' = v_2'$, etc. but $v_p' = v_p'' \pm 1$.

For the Stokes case $v_p' = v_p'' + 1$ and the second term in (2.86) then has the value

$$(v_p'' + 1)^{1/2} \left(\frac{\partial \beta_{xyz}}{\partial Q_p}\right)_0 C_p \quad (2.88)$$

For the anti-Stokes case $(v_p' = v_p'' - 1)$ the value is

$$(v_p'')^{1/2} \left(\frac{\partial \beta_{xyz}}{\partial Q_p}\right)_0 C_p \quad (2.89)$$

where

$$C_p = \left(\frac{h}{8\pi^2 \nu_p}\right)^{1/2} \quad (2.90)$$

In these formulae and elsewhere it is assumed that the normal coordinate Q_p is mass-adjusted.

Thus in the harmonic oscillator approximation only fundamental vibrations $(\Delta v = \pm 1)$ arise in the hyper Raman effect, provided derivatives of β higher than the first are neglected in the expansion of β in terms of

Q. If second derivatives of β are taken into account, there arises a third term in (2.86) of the general form.

$$\frac{1}{2}\left(\frac{\partial^2 \beta_{xyz}}{\partial Q_p \partial Q_q}\right)_0 \langle v'|Q_p Q_q|v''\rangle \tag{2.91}$$

In the harmonic oscillator approximation, this term can be different from zero only if $v' = v''$ and $p = q$, or for one normal vibration only if $v_p' = v_p'' \pm 2$ or for two normal vibrations $v_p' = v_p'' \pm 1$ and $v_q' = v_q'' \pm 1$ (when $p \neq q$ in (2.91)).

When $v' = v''$ and $p = q$ (2.91) has the value

$$\tfrac{1}{2}(v_p'' + 1)\left(\frac{\partial^2 \beta_{xyz}}{\partial Q_p^2}\right)_0 C_p^2 \tag{2.92}$$

and forms an additional contribution to hyper Rayleigh scattering.

When $v_p' = v_p'' + 2$ and $p = q$ (2.91) has the value

$$\tfrac{1}{2}\{(v_p'' + 1)(v_p'' + 2)\}^{1/2}\left(\frac{\partial^2 \beta_{xyz}}{\partial Q_p^2}\right)_0 C_p^2 \tag{2.93}$$

and this controls the intensity of first overtones in the Stokes hyper Raman effect.

When $v_p' = v_p'' + 1$, $v_q' = v_q'' + 1$, and $p \neq q$ then (2.91) has the value

$$\tfrac{1}{2}\{(v_p'' + 1)(v_q'' + 1)\}^{1/2}\left(\frac{\partial^2 \beta_{xyz}}{\partial Q_p \partial Q_q}\right)_0 C_p C_q \tag{2.94}$$

and this controls the intensity of combination tones in the Stokes hyper Raman effect. Similar formulae can be derived for the anti-Stokes cases.

Overtones and combination bands can also arise in the hyper Raman effect as a result of mechanical anharmonicity. Bands arising in this way depend on the first derivative of β. We shall not consider these further here.

It must be emphasized that although the restrictions on changes in the vibrational quantum numbers given above are *necessary* conditions which must be satisfied for the appearance of vibrational bands in the hyper Raman effect they are not *sufficient* conditions. Unless certain symmetry conditions are also satisfied the vibrational matrix element will be zero even when the quantum numbers change only as specified. We shall consider these symmetry requirements in the next section. These symmetry requirements determine whether the derivatives of β in the expansion (2.86) are zero or non-zero.

(*b*) *Transformation properties and hyper Raman activity*

For $[\beta_{xyz}]_{v'v''}$, a vibrational matrix element of a component of the symmetric hyperpolarizability tensor to be non-zero the triple direct product of the species of the two states defined by the sets of vibrational quantum numbers v' and v'' and β_{xyz} must contain the totally symmetric species. The symmetry properties of vibrational states are known for all the molecular point groups.

The hyperpolarizability components β_{xyz} have the transformation properties of the product xyz of Cartesian vectors. These transformation properties have been treated very fully by Cyvin, Rauch and Decius[8] and the following account is based closely upon their work. These authors chose as the ten distinct components of β to be listed β_{xxx}, β_{yyy}, β_{zzz}, β_{xxy}, β_{xyy}, β_{yyz}, β_{zxx}, β_{zzx}, β_{yzz} and β_{xyz}. This choice will be adhered to here although it is to some extent arbitrary since $\beta_{xyy} = \beta_{yyx} = \beta_{yxy}$, etc. It may be shown that the character $\chi_\beta(\mathrm{R})$ of a reducible representation based on the ten distinct components of β under a group operation R is given by

$$\chi_\beta(\mathrm{R}) = 2\cos\phi(4\cos^2\phi \pm 2\cos\phi - 1) \qquad (2.93)$$

The arbitrarily chosen Cartesian coordinate system has Z as the rotation symmetry axis. In (2.93) ϕ is the angle of rotation. For proper rotations C_n^k (including the identity E), the + sign is taken, and for improper rotations S_n^k (including σ and i) the − sign is taken; $\phi = 2\pi k/n$. When the character of the reducible representation is known, standard methods of group theory may be used to find its structure, that is the number of times it contains each irreducible representation for the group in question. If a given irreducible representation is contained n times in the reducible representation then n linear combinations of the components of β may be found which transform as the irreducible representation and so belong to a given symmetry species.

Table 2.1 gives for all the molecular point groups the proper linear combinations of β components together with the more familiar results for the polarizability components α_{ij} and the dipole components μ_i. This table is reproduced from the paper of Cyvin, Decius and Rauch.[8] The table uses the species designation found in Wilson, Decius and Cross.[15] Attention must be drawn to the fact that in some instances the distribution of β combinations depends on the orientation of the Cartesian axes with respect to the symmetry elements. This sort of ambiguity is not encountered when only the μ's and α's are considered; it is therefore appropriate to give a more detailed discussion. One typical example is found in the D_{3h} group. A molecule of this symmetry is a symmetric top where the x and y axes are equivalent. It is therefore not immediately understood why the combination $\beta_{xxx} - 3\beta_{xyy}$ should be totally symmetric (see Table 2.1) rather than $\beta_{yyy} - 3\beta_{xxy}$, since one of these combinations goes into the other on interchanging x and y. As a matter of fact, all the four components β_{xxx}, β_{yyy}, β_{xxy}, and β_{xyy} belong to the same symmetric equivalent set in the

TABLE 2.1. *Symmetry properties of dipole moment* (μ), *polarizability* (α) *and hyperpolarizability* (β) *components.* (*Reproduced with the permission of Cyvin, Decius and Rauch*[8] *and the editorial board of the* Journal of Chemical Physics.)

		A. The groups C_s, C_i, and the cyclic groups C_n ($n = 2, 3, 4, 5, 6$)		
C_s	a′	μ_x, μ_y	$\alpha_{xx}, \alpha_{yy}, \alpha_{zz}, \alpha_{xy}$	$\beta_{xxx}, \beta_{yyy}, \beta_{xxy}, \beta_{zzx}, \beta_{xyy}, \beta_{yzz}$
	a″	μ_z	α_{yz}, α_{zx}	$\beta_{zzz}, \beta_{yyz}, \beta_{zxx}, \beta_{xyz}$
C_i	a_g		$\alpha_{xx}, \alpha_{yy}, \alpha_{zz}, \alpha_{xy}, \alpha_{yz}, \alpha_{zx}$	
	a_u	μ_x, μ_y, μ_z		$\beta_{xxx}, \beta_{yyy}, \beta_{zzz}, \beta_{xxy}, \beta_{yyz}, \beta_{zzx}, \beta_{xyy}, \beta_{yzz}, \beta_{zxx}, \beta_{xyz}$
C_2	a	μ_z	$\alpha_{xx}, \alpha_{yy}, \alpha_{zz}, \alpha_{xy}$	$\beta_{zzz}, \beta_{yyz}, \beta_{zxx}, \beta_{xyz}$
	b	μ_x, μ_y	α_{yz}, α_{zx}	$\beta_{xxx}, \beta_{yyy}, \beta_{xxy}, \beta_{zzx}, \beta_{xyy}, \beta_{yzz}$
C_3	a	μ_z	$\alpha_{xx} + \alpha_{yy}, \alpha_{zz}$	$\beta_{xxx} - 3\beta_{xyy}, \beta_{yyy} - 3\beta_{xxy}, \beta_{yyz} + \beta_{zxx}, \beta_{zzz}$
	e	(μ_x, μ_y)	$(\alpha_{xx} - \alpha_{yy}, \alpha_{xy}), (\alpha_{yz}, \alpha_{zx})$	$(\beta_{xxx} + \beta_{xyy}, \beta_{yyy} + \beta_{xxy}), (\beta_{yyz} - \beta_{zxx}, \beta_{xyz}), (\beta_{zzx}, \beta_{yzz})$
C_4	a	μ_z	$\alpha_{xx} + \alpha_{yy}, \alpha_{zz}$	$\beta_{yyz} + \beta_{zxx}, \beta_{zzz}$
	b		$\alpha_{xx} - \alpha_{yy}, \alpha_{xy}$	$\beta_{yyz} - \beta_{zxx}, \beta_{xyz}$
	e	(μ_x, μ_y)	$(\alpha_{yz}, \alpha_{zx})$	$(\beta_{xxx}, \beta_{yyy}), (\beta_{xyy}, \beta_{xxy}), (\beta_{zzx}, \beta_{yzz})$
C_5	a	μ_z	$\alpha_{xx} + \alpha_{yy}, \alpha_{zz}$	$\beta_{yyz} + \beta_{zxx}, \beta_{zzz}$
	e_1	(μ_x, μ_y)	$(\alpha_{yz}, \alpha_{zx})$	$(\beta_{xxx} + \beta_{xyy}, \beta_{yyy} + \beta_{xxy}), (\beta_{zzx}, \beta_{yzz})$
	e_2		$(\alpha_{xx} - \alpha_{yy}, \alpha_{xy})$	$(\beta_{xxx} - 3\beta_{xyy}, \beta_{yyy} - 3\beta_{xxy}), (\beta_{yyz} - \beta_{zxx}, \beta_{xyz})$
C_6	a	μ_z	$\alpha_{xx} + \alpha_{yy}, \alpha_{zz}$	$\beta_{yyz} + \beta_{zxx}, \beta_{zzz}$
	b			$\beta_{xxx} - 3\beta_{xyy}, \beta_{yyy} - 3\beta_{xxy}$
	e_1	(μ_x, μ_y)	$(\alpha_{yz}, \alpha_{zx})$	$(\beta_{xxx} + \beta_{xyy}, \beta_{yyy} + \beta_{xxy}), (\beta_{zzx}, \beta_{yzz})$
	e_2		$(\alpha_{xx} - \alpha_{yy}, \alpha_{xy})$	$(\beta_{yyz} - \beta_{zxx}, \beta_{xyz})$
		B. The dihedral groups D_n ($n = 2, 3, 4, 5, 6$)		
D_2	a		$\alpha_{xx}, \alpha_{yy}, \alpha_{zz}$	β_{xyz}
	b_1	μ_z	α_{xy}	$\beta_{zzz}, \beta_{yyz}, \beta_{zxx}$
	b_2	μ_y	α_{zx}	$\beta_{yyy}, \beta_{xxy}, \beta_{yzz}$
	b_3	μ_x	α_{yz}	$\beta_{xxx}, \beta_{zzx}, \beta_{xyy}$

TABLE 2.1 (*Continued*)

B. The dihedral groups D_n ($n=2,3,4,5,6$) (*continued*)				
D_3	a_1[a]		$\alpha_{xx}+\alpha_{yy}, \alpha_{zz}$	$\beta_{xxx}-3\beta_{xyy}$
	a_2[a]	μ_z		$\beta_{yyy}-3\beta_{xxy}, \beta_{yyz}+\beta_{zxx}, \beta_{zzz}$
	e	(μ_x, μ_y)	$(\alpha_{xx}-\alpha_{yy}, \alpha_{xy}), (\alpha_{yz}, \alpha_{zx})$	$(\beta_{xxx}+\beta_{xyy}, \beta_{yyy}+\beta_{xxy}), (\beta_{yyz}-\beta_{zxx}, \beta_{xyz}), (\beta_{zzx}, \beta_{yzz})$
D_4	a_1		$\alpha_{xx}+\alpha_{yy}, \alpha_{zz}$	
	a_2	μ_z		$\beta_{yyz}+\beta_{zxx}, \beta_{zzz}$
	b_1		$\alpha_{xx}-\alpha_{yy}$	β_{xyz}
	b_2		α_{xy}	$\beta_{yyz}-\beta_{zxx}$
	e	(μ_x, μ_y)	$(\alpha_{yz}, \alpha_{zx})$	$(\beta_{xxx}, \beta_{yyy}), (\beta_{xyy}, \beta_{xxy}), (\beta_{zzx}, \beta_{yzz})$
D_5	a_1		$\alpha_{xx}+\alpha_{yy}, \alpha_{zz}$	
	a_2	μ_z		$\beta_{yyz}+\beta_{zxx}, \beta_{zzz}$
	e_1	(μ_x, μ_y)	$(\alpha_{yz}, \alpha_{zx})$	$(\beta_{xxx}+\beta_{xyy}, \beta_{yyy}+\beta_{xxy}), (\beta_{zzx}, \beta_{yzz})$
	e_2		$(\alpha_{xx}-\alpha_{yy}, \alpha_{xy})$	$(\beta_{xxx}-3\beta_{xyy}, \beta_{yyy}-3\beta_{xxy}), (\beta_{yyz}-\beta_{zxx}, \beta_{xyz})$
D_6	a_1		$\alpha_{xx}+\alpha_{yy}, \alpha_{zz}$	
	a_2	μ_z		$\beta_{yyz}+\beta_{zxx}, \beta_{zzz}$
	b_1[b]			$\beta_{xxx}-3\beta_{xyy}$
	b_2[b]			$\beta_{yyy}-3\beta_{xxy}$
	e_1	(μ_x, μ_y)	$(\alpha_{yz}, \alpha_{zx})$	$(\beta_{xxx}+\beta_{xyy}, \beta_{yyy}+\beta_{xxy}), (\beta_{zzx}, \beta_{yzz})$
	e_2		$(\alpha_{xx}-\alpha_{yy}, \alpha_{xy})$	$(\beta_{yyz}-\beta_{zxx}, \beta_{xyz})$
C. The groups C_{nv} ($n = 2, 3, 4, 5, 6$)				
C_{2v}	a_1	μ_z	$\alpha_{xx}, \alpha_{yy}, \alpha_{zz}$	$\beta_{zzz}, \beta_{yyz}, \beta_{zxx}$
	a_2		α_{xy}	β_{xyz}
	b_1	μ_x	α_{zx}	$\beta_{xxx}, \beta_{zzx}, \beta_{xyy}$
	b_2	μ_y	α_{yz}	$\beta_{yyy}, \beta_{xxy}, \beta_{yzz}$
C_{3v}	a_1[c]	μ_z	$\alpha_{xx}+\alpha_{yy}, \alpha_{zz}$	$\beta_{xxx}-3\beta_{xyy}, \beta_{yyz}+\beta_{zxx}, \beta_{zzz}$

	a_2[c]			$\beta_{yyy} - 3\beta_{xxy}$
	e	(μ_x, μ_y)	$(\alpha_{xx} - \alpha_{yy}, \alpha_{xy}), (\alpha_{yz}, \alpha_{zx})$	$(\beta_{xxx} + \beta_{xyy}, \beta_{yyy} + \beta_{xxy}), (\beta_{yyz} - \beta_{zxx}, \beta_{xyz}), (\beta_{zzx}, \beta_{yzz})$
C_{4v}	a_1	μ_z	$\alpha_{xx} + \alpha_{yy}, \alpha_{zz}$	$\beta_{yyz} + \beta_{zxx}, \beta_{zzz}$
	a_2			
	b_1		$\alpha_{xx} - \alpha_{yy}$	$\beta_{yyz} - \beta_{zxx}$
	b_2		α_{xy}	β_{xyz}
	e	(μ_x, μ_y)	$(\alpha_{yz}, \alpha_{zx})$	$(\beta_{xxx}, \beta_{yyy}), (\beta_{xyy}, \beta_{xxy}), (\beta_{zzx}, \beta_{yzz})$
C_{5v}	a_1	μ_z	$\alpha_{xx} + \alpha_{yy}, \alpha_{zz}$	$\beta_{yyz} + \beta_{zxx}, \beta_{zzz}$
	a_2			
	e_1	(μ_x, μ_y)	$(\alpha_{yz}, \alpha_{zx})$	$(\beta_{xxx} + \beta_{xyy}, \beta_{yyy} + \beta_{xxy}), (\beta_{zzx}, \beta_{yzz})$
	e_2		$(\alpha_{xx} - \alpha_{yy}, \alpha_{xy})$	$(\beta_{xxx} - 3\beta_{xyy}, \beta_{yyy} - 3\beta_{xxy}), (\beta_{yyz} - \beta_{zxx}, \beta_{xyz})$
C_{6v}	a_1	μ_z	$\alpha_{xx} + \alpha_{yy}, \alpha_{zz}$	$\beta_{yyz} + \beta_{zxx}, \beta_{zzz}$
	a_2			
	b_1[d]			$\beta_{xxx} - 3\beta_{xyy}$
	b_2[d]			$\beta_{yyy} - 3\beta_{xxy}$
	e_1	(μ_x, μ_y)	$(\alpha_{yz}, \alpha_{zx})$	$(\beta_{xxx} + \beta_{xyy}, \beta_{yyy} + \beta_{xxy}), (\beta_{zzx}, \beta_{yzz})$
	e_2		$(\alpha_{xx} - \alpha_{yy}, \alpha_{xy})$	$(\beta_{yyz} - \beta_{zxx}, \beta_{xyz})$

D. The groups C_{nh} ($n = 2, 3, 4, 5, 6$)

C_{2h}	a_g		$\alpha_{xx}, \alpha_{yy}, \alpha_{zz}, \alpha_{xy}$	
	b_g		α_{yz}, α_{zx}	
	a_u	μ_z		$\beta_{zzz}, \beta_{yyz}, \beta_{zxx}, \beta_{xyz}$
	b_u	μ_x, μ_y		$\beta_{xxx}, \beta_{yyy}, \beta_{xxy}, \beta_{zzx}, \beta_{xyy}, \beta_{yzz}$
C_{3h}	a′		$\alpha_{xx} + \alpha_{yy}, \alpha_{zz}$	$\beta_{xxx} - 3\beta_{xyy}, \beta_{yyy} - 3\beta_{xxy}$
	e′	(μ_x, μ_y)	$(\alpha_{xx} - \alpha_{yy}, \alpha_{xy})$	$(\beta_{xxx} + \beta_{xyy}, \beta_{yyy} + \beta_{xxy}), (\beta_{zzx}, \beta_{yzz})$
	a″	μ_z		$\beta_{yyz} + \beta_{zxx}, \beta_{zzz}$
	e″		$(\alpha_{yz}, \alpha_{zx})$	$(\beta_{yyz} - \beta_{zxx}, \beta_{xyz})$
C_{4h}	a_g		$\alpha_{xx} + \alpha_{yy}, \alpha_{zz}$	
	b_g		$\alpha_{xx} - \alpha_{yy}, \alpha_{xy}$	
	e_g		$(\alpha_{yz}, \alpha_{zx})$	
	a_u	μ_z		$\beta_{yyz} + \beta_{zxx}, \beta_{zzz}$
	b_u			$\beta_{yyz} - \beta_{zxx}, \beta_{xyz}$
	e_u	(μ_x, μ_y)		$(\beta_{xxx}, \beta_{yyy}), (\beta_{xyy}, \beta_{xxy}), (\beta_{zzx}, \beta_{yzz})$

TABLE 2.1 (*Continued*)

		D. The groups C_{nh} ($n = 2, 3, 4, 5, 6$) (*continued*)		
C_{5h}	a′		$\alpha_{xx} + \alpha_{yy}, \alpha_{zz}$	
	e_1'	(μ_x, μ_y)		$(\beta_{xxx} + \beta_{xyy}, \beta_{yyy} + \beta_{xxy}), (\beta_{zzx}, \beta_{yzz})$
	e_2'		$(\alpha_{xx} - \alpha_{yy}, \alpha_{xy})$	$(\beta_{xxx} - 3\beta_{xyy}, \beta_{yyy} - 3\beta_{xxy})$
	a″	μ_z		$\beta_{yyz} + \beta_{zxx}, \beta_{zzz}$
	e_1''		$(\alpha_{yz}, \alpha_{zx})$	
	e_2''			$(\beta_{yyz} - \beta_{zxx}, \beta_{xyz})$
C_{6h}	a_g		$\alpha_{xx} + \alpha_{yy}, \alpha_{zz}$	
	b_g			
	e_{1g}		$(\alpha_{yz}, \alpha_{zx})$	
	e_{2g}		$(\alpha_{xx} - \alpha_{yy}, \alpha_{xy})$	
	a_u	μ_z		$\beta_{yyz} + \beta_{zxx}, \beta_{zzz}$
	b_u			$\beta_{xxx} - 3\beta_{xyy}, \beta_{yyy} - 3\beta_{xxy}$
	e_{1u}	(μ_x, μ_y)		$(\beta_{xxx} + \beta_{xyy}, \beta_{yyy} + \beta_{xxy}), (\beta_{zzx}, \beta_{yzz})$
	e_{2u}			$(\beta_{yyz} - \beta_{zxx}, \beta_{xyz})$
		E. The groups D_{nh} ($n = 2, 3, 4, 5, 6$)		
D_{2h}	a_g		$\alpha_{xx}, \alpha_{yy}, \alpha_{zz}$	
	b_{1g}		α_{xy}	
	b_{2g}		α_{zx}	
	b_{3g}		α_{yz}	
	a_u			β_{xyz}
	b_{1u}	μ_z		$\beta_{zzz}, \beta_{yyz}, \beta_{zxx}$
	b_{2u}	μ_y		$\beta_{yyy}, \beta_{xxy}, \beta_{yzz}$
	b_{3u}	μ_x		$\beta_{xxx}, \beta_{zzx}, \beta_{xyy}$

D_{3h}	a_1'[e]		$\alpha_{xx} + \alpha_{yy}, \alpha_{zz}$	$\beta_{xxx} - 3\beta_{xyy}$
	a_2'[e]			$\beta_{yyy} - 3\beta_{xxy}$
	e'	(μ_x, μ_y)	$(\alpha_{xx} - \alpha_{yy}, \alpha_{xy})$	$(\beta_{xxx} + \beta_{xyy}, \beta_{yyy} + \beta_{xxy}), (\beta_{zzx}, \beta_{yzz})$
	a_1''			
	a_2''	μ_z		$\beta_{yyz} + \beta_{zxx}, \beta_{zzz}$
	e''		$(\alpha_{yz}, \alpha_{zx})$	$(\beta_{yyz} - \beta_{zxx}, \beta_{xyz})$
D_{4h}	a_{1g}		$\alpha_{xx} + \alpha_{yy}, \alpha_{zz}$	
	a_{2g}			
	b_{1g}		$\alpha_{xx} - \alpha_{yy}$	
	b_{2g}		α_{xy}	
	e_g		$(\alpha_{yz}, \alpha_{zx})$	
	a_{1u}			
	a_{2u}	μ_z		$\beta_{yyz} + \beta_{zxx}, \beta_{zzz}$
	b_{1u}			β_{xyz}
	b_{2u}			$\beta_{yyz} - \beta_{zxx}$
	e_u	(μ_x, μ_y)		$(\beta_{xxx}, \beta_{yyy}), (\beta_{xyy}, \beta_{xxy}), (\beta_{zzx}, \beta_{yzz})$
D_{5h}	a_1'		$\alpha_{xx} + \alpha_{yy}, \alpha_{zz}$	
	a_2'			
	e_1'	(μ_x, μ_y)		$(\beta_{xxx} + \beta_{xyy}, \beta_{yyy} + \beta_{xxy}), (\beta_{zzx}, \beta_{yzz})$
	e_2'		$(\alpha_{xx} - \alpha_{yy}, \alpha_{xy})$	$(\beta_{xxx} - 3\beta_{xyy}, \beta_{yyy} - 3\beta_{xxy})$
	a_1''			
	a_2''	μ_z		$\beta_{yyz} + \beta_{zxx}, \beta_{zzz}$
	e_1''		$(\alpha_{yz}, \alpha_{zx})$	
	e_2''			$(\beta_{yyz} - \beta_{zxx}, \beta_{xyz})$

TABLE 2.1 (*Continued*)

E. The groups D_{nh} ($n=2, 3, 4, 5, 6$) (*continued*)

D_{6h}	a_{1g}		$\alpha_{xx} + \alpha_{yy}, \alpha_{zz}$	
	a_{2g}			
	b_{1g}			
	b_{2g}			
	e_{1g}		$(\alpha_{yz}, \alpha_{zx})$	
	e_{2g}		$(\alpha_{xx} - \alpha_{yy}, \alpha_{xy})$	
	a_{1u}			
	a_{2u}	μ_z		$\beta_{yyz} + \beta_{zxx}, \beta_{zzz}$
	b_{1u}[f]			$\beta_{xxx} - 3\beta_{xyy}$
	b_{2u}[f]			$\beta_{yyy} - 3\beta_{xxy}$
	e_{1u}	(μ_x, μ_y)		$(\beta_{xxx} + \beta_{xyy}, \beta_{yyy} + \beta_{xxy}), (\beta_{zzx}, \beta_{yzz})$
	e_{2u}			$(\beta_{yyz} - \beta_{zxx}, \beta_{xyz})$

F. The groups D_{nd} ($n = 2, 3, 4, 5, 6$)

D_{2d}	a_1		$\alpha_{xx} + \alpha_{yy}, \alpha_{zz}$	β_{xyz}
	a_2			$\beta_{yyz} - \beta_{zxx}$
	b_1		$\alpha_{xx} - \alpha_{yy}$	
	b_2	μ_z	α_{xy}	$\beta_{yyz} + \beta_{zxx}, \beta_{zzz}$
	e	(μ_x, μ_y)	$(\alpha_{yz}, \alpha_{zx})$	$(\beta_{xxx}, \beta_{yyy}), (\beta_{xyy}, \beta_{xxy}), (\beta_{zzx}, \beta_{yzz})$
D_{3d}	a_{1g}		$\alpha_{xx} + \alpha_{yy}, \alpha_{zz}$	
	a_{2g}			
	e_g		$(\alpha_{xx} - \alpha_{yy}, \alpha_{xy}), (\alpha_{yz}, \alpha_{zx})$	
	a_{1u}[g]			$\beta_{xxx} - 3\beta_{xyy}$
	a_{2u}[g]	μ_z		$\beta_{yyy} - 3\beta_{xxy}, \beta_{yyz} + \beta_{zxx}, \beta_{zzz}$
	e_u	(μ_x, μ_y)		$(\beta_{xxx} + \beta_{xyy}, \beta_{yyy} + \beta_{xxy}), (\beta_{yyz} - \beta_{zxx}, \beta_{xyz}), (\beta_{zzx}, \beta_{yzz})$

D_{4d}	a_1		$\alpha_{xx} + \alpha_{yy}, \alpha_{zz}$	
	a_2			
	b_1			
	b_2	μ_z		$\beta_{yyz} + \beta_{zxx}, \beta_{zzz}$
	e_1	(μ_x, μ_y)		$(\beta_{xxx} + \beta_{xyy}, \beta_{yyy} + \beta_{xxy}), (\beta_{zzx}, \beta_{yzz})$
	e_2		$(\alpha_{xx} - \alpha_{yy}, \alpha_{xy})$	$(\beta_{yyz} - \beta_{zxx}, \beta_{xyz})$
	e_3		$(\alpha_{yz}, \alpha_{zx})$	$(\beta_{xxx} - 3\beta_{xyy}, \beta_{yyy} - 3\beta_{xxy})$
D_{5d}	a_{1g}		$\alpha_{xx} + \alpha_{yy}, \alpha_{zz}$	
	a_{2g}			
	e_{1g}		$(\alpha_{yz}, \alpha_{zx})$	
	e_{2g}		$(\alpha_{xx} - \alpha_{yy}, \alpha_{xy})$	
	a_{1u}			
	a_{2u}	μ_z		$\beta_{yyz} + \beta_{zxx}, \beta_{zzz}$
	e_{1u}	(μ_x, μ_y)		$(\beta_{xxx} + \beta_{xyy}, \beta_{yyy} + \beta_{xxy}), (\beta_{zzx}, \beta_{yzz})$
	e_{2u}			$(\beta_{xxx} - 3\beta_{xyy}, \beta_{yyy} - 3\beta_{xxy}), (\beta_{yyz} - \beta_{zxx}, \beta_{xyz})$
D_{6d}	a_1		$\alpha_{xx} + \alpha_{yy}, \alpha_{zz}$	
	a_2			
	b_1			
	b_2	μ_z		$\beta_{yyz} + \beta_{zxx}, \beta_{zzz}$
	e_1	(μ_x, μ_y)		$(\beta_{xxx} + \beta_{xyy}, \beta_{yyy} + \beta_{xxy}), (\beta_{zzx}, \beta_{yzz})$
	e_2		$(\alpha_{xx} - \alpha_{yy}, \alpha_{xy})$	
	e_3			$(\beta_{xxx} - 3\beta_{xyy}, \beta_{yyy} - 3\beta_{xxy})$
	e_4			$(\beta_{yyz} - \beta_{zxx}, \beta_{xyz})$
	e_5		$(\alpha_{yz}, \alpha_{zx})$	

G. The groups S_n ($n = 4, 6, 8$)

S_4	a		$\alpha_{xx} + \alpha_{yy}, \alpha_{zz}$	$\beta_{yyz} - \beta_{zxx}, \beta_{xyz}$
	b	μ_z	$\alpha_{xx} - \alpha_{yy}, \alpha_{xy}$	$\beta_{yyz} + \beta_{zxx}, \beta_{zzz}$
	e	(μ_x, μ_y)	$(\alpha_{yz}, \alpha_{zx})$	$(\beta_{xxx}, \beta_{yyy}), (\beta_{xyy}, \beta_{xxy}), (\beta_{zzx}, \beta_{yzz})$

TABLE 2.1 (*Continued*)

G. The groups S_n ($n=4, 6, 8$) (*continued*)				
S_6	a_g		$\alpha_{xx}+\alpha_{yy}, \alpha_{zz}$	
	e_g		$(\alpha_{xx}-\alpha_{yy}, \alpha_{xy}), (\alpha_{yz}, \alpha_{zx})$	
	a_u	μ_z		$\beta_{xxx}-3\beta_{xyy}, \beta_{yyy}-3\beta_{xxy}, \beta_{yyz}+\beta_{zxx}, \beta_{zzz}$
	e_u	(μ_x, μ_y)		$(\beta_{xxx}+\beta_{xyy}, \beta_{yyy}+\beta_{xxy}), (\beta_{yyz}-\beta_{zxx}, \beta_{xyz}), (\beta_{zzx}, \beta_{yzz})$
S_8	a		$\alpha_{xx}+\alpha_{yy}, \alpha_{zz}$	
	b	μ_z		$\beta_{yyz}+\beta_{zxx}, \beta_{zzz}$
	e_1	(μ_x, μ_y)		$(\beta_{xxx}+\beta_{xyy}, \beta_{yyy}+\beta_{xxy}), (\beta_{zzx}, \beta_{yzz})$
	e_2		$(\alpha_{xx}-\alpha_{yy}, \alpha_{xy})$	$(\beta_{yyz}-\beta_{zxx}, \beta_{xyz})$
	e_3		$(\alpha_{yz}, \alpha_{zx})$	$(\beta_{xxx}-3\beta_{xyy}, \beta_{yyy}-3\beta_{xxy})$
H. The cubic groups τ, τ_h, τ_d, θ, θ_h				
τ	a		$\alpha_{xx}+\alpha_{yy}+\alpha_{zz}$	β_{xyz}
	e		$(\alpha_{xx}+\alpha_{yy}-2\alpha_{zz}, \alpha_{xx}-\alpha_{yy})$	
	f	(μ_x, μ_y, μ_z)	$(\alpha_{xy}, \alpha_{yz}, \alpha_{zx})$	$(\beta_{xxx}, \beta_{yyy}, \beta_{zzz}), (\beta_{xyy}, \beta_{yzz}, \beta_{zxx}), (\beta_{zzx}, \beta_{xxy}, \beta_{yyz})$
τ_h	a_g		$\alpha_{xx}+\alpha_{yy}+\alpha_{zz}$	
	e_g		$(\alpha_{xx}+\alpha_{yy}-2\alpha_{zz}, \alpha_{xx}-\alpha_{yy})$	
	f_g		$(\alpha_{xy}, \alpha_{yz}, \alpha_{zx})$	
	a_u			β_{xyz}
	e_u			
	f_u	(μ_x, μ_y, μ_z)		$(\beta_{xxx}, \beta_{yyy}, \beta_{zzz}), (\beta_{xyy}, \beta_{yzz}, \beta_{zxx}), (\beta_{zzx}, \beta_{xxy}, \beta_{yyz})$
τ_d	a_1		$\alpha_{xx}+\alpha_{yy}+\alpha_{zz}$	β_{xyz}
	a_2			
	e		$(\alpha_{xx}+\alpha_{yy}-2\alpha_{zz}, \alpha_{xx}-\alpha_{yy})$	
	f_1			$(\beta_{xyy}-\beta_{zzx}, \beta_{yzz}-\beta_{xxy}, \beta_{zxx}-\beta_{yyz})$
	f_2	(μ_x, μ_y, μ_z)	$(\alpha_{xy}, \alpha_{yz}, \alpha_{zx})$	$(\beta_{xxx}, \beta_{yyy}, \beta_{zzz}), (\beta_{xyy}+\beta_{zzx}, \beta_{yzz}+\beta_{xxy}, \beta_{zxx}+\beta_{yyz})$

θ	a_1		$\alpha_{xx} + \alpha_{yy} + \alpha_{zz}$	
	a_2			β_{xyz}
	e		$(\alpha_{xx} + \alpha_{yy} - 2\alpha_{zz}, \alpha_{xx} - \alpha_{yy})$	
	f_1	(μ_x, μ_y, μ_z)		$(\beta_{xxx}, \beta_{yyy}, \beta_{zzz}), (\beta_{xyy} + \beta_{zzx}, \beta_{yzz} + \beta_{xxy}, \beta_{zxx} + \beta_{yyz})$
	f_2		$(\alpha_{xy}, \alpha_{yz}, \alpha_{zx})$	$(\beta_{xyy} - \beta_{zzx}, \beta_{yzz} - \beta_{xxy}, \beta_{zxx} - \beta_{yyz})$
θ_h	a_{1g}		$\alpha_{xx} + \alpha_{yy} + \alpha_{zz}$	
	a_{2g}			
	e_g		$(\alpha_{xx} + \alpha_{yy} - 2\alpha_{zz}, \alpha_{xx} - \alpha_{yy})$	
	f_{1g}			
	f_{2g}		$(\alpha_{xy}, \alpha_{yz}, \alpha_{zx})$	
	a_{1u}			
	a_{2u}			β_{xyz}
	e_u			
	f_{1u}	(μ_x, μ_y, μ_z)		$(\beta_{xxx}, \beta_{yyy}, \beta_{zzz}), (\beta_{xyy} + \beta_{zzx}, \beta_{yzz} + \beta_{xxy}, \beta_{zxx} + \beta_{yyz})$
	f_{2u}			$(\beta_{xyy} - \beta_{zzx}, \beta_{yzz} - \beta_{xxy}, \beta_{zxx} - \beta_{yyz})$

I. The groups $C_{\infty v}$ and $D_{\infty h}$

$C_{\infty v}$	σ^+	μ_z	$\alpha_{xx} + \alpha_{yy}, \alpha_{zz}$	$\beta_{yyz} + \beta_{zxx}, \beta_{zzz}$
	σ^-			
	π	(μ_x, μ_y)	$(\alpha_{yz}, \alpha_{zx})$	$(\beta_{xxx} + \beta_{xyy}, \beta_{yyy} + \beta_{xxy}), (\beta_{zzx}, \beta_{yzz})$
	δ		$(\alpha_{xx} - \alpha_{yy}, \alpha_{xy})$	$(\beta_{yyz} - \beta_{zxx}, \beta_{xyz})$
	ϕ			$(\beta_{xxx} - 3\beta_{xyy}, \beta_{yyy} - 3\beta_{xxy})$

TABLE 2.1 (*Continued*)

I. The groups $C_{\infty v}$ and $D_{\infty h}$ (*continued*)				
$D_{\infty h}$	σ_g^+		$\alpha_{xx} + \alpha_{yy}, \alpha_{zz}$	
	σ_g^-			
	π_g		$(\alpha_{yz}, \alpha_{zx})$	
	δ_g		$(\alpha_{xx} - \alpha_{yy}, \alpha_{xy})$	
	ϕ_g			
	σ_u^+	μ_z		$\beta_{yyz} + \beta_{zxx}, \beta_{zzz}$
	σ_u^-			
	π_u	(μ_x, μ_y)		$(\beta_{xxx} + \beta_{xyy}, \beta_{yyy} + \beta_{xxy}), (\beta_{zzx}, \beta_{yzz})$
	δ_u			$(\beta_{yyz} - \beta_{zxx}, \beta_{xyz})$
	ϕ_u			$(\beta_{xxx} - 3\beta_{xyy}, \beta_{yyy} - 3\beta_{xxy})$

[a] x axis along C_2. [b] x axis along $C_{2'}$. [c] x axis in σ_v [d] x axis in σ_v [e] x axis along C_2 [f] x axis along $C_{2'}$ [g] x axis along C_2

case considered, and one could form infinitely many different combinations of them belonging to species a_1' and a_2', depending on the orientation of x and y with respect to the C_2 symmetry axes. In particular, the combinations assigned to a_1' and a_2' in the table could be interchanged if the y axis was chosen along one of the three C_2 axes of symmetry. The presently chosen orientation is such that the x axis is coincident with one of the C_2. In most of the cases of symmetric tops, however, the interchanging of x and y does not affect the species designation. In all of the cases where it does matter, the appropriate species are identified by footnotes.

With the species for the hyperpolarizability components established for all the molecular point groups the previously stated general symmetry condition for hyper Raman activity may be readily applied to specific vibrational transitions. For a fundamental vibration, in which all the initial vibrational quantum numbers are zero and only one vibrational quantum number changes from zero to unity, in the transition to the upper state the symmetry condition takes on a simple form. The pth fundamental vibration will be hyper Raman active if the pth vibrational wave function with $v_p' = 1$ belongs to the same species as a component or linear combination of components of β. Other conditions can be deduced for the hyper Raman activity of overtones, combinations and so on, following the procedures established for infrared and Raman activity, but we shall not consider these further here.

Examination of Table 2.1 reveals that many vibrations inactive in both infrared and Raman may be hyper Raman active. Some specific examples are the twisting frequency (a_u) in ethylene, the CH_2 wagging frequency (a_2') in cyclopropane, the torsional frequency (a_{1u}) in ethane, six modes ($2b_{1u} + 2b_{2u} + 2e_{2u}$) in benzene and a triply degenerate mode (f_{2u}) in octahedral MX_6 molecules like SF_6.

Some important generalizations may be made concerning activity in the hyper Raman effect. *All infrared-active modes will be hyper Raman active. In molecules with a centre of symmetry any hyper Raman active modes belong to the gerade species.*

Hyper Raman spectroscopy is thus of very considerable importance for it is the first new optical spectroscopic technique for the study of otherwise inaccessible energy levels since the discovery of the Raman effect in 1928. Furthermore it yields all the *frequency* information contained in the infrared.

(c) *Directional and polarization properties*

The directional and polarization properties of light scattering controlled by the hyperpolarizability tensor may be calculated from a form of equation (2.80) in which hyperpolarizability tensor components replace third-order scattering tensor components and space fixed axes XYZ replace the general axes ijk. The required relationship is given by

$$[B_X]_{nk} = \tfrac{1}{2} \sum_{X,Y} [\beta_{XYZ}]_{nk} A_X A_Y \tag{2.94}$$

It must be remembered that there is permutation symmetry of the spatial suffices so that there are only ten distinct components of β. n and k are sets of nuclear quantum numbers of the initial and final states.

In general terms the application of this formula is straightforward. For example, for illumination along the Y direction with light polarized in the Z direction ($A_Z \neq 0$; $A_X = A_Y = 0$) and observation along the X direction, the intensity of scattered radiation arising from a transition $k \to n$ polarized in the Z direction is proportional to

$$[\beta_{ZZZ}]^2_{nk} A^2_Z \tag{2.95}$$

and that polarized in the Y direction is proportional to

$$[\beta_{YZZ}]^2_{nk} A^2_Z \tag{2.96}$$

and the total intensity is proportional to

$$\{[\beta_{ZZZ}]^2_{nk} + [\beta_{YZZ}]^2_{nk}\} A^2_Z \tag{2.97}$$

The situation becomes more complicated when it is required to move from β_{XYZ} to β_{xyz}, for which the general transformation is given by (2.85). Two specific cases will be considered here, liquids and crystals.

(*i*) *Liquids*

Assuming random orientation of the molecules we may follow the treatment used for the normal Raman effect and average classically over all directions. A detailed analysis of this problem has been given by Cyvin, Rauch and Decius[8] and we shall only concern ourselves with the results in so far as they determine depolarization ratios. For illumination with natural light the depolarization ratio ρ_{HR} for fundamental vibrations lies between zero and $\frac{4}{5}$. Vibrations with $\rho_{HR} = \frac{4}{5}$ are said to be depolarized and those with $\rho_{HR} < \frac{4}{5}$ are said to be polarized. For plane polarized incident radiation the depolarization ratio lies between zero and $\frac{2}{3}$. In the hyper Raman effect, unlike the normal Raman effect, it is not possible to make generalizations relating depolarization values and the symmetry of vibrations. Depolarized lines are to be found amongst symmetric and non-symmetric vibrations in the hyper Raman effect and polarized lines are likewise not confined to symmetric modes. However it can be shown that depolarized lines are always those which are infrared inactive. For selected point groups more specific rules may be established. For example for e′ and a_2'' vibrations in molecules of the D_{3h} point group $\frac{1}{5} \leqslant \rho_{HR} \leqslant \frac{4}{5}$ for natural radiation and $\frac{1}{9} \leqslant \rho_{HR} \leqslant \frac{2}{3}$ for plane polarized illumination.

(*ii*) *Crystals*

Here the situation can be less complicated. The direction cosines in (2.85) are constants for a given experiment and it is often possible to arrange matters so that the *XYZ* and *xyz* axes are coincident. In such cases those components of the hyperpolarizability tensor matrix element for a vibrational transition of a given symmetry which are non-zero are obtainable from Table 2.1. The pattern of non-zero entries in the tensor may be determined experimentally by orientation studies just as for normal Raman spectroscopy and so used to assign frequencies unambiguously to symmetry classes.

F. Experimental

The first observation of the hyper Raman effect was reported by Terhune, Maker and Savage.[7] Their initial experimental arrangement is shown diagrammatically in Figure 2.1. Spectra were excited with a Q-switched

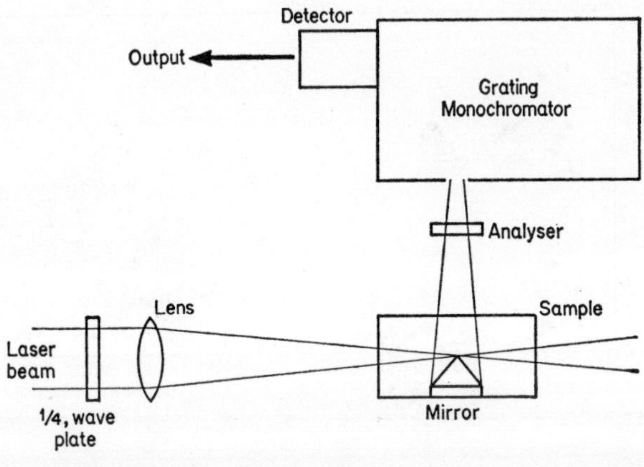

FIGURE 2.1. *Diagrammatic representation of apparatus used by Terhune, Maker and Savage in their first studies of non-linear light scattering.*

ruby laser with the following characteristics: plane polarized radiation, up to 10^6 watt peak power, 80×10^{-9} s half-width, and 10^{-3} radian beam divergence. A repetition rate of 2 pulses per minute was used. The laser was focused into the sample with an $f/10$ lens. The scattered radiation was

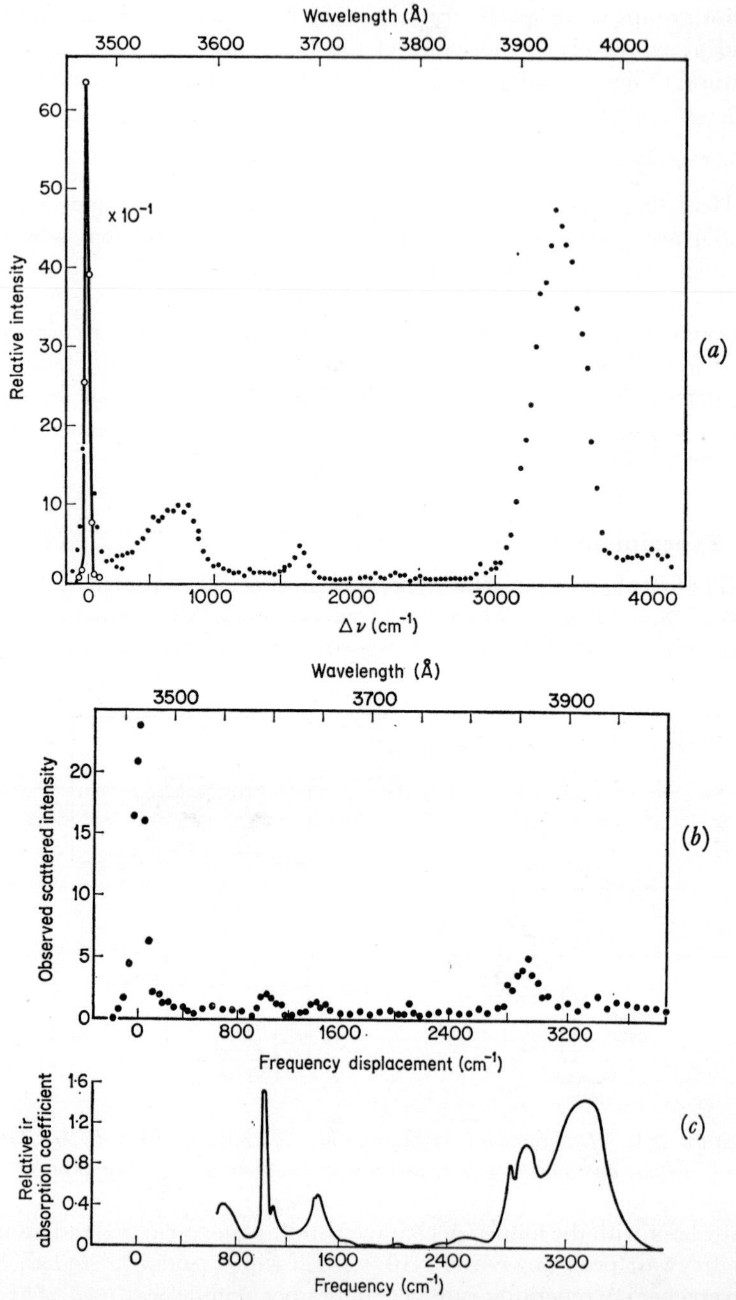

FIGURE 2.2. *(a) Stokes hyper Raman spectrum of water. (b) Stokes hyper Raman spectrum of methanol. (c) Infrared absorption spectrum of methanol.*

collected at $f/1$ and dispersed in a grating monochromator. The spectrum was recorded spectral element by spectral element. With the monochromator set to pass radiation at $\nu \pm \Delta\nu$ (where $2\Delta\nu$ is the spectral band pass) the intensity of the scattered radiation in this spectral element was found by determining the average over 25 laser pulses of the number of photoelectrons produced by the photomultiplier detector. This procedure was repeated spectral element by spectral element over the whole spectral region of interest. All the measurements were carried out just below threshold for dielectric breakdown or onset of the stimulated Raman effect.

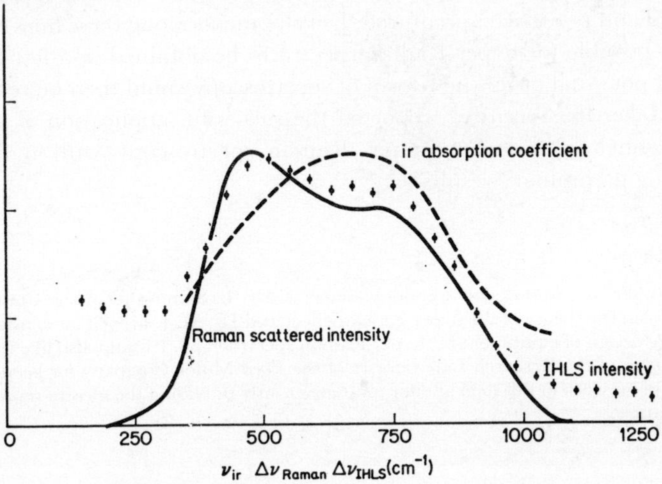

FIGURE 2.3 *Infrared absorption, normal Raman and hyper Raman spectra of water in the librational region, 200–1200 cm*$^{-1}$.

Subsequently Maker[12] has made considerable improvements in his experimental procedure. Using a Q-switched laser with a repetition rate of 1 pulse per second and a fully automated digital detection system he has been able to average over a much larger number of laser shots per spectral element. However, despite the increased repetition rate of the laser the process of acquiring information spectral element by spectral element has meant that one to two days are needed to produce good spectra.

Figure 2.2 shows hyper Raman spectra obtained in this way by Maker[12] for water and methanol. Figure 2.3 shows details of the hyper Raman spectrum of water[12] in the librational region 250–1000 cm^{-1}. The infrared and normal Raman spectra of water in this region are also included. Walrafen[16] has discussed the significance of these results. Briefly the observation that selection rules are operative for these intermolecular motions implies some finite degree of short range intermolecular symmetry or order and enables the various models proposed for water structure to be tested. It is found that only a model with C_{2v} local symmetry can predict the observed spectra.

The time needed for obtaining spectra could be markedly reduced by using a multichannel system in which information from all the spectral elements in the desired spectral region is recorded for each laser shot. An image intensifier could be used to produce on the output phosphor an intensified spectrum which could be preserved long enough for scanning by a television camera. The output of the television camera could then be stored and processed as required. Developments along these lines should make it possible for hyper Raman spectra to be obtained in a few hours. The full potential of this new form of spectroscopy could then be realized. P. D. Maker has recently[17] reported the successful application of such a multichannel technique to hyper Raman spectroscopy with most encouraging preliminary results.

Acknowledgement

I would like to acknowledge my indebtedness to Dr. L. Stanton for many stimulating discussions on the theory of the hyper Raman effect and Dr. M. J. French for considerable help in the design of equipment for hyper Raman spectroscopy. I would also like to thank Dr. P. Maker of the Research Laboratories of the Ford Motor Company for keeping me informed of his work in this field and for most generously providing the spectra reproduced in this chapter.

APPENDIX

The inclusion of the permittivity of free space ϵ_0 in equation (2.1) ensures that in the S.I. system the units of the polarizabilities, α, β and γ are not over-cumbersome. Since ϵ_0 is dimensionless in the e.s.u. system, the units for the polarizabilities in this system are unchanged. The following table summarizes the appropriate units in the two systems.

System	ϵ_0	α	β	γ	μ	E
e.s.u.	$4\pi\epsilon_0 = 1$	cm^3	$e.s.u.^{-1}\ cm^5$	$e.s.u.^{-2}\ cm^7$	e.s.u. cm	$e.s.u.\ cm^{-2}$
S.I. (derived)	$CV^{-1}\ m^{-1}$	m^3	$m^4\ V^{-1}$	$m^5\ V^{-2}$	C m	Vm^{-1}
S.I. (base)	$kg^{-1}\ s^4\ m^{-3}\ A^2$	m^3	$kg\ s^3\ m^2\ A$	$kg^{-2}\ s^6\ mA^2$	Asm	$kg\ s^{-3}\ A^{-1}\ m$

References

1. Placzek, G., *Handbuch der Radiologie*, 1934, **6**, (ii), 205.
2. Woodward, L. A., and Long, D. A., *Trans. Faraday Soc.*, 1949, **45**, 1131.
3. Buckingham, A. D., and Orr, B. J., *Quart. Rev.*, 1967, **21**, 195.
4. Decius, J. C., and Rauch, J. E., 'Ohio State Symposium of Molecular Structure and Spectroscopy', 1959, Paper 48.
5. Kielich, S., *Bull. Acad. Polon. Sci.*, 1964, **12**, 53.
6. Li, Y. Y., *Acta Phys. Sinica*, 1964, **20**, 164.
7. Terhune, R. W., Maker, P. D., and Savage, C. M., *Phys. Rev. Letters*, 1965, **14**, 681.
8. Cyvin, S. J., Rauch, J. E., and Decius, J. C., *J. Chem. Phys.*, 1965, **43**, 4083.
9. Fanconi, B., Nafie, L. A., Small, W., and Peticolas, W. L., *J. Chem. Phys.*, 1969, **51**, 3993.
10. Fanconi, B., and Peticolas, W. L., *J. Chem. Phys.*, 1969, **50**, 2244.
11. Long, D. A., and Stanton, L., *Proc. Roy. Soc. A.*, 1970, **A318**, 441.
12. Maker, P. D., personal communication.
13. Stanton, L., *Molec. Phys.*, 1969, **16**, 37.
14. Bass, M., Franken, P. A., Ward, J. F., and Wienrich, G., *Phys. Rev. Letters*, 1962, **9**, 446.
15. Wilson, E. B., Decius, J. C., and Cross, P. C., 'Molecular Vibrations', McGraw-Hill, New York, 1955.
16. Walrafen, G. E., personal communication.
17. Maker, P. D., Second International Raman Conference, Oxford, 1970.

CHAPTER 3

Vibrational Assignments in Small Molecules

W. H. FLETCHER

A. Introduction

The assignment of the vibrational bands which are observed in the infrared and Raman spectra of small molecules is a relatively simple process for molecules having only three or four atoms and for some larger molecules if they possess sufficient symmetry. However, the complexity of the spectrum increases rapidly with each added atom, and unequivocal assignments are still sought for some molecules with as few as four and five atoms. The object is to identify the vibrational bands which are observed in the infrared and/or Raman spectrum and to associate each one, insofar as possible, with particular vibrational modes or normal coordinates of the molecule. Our primary interest is in identifying the fundamental frequencies, and our attention will be focused in this direction. However, in most spectra many overtone and combination bands are observed and assigned also, although these assignments are frequently very uncertain. The tools which can be brought to bear in making a vibrational assignment include empirical correlations and the group frequency concept, infrared and Raman selection rules, molecular symmetry and group theory, theoretical band shapes for partially resolved bands, frequency shifts with isotopic substitution, frequency calculation with approximate force fields, the polarization characteristics of Raman bands, and (perhaps too often) the investigator's intuition and prejudice.

B. General principles

The assignment of fundamental frequencies is made easier if the molecular configuration is known from other structural determinations. The point group is then known and hence the number of infrared- and Raman-active fundamentals are deduced by standard methods of group theory.[1, 2] The expected frequencies for the different fundamentals are initially estimated by making use of group frequency correlations.[3] The actual assignments of the observed spectra can then be made by employing the infrared and Raman selection rules, predicted band shapes in the case of gases, polarization data in the Raman spectrum, frequency shifts for isotopic substitutions, and correlations which may be pertinent to the problem.

If the molecular configuration is not known it is often possible to deduce it from vibrational spectra. There are usually only two or three plausible configurations to be considered and the selection rules for these structures may be sufficiently different for them to be distinguished on the basis of the observed infrared and Raman spectra, particularly when the fullest use is made of band shapes, polarization characteristics and isotopic substitution. Essentially in such cases deduction of configuration and assignment of fundamentals proceed simultaneously.

A non-linear molecule containing N atoms has $3N$-6 vibrational modes of motion and $3N$-6 fundamental frequencies if all vibrational modes are non-degenerate. Overtones and combinations of these fundamental frequencies may be observed also, and the task of unravelling an observed spectrum would rapidly become formidable if some simplifying factors did not exist. The two most important simplifying factors are molecular symmetry and the concept of 'group frequencies'. If the molecule possesses some symmetry the normal vibrations are divided among the symmetry species or irreducible representations of the group.[1] The classification of a given vibration depends on the symmetry of the distorted molecule. For instance, the methyl halides belong to point group C_{3v}, and three of their nine normal modes of vibration preserve the symmetry of the molecule. These are called totally symmetric vibrations and belong to species a_1 of the group. The remaining six vibrations form three doubly degenerate pairs, and these destroy the symmetry of the molecule. They belong to species e of the group. For a molecule like methane the nine normal vibrations fall into three species, one totally symmetric (a_1) stretching vibration, one doubly degenerate (e) bending vibration, and two triply degenerate (f_2) vibrations. The nine normal modes give rise to only four fundamental frequencies in methane and similar tetrahedral molecules, while there are six distinct fundamentals in methyl halides.

The selection rules for infrared and Raman activity provide one of the most powerful tools for vibrational analysis. These selection rules depend upon molecular symmetry and are easily deducible for any molecule by use of group theory. Basically a change in dipole moment must accompany a vibration in order for it to be infrared active, and a change in polarizability must occur with the vibration in order for it to be Raman active. In a linear molecule without a centre of symmetry, all vibrations are both infrared and Raman active, but in a linear molecule with a centre of symmetry, vibrations are *either* Raman active *or* infrared active. Those distortions which destroy the centre of symmetry produce a change in dipole moment and are infrared active, while those which retain the centre of symmetry in the distorted configuration cause a change in the polarizability and are therefore Raman active. This mutual exclusion rule[2] for infrared and Raman activity applies to all molecules with a centre of symmetry. No fundamental is both infrared and Raman active unless there is a breakdown in selection rules. This may occur in the liquid state, but the forbidden band is normally very weak. Some vibrations are inactive in both the infrared and Raman spectra. The torsional mode of ethylene is 'silent', and in benzene only eleven of the twenty distinct frequencies are theoretically active.[3]

Most molecules contain characteristic groups such as CH_3, C═O, C═C, and ≡C—H, and these virtually always produce a characteristic 'group frequency'† which can be used for identification and for making assignments.[3] Some of these group frequencies must be used with care, since mixing of group modes with skeletal vibrations can occur and cause a significant and unexpected change in frequency. Examples of such changes will be pointed out later.

Until relatively recently most of the Raman spectra which have been used in vibrational analyses have been observed for the liquid or powdered solid state. In the case of liquids polarization data provide a means of identifying the totally symmetric vibrations. The non-totally symmetric vibrations exhibit a depolarization ratio of 6/7 when the incident radiation is natural (or 3/4 when the incident radiation is plane polarized). The ratio for the totally symmetric vibrations lies between zero and 6/7 (or 3/4).[2] For spherical tops the ratio for polarized bands is theoretically zero. Any degree of depolarization which is less than 6/7 (or 3/4) is considered adequate evidence that the vibration is 'polarized' and belongs to the totally symmetric species. A depolarization ratio of 6/7 (or 3/4) can in principle arise from either a non-totally symmetric vibration or a totally symmetric vibration for which the ratio is so close to 6/7 (or 3/4) that it is indistinguishable experimentally from the limiting value. The second possibility is not a common occurrence but cases are known where totally symmetric bands have depolarization ratios indistinguishable experimentally from 6/7 (or 3/4). It is therefore important to realize that whereas

† If the reader is not familiar with these 'group frequencies', he should consult one of the references in (3) for their magnitudes and reliability.

a depolarization ratio less than 6/7 (or 3/4) is unambiguous evidence for a totally symmetric vibration, an observed ratio of 6/7 (or 3/4) is not necessarily indicative of a non-totally symmetric vibration. Generally speaking, the use of powdered solid samples precludes polarization measurements since the numerous reflections suffered by the scattered radiation destroy the polarization characteristics impressed by the scattering act. It has, however, recently been shown that depolarization ratios of powders can be successfully determined if the powder is suspended in a liquid having the same refractive index as the solid.[4]

Laser sources give access to further valuable information with direct bearing on assignments. Always determinable in principle, this information could not be obtained easily in practice with previously available sources. From studies of oriented single crystals the form of the scattering tensor for each vibrational mode can be determined and this is usually uniquely characteristic of the symmetry of the mode. Additionally, gas-phase Raman spectra can be obtained relatively easily with laser sources and such observations also greatly assist vibrational assignments. Recent work on single crystal and gas-phase Raman spectra are the subject of Chapter 5 and will not be further considered here.

The shapes of infrared absorption bands of gases can yield a great deal of information about a molecule even if the rotational structure is not completely resolved. For instance, the stretching vibrations of linear molecules give rise to bands which look exactly like the bands of a diatomic molecule. The spacing of resolved rotational lines is approximately $2B$, and if the structure is not resolved, the spacing of the P and R maxima in the band envelope is given by the following expression (reference 2b, p. 391)

$$\Delta\tilde{\nu} = (8kTB/hc)^{1/2}$$

Corresponding to the bending vibrations of a linear molecule the selection rule for rotational transitions is $\Delta J = 0, \pm 1$. The $\Delta J = 0$ transitions produce a strong Q branch at the band centre, while the P and R structure of the band is analogous to that seen in the stretching bands. The two band types are usually readily recognizable whether the rotational lines are resolved or not, and the observance of a spectrum which has only these two types of bands would provide strong evidence that the molecule is linear.

Symmetric tops have two distinct types of fundamental also, which give rise to easily recognizable band shapes. The parallel bands are those for which the dipole moment change is parallel to the symmetry axis of the molecule. The resulting bands look qualitatively like the perpendicular (bending) vibrations of linear molecules. If the rotational lines of the P and R branches are resolved, the spacing is approximately $2B$, as in linear molecules, otherwise a distinct P—Q—R band contour may be observed.

The Q branch is very strong for oblate tops, but may become very weak in the prolate tops. An additional bit of information used to check symmetric top spectra is a calculation of the separation of the P and R maxima in partially resolved parallel bands. The equations necessary for this are given by Gerhard and Dennison[5] in a classic paper on the shapes of infrared bands of symmetric tops.

The perpendicular bands of symmetric tops are always doubly degenerate and have alternating dipole moments which are perpendicular to the figure axis. The rotational selection rules are $\Delta K = \pm 1$ and $\Delta J = 0, \pm 1$. Because $\Delta K \neq 0$, the sub-bands do not coincide, and the Q branches of the sub-bands form the most prominent feature of the perpendicular bands of a prolate symmetric top. These Q branches have a spacing of $2(A' - B')$ if Coriolis effects are ignored. If the molecule has a 3-fold axis of symmetry, as in methyl halides, methyl cyanide and ethane, the Q branches exhibit an intensity alternation which goes strong, weak, weak, strong.[2b] The strong lines are those identified by values of K divisible by three in the ground state, that is $K = 0, 3, 6, \ldots$. The ratio of the intensities of the strong and weak lines depends on the nuclear statistics of the identical nuclei around the 3-fold axis, but the alternation itself is independent of the statistics. Figure 3.1 shows typical parallel and perpendicular band envelopes for prolate and oblate symmetric tops.

If the Coriolis coupling coefficient ζ_i differs from zero in a perpendicular vibration, the band shape may be appreciably different from those shown in Figure 3.1. Since the Q branch spacing is $2[A(1 - \zeta_i) - B']$ we find the sub-band origins coincide if $\zeta_i = 1 - B'/A'$ and the resulting band is indistinguishable from a parallel band. This is a frequent occurrence in oblate tops. The perpendicular bands of oblate tops[6] may vary greatly in shape with different values of ζ. In a prolate top, if the sub-band origins are close but not coincident, a single peak is seen, as in Figure 3.1(b).

Asymmetric tops exhibit extremely complex resolved spectra, but the envelopes of unresolved bands are often well formed and characteristic of the type of vibration involved. The bands of an asymmetric top are classified according to the direction of the dipole moment change for the vibration. Type A bands are those in which the vibration produces an alternating dipole moment which is parallel to the axis of least moment of inertia. Type B bands have an alternating dipole moment parallel to the axis of the intermediate moment of inertia, and the alternating dipole moment for type C bands is parallel to the axis of the largest moment of inertia. In many asymmetric tops which exhibit only a band contour the three types are easily distinguished. The type A bands have well-formed P, Q, and R branches; type B bands are characterized by a doublet structure with a pronounced central minimum, while type C bands have a very strong central Q branch with the P and R branches showing only as rather weak shoulders on the Q branch. Examples of all three of these are seen in the spectrum of dimethyl ketene which appears in Figure 3.6.

The contours of asymmetric top bands were first discussed by Badger and Zumwalt,[7] and this classic paper is still an important one. The subject of band shapes of symmetric and asymmetric tops has been reviewed

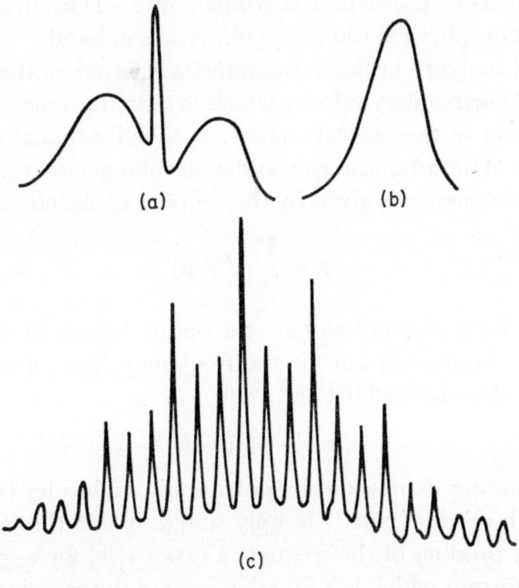

FIGURE 3.1. *(a) Parallel band envelope of an oblate top. Parallel bands of prolate tops have similar shapes, but with much weaker Q branches. (b) Perpendicular band envelope of a prolate top. (c) Resolved perpendicular band of a prolate top* (CH_3Cl).

recently by Seth-Paul.[8] Typical contours are shown for type A, B, and C bands of asymmetric tops with varying degrees of asymmetry and for the parallel and perpendicular bands of symmetric tops ranging from oblate tops to linear molecules. Unfortunately, the shapes shown for perpendicular bands of prolate symmetric tops are somewhat misleading, and not all oblate tops have perpendicular bands with these well formed P, Q and R branches. It is true that the envelopes of perpendicular bands of spherical tops and linear molecules do look alike, but prolate tops give the structure shown in Figure 3.1(c), and if I_A is large the band envelope shows a single central maximum as in Figure 3.1(b). In the case of oblate tops, the perpendicular band shapes are usually distorted by the effects of Coriolis coupling, which are not considered by Seth-Paul. The parallel bands of symmetric tops and the type A and C bands look so much alike in these diagrams that the inexperienced reader will get little help from them. The type B diagrams are characteristic and satisfactory. The reader who wishes to make use of observed band contours would be well advised to study Badger and Zumwalt's article and refer to observed asymmetric top spectra for examples. Unfortunately, the effects of Coriolis coupling are beyond the scope of this discussion, too, so the interested reader is referred to papers by Edgell and Moynihan, by McDowell, and by Hoskins[9] on the shapes of spherical top and symmetric top bands.

A powerful tool for checking assignments is found in the isotope effect. Except for anharmonicity effects, which are small, molecular force fields are independent of the nuclear masses, so the vibrational frequencies are functions only of the atomic masses and molecular geometry. For a diatomic molecule the frequency is given by the simple expression

$$\nu = \frac{1}{2\pi}\sqrt{(k/\mu)}$$

where k is the force constant and μ is the reduced mass. From this one finds the following simple relation for the frequency ratio of two isotopically related molecules such as HCl and DCl.

$$\nu/\nu' = \sqrt{(\mu'/\mu)}$$

The corresponding expression for polyatomic molecules is known as the Teller-Redlich product rule.[10] Simply stated, it says that for two isotopic molecules the product of the frequency ratios ν_i/ν_i' for a given symmetry species is a constant which is a function only of the atomic masses and the moments of inertia of the molecule.

$$\prod_i \nu_i/\nu_i' = f(m_1, m_1', m_2, m_2', \ldots I_A, I_A' \ldots)$$

This rule is based on the harmonic oscillator approximation, and it does not hold exactly for observed frequencies. The discrepancy between calculated and observed ratios increases with the number of frequencies in the symmetry species.

After the fundamentals have been selected and assigned in the observed spectrum, the remaining bands, the overtones and combinations, should be assignable in terms of the proposed fundamentals. Some anharmonicity is expected, and this usually causes the observed overtone or combination to be slightly less than the simple multiple or sum of the fundamentals, but in a few instances the observed frequencies are higher. Assignments of these bands, which must be done with due regard to symmetry and selection rules, frequently confirm the fundamental assignments or show that a proposed assignment is in error. Fermi resonances, of course, displace overtones (or combinations) as well as fundamentals, and due allowance must be made for this. Such resonances are generally recognized by the unusually high intensity of the overtone (or combination). If low frequency fundamentals (usually 400 cm^{-1} or less) are present some difference tones such as $\tilde{\nu}_i - \tilde{\nu}_j$ may be expected. If a difference band is seen, the corresponding sum band $\tilde{\nu}_i + \tilde{\nu}_j$ should also be seen with an even greater intensity, since the population of the lower state of the 'hot band' is reduced, relative to the ground state, by the Boltzmann factor, $e^{-h\nu_j/kT}$

Some examples of the application of these principles are given in the following pages. The interested reader is referred to the original papers for details which cannot be included here.

C. Linear molecules

A linear molecule containing N atoms has $3N$-5 fundamental modes of vibration or normal coordinates. Of these N-1 are non-degenerate stretching modes, and $2N$-4 are doubly degenerate bending vibrations. There are, therefore, just N-2 distinct frequencies representing bending vibrations. Most linear molecules have only five atoms or less, so it is really not necessary to employ group theory in making the vibrational assignment. Nevertheless, the mutual exclusion in the Raman and infrared selection rules provides gratifying confirmation of chosen assignments for those molecules having $D_{\infty h}$ symmetry.

The bond-stretching vibrations are parallel in nature and the corresponding infrared bands are identical in appearance with the vibration-rotation bands of diatomic molecules. The rotational selection rule is $\Delta J = \pm 1$, and the spacing of adjacent lines in the P and R branches is approximately $2B$. The bending vibrations are perpendicular in nature, and because of the degenerate nature of the motion, there is a resultant vibrational angular momentum in the first excited state. Consequently the selection rule is $\Delta J = 0, \pm 1$, and these bands have a strong Q branch as well as the P—R structure. All fundamentals are both infrared- and Raman-active in molecules with $C_{\infty v}$ symmetry such as N_2O and COS, but in molecules with $D_{\infty h}$ symmetry (i.e. CO_2 and C_2H_2) only those vibrations which destroy the centre of symmetry, and therefore cause a change in dipole moment, are active in the infrared.

(a) *Carbon dioxide and carbon disulphide*

The infrared and Raman spectra of CS_2 provide a convenient example which is relatively free of confusing perturbations. The Raman spectrum shows one strong, polarized band at 656 cm^{-1} in the liquid and at

659 cm^{-1} in the gas. This is clearly due to the symmetric C—S stretching vibration. The infrared spectrum shows a strong parallel band at 1537·6 cm^{-1} and a perpendicular band at 396·6 cm^{-1}. These are the antisymmetric C—S stretching mode and the doubly degenerate bending mode, respectively. All three bands are in different symmetry species, and their assignment is unequivocal.

The fundamentals of CO_2 fall in a similar pattern: the symmetric C—O stretch $\nu_1 = 1388$ cm^{-1}, the bending vibration $\nu_2 = 668$ cm^{-1}, and the antisymmetric stretch $\nu_3 = 2349$ cm^{-1}. The last two appear as perpendicular and parallel bands, respectively, in the infrared spectrum. As in the CS_2 spectrum, only one fundamental ν_1 is Raman active, but the Raman spectrum of CO_2 shows two strong bands of nearly equal intensity at 1285·5 and 1388·2 cm^{-1}. This phenomenon, first correctly interpreted by Fermi and discussed in detail by Adel and Dennison,[12] affords a good example of Fermi resonance,[13] which arises because the overtone $2\nu_2$ belongs to the same symmetry species as ν_1 and lies sufficiently close to it that the wave functions can mix. The unperturbed levels ν_1 and $2\nu_2$ are separated by approximately 5 cm^{-1} in CO_2, while the observed levels are separated by 103 cm^{-1}. The result of the interaction and mixing of the wave functions is a 'borrowing of intensity' by the overtone from the strong fundamental so that both bands have similar intensities. In addition, the lower of the two levels $2\nu_2$ is pushed downward and the upper level ν_1 is pushed up by the same amount.† This same perturbation is observed in CS_2, but it is much weaker than in CO_2, and the overtone $2\nu_2$ is only slightly stronger than it would be with no interaction.

This type of interaction depends on the anharmonic terms of the potential function, and it may occur whenever two levels (i.e., a fundamental ν_i and an overtone $n\nu_j$) belonging to the same species are accidentally nearly degenerate. However, as there are usually insufficient data to make an analysis of the perturbation, the postulation of Fermi resonance should be made with care.

(*b*) *Acetylene*

Acetylene also has $D_{\infty h}$ symmetry, but its spectrum demonstrates some additional complications. The observed frequencies, their assignments, symmetry species, and approximate description of the modes for C_2H_2, C_2D_2 and C_2HD are given in Table 3.1. The symmetry coordinates appropriate for these modes are defined in Table 3.2. Note that each of the

† G. Amat and M. Pimbert[12] have considered higher order perturbations in the analysis of high resolution data for CO_2 overtones observed by C. P. Courtoy, and they suggest that the assignment of ν_1 and ν_2 should be reversed.

TABLE 3.1. *Fundamental frequencies of acetylenes*

Species	Mode	Approximate description	Frequency (cm^{-1}) C_2H_2	C_2D_2	C_2HD
Σ_g^+	ν_1	C—H stretch	3372·5	2703·8	3334·0
	ν_2	C≡C stretch	1973·5	1763·5	1852·6
Σ_u^+	ν_3	C—H stretch	3284·2	2439·2	2583·6
Π_g	ν_4	C≡C—H bend	613·5	511·1	518·8
Π_u	ν_5	C≡C—H bend	730·2	537·8	683

TABLE 3.2. *Symmetry coordinates of C_2H_2*

Σ_g^+ $$S_1 = \frac{1}{\sqrt{2}}(\Delta r_1 + \Delta r_2)$$

$$S_2 = \Delta R$$

Σ_u^+ $$S_3 = \frac{1}{\sqrt{2}}(\Delta r_1 - \Delta r_2)$$

Π_g $$S_{4a} = \frac{1}{\sqrt{2}}(\Delta\theta_1 - \Delta\theta_2)$$

$$S_{4b} = \frac{1}{\sqrt{2}}(\Delta\phi_1 - \Delta\phi_2)$$

Π_u $$S_{5a} = \frac{1}{\sqrt{2}}(\Delta\theta_1 + \Delta\theta_2)$$

$$S_{5b} = \frac{1}{\sqrt{2}}(\Delta\phi_1 + \Delta\phi_2)$$

$\Delta r_1, \Delta r_2$ are C—H bonds

ΔR is C≡C bond

$\Delta\theta_1, \Delta\theta_2$ are H—C≡C angles in xz plane

$\Delta\phi_1, \Delta\phi_2$ are H—C≡C angles in yz plane

last three vibrations is in a species by itself, so the symmetry coordinates are also normal coordinates of the molecule. The same is true for the vibrations of CO_2 and CS_2. Apart from a small anharmonicity (1–2%) the frequency changes observed in the deuterated molecule comply with those calculated from the reduced mass ratios.[14] These simple ratios do not apply individually to the two Σ_g^+ vibrations because the symmetry coordinates of vibrations belonging to a single species mix to form the normal coordinates. Consequently, the isotope effect which may appear to influence only one vibration may in fact be distributed over several vibrations. The extent of this mixing can be calculated from the eigenvectors (assuming we have been able to find the correct force constants), but a good qualitative picture may be seen by examining the frequency ratios for ν_1 and ν_2 in C_2H_2 and C_2D_2 and comparing them with the appropriate, corresponding diatomic molecules. For ν_1 the observed frequency ratio is 1·25 while the calculated ratio for C—H and C—D is 1·37. For ν_2 the appropriate representation is a hypothetical homonuclear diatomic molecule with masses 13 in one case and 14 in the other. The observed and calculated frequency ratios are 1·12 and 1·04, respectively. It is clear that the normal coordinate of ν_1 includes a significant amount of the C≡C stretch, and likewise, the normal coordinate of ν_2 contains a significant contribution from the C—H stretch.

(c) Diacetylene, carbon suboxide, dicyanoacetylene and related species

The spectra of the larger linear molecules with $D_{\infty h}$ symmetry, such as carbon suboxide, diacetylene, dicyanoacetylene, and dicyanodiacetylene, are much more complex, and their assignments have been subject to controversy. The spectrum of diacetylene was the first of these to be assigned correctly.[15] Diacetylene has three Σ_g^+ stretching frequencies, two Σ_u^+ stretching modes, and two each of the Π_g and Π_u bending modes. The interesting point of this case is that the early assignments placed the single bond C—C stretching vibration at 642 cm^{-1}, and normal coordinate calculations with this frequency gave a C—C force constant of 3·2 to 3·6 mdyne $Å^{-1}$, which is much too low and is not consistent with the known C—C bond length (1·36 Å). Because of this the spectrum of diacetylene was re-examined in 1951 by A. V. Jones, who found a previously unobserved polarized line at 877 cm^{-1} in the Raman spectrum of the liquid. Assigning this to the C—C stretch, ν_3, leads to a force constant of 7·15 mdyne $Å^{-1}$, which is quite acceptable since the bond is anomalously short for a single bond. That this bond should have some double bond character is not surprising since it is between two triple bonds. A similar situation exists for the single bond in cyanogen, cyanoacetylene, and dicyanoacetylene.

Jones was not able to observe the lowest frequency fundamental which is not Raman-active, but he did deduce it correctly from overtones. The fundamental was later observed directly by Miller, Lemmon, and Witkowski,[15] who found it within 1·5 cm^{-1} of Jones' value of 220 cm^{-1}. The fundamentals of diacetylene are listed in Table 3.3.

Carbon suboxide has proved to have the most difficult spectrum to interpret, and it has been studied by several investigators.[16] Long, Murfin and Williams examined the infrared and Raman spectrum, and they observed and assigned correctly six of the seven fundamentals on the basis of a $D_{\infty h}$ structure, which is predicted by valence theory. The assignment of the observed bands is unequivocal since no more than two vibrations

TABLE 3.3. *Fundamental frequencies of carbon suboxide, cyanogen, dicyanodiacetylene, dicyanoacetylene, and diacetylene*

		Frequency (cm^{-1})				
Species	Approximate description	C_3O_2	C_2N_2	C_6N_2	C_4N_2	C_4H_2
Σ_g^+	C=O, C≡N, or C—H stretch	2200	2330	2235	2290	3293
	C≡C or C=C stretch	830		2183	2119	2184
	C—C stretch		848	1287·5	692	874
	C—C stretch			571		
Σ_u^+	C=O, C≡N, or C—H stretch	2258	2150	2266	2241	3329
	C≡C or C=C stretch	1573		2097		2020
	C—C stretch			717	1154	
Π_g	C=C=O, C—C≡N, or C≡C—H bend	577	506	501	504	647
	C≡C—C bend			455	263	484
	C≡C—C bend			170		
Π_u	C=C=O, C—C≡N, or C≡C—H bend	550	240	490·5	472	500
	C=C=C or C≡C—C bend	63		276	107	221·5
	C≡C—C bend			61·5		

belong to a single species. In the Raman spectrum polarization measurements identify the Π_g vibrations, and in the infrared the Σ_u^+ and Π_u vibrations are distinguished by their parallel and perpendicular character, respectively. The C=O and C=C stretches are distinguishable by using group frequency correlations. The lowest frequency ν_7, a Π_u bending mode, was not observed by the early investigators because it was out of the range of their instruments, but it was estimated to be about 200 cm^{-1}. While this is a reasonable value and was supported by some combination assignments, it is in fact much too high.

The major problem with the data for carbon suboxide was that neither the band shapes nor the selection rules established whether the molecule is linear or not. The infrared band envelopes should have shown distinct P—R or P—Q—R structure with a P—R separation of about 11 cm^{-1}, but no such contours were observed. The absence of these led Rix to believe the molecule was bent and had C_{2h} symmetry. Miller and Fateley were also nearly convinced it was non-linear until Lafferty, Maki and Plyler analyzed an overtone and its 'hot bands' in the region above 3100 cm^{-1}. Their analysis demonstrated conclusively that C_3O_2 is a linear molecule. Miller and Fateley re-examined the infrared spectrum from 70 to 4000 cm^{-1}, taking exceptional care to insure that their samples contained no impurities. Samples prepared by three different methods gave identical spectra, but the lowest frequency fundamental was still unobserved.

A special study of the low frequency bending modes of carbon suboxide, dicyanoacetylene, diacetylene and dimethylacetylene was made by Miller, Lemmon and Witkowski,[15] and this time a very weak band was found at 63 cm^{-1} and identified as ν_7 of C_3O_2. (The fundamental assignments of C_3O_2 are given in Table 3.3.) Two important questions arise concerning this band. First, why is the frequency so much lower than the other two bending modes, and second, why is the intensity so low when such a bending mode is usually intense? The answer to both of these lies in a consideration of the resonance structures II and III below

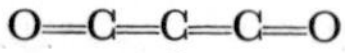
I

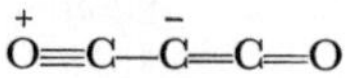
II

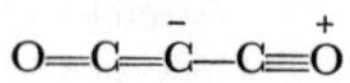
III

The central carbon ($-\bar{C}=$) is isoelectronic with $-N=$ which normally has a bent configuration, and if the forms II and III make a significant contribution to the structure, the bending mode ν_7 should have a low frequency. The other two bending modes involve bending at the outer carbons, and the $\overset{+}{O}{\equiv}C-$ and $C{=}C{=}O$ groups are linear, so these modes are not influenced by the resonance in the same way as ν_7.

The Π_u bending mode ν_6 is slightly lower in frequency than the Π_g mode ν_5, but the simple theory of coupled oscillators predicts the opposite. The nature of the system can be easily understood by referring to Figure 3.2. In a linear chain of identical coupled oscillators the highest frequency mode is the one with the most nodes, the second highest frequency has the second largest number of nodes, and the lowest frequency has the smallest number of nodes. This predicts that $\nu_6 > \nu_5 > \nu_7$ should be the order of frequencies, while the observed order is $\nu_5 > \nu_6 > \nu_7$, and this assignment is certainly correct. Note that ν_6 involves a bending at the central carbon while ν_5, by virtue of the symmetry of the mode, does not. Therefore, the contribution of the resonance forms II and III can decrease the potential energy required for distortion in the ν_6 mode, but the ν_5 mode cannot be affected. It would be interesting to assess the approximate magnitude of the effect of the resonance on ν_6 by computing its unperturbed value with a simple force field.

The abnormally weak intensity of ν_7 is also explainable in terms of the resonance forms. By analogy with other molecules, the distortion apparently involves the displacement of large dipoles at the ends of long lever arms, and $d\mu/dQ$ is expected to be large. However, in form I the oxygen atoms are negative ends of the dipoles, while in II and III the charge is sharply reduced by the positive charge and $d\mu/dQ$ is reduced nearly to zero.

The existence of this very low frequency, $\nu_7 = 63\ \mathrm{cm}^{-1}$, also provides the explanation for the absence of the clearly defined parallel and perpendicular band contours which are expected in a linear molecule. The

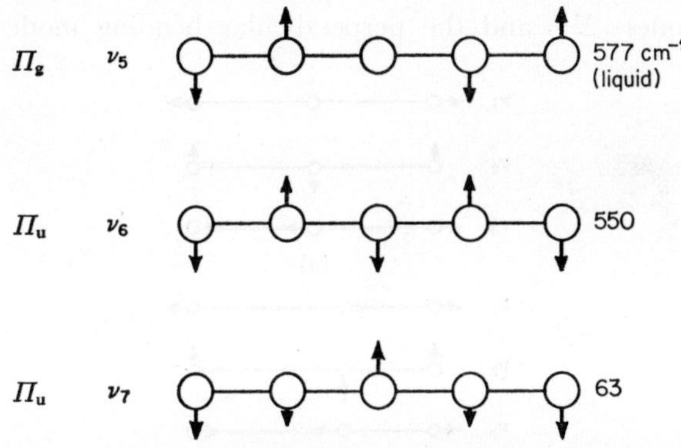

FIGURE 3.2. *Bending distortions of carbon suboxide. (Reproduced with permission from ref. 15.)*

upper states of this vibrational mode are well populated at room temperature, and every fundamental has its band contour distorted by underlying hot bands which are necessarily rather closely spaced.

All fundamentals of dicyanoacetylene have been observed, [15] and there is no problem involving their assignments. Dicyanodiacetylene is the longest linear molecule for which the infrared and Raman spectra have been observed and analyzed.[17] The thirteen fundamentals are distributed among four symmetry species, and all have been observed and assigned.

$$\Gamma_{\mathrm{vib}} = 4\Sigma_g^+ + 3\Sigma_u^+ + 3\Pi_g + 3\Pi_u$$

The data strongly indicate that the molecule is linear, and while no controversy exists over the assignments, the data do not provide a basis for firm conclusions. For instance, the low frequency C—C stretching band ν_4 should be polarized, but the band ($571\ \mathrm{cm}^{-1}$) assigned to it is depolarized. A totally symmetric vibration may be virtually completely depolarized, however, and this is the only logical choice for this fundamental. The two parallel infrared bands ν_5 and ν_6 are assigned to bands which do not have the expected structure, but alternative assignments are not acceptable. The absence of the doublet structure is no doubt due to the same cause as was discussed in the case of carbon suboxide. Dicyanodiacetylene has a strong bending mode ν_3 at 61·5 cm^{-1}, and the well populated excited levels of this vibration produce hot bands which obscure the doublet structure of the stretching fundamentals. The fundamentals of dicyanoacetylene and dicyanodiacetylene are listed in Table 3.3.

(d) Nitrous oxide, hydrogen cyanide, cyanoacetylene and related species

Linear molecules having symmetry $C_{\infty v}$, such as N_2O, COS, and monosubstituted acetylenes have only two species of fundamental vibrations, the parallel modes (Σ^+) and the perpendicular bending modes (Π). The

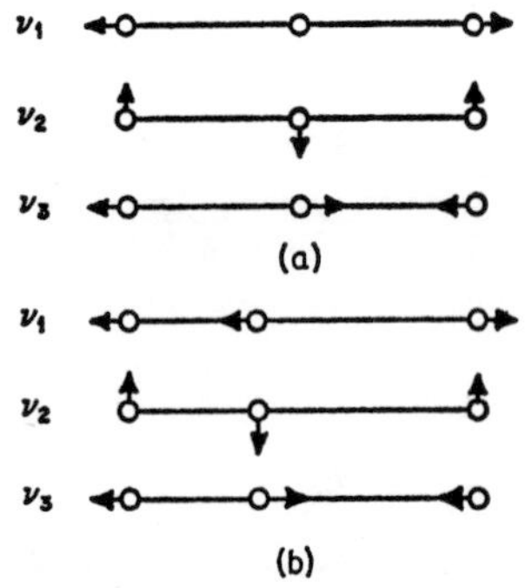

FIGURE 3.3. *Normal vibrations (a) CO_2, (b) N_2O.*

symmetric and antisymmetric character (g and u) of the vibrations does not exist, but the actual molecular distortions may resemble the corresponding modes in $D_{\infty h}$ molecules. Let us compare the vibrations of N_2O and CO_2, which are shown in Figure 3.3. The distortions ν_1 and ν_3 in CO_2 belong to different species, and the symmetry coordinates are also normal coordinates. The vibrations are precisely described as symmetric and antisymmetric C=O stretching vibrations, respectively. In N_2O the corresponding modes belong to the same symmetry species, and they are descriptively called N=O and N=N stretching vibrations, respectively. This is not an accurate description because the symmetry coordinates Δr_{NN} and Δr_{NO} both contribute to each of the normal coordinates. In fact, the actual motions may quite properly be called 'pseudo-symmetric' and 'pseudo-antisymmetric' vibrations.[18] In ν_1 the N=N and N=O bonds stretch simultaneously as in CO_2, but unlike CO_2 the central atom does move slightly because N_2O has no centre of symmetry. The ν_3 mode of N_2O resembles that of CO_2 in the fact that the central atom executes the largest displacement. In CO_2 the O—O distance remains invariant in this motion, with the carbon atom oscillating between the two oxygen atoms which move together just enough to keep the centre of mass invariant. The similarity of the ν_3 mode in N_2O is easily seen. The fundamental frequencies of N_2O and COS are listed in Table 3.4, with the frequencies of CO_2 and

TABLE 3.4. *Fundamental frequencies of linear triatomic molecules*

		$D_{\infty h}$		$C_{\infty v}$	
Mode	Description	CS_2	CO_2	N_2O	COS
ν_1	Symmetric stretch or pseudo sym. stretch	658	1388	1285	859
ν_3	Antisym. stretch or pseudo antisym. stretch	1533	2349	2223	2062
ν_2	Bending	397	667	589	527

CS_2 for comparison. All other bands in the spectra of these molecules can be assigned as overtones and combinations of these. The references should be consulted for details of the complete assignments.

Hydrogen cyanide has the same symmetry as N_2O and COS, but in this case the two bonds have characteristic stretching frequencies sufficiently different that there is very little mixing in the normal coordinates. The observed bands are properly called C—H (3312 cm^{-1}) and C≡N (2089 cm^{-1}) stretching bands. Another interesting point of behaviour appears in the infrared spectrum. The C≡N stretching mode is expected to be quite intense, and indeed these modes are very intense in some molecules, but it is only weakly absorbing in HCN. The reason is that the C—H group is isoelectronic with N, and $d\mu/dQ$ is very small for this vibration.

The infrared spectra of fluoro-, chloro-, and bromo-acetylenes have been observed and analyzed by Hunt and Wilson.[19] The three parallel bands in each molecule are easily assigned to C—H, C≡C, and C—X stretching modes on the basis of group frequencies. In the case of the two perpendicular bands, the C≡C—H band is readily identified by the isotope effect. The frequency assignments are given in Table 3.5.

Cyanoacetylene has four parallel stretching modes and three doubly degenerate bending modes. Since all modes are both infrared and Raman active, it should be possible to make a complete assignment from either spectrum. The infrared spectrum was first observed by Turrell, Jones and Maki.[20] The gas-phase spectrum shows two perpendicular bands and five parallel bands in the range of 400 to 3400 cm^{-1}. The three highest-frequency

TABLE 3.5. *Fundamental frequencies of haloacetylenes and cyanoacetylene*

Approximate description	Mode	Frequency (cm^{-1}) FC≡CH	FC≡CD	ClC≡CH	ClC≡CD	BrC≡CH	BrC≡CD	HC≡CCN	DC≡CCN
C—H(C—D) stretch	ν_1	3335	2645	3340	2612	3325	2600	3256	2608
C≡N stretch								2270	2245
C≡C stretch	ν_2	2255	2065	2110	1980	2085	1950	2070	1952
C—X stretch	ν_3	1055	1048	756	742	618	606	877	860
C≡C—H bend	ν_4	578	439	604	472	618	480	684	545
C≡C—X bend	ν_5	367	(364)	326	(312)	295	(283)		
C—C≡N bend								502	498
C≡C—C bend								230	219

Figures in parentheses were calculated.

parallel bands which are at 3328, 2271 and 2078 cm^{-1} are readily assigned to the C—H, C≡N, and C≡C stretches, respectively, on the basis of group frequency correlations. A strong parallel band at 1314 cm^{-1} and a weak one at 720 cm^{-1} remain as possible choices for the C—C stretch. Neither is acceptable, either on the basis of group frequency correlations (cf. cyanogen and diacetylene) or on the basis of the force constants computed from the assignment. It was concluded that the C—C stretching frequency is simply not observed in the gas phase. The liquid shows a medium strength band at 875 cm^{-1}, and the solid shows a strong band at 879 cm^{-1}. These fit well with the Raman-active C—C stretching band which was observed at 874 cm^{-1} in diacetylene. The weakness of this band is not surprising, since it is analogous to the infrared-inactive C—C stretch in diacetylene.

The two perpendicular bands are assigned to the C≡C—H bend (663 cm^{-1}) and to the C—C≡C bend (502 cm^{-1}). The third bending mode was not observed directly, but was placed at 230 cm^{-1} on the basis of observed combinations. That these deductions are basically correct was demonstrated by Job and King[21] who examined the infrared and Raman spectra of cyanoacetylene and deuterocyanoacetylene. The C—C stretch was observed directly in the Raman spectrum, where it produces a strong band at 877 cm^{-1}. The third bending vibration was also observed in the Raman spectrum at 230 cm^{-1} in the light molecule and at 219 cm^{-1} in the heavy molecule. Job and King assign the C≡C—C bend to 230 cm^{-1} and the C—C≡N bend to 502 cm^{-1}, while Turrell, Jones and Maki chose the converse assignment. On the basis of their data, there was no way of making a definite choice between the two possibilities, but the data for the deuterated molecule provide another bit of information. Examination of the assignments in Table 3.5 shows that the low frequency mode has a considerably larger isotope shift, and the substitution of the hydrogen by deuterium would indeed have a larger effect on the C≡C—C bending mode than on the C—C≡N bending mode. This conclusion could be confirmed very nicely by introducing ^{15}N into the molecule.

D. Prolate symmetric tops

(a) *Methyl halides and related molecules*

The simplest prolate symmetric tops are the methyl halides. Their symmetry is C_{3v}, and the nine fundamental vibrations are divided into three parallel vibrations (species a_1) and three doubly degenerate perpendicular vibrations (species e). The parallel bands have typical PQR contours under moderate resolution, and the perpendicular bands all show the expected strong and nearly uniformly spaced Q branches with a strong–weak, weak–strong intensity alternation. The fundamentals are the strongest bands in the spectrum, and the overtones are readily accounted for in terms of these six strongest bands. The three parallel vibrations are a symmetric C—H stretch, a symmetric CH_3 deformation (umbrella motion) and a C—X stretch. The degenerate vibrations are an antisymmetric C—H stretch, an antisymmetric CH_3 deformation, and a CH_3 rocking mode. The detailed assignment of frequencies to these modes is readily made on the basis of group frequencies.

All vibrations are both infrared and Raman active, so a complete assignment can be made from either spectrum. The a_1 Raman lines are readily recognized by their intensity and the fact that they are polarized. As a general rule the totally symmetric Raman lines in liquids are strong and narrow while the lines for other vibrations are medium to weak and broader. The fundamental assignments for methyl halides, CH_3X and CD_3X are given in Table 3.6. It is readily observed that the C—H (and

TABLE 3.6. *Fundamental frequencies of methyl halides*

Mode	Approximate description	Frequency (cm^{-1})			
		CH_3F	CH_3Cl	CH_3Br	CH_3I
ν_1	C—H stretch (sym.)	2964	2966	2972	2970
ν_2	CH_3 def. (sym.)	1464	1355	1305	1252
ν_3	C—X stretch	1049	732	611	533
ν_4	C—H stretch (antisym.)	3006	3042	3057	3060
ν_5	CH_3 def. (antisym.)	1467	1455	1444	1440
ν_6	CH_3 rock	1182	1015	954	880
		CD_3F	CD_3Cl	CD_3Br	CD_3I
ν_1	C—D stretch (sym.)	2090	2161	2157	2155
ν_2	CD_3 def. (sym.)	1136	1029	993	951
ν_3	C—X stretch	991	695	578	501
ν_4	C—D stretch (antisym.)	2258	2286	2293	2300
ν_5	CD_3 def. (antisym.)	1072	1058	1056	1049
ν_6	CD_3 rock	903	775	712	662
Product rule ratios					
a_1:	observed	0·5221	0·5157	0·5099	0·5144
	calculated	0·5169	0·5153	0·5084	0·5057
e:	observed	0·4130	0·40356	0·3995	0·3991
	calculated	0·4193	0·40354	0·4005	0·3986

C—D) stretching motions and the antisymmetric deformations ν_5 have frequencies which are almost independent of the halogen. Furthermore the isotope effects show that the distortions are almost entirely hydrogen displacements. These are, therefore, 'good' group frequencies. On the other hand the symmetric CH_3 deformation and the rocking mode frequencies have a distinct dependence on the atom attached to the carbon. In the case of the deformation this dependence arises because the normal coordinate of the motion is a mixture of the CH_3 deformation and the C—X stretch.

Methyl cyanide[22] and methyl isocyanide[23] are prolate symmetric tops with one more atom than the methyl halides. The observed bands resemble those of the methyl halides except for the low frequency skeletal bending vibration. The three higher-frequency perpendicular fundamentals have easily resolvable sub-band Q branches, but the skeletal bending modes show only a band envelope which has the classical shape described by Gerhard and Dennison.[6] The reason the fine structure has not been resolved is that $\zeta_8 = 0{\cdot}93$, and the Q branch spacing computed from $2[A'(1-\zeta_8)-B']$ is 0·13 cm^{-1} in CH_3CN,[24] and most spectrometers cannot resolve lines as close together as this. The situation is almost identical in methyl isocyanide.

(*b*) *Methylmercuric halides*

The infrared and Raman spectra of methylmercuric halides have been observed by Goggin and Woodward,[25] who used mulls and oriented crystals for the infrared measurements and benzene and nitromethane solutions for the Raman spectra. The normal coordinates of these molecules are similar to those of methyl cyanide, so the group frequency correlations are similar. Three polarized lines are seen in the Raman spectra of nitromethane solutions for each molecule, and these are clearly species a_1 vibrations. Confirmation of these frequencies is seen in the infrared except for two low-frequency fundamentals (226 cm^{-1} in CH_3HgBr and 181 cm^{-1} in CH_3HgI) which were beyond the range of the infrared spectrometer used in this work. Group frequency correlations were used to assign the observed infrared bands. Table 3.7 shows the observed fundamentals of the three methylmercuric halides. Goggin and Woodward's assignment for CH_3HgBr was confirmed by Meić and Randić,[26] who examined this spectrum as well as that of CD_3HgBr.

E. Oblate symmetric tops

(*a*) *Boron trifluoride*

For most oblate symmetric tops all fundamentals are both infrared and Raman active, so a vibrational assignment might be made with either

TABLE 3.7. *Fundamental frequencies of methylmercuric halides*

Mode	Approximate description	Frequency (cm^{-1})			
		CH_3HgCl	CH_3HgBr	CH_3HgI	CD_3HgBr
ν_1	C—H stretch (sym.)	2930	2925	2912	2126
ν_2	CH_3 deformation (sym.)	1203	1193	1182	916
ν_3	Hg—C stretch	556	546	535	495
ν_4	Hg—X stretch	334	228	181	201
ν_5	C—H stretch (antisym.)	3015	3015	3000	2262
ν_6	CH_3 def. (antisym.)	1410	1410	1410	1027
ν_7	CH_3 rock	790	790	785	597
ν_8	C—Hg—X bend		(100)		(92)

All frequencies were observed by Goggin and Woodward,[25] except those for CD_3HgBr which were reported by Meić and Randić.[26] The figures in parentheses are calculated values given by the latter authors.

spectrum. Molecules with D_{nh} symmetry like BF_3 and cyclopropane are notable exceptions, and the infrared bands of the former illustrate the difficulty of determining a band type from a band contour alone. The six fundamental vibrations of BF_3 give rise to four distinct frequencies.

$$\Gamma_{vib} = a_1' + a_2'' + 2e'$$

The a_1' vibration is Raman active only, the a_2'' mode is infrared active only, while the e′ modes are active in both. The symmetric B—F stretch is easily recognized by its strong, polarized Raman line and the fact that it has no boron isotope effect. The latter shows that the boron atom is stationary in this vibration. The bending mode ν_2 is recognized by the fact that it appears only in the infrared. It has a typical parallel band contour, and the assignment is confirmed by the fact that two overlapping bands are present with Q branches at 691·3 cm^{-1} and 719·5 cm^{-1}.[27] These frequencies fit the calculated isotope splitting[28] for ^{11}B and ^{10}B.

The two e′ modes are recognized easily by the fact that they are both infrared and Raman active. The infrared bands would be expected to show a perpendicular type contour, but ν_4 actually looks like a parallel band, and ν_3 has an unusual doublet contour. The reason for this is the magnitude of the Coriolis coupling coefficients. For the degenerate bending vibration $\zeta_4 = -0{\cdot}80$ and $2[A'(1 - \zeta_4) - B'] = 0{\cdot}07$ cm^{-1}. Consequently, Q branches for the different sub-bands are nearly coincident, and the resulting contour looks like a parallel band. The degenerate stretching band ν_3 has the appearance of distorted P and R contours. In this case $\zeta_3 = +0{\cdot}80$ and

$2[A'(1 - \zeta_3) - B] = -0{\cdot}318$, and the Q branches are dispersed among the J lines of the P and R branches. The Q branches are weak so their presence among the J lines is not apparent.[2b]

(b) Nitrogen and phosphorus trifluorides

The infrared spectra of NF_3 and PF_3[29] shown in Figure 3.4 illustrate again the effects of Coriolis coupling on band shapes. No Raman polariza-

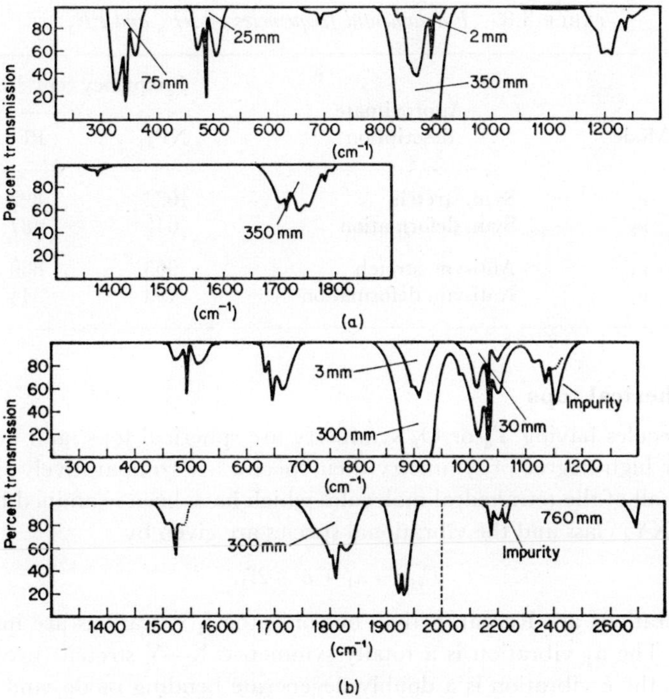

FIGURE 3.4. *Infrared spectra of NF_3 and PF_3.*
(a) PF_3 (gas, room temp, 10 cm cell)
(b) NF_3 (gas, room temp, 10 cm cell)
(*Reproduced with permission from ref. 29.*)

tion data were available when these spectra were observed, so the assignment had to be made by using correlations with similar molecules. In both molecules the two lowest-frequency bands are the bending fundamentals, but the contours give no clue as to which is the degenerate fundamental. This was assigned to the lowest frequency in each case by analogy with PCl_3, PBr_3, OPF_3, $OPCl_3$, $OPBr_3$ and HCF_3. In all of these the degenerate bend is the lowest frequency. A recent report of high resolution work on NF_3 by Popplewell, Masri and Thompson[30] confirms this assignment. The two stretching vibrations are less puzzling because the bands at 860 cm^{-1} in PF_3 and 905 cm^{-1} in NF_3 have contours which do resemble that of a partially resolved perpendicular band. In each case there is a band at higher frequency with distinct parallel contours, which is assigned to the symmetric stretching vibration. The assignment of all fundamentals is given in Table 3.8. A report of Raman polarization data on NF_3 has since confirmed these assignments.[31]

TABLE 3.8. *Fundamental frequencies of NF_3 and PF_3*

Mode	Approximate description	Frequency (cm^{-1})	
		NF_3	PF_3
ν_1	Sym. stretch	1032	892
ν_2	Sym. deformation	647	487
ν_3	Antisym. stretch	905	860
ν_4	Antisym. deformation	493	344

F. Spherical tops

Molecules having T_d or O_h symmetry are spherical tops, and because of their high degree of symmetry their spectra are comparatively simple. Nearly all of the tetrahedral molecules which have been examined belong to the XY_4 class and the vibrational species are given by

$$\Gamma_{vib} = a_1 + e + 2f_2.$$

All vibrations are Raman active, but only the f_2 vibrations are infrared active. The a_1 vibration is a totally symmetric X—Y stretch (breathing mode), the e vibration is a doubly degenerate bending mode, and the f_2 vibrations consist of a triply degenerate stretch and a triply degenerate bending mode. Positive identification of the bands is achievable because the a_1 band is the only polarized fundamental in the Raman spectrum, and the two f_2 vibrations appear in the infrared as well as in the Raman spectrum. The infrared bands have easily recognized PQR contours. The P and R branches may be slightly distorted by Coriolis coupling.

(a) CF_4

The spectrum of CF_4 is an example in which more than ample data are available. Claassen[32] first observed the Raman spectrum of the gas under moderate resolution. This spectrum and the infrared data of Woltz and Nielsen[33] yield a positive identification of the fundamentals. A Raman spectrum of the gas was photographed under higher resolution by Monostori and Weber,[34] and this provides complete and independent data for an unequivocal assignment (see Figure 3.5). In addition to being polarized the ν_1 vibration appears as a sharp line. There is no rotational structure because no rotation of the spherical top produces a change in the polarizability. The e vibration ν_2 is easily identified by its measured P—R spacing which is 33 cm^{-1}. This band is not perturbed by Coriolis coupling and the P—R spacing is predicted to be about 30 cm^{-1}. The spacing of the f_2

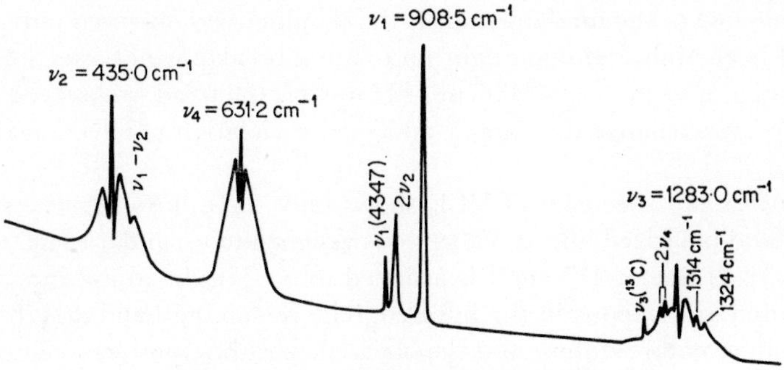

FIGURE 3.5. *Raman spectrum of CF_4.*
(*Reproduced with permissiom from ref. 34.*)

fundamentals is much less, and while the actual spacing is dependent on the zeta values, there is no question as to which vibration belongs to species e. The fundamental assignments are: $\nu_1 = 908{\cdot}5$ cm^{-1}, $\nu_2 = 435{\cdot}0$ cm^{-1}, $\nu_3 = 1283{\cdot}0$ cm^{-1}, and $\nu_4 = 631{\cdot}2$ cm^{-1}.

(*b*) *$TiCl_4$ and VCl_4*

The infrared and Raman spectra of $TiCl_4$ and VCl_4 were examined by Dove, Creighton and Woodward.[35] The liquid state was used for both except in the case of the Raman spectrum of VCl_4, which was observed in CCl_4 solution. Four Raman lines were observed for $TiCl_4$, one very strong, polarized line at 389 cm^{-1}, which is clearly ν_1, and three depolarized lines at 490, 140 and 120 cm^{-1}. The 490 cm^{-1} band can be assigned to ν_3, the triply degenerate stretch, because of its frequency and because the corresponding infrared band is observed at 485 cm^{-1} in the liquid. The two bending fundamentals are in a frequency range too low for the infrared spectrometer used, so the infrared coincidence could not be used to identify ν_4. They were assigned $\nu_4 = 140$ cm^{-1} and $\nu_2 = 120$ cm^{-1} because $\nu_4 > \nu_2$ in many other tetrahedral molecules for which the assignment is fairly certain (i.e. SiF_4, GeF_4 and $SnCl_4$). There is little doubt that the spectra indicate correctly that $TiCl_4$ is tetrahedral and that the assignments are correct. The ν_3 band (490 cm^{-1} in the Raman) is part of a doublet, for which the higher frequency and slightly weaker component appears at 506 cm^{-1} in the Raman (liquid) and in the infrared in both the liquid and vapour state. This upper component does not appear in the infrared spectrum of CCl_4 solutions. The band is best assigned to the unlikely, but entirely possible combination $\nu_3 + \nu_4 - \nu_2 = 505$ cm^{-1}. Fermi resonance cannot be involved here because this interaction occurs between closely spaced energy levels (the wave functions mix), and the levels involved in a difference band are not close to the fundamental. If the shoulder was observed only in the liquid in the infrared one might postulate a breakdown of selection rules and assign it to $\nu_2 + \nu_3 = 509$ cm^{-1}. However, the band is observed in the infrared spectrum of the vapour and such a violation of selection rules is unlikely.†

The Raman spectrum of VCl_4 shows only three lines. A very strong, sharp and polarized line at 383 cm^{-1} is assigned to ν_1, and a weak, broad, depolarized line at 475 cm^{-1} is assigned to ν_3. This band appears also in the liquid and vapour in the infrared. The remaining band at 128 cm^{-1}, medium strength, diffuse and depolarized, possibly represents ν_2 and ν_4. The coincidence of ν_2 and ν_4 is unusual,[36] but in this case it appears to occur in the pure liquid and in the vapour state. The infrared spectrum of the vapour shows four lines: 490 cm^{-1}, very strong, assigned to ν_3; 509 cm^{-1}, strong; 620 cm^{-1}, very strong; and 640 cm^{-1}, medium. If we assign the 620 cm^{-1} band to $\nu_2 + \nu_3$ and ignore anharmonicity, which is small anyway, we find $\nu_2 = 130$ cm^{-1}. The 509 cm^{-1} band can be assigned to $\nu_1 + \nu_4$, and this gives $\nu_4 \approx 130$ cm^{-1}. The medium strength band at 640 cm^{-1} can be assigned to $\nu_1 + 2\nu_4 \approx 638$ cm^{-1}. The same bands are observed in the infrared spectrum of the liquid. A more detailed account of the vibrational spectrum of VCl_4 embodying an alternative explanation of the data appears in Chapter 14 (pp. 366–368).

(c) OsO_4

The OsO_4 molecule is another example in which ν_2 and ν_4 appear to be nearly coincident. Woodward and Roberts[37] found only three Raman lines in the spectrum: 965 cm^{-1}, very strong, sharp and polarized, assigned to ν_1; 954 cm^{-1}, diffuse, weak, and depolarized, assigned to ν_3; and 335 cm^{-1},

† The infrared-inactive ν_2 band has been observed weakly in CF_4 at 2 atmospheres in a one-metre cell. It is also observed in CH_4 and SiH_4, where the breakdown in selection rules is due to Coriolis coupling with the ν_4 vibration. The band at 506 cm^{-1} in $TiCl_4$ appears to be too strong to be explained as $\nu_1 + \nu_2$, and the violation of selection rules attributed to Coriolis coupling, but this possibility cannot be dismissed with present evidence.

broad, strong, and depolarized, assigned to ν_2 and ν_4. The assignment of the latter two bands to ν_3 and ν_4 is confirmed by their appearance in the infrared spectrum. The assumption that the broad, low frequency band can be assigned to ν_2 as well as ν_4 merely means that two vibrations are too close to be resolved, not necessarily coincident. One could postulate that ν_2 is simply too weak to be observed in the Raman spectrum, but this vibration is observed in most tetrahedral molecules and there are numerous examples in which ν_2 and ν_4 are only 15–20 cm^{-1} apart.[36]

G. Asymmetric tops

All vibrations of an asymmetric top are non-degenerate (except for accidental coincidences), so the spectrum of a molecule with a given number of atoms is more complex than it is for symmetric and spherical tops. However, for many molecules the type A, B and C infrared bands have easily recognizable contours which are extremely useful in making assignments. If the molecule has a centre of symmetry as in point groups C_{2h} and D_{2h}, no more than half of the fundamental vibrations are infrared active, and most of the other fundamentals are Raman active.

(a) *Ethylene*

Ethylene is an asymmetric top with symmetry D_{2h}, and it provides a good illustration of the use of the characteristic type A, B and C band shapes. The mutual exclusion rule applies to the infrared and Raman activity, and one would expect the spectra to provide a fundamental assignment rather easily. Actually the correct assignments for C_2H_4 and C_2D_4 were not proposed until 1953 when Crawford, Lancaster and Inskeep[38] examined the spectra of several partially deuterated ethylenes and computed an ethylene potential function based on the frequencies of seven isotopically substituted ethylenes. Some of their frequencies have since been corrected as a result of analysis of high resolution data, such as that reported by Feldman, Romanko and Welsh,[39] but no change in their assignments has been made.

The fundamental vibrations of ethylene comprise the following set of irreducible representations,

$$\Gamma_{vib} = 3a_g + a_u + 2b_{1g} + b_{1u} + b_{2g} + 2b_{2u} + 2b_{3u}$$

The a_g, b_{1g} and b_{2g} vibrations are Raman active, the b_{1u}, b_{2u} and b_{3u} modes are infrared active, and the a_u vibration is inactive. The assignments of the fundamentals of C_2H_4 and C_2D_4 from the best data which are presently available[38, 39] are given in Table 3.9. Of the four Raman active bands observed by Feldman, Romanko and Welsh,[39] three are polarized

TABLE 3.9. *Fundamental frequencies of C_2H_4 and C_2D_4*

Species	Mode	Approximate description	Frequency (cm^{-1}) C_2H_4	Frequency (cm^{-1}) C_2D_4	Product rule ratios Observed	Product rule ratios Calculated
a_g	ν_1	C—H stretch	3026·4	2260	0·5122	0·5003
	ν_2	C═C stretch	1622·6	1517·5		
	ν_3	CH_2 deformation	1342·2	984·5		
a_u	ν_4	Torsion	(1027)	(725)		
b_{1g}	ν_5	C—H stretch	3102·5	2305	0.6077	0.6075
	ν_6	CH_2 rock	1236	1011		
b_{1u}	ν_7	CH_2 wag	949·2	720	0.7585	0.7565
b_{2g}	ν_8	CH_2 wag	950	780	0.8211	0.8277
b_{2u}	ν_9	C—H stretch	3105·5	2345	0·5339	0·5351
	ν_{10}	CH_2 rock	826	584		
b_{3u}	ν_{11}	C—H stretch	2989·5	2200·2	0·5496	0·5351
	ν_{12}	CH_2 deformation	1443·5	1077·9		

Frequencies for the inactive a_u vibrations were computed.

and can immediately be assigned to the a_g vibrations, the symmetric C—H stretch, the C═C stretch and the symmetric CH_2 deformation. The depolarized band at 3102·5 cm^{-1} can be assigned only to the b_{1g} C—H stretch, ν_5. The b_{1g} CH_2 rocking fundamental, ν_6, has been assigned on the basis of a rather weak depolarized line[41] observed in the liquid at 1236 cm^{-1}. This assignment is supported by Arnett and Crawford's[41] observation of the combination $\nu_6 + \nu_{10}$ at 2046·5 cm^{-1}. The remaining Raman-active fundamental ν_8 (b_{2g} wagging mode) was observed by Stoicheff[42] in the gas phase. The band is weak and unresolved, but it has the expected band contour.

Five of the fundamentals of C_2H_4 are infrared active. The species b_{1u}, b_{2u} and b_{3u} vibrations give rise to type C, type B, and type A bands, respectively. The A axis is parallel to the C═C bond, the B axis is perpendicular to the C═C bond and in the plane of the molecule, and the C axis is perpendicular to the molecular plane. The high resolution work of Galloway and Barker[43] on C_2H_4 shows three strong vibration-rotation bands which were correctly assigned as fundamentals. The rotational spacing of the resolved band at 2989·5 cm^{-1} and the unresolved envelope[38] of the 1443·5 cm^{-1} band shows that they are type A bands, and they are assigned to ν_{11}, the C—H stretch, and ν_{12}, the CH_2 deformation, respectively. The strongest Q branch in the spectrum is found at 949·2 cm^{-1}, and it is accompanied by rotational structure which shows it to be a type C band. This is assigned to ν_7, the b_{1u} wagging motion. A type B band in the C—H stretching region is unquestionably due to ν_9 in species b_{2u}. The remaining infrared-active fundamental is ν_{10}, the b_{2u} rocking mode, which was assigned to 810·3 cm^{-1} by Arnett and Crawford[41] on the basis of a set of lines, with a strong–weak–strong intensity alternation, appearing on the side of the much stronger ν_7 band. The line spacing identifies these lines as part of a type B band, and it was interpreted as the P branch of ν_{10}. These bands have been examined more recently by Smith and Mills[44] under higher resolution, and the analysis indicates a Coriolis perturbation coupling of ν_7 and ν_{10}; the origin of ν_7 is found to be at 826 cm^{-1}. This conclusion is supported by an observed Raman band at 1656 cm^{-1}, which is readily assigned as $2\nu_7 = 1652$ cm^{-1}. A weak Fermi resonance with ν_2 accounts for the observed frequency being a little higher.

The torsional motion of the CH_2 groups is in species a_u and is inactive in C_2H_4 and C_2D_4, but it is Raman active in *cis*-$C_2H_2D_2$ and *asym*-$C_2H_2D_2$. It is infrared active in *trans*-$C_2H_2D_2$, while it is both infrared and Raman active in C_2H_3D and C_2HD_3. It has been observed at 999·4 cm^{-1} in C_2H_3D, 987 cm^{-1} in *trans*-$C_2H_2D_2$, and at 764 cm^{-1} in C_2HD_3 by Arnett and Crawford. Application of the product rule to these data yield 1027 cm^{-1} and 726 cm^{-1} for ν_4 in C_2H_4 and C_2D_4, respectively.[41]

The observed spectra of C_2D_4 will not be discussed here, but assignments can be made rather independently by using arguments similar to those used for C_2H_4. The observed and calculated product rule ratios aid in settling some uncertain points in assigning the C_2D_4 fundamentals. Inspection of Table 3.9 shows good agreement in the product rule ratios in all species.

(*b*) *Nitryl fluoride*

The infrared and Raman spectra of nitryl fluoride, NO_2F, were examined and assigned by Dodd, Rolfe and Woodward[45] on the basis of C_{2v} symmetry. There are three a_1 vibrations, the symmetric N=O stretch, the N—F stretch, and the NO_2 deformation, two b_1 modes, the antisymmetric N=O stretch and the NO_2 rocking mode, and one b_2 out-of-plane bend. The Raman spectrum shows eight lines, five of which are polarized. Only three polarized fundamentals exist, so the weakest of these bands are assigned to overtones. The species a_1 fundamentals are: ν_1 (N=O stretch) = 1306 cm^{-1}, ν_2 (N—F stretch) = 821 cm^{-1}, and ν_3 (NO_2 deformation) = 466 cm^{-1}. The three depolarized lines must be fundamentals, so they are assigned ν_4 (antisymmetric N=O stretch) = 1797 cm^{-1}, ν_5 (NO_2 rock) = 555 cm^{-1}, and $\nu_6 = 742$ cm^{-1}. This last figure is from the gas-phase infrared data.

The assignments are confirmed very well by the infrared data. The geometry of NO_2F was not known at the time of the investigation, but the band contours show that it is an asymmetric top with the I_A axis coinciding with the N—F bond. The three polarized Raman bands are nearly coincident with the three strong infrared bands which have definite type A contours. These bands in the vapour-state spectrum are found at 1312, 822 and 460 cm^{-1}. There is only one strong type C band in the spectrum, which is clearly produced by the out-of-plane vibration ν_6. In addition, there are two type B bands at frequencies corresponding to the depolarized Raman lines at 1797 and 555 cm^{-1}. The infrared bands are observed at 1793 and 570 cm^{-1}; the 15 cm^{-1} difference observed between the liquid- and gas-phase frequencies is not uncommon for bending modes. The overtones observed in the Raman and infrared spectra are readily explained in terms of the fundamentals assigned above.

(c) *Dimethyl ketene*

Dimethyl ketene[46] may hardly qualify as a small molecule since it has eleven atoms, but it provides a good example of a case in which the use of the characteristic band shapes of an asymmetric top, plus some simple force-field calculations, has yielded an acceptable assignment even without isotopic data. Dimethyl ketene has C_{2v} symmetry and its twenty-seven genuine vibrations fall in the irreducible representations as follows.

$$\Gamma_{vib} = 9a_1 + 4a_2 + 8b_1 + 6b_2.$$

The moments of inertia, calculated from assumed bond lengths and distances transferred from ketene and acetone, are: $96{\cdot}5 \times 10^{-40}$, 221×10^{-40} and 307×10^{-40} g cm^2 for I_A, I_B and I_C, respectively, and the degree of asymmetry is large enough that the type A, B and C bands are easily distinguishable. These band types are associated with species a_1, b_1 and b_2, respectively. All vibrations are Raman active, but only the a_1, b_1 and b_2 vibrations are infrared active. Only infrared data are available for analysis, and of the twenty-three allowed infrared fundamentals, the b_2 torsional mode of the CH_3 groups is too weak to be observed. The equations of Badger and Zumwalt[7] predict a P—R separation of 16–18 cm^{-1} in the type A bands and 12–13 cm^{-1} in the type B bands.

A detailed discussion of the assignments will not be given here, but the essential points can easily be seen by examining the observed spectra in Figure 3.6 and the tabulated fundamentals in Table 3.10. Noting that the overlapping of bands may tend to obscure the expected band envelopes, one can pick out six type A bands with well formed PQR structure and the expected P—R separation. These are at 2983, 2928, 2134, 1392, 1013·5

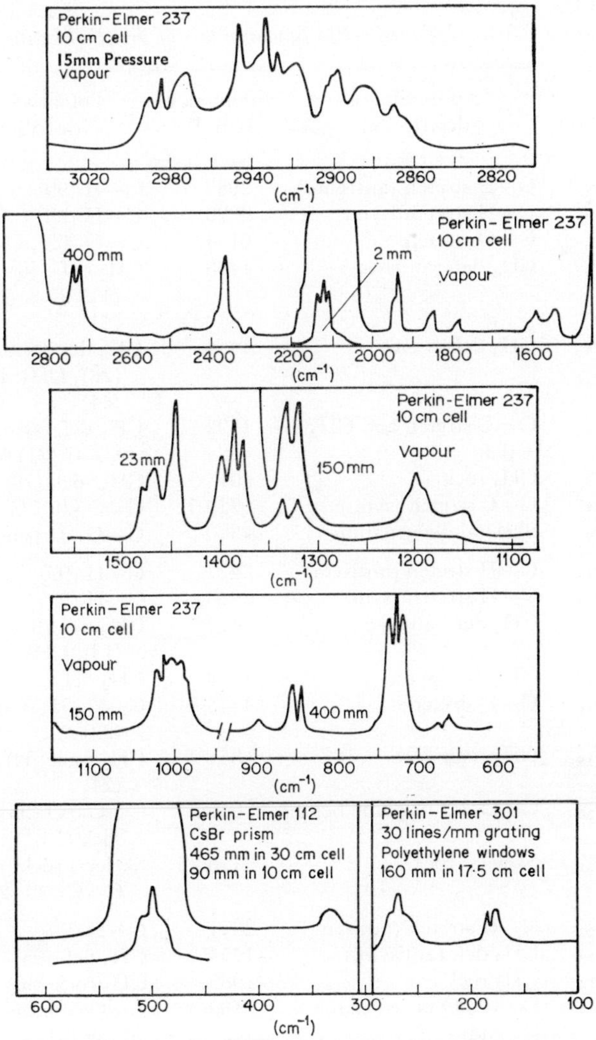

FIGURE 3.6. *Infrared spectrum of dimethyl ketene.*

and 729·5 cm^{-1} (see Table 3.10). The two remaining species a_1 fundamentals are CH_3 deformation modes which are expected to be in the range of 1420 to 1480 cm^{-1}. These two modes are symmetric and antisymmetric with respect to the threefold axes of the methyl groups.

Of the eight expected type b_1 fundamentals, five are rather easily identified at 2940, 1335, 851, 676 and 178 cm^{-1}. Two additional ones can be picked out tentatively from the overlapping contours, and these appear

TABLE 3.10. *Infrared active fundamentals of dimethyl ketene*

Species	Mode	Approximate description	Frequency (cm^{-1})	Potential energy distribution
a_1	ν_1	C—H stretch (antisym.)	2983	C—H (99)
	ν_2	C—H stretch (sym.)	2928	C—H (99)
	ν_3	C=O stretch	2134	C=O (62), C=C (27)
	ν_4	CH_3 def. (antisym.)	1468	CH_3 def. (46), C=C (11), rock (13), C—C (11)
	ν_5	CH_3 def. (sym.)	1454	CH_3 def. (antisym.) (28), CH_3 def. (sym.) (44)
	ν_6	C=C stretch and CH_3 def.	1392	CH_3 def. (sym.) (46), C=C (21)
	ν_7	CH_3 rock	1013·5	CH_3 rock (78)
	ν_8	C—C stretch (sym.)	729·5	C=C (26), C—C (64)
	ν_9	Skeletal deformation	~320	C—C—C bend (93)
b_1	ν_{14}	C—H stretch (antisym.)	2977	C—H (99)
	ν_{15}	C—H stretch (sym.)	2940	C—H (99)
	ν_{16}	CH_3 def. (antisym.)	1459	CH_3 def. (69), C—C (13), CH_3 rock (12)
	ν_{17}	CH_3 def. (sym.)	~1390	CH_3 def. (sym.) (80)
	ν_{18}	C—C stretch	1335·5	C—C (51), CH_3 def. (22)
	ν_{19}	CH_3 rock	851·2	CH_3 rock (53), C—C (24)
	ν_{20}	C=C=O in-plane bend	676·2	C=C=O (53), skeletal rock (18)
	ν_{21}	Skeletal rock	178·2	Skeletal rock (49), C=C=O (27)
b_2	ν_{22}	C—H stretch (antisym.)	2971	C—H (99)
	ν_{23}	CH_3 def. (antisym.)	1454	CH_3 def. (93)
	ν_{24}	CH_3 rock	989	CH_3 rock (90)
	ν_{25}	C=C=O out-of-plane bend	499·5	C=C=O (56), skeletal wag (21)
	ν_{26}	Skeletal wag	263·5	Skeletal wag (67), C=C=O (17)
	ν_{27}	CH_3 torsion	—	

to be at approximately 2977 and 1390 cm^{-1}. (One might say that the authors' prejudice plays a role here!) The remaining species b_1 fundamental is again a CH_3 deformation which is expected to be near 1450 cm^{-1}.

Only five type C fundamentals are expected to be observed, and four of these are discernible at 1454, 989, 499·5, and 263·5 cm^{-1}. The 989 cm^{-1} band is nearly obscured by an overlapping type A band. The remaining type C fundamental is an antisymmetric C—H stretch, which is assumed to be at 2971 cm^{-1} on the basis of the distorted envelope of the type A band at 2983 cm^{-1}.

After making specific assignments of the characteristic type A, B, and C bands listed above, there are three CH_3 deformations left to be assigned. One can do little more than guess the actual positions of these by fitting the expected band shapes into the distorted absorption contour in the 1450–1500 cm^{-1} region.

Supporting evidence for the assignments given in Table 3.10 was provided by using a set of force constants transferred from other molecules with similar structure, such as acetone and ketene, to calculate the observed frequencies. Since isotopic data were not available, only the principal force constants were allowed to vary in the least squares adjustment of the force field to give a best fit of the observed frequencies. The figures for the principal contributions to the potential energy distribution show that strong mixing of the symmetry coordinates occurs in the frequencies which were hardest to identify and assign. Furthermore, no single frequency seems to be assignable to the C=C stretching motion. It should be noted that the apparently successful calculation does not prove that the assignment is correct, but only that it is acceptable.

H. Force fields and vibrational assignments

Molecular force fields have been used extensively as an aid in making vibrational assignments, and they are indeed very useful, but their limitations should be fully realized. The fact that one calculates a reasonable set of force constants for a molecule does not prove that the assignment of fundamentals is correct, although the successful calculation may be very comforting to the investigator. In small organic molecules where the force constants are well known to be transferable, such calculations may be given more weight, but the assignments can often be established firmly with other arguments. One of the most useful applications of molecular force fields is in establishing satisfactory assignments and correlations in isotopically related sets of molecules. An example of this was seen in the case of C_2H_4 and C_2D_4, for which a correct assignment was not made until observed frequencies for several isotopic species were fitted to a force field.[38] In this instance the calculation was essentially equivalent to using product- and sum-rule relations,[47] which can easily be applied when the number of vibrations in each symmetry species is no more than two or three. In larger molecules, however, the calculation of frequencies and adjustment of a set of force constants to fit the observations is much more informative.

The calculation of force constants from observed frequencies may show up an incorrect assignment, as it did in the case of the C—C stretching vibration in diacetylene.[15] Useful correlations of assignments may be made by comparing force constants in a series of molecules such as $XeOF_4$ and the pentafluorides of the heavier halogens,[48] or the phosphoryl and thiophosphoryl halides,[36] after the assignments have been fairly well established by other arguments, but the use of force fields as a primary tool in determining assignments must be viewed with scepticism.

I. Conclusions

The principles employed in making vibrational assignments in small molecules have been outlined, and the use of such tools as molecular symmetry, characteristic group frequencies, selection rules, and band shapes has been described. Illustrations of the interpretation of observed data have been given for a number of linear molecules, prolate symmetric tops, oblate symmetric tops, spherical tops and asymmetric tops.

The characteristic shapes of both the resolved bands and the band envelopes in the infrared spectra of gases provide valuable and frequently unequivocal information concerning the species of vibrations in symmetric and asymmetric tops. These band shapes may be severely perturbed by Coriolis coupling in the perpendicular vibrations of symmetric tops. These effects are most severe in oblate tops, and some perpendicular vibrations of oblate tops produce bands which closely resemble parallel bands.

Characteristic types of band contours observed in the Raman spectra of gases may be expected to yield information concerning species of vibrations, but data are available on Raman band shapes for only a few molecules. A recent study of Raman bands of spherical and symmetric tops by F. N. Masri shows that, while symmetry information and Coriolis coefficients may be obtained from Raman bands, the band structure is more complex than in the infrared. This is because Raman selection rules are less restrictive. The utility of Raman band shapes will be greatest for symmetrical molecules which have some infrared-inactive vibrations, such as cyclopropane. The development of laser excitation and photon counting techniques with double monochromators has made the observation of Raman band shapes more feasible than it has been in previous years, and we may expect to see an increasing number of reports of observed band contours in the Raman spectra of gases.

Acknowledgement

The author wishes to express his appreciation to Dean Alvin H. Nielsen and Dr. Frederick N. Masri for their critical and constructive reviews of the manuscript.

References

1. Cotton, F. A., 'Chemical Applications of Group Theory', Wiley, 1963.
2. (a) Wilson, E. B., Decius, J. C., and Cross, P. C., 'Molecular Vibrations', McGraw-Hill, 1955; (b) Herzberg, G., 'Infrared and Raman Spectra', Van Nostrand, 1945.
3. Bellamy, L. J., 'The Infrared Spectra of Complex Molecules', Wiley, 1945; Szymanski, H. A., 'Theory and Practice of Infrared Spectroscopy', Plenum Press, 1964, Chapter 5; King, W. J., and Crawford, B. L., Jr., *J. Mol. Spectroscopy*, 1960, **5**, 421.
4. Bulkin, B. J., *J. Opt. Soc. Amer.*, 1969, **59**, 1387.
5. Gerhard, S. L., and Dennison, D. M., *Phys. Rev.*, 1933, **43**, 197.
6. Masri, Frederick N., D.Phil. thesis, University of Oxford, 1968.
7. Badger, R. M., and Zumwalt, L. R., *J. Chem. Phys.*, 1938, **6**, 711.
8. Seth-Paul, W. A., *J. Mol. Structure*, 1969, **3**, 403.
9. Hoskins, L. C., *J. Chem. Phys.*, 1966, **45**, 4594; Edgell, W. F., and Moynihan, R. E., *J. Chem. Phys.*, 1966, **45**, 1205; 1957, **27**, 155; McDowell, R. S., *J. Chem. Phys.*, 1965, **43**, 319.
10. Reference 2b, p. 231.
11. Stoicheff, B. P., *Canad. J. Phys.*, 1958, **36**, 218.
12. Fermi, E., *Z. Physik*, 1931, **71**, 250; Adel, A., and Dennison, D. M., *Phys. Rev.*, 1933, **43**, 716; Amat, G., and Pimbert, M., *J. Mol. Spectroscopy*, 1965, **16**, 278.
13. Reference, 2b, p. 215.
14. Reference 2b, p. 181.
15. Miller, F. A., Lemmon, D. H., and Witkowski, R. E., *Spectrochim. Acta*, 1965, **21**, 1709; Jones, A. V., *Proc. Roy. Soc.*, 1952, **221A**, 285.
16. Miller, F. A., and Fateley, W. G., *Spectrochim. Acta*, 1964, **20**, 253; Lafferty, W. J., Maki, A. G., and Plyler, E. K., *J. Chem. Phys.*, 1964, **40**, 224; Long, D. A., Murfin, F. S., and Williams, R. L., *Proc. Roy. Soc.*, 1954, **A223**, 251; Rix, H. D., *J. Chem. Phys.*, 1954, **22**, 429.
17. Miller, F. A., and Lemmon, D. H., *Spectrochim. Acta*, 1967, **23A**, 1415.
18. Orville-Thomas, W. J., *J. Chem. Phys.*, 1951, **19**, 1162.
19. Hunt, G. R., and Wilson, M. K., *J. Chem. Phys.*, 1961, **34**, 1301.
20. Turrell, G. C., Jones, W. D., and Maki, A., *J. Chem. Phys.*, 1957, **26**, 1544.
21. Job, V. A., and King, G. W., *Canad. J. Chem.*, 1963, **41**, 3132.
22. Nakagawa, I., and Shimanouchi, T., *Spectrochim. Acta*, 1962, **18**, 513; Fletcher, W. H., and Shoup, C. S., *J. Mol. Spectroscopy*, 1963, **10**, 300.
23. Fletcher, W. H., and Mottern, J. G., *Spectrochim. Acta*, 1962, **18**, 995.
24. Parker, F. W., Nielsen, A. H., and Fletcher, W. H., *J. Mol. Spectroscopy*, 1957, **1**, 107.
25. Goggin, P. L., and Woodward, L. A., *Trans. Faraday Soc.*, 1966, **62**, 1423.

26. Meić, Z., and Randić, M., *Trans. Faraday Soc.*, 1968, **64**, 1438.
27. Nielsen, A. H., *J. Chem. Phys.*, 1954, **22**, 659.
28. Reference 2b, p. 178.
29. Wilson, M. K., and Polo, S. R., *J. Chem. Phys.*, 1952, **20**, 1716.
30. Popplewell, R. J. L., Masri, F. N., and Thompson, H. W., *Spectrochim. Acta*, 1967, **23A**, 2797.
31. Shamir, J., and Hyman, H. H., *Spectrochim. Acta*, 1967, **23A**, 1899.
32. Claassen, H. H., *J. Chem. Phys.*, 1954, **22**, 50.
33. Woltz, P. J. H., and Nielsen, A. H., *J. Chem. Phys.*, 1952, **20**, 307.
34. Monostori, B., and Weber, A., *J. Chem. Phys.*, 1960, **33**, 1867.
35. Dove, M. F. A., Creighton, J. A., and Woodward, L. A., *Spectrochim. Acta*, 1962, **18**, 267.
36. Nakamoto, K., 'Infrared Spectra of Inorganic and Coordination Compounds', Wiley, 1963, p. 106.
37. Woodward, L. A., and Roberts, H. L., *Trans. Faraday Soc.*, 1956, **52**, 615; Hawkins, N. J., and Sabol, W. W., *J. Chem. Phys.*, 1956, **25**, 775.
38. Crawford, B. L., Jr., Lancaster, J. E., and Inskeep, R. G., *J. Chem. Phys.*, 1953, **21**, 678.
39. Feldman, T., Romanko, J., and Welsh, H. L., *Canad. J. Phys.*, 1956, **34**, 737.
40. Rank, D. H., Shull, E. R., and Oxford, D. W. E., *J. Chem. Phys.*, 1950, **18**, 116.
41. Arnett, R. L., and Crawford, B. L., Jr., *J. Chem. Phys.*, 1950, **18**, 118.
42. Stoicheff, B. P., *J. Chem. Phys.*, 1953, **21**, 755.
43. Galloway, W. S., and Barker, E. F., *J. Chem. Phys.*, 1942, **10**, 88.
44. Smith, W. L., and Mills, I. M., *J. Chem. Phys.*, 1964, **40**, 2095.
45. Dodd, R. E., Rolfe, J. A., and Woodward, L. A., *J. Chem. Phys.*, 1956, **52**, 1956.
46. Fletcher, W. H., and Barish, W. B., *Spectrochim. Acta*, 1965, **21**, 1647.
47. Brodersen, S., and Langseth, A., Kgl. Danske Videnskab, *Mat-Fys. Skrift.*, 1958, **1**, No. 5.
48. Begun, G. M., Smith, D. F., and Fletcher, W. H., *J. Chem. Phys.*, 1965, **42**, 2236.

CHAPTER 4

Interferometry: Experimental Techniques and Applications to Inorganic Structures

G. W. CHANTRY

A. Introduction

The principles underlying the assignment of vibrational frequencies in small molecules have been discussed and illustrated in Chapter 3. Until relatively recently, experimental difficulties have prevented the determination of all the information which is, in principle, available from infrared and Raman spectra. In Raman spectroscopy the main restriction has been on the states of matter that could be studied; using the mercury arc as a source, only solutions, liquids and powdered solids could be easily dealt with. In infrared spectroscopy the principal restriction has been on the frequency range that could be investigated; with conventional spectrometers the lower limit has been around 200 cm^{-1}. Chapter 5 deals with recent work on the Raman spectroscopy of single crystals and vapours using laser excitation and illustrates the new information that can be obtained. The present chapter describes the experimental determination of low-lying infrared frequencies using interferometric techniques and discusses the contributions such observations have made to the understanding of the vibrational spectra of selected species.

B. Interferometric methods in the far-infrared region

(a) *Background*

The difficulties encountered as one attempts to extend the operation of conventional spectrometers to longer and longer wavelengths are of two different kinds. The first, and relatively trivial difficulty arises from 'reststrahlen' absorption. Many ionic crystals show strong absorption of this kind below 500 cm^{-1} and this tends to obscure true intramolecular absorptions. In practice this means that sodium chloride prisms and windows cannot be used below 700 cm^{-1} and that no alkali halide material is useable below 200 cm^{-1}. This difficulty can be avoided by using reflection gratings as the dispersing elements and materials such as polyethylene for the windows. The more profound difficulties arise from the spectral distribution function of a black body source. As is well known, the total power radiated by a black body is proportional to the fourth power of the temperature but the total power radiated in the far-infrared (below 200 cm^{-1}) is, to a good approximation, proportional only to the first power of the temperature. The spectral distribution function shows a sharp peak at a wavenumber roughly twice the source temperature in degrees Kelvin so that, whereas it pays to run the source as hot as possible (and 1000 K seems the best readily achievable), it nevertheless remains true that the bulk of the radiated power lies in the near-infrared, peaking near 3 μ (3000 cm^{-1}). With a source temperature of 1000 K it is found that only 10^{-4} of the total power is radiated in the region below 200 cm^{-1} and within this interval the spectral radiance is proportional to the square of the frequency. Another related difficulty arises when one considers the performance of the detector. Visible and near-infrared detectors operate by promoting electrons between allowed levels of a solid or by photo-emission. The chance that electrons may be so promoted by random thermal processes is very small since the energy interval is many times greater than kT, and therefore the signal-to-noise performance of the detectors is good. In the far-infrared, a photon of wavenumber 200 cm^{-1} has the same energy as a room temperature (300 K) degree of freedom. It is clear that any detector operating on quantum principles will have to be cooled to liquid helium temperature if it is to work satisfactorily, and if we decide to use room temperature detectors, then these will have to be simple thermal devices. The best thermal detector available, namely the Golay cell, can detect 5×10^{-11} watt with unit signal-to-noise ratio in an integrating time of one second; this performance compares poorly with that of visible and near-infrared detectors.

(*b*) *The interferometric method*

Faced with these difficulties, physicists started, some ten years ago, to investigate the best way to do spectroscopy in energy-limited circumstances. The results of these enquiries are now well known, the answers, basically, being given in terms of the information throughput of the system. In spectroscopy we are studying a time-varying electromagnetic field. Whatever form this takes, from the almost pure cosine waves from a laser to the totally random fluctuations from an incoherent black body source, the field at a fixed point in space can be resolved into an infinite number of pure cosine waves (i.e. Fourier components). The spectrum is then merely a plot of the total intensity carried by components having frequencies between ν and $d\nu$ as a function of ν. If we have coherent sources, and detectors capable of following the rapid fluctuations of the field, as, for example, in the radio-frequency region, then the spectroscopy is straightforward: we merely attach a wave analyser to the detector and plot its output as a function of the maximum frequency of its pass band.

In the infrared region, we have to do spectroscopy with incoherent sources, and since the detectors give a d.c. output proportional to the mean power arriving, direct wave analysis is not possible. The only method open to us is based on the delay principle. The incident wave front is divided into a number of beams. These are subjected to varying time delays and then the beams are allowed to recombine with interference in the image plane of the spectrometer. A prism spectrometer divides the beam into an infinite number of beams and these recombine in the image plane to give a unique spectral record; each point in the image plane corresponds to one and only one frequency and *vice versa*. A grating instrument divides the beam into a finite number of beams and the resulting intensity distribution in the image plane is no longer unique. Monochromatic incident radiation yields maxima of intensity in several places in the image plane, corresponding to the various grating orders.

A two-beam interferometer splits the beam into two. The operation of a two-beam interferometer brings out the delay principle in its simplest form. Consider one of the Fourier components of time frequency ν divided into two beams one of which traverses an extra path length X. The electric field of the recombined beams at the detector is given by

$$E(t) = \tfrac{1}{2}E_0 \cos 2\pi\nu t + \tfrac{1}{2}E_0 \cos(2\pi\nu t + 2\pi\bar{\nu}X) \tag{4.1}$$

where $\bar{\nu}$ is the wave number equal to ν/c. The detector reports the mean power in the radiation arriving, that is the time average of $E(t)^2$. This can be readily shown to be

$$P(t) = \tfrac{1}{4}E_0^2[1 + \cos 2\pi\bar{\nu}X] \tag{4.2}$$

If we are studying monochromatic radiation, the output of the detector as the path difference is varied is a cosine function, from the maxima and minima of which $\bar{\nu}$ can be calculated. Thus, although through the operation of the detector we lose all information about the time-dependence of the field, we can recover this information by varying the path difference and determining the wavenumber. If on the other hand we are studying a broad band of radiation, then we can imagine this to be divided up into monochromatic elements for which the output signal of the detector is given by

$$V(X) \approx \frac{1}{2} \int_0^\infty P(\bar{\nu}) [1 + \cos 2\pi\bar{\nu}X] \, d\bar{\nu} \quad (4.3)$$

The form of this function for a suitable choice of $P(\bar{\nu})$ is shown in Figure 4.1. From the form of the interferogram function $V(X)$ one immediate

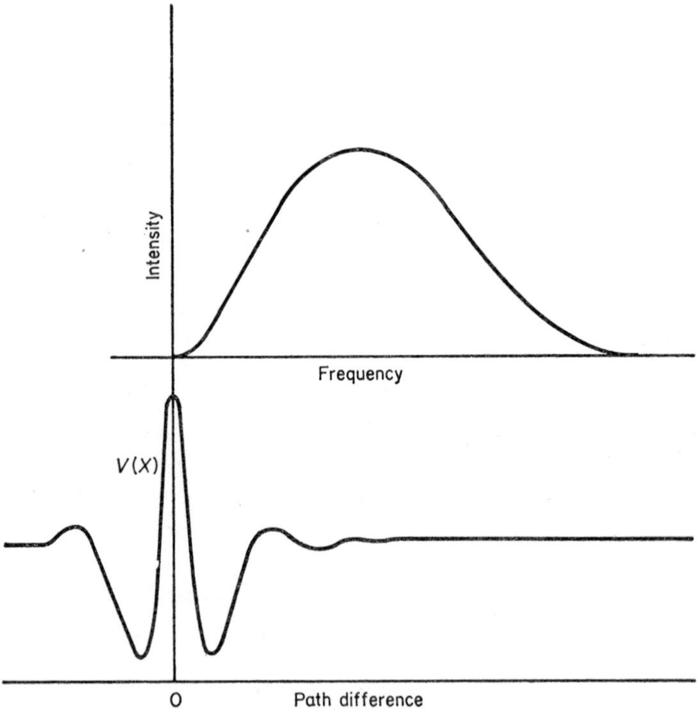

FIGURE 4.1. *Variation of output signal of detector in an interference experiment.*

conclusion can be drawn: whereas the spectrum produced by a prism instrument or a grating instrument (provided overlapping orders are avoided) is immediately intelligible in terms of frequency, the output of a two-beam interferometer is not. However, an information throughput analysis shows that the energy transfer from source to detector, normalized to some parameter such as resolution, is least for the prism, intermediate for the grating, and highest for the two-beam interferometer. It follows that in the far-infrared region, the two-beam interferometer is the natural choice for spectroscopy, provided we can find a way of unscrambling the spectral information. Equation (4.3) is a Fourier integral and may be inverted to read

$$P(\bar{\nu}) \approx \int_{-\infty}^{+\infty} V'(X) \cos 2\pi\bar{\nu}X \cdot \mathrm{d}X \tag{4.4}$$

where $V'(X) = V(X) - \overline{V(X)}$ and $\overline{V(X)}$ is the average value of $V(X)$. The transformation represented by equation (4.4) is readily achieved in even a small computer, and the computer output is the desired spectrum.

(c) *Technical aspects*

There are two distinct ways of realizing two-beam interferometry in practice and the corresponding instruments are the Michelson interferometer and the lamellar grating interferometer. To a certain extent the two instruments are complementary, the lamellar grating working best at frequencies below 70 cm^{-1} and the Michelson interferometer best at those above 70 cm^{-1}. However, since it is perfectly feasible with beam divider changes (see below) to operate Michelson interferometers down to 2 cm^{-1}, and since few inorganic systems have been studied using a lamellar grating instrument, we will confine ourselves to the former instrument. The optical layout of a Michelson interferometer is shown in Figure 4.2. The incoming radiation is partly reflected and partly transmitted by a thin (6 μm) stretched-film, dielectric beam-divider (B) made from polyethylene terephthalate. After reflection from the fixed (F) and moving (M) mirrors the radiation re-encounters the beam divider where it again suffers division of amplitude, some of the radiation returning to the source (S) and some going on via the condensing lens (L) to the detector (G). The output of the detector as a function of distance of the moving mirror from the zero path position (i.e. when both mirrors are equidistant from the beam divider) is the interferogram function. Transmission spectra of samples can be obtained by computing spectra with and without the sample in the beam and then taking the ratio of the two spectra in the computer.

All spectroscopic systems have fundamental features in common. For example, ultimate resolution is determined by two factors: the maximum path length difference introduced between the split beams and the physical size of the limiting aperture stop in the system. For a prism instrument these two quantities would correspond to the base length of the prism and the width of the entrance or exit slit. For a Michelson interferometer the two quantities are the maximum mirror travel available and the size of the detector window. The fact that the limiting stop in the Michelson instrument is circular, whilst in the prism instrument it is a long but very narrow slit, is the physical origin of the superior luminosity of the interferometer. There are other features, which may or may not be present in any given system, and the most important of these has come to be known as the multiplex principle. Imagine that we divide up the spectral band under investigation into N regions, each one resolution limit wide. If,

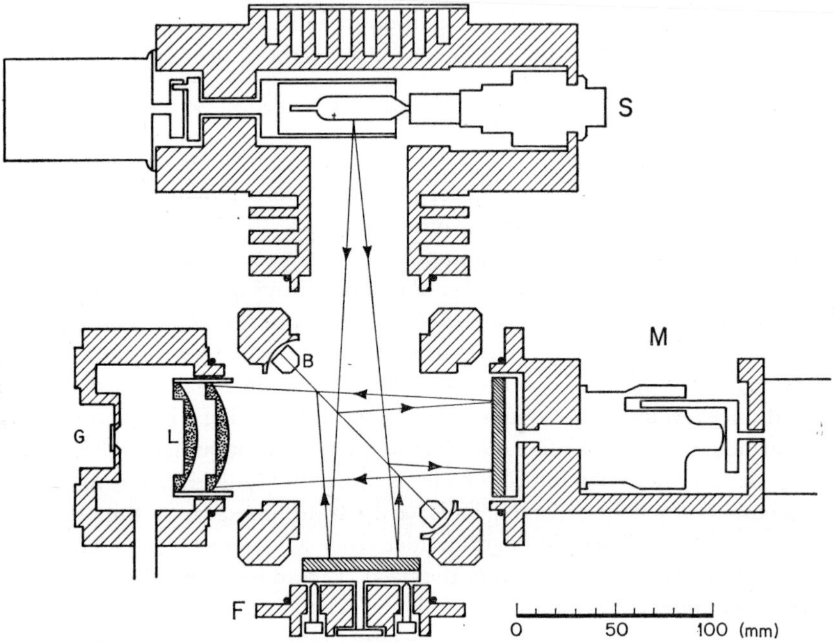

FIGURE 4.2. *Optical layout of a Michelson interferometer.*

further, we have a time T available to do the experiment, then for a conventional spectrometer each interval is studied for a time T/N, and the signal-to-noise ratio is proportional to $\sqrt{(T/N)}$. For an interferometer, on the other hand, every region is being studied all the time—in other words the interferometer is a multiplex instrument—and the signal-to-noise ratio is proportional to $\sqrt{T}$. It follows that the interferometer has an advantage in signal-to-noise by the factor $\sqrt{N}$, which can be quite large.

These two principal advantages of interferometry inspired the successful development of apparatus at the National Physical Laboratory by the group led by Dr. H. A. Gebbie. Several commercial instruments based on N.P.L. prototypes are now available and are in use throughout the world.

One difficulty with dielectric thin-film beam dividers is that multiple beam interference in the divider leads to minima in the instrumental transmission function at frequencies given by

$$\bar{\nu} = \frac{m}{2nt\cos\theta'} \qquad m = 0, 1, 2, \text{etc.} \tag{4.5}$$

where n is the refractive index of the beam divider, t its thickness and θ' the angle of refraction in the material. These minima are not troublesome at high frequencies ($70\ \text{cm}^{-1} < \bar{\nu} < 500\ \text{cm}^{-1}$), but become progressively so at lower frequencies where thicker beam dividers are required. In fact, it turns out that several beam dividers are required to cover the range 2–70 cm^{-1}, and modern interferometers feature rapid and reproducible change of pre-mounted beam dividers. Manufacturers of commercial interferometers supply complete electronic equipment so that the instrument yields a digitized interferogram on punched tape, punched cards or magnetic tape ready for immediate processing in the computer. Additionally, the Research and Industrial Instrument Company supplies an analogue computer which can process the data in analogue (i.e. d.c. voltage) form and provide true transmission spectra by taking the ratio of the spectra with and without the sample in the beam.

Far-infrared spectra of inorganic materials are usually observed with the solid specimen finely ground and dispersed in a nujol mull or in a pressed disc of polyethylene. Studies of aqueous systems are virtually ruled out by the enormous absorption coefficients (approximately 500 neper/cm) of liquid water in this spectral region. Raman spectra, on the other hand, are more commonly observed using aqueous solutions. Because of this difference of approach, there are four factors to be borne in mind which complicate the interpretation of far-infrared spectra, and which can make direct comparison with Raman spectra rather hazardous.

The first of these is the reststrahlen phenomenon briefly mentioned earlier. In the vicinity of the reststrahlen maximum the absorption coefficients are very high, but the reflection coefficients are also large, often being greater than 90%. It follows that radiation may traverse the sample, not by transmission through the crystallites, but by multiple reflection from them. The observed transmission spectrum will therefore be a complicated mixture of absorption and reflection. At frequencies much greater than the reststrahlen maximum the spectrum will be more or less a true transmission spectrum, but in the heavily absorbing region peculiar effects may be observed. The ideal way round this difficulty is to observe reflection spectra of large single crystals in all relevant orientations and to derive the optical constants (n the refractive index and α the absorption coefficient) by Kramers–Kronig analysis of the reflection data. Such a regimen is very time consuming; large single crystals may not even be available for many compounds, and, although it may appeal to physicists, it is less attractive to the chemist who usually wishes to study a great many compounds fairly rapidly. The mathematical problem of studying the transmission of a nujol mull made up of crystallites of varying sizes and shapes, and with marked frequency-dependence of both n and α, is just too formidable for serious consideration. The variation of the refractive index can lead to strange effects, for, whereas normally a great deal of the incident energy is lost by scattering, when the refractive index of the crystallites becomes equal to that of the nujol, this scattering loss will be virtually eliminated. Spurious maxima of transmission may be observed which bear no relationship to any fundamental change in the optical constants of the crystallite. Band shapes may be distorted and slight maxima of transmission may be observed at the absorption maximum. Clearly, great care is very necessary in interpreting spectra.

The second difficulty is akin to this, in that the lattice modes (including the reststrahlen type) appear in the spectrum; one then has the problem of picking out the true intramolecular bands from a rather rich spectrum. Theoretical predictions based on approximate force fields can be helpful, but the best approach is to change the physical state of the system in such a manner as to shift the lattice modes whilst leaving the intramolecular modes invariant. Either lowering the temperature, or subjecting the specimen to high pressure, shifts lattice modes to higher frequency, and these methods have proved useful. For ionic crystals, however, one has another variable; if one is interested in the intramolecular modes of an anion, one may vary the cation and note those modes which are relatively invariant. This problem of identifying intramolecular bands is quite acute for inorganic specimens; whereas in organic compounds the intramolecular modes are usually much higher in frequency than the lattice modes, for inorganic crystals they may have comparable or even lower frequencies. As an example, the fundamental vibration of lowest frequency of the PtI_6^{2-} ion occurs at 46 cm^{-1}, well below the lattice modes for K_2PtI_6 which occur in the 80–100 cm^{-1} region.

The third problem which is again related in a way to the lattice vibrations, arises from crystal field splittings and relaxation of selection rules. Lattice modes can arise only if the unit cell contains more than one entity. Consider, as an example, NaCl. The fundamental selection rule for a crystal is that only those modes in which corresponding atoms in all unit cells are vibrating in phase may lead to absorption of radiation. One can therefore think of two possible modes for NaCl (both triply degenerate), one in which the two atoms in the unit cell move in the same direction, and one in which they move in opposite directions. This is tantamount to combining the six translational degrees of freedom, in-phase and out-of-phase. The first of these corresponds to physical translation of the crystal and has therefore zero frequency whilst the second is the reststrahlen lattice mode occurring at 160 cm^{-1} for room temperature. For more complicated cases the lattice modes can be predicted from factor group analysis of the crystal space group. The important point is that, just as one may combine the translational degrees of freedom, one may also combine the intramolecular modes and in the general case if there are N molecules of a given kind in the unit cell each fundamental frequency may split into N components. The actual number of components and their spectral activity may be worked out by standard group-theoretical methods, provided the crystal structure is known, but we must always be on the alert for this phenomenon and especially for the more subtle consequences, when some components are infrared active and others Raman active and the two spectra look different. The selection rules for a molecule in a crystal lattice are determined by the site symmetry about the molecule rather than by the molecular symmetry itself, and, as a consequence, the spectrum of a molecule in a crystal may show lines which are forbidden for the isolated molecule. While this phenomenon makes the interpretation of spectra more difficult, it is not unwelcome, since it does give us the missing frequencies, a particularly important consideration if we are trying to derive reasonably accurate force fields for the molecule.

The fourth problem is that, in general, frequencies shift with change of phase. It is not uncommon for these shifts to be as much as 20 cm^{-1} in going from an aqueous solution to the crystalline phase. Since Raman and far-infrared spectra are commonly observed in different phases, this makes comparison of frequencies unreliable. Fortunately, considerable progress has been made in the last five years in recording the Raman spectra of polycrystalline samples, and it is now possible to derive sets of frequencies which correspond to one and the same phase. Having mentioned the difficulties of the technique, we will now go on to describe some interesting far-infrared spectra, obtained using interferometers, which are relevant to structural chemistry.

C. Some practical examples

(*a*) *Octahedral ions* XY_6

Octahedral ions and molecules of the formula XY_6 are particularly attractive to the chemical physicist because of their high symmetry (O_h).

The representation of the point group O_h formed by the fifteen normal coordinates can be reduced as

$$\Gamma(Q) = A_g + E_g + 2F_{1u} + F_{2g} + F_{2u}$$

There will thus be six normal frequencies of which three (A_g, E_g and F_{2g}) occur in the Raman spectrum, two ($2F_{1u}$) in the infrared spectrum and one (F_{2u}) is inactive in both. Raman spectra of a large number of octahedral ions have been reported, principally by Woodward and his colleagues.[1, 2] In recent years far-infrared spectra of an even larger array have been reported, principally by Adams and his co-workers,[3–9] who have also provided Raman data for those not reported by Woodward. The very high symmetry of the ions implies a large degree of factoring of the potential energy matrix, and in fact the harmonic force field requires only seven independent force constants. Nevertheless, since only five frequencies are normally available, there are still too many unknown terms. Many authors have favoured a modified form of the Urey-Bradley force field, which, because of its physical assumptions, requires only five independent parameters. Adams and Morris[6] give as the elements of the factored $\mathscr{F}$ matrix

$$A_g: \mathscr{F}_{11} = K + A + 4\,F$$

$$E_g: \mathscr{F}_{22} = K + A + 0{\cdot}7\,F$$

$$F_{1u}: \mathscr{F}_{33} = K - A + 1{\cdot}8\,F,\ \mathscr{F}_{34} = 0{\cdot}9\,F,\ \mathscr{F}_{44} = H + 0{\cdot}55\,F + 2B$$

$$F_{2g}: \mathscr{F}_{55} = H + 0{\cdot}55\,F$$

$$F_{2u}: \mathscr{F}_{66} = H + 0{\cdot}55\,F - 2B,$$

where K, H and F are X—Y stretching, Y—X—Y bending and X ... X repulsive constants, respectively, and A and B are interaction constants between two collinear bonds and two adjacent non coplanar angles respectively. The factored $\mathscr{G}$ matrix elements are

$$A_g: \mathscr{G}_{11} = \mu_y$$

$$E_g: \mathscr{G}_{22} = \mu_y$$

$$F_{1u}: \mathscr{G}_{33} = \mu_y + 2\mu_x,\ \mathscr{G}_{34} = -4\mu_x,\ \mathscr{G}_{44} = 2\mu_y + 8\,\mu_x$$

$$F_{2g}: \mathscr{G}_{55} = 4\mu_y$$

$$F_{2u}: \mathscr{G}_{66} = 2\mu_y \text{ where } \mu_x = m_x^{-1}$$

The spectra (and the force constants calculated from them) show some very interesting dependences on the position of X in the periodic table. Woodward and Creighton[2] pointed out that, whereas for $SnCl_6^{2-}$ ν_1 and ν_2 are widely separated and the band due to ν_1 is much more intense than that due to ν_2, for $PtCl_6^{2-}$ ν_1 and ν_2 are close together and the bands are of approximately equal intensity. Both ν_1 and ν_2 are pure X—Y stretching modes and have identical $\mathcal{G}$ matrix elements so that the difference in frequency arises purely from force constant effects. The varying quantity is the contribution from the parameter F given above, or its equivalent $f_{rr'}$ the interaction constant between bonds at 90° in the valence force field. For $SnCl_6^{2-}$ F is 0·27 whereas for $PtCl_6^{2-}$ it is only 0·13 mdyne Å^{-1}. The effect appears not to be a function of ionic size, and Woodward and Creighton's suggestion of electronic effects involving the 5d electrons of platinum seems plausible. The principal stretching force constant K is also sensitive to electronic effects. The fundamental ν_3 for the $CrCl_6^{3-}$ ion occurs at 315 cm^{-1} whereas that for $FeCl_6^{3-}$ is found at 248 cm^{-1}. The mass change is slight, so the large drop in frequency must come from a diminution of the stretching force constant, which, it has been suggested, arises from the occupation of antibonding levels in the latter ion.[9] The $MnCl_6^{3-}$ ion is particularly interesting, since this should suffer Jahn-Teller distortion, but only one band attributable to an Mn—Cl stretching mode is observed in the far-infrared spectrum at 342 cm^{-1}. The suggestion is that the long bonds have a much lower force constant which shifts the absorption involving their stretching out of the 300 cm^{-1} region. It is interesting that the single observed band occurs at a higher frequency than for $CrCl_6^{3-}$ and is only a little lower than that (358 cm^{-1}) for $MnCl_6^{2-}$, which should not be distorted and in which the manganese has a higher oxidation state. Adams and Morris[9] report Raman and infrared spectra for the rare earth anion $CeCl_6^{2-}$. The results appear quite normal, leading to force constants $K = 1{\cdot}14, F = 0{\cdot}11, H = 0{\cdot}02, A = 0{\cdot}26$ and $B = 0{\cdot}01$ mdyne Å^{-1}. The ions SeX_6^{2-} and TeX_6^{2-} (X = halogen) are particularly interesting because of the presence of the lone pair of electrons on the central atom. It has been thought that this lone pair might be stereochemically active and lead to structures for the anion based essentially on seven-coordination. The molecule XeF_6, isoelectronic with TeX_6^{2-}, as far as valency electrons are concerned, is thought to be distorted in this fashion. The far-infrared spectra reported by Greenwood and Straughan[10] and subsequently presented, together with Raman data by Adams and Morris[8] do not support the anticipated distortion of the SeX_6^{2-} and TeX_6^{2-} ions. The spectra are completely consistent with regular octahedral symmetry for the ions, and the splitting of the ν_3 fundamental observed in some cases can be explained by site symmetries lower than O_h. The band due to the ν_3 vibration for $TeCl_6^{2-}$ is remarkably broad compared say with the ν_1 and ν_2 Raman bands. Adams and Morris[8] point out that this is the only stretching mode in which the central atom moves, and in which therefore transient directional effects are possible. Such effects involve the mixing of the lone pair, initially in an a_g^* orbital, with the bonding orbitals. The force constants in this series of anions are rather anomalous. The values of K in $TeCl_6^{2-}$ and $TeBr_6^{2-}$ are very similar (0·92 and 0·90 mdyne Å^{-1} respectively), whereas this quantity usually falls in going from a chlorine to a bromine compound. Also the ν_3 fundamentals of SeI_6^{2-} and TeI_6^{2-} are closely similar in frequency (142 cm^{-1}).

The effects of varying the cation on the spectrum of XY_6 anions are of two distinct types. Firstly, we may have a lowering of the unit cell symmetry with a corresponding relaxation of selection rules; secondly, we may have frequency shifts without the appearance of any extra lines in the spectrum. The most spectacular illustrations of the first effect occur in diphenyliodonium hexachloroplatinate[6] where ν_3 splits and the forbidden ν_1 appears in the infrared spectrum, as does the inactive fundamental ν_6, which is observed as a medium-strong band at 86 cm^{-1}. The hexachloroplatinate ion also affords a good example of the second effect, the value of ν_3 varying between 353 (Tl^+) and 335 (Ph_2I^+) cm^{-1}. A possible explanation of these shifts is that the extent of d_π—d_π bonding in the ion, suggested by Woodward and Ware,[11] is affected by the electrostatic field of the nearby cation, the smaller cations having the larger effects. Bands due to lattice modes are observed in the 100 cm^{-1} region for nearly all the compounds studied. As an example, a band observed at 84 cm^{-1} for K_2IrCl_6 was proved[3] to be due to a lattice mode by observing its shift to 129 cm^{-1} for $(NH_4)_2IrCl_6$. Some care has to be taken where one is loosely distinguishing between intramolecular modes and intermolecular modes for crystals of very high symmetry, and for which the two types are closely similar in frequency. Thus, Adams and Morris[8] point out that ν_4 and the cation lattice mode ν_L for K_2TeBr_6 and K_2TeI_6 are of the same symmetry species and are expected to be roughly equal in frequency. Strong interaction is expected and distinguishing labels are probably meaningless.

(b) XY_4 *molecules and ions*

Not much work has been published on the far-infrared spectra of strictly tetrahedral molecules and ions, probably because, for most of these Raman spectra are readily obtained and, since all fundamentals are Raman active, the far-infrared spectra contain no fresh information. An early study (1963) by Butcher *et al.*[12] reported the ν_3 and ν_4 fundamentals for the tin tetrahalides in cyclohexane solution. These authors also report and assign the far-infrared spectra of some dialkyltin dihalides on the basis of effective C_{2v} symmetry. Tin tetraiodide is quite interesting because of its ability to form adducts with various ligands, in which the tin atom achieves six-coordination.[13] For a monodentate ligand the molecule

L_2SnI_4 can have two isomeric forms, *cis* and *trans*. The latter has a centre of symmetry so that no coincidences are expected in the infrared and Raman spectra. The adduct with trimethylamine seems to adopt this stereochemical form; there are six bands in the infrared spectrum below 500 cm^{-1} and none coincident with the solitary Raman line at 148 cm^{-1}. Far infrared spectra have been reported[14] for complex ions of the type $[LMX_3]^-$ (where L = triphenylphosphine or pyridine; M = cobalt, nickel or zinc; X = chlorine or bromine). On the basis of effective C_{3v} symmetry six skeletal vibrations are predicted to be infrared active; for most of the compounds studied this prediction has been realized. The spectra for $[LMX_3]^-$ differ considerably from those for L_2MX_2 or MX_4^{2-}, and thus one can distinguish by inspection the $[LMX_3]^-$ ion from an equimolecular mixture of L_2MX_2 and MX_4^{2-}. Within a series, the cobalt complex shows the highest metal-halogen stretching frequency and zinc the lowest, a fact which may reflect the increasing degree of ionic character in the bonds.

One of the most interesting XY_4 ion systems is provided by the halo-complexes of copper(II). The ion $CuCl_4^{2-}$ is known to depart significantly from T_d symmetry in the solid state, generally taking on a flattened tetrahedral structure (D_{2d}) with all bond lengths equal to 2·22 Å. There is even some evidence that with certain cations the $CuCl_4^{2-}$ anion may adopt square planar (D_{4h}) symmetry. The infrared spectrum[15] of $CuCl_4^{2-}$ generally shows two bands attributable to copper-chlorine stretching modes, consistent with D_{2d} symmetry; with very large cations, such as $[Ph_3MeAs]^+$, however, only a single band is observed. The crystal Cs_2CuBr_4 is particularly interesting since the electronic spectrum suggests that the anion is even further distorted, having only C_s symmetry. In agreement with this the far-infrared spectrum[15] shows three bands (four expected) associated with copper-bromine stretching modes, at 224, 189 and 172 cm^{-1}. Molecules formally written $[Cr(NH_3)_6]$ $[CuCl_5]$ have been thought to contain trigonal bipyramidal $CuCl_5^{3-}$ ions. The far-infrared spectrum does not support this, featuring as it does just a single copper-chlorine stretching band at 268 cm^{-1}. Adams and Lock[15] suggest that the crystal contains tetrahedral or square planar $CuCl_4^{2-}$ ions together with Cl^- ions. In cupric chloride itself in the solid phase, the copper atoms form long chains with successive atoms bridged by pairs of chlorine atoms, giving a planar polymer. The situation is most reminiscent of that in $PdCl_2$, which is essentially a molecular crystal of an inorganic polymer. However, each copper atom in $CuCl_2$ is also associated with two chlorine atoms in the chains immediately above and below their own chain. The in-chain CuCl distance is 2·30 Å and the interchain distance 2·95 Å. The appropriate procedure for dealing with the molecular vibrations of linear chains[16] is to refer to the line group, together with its associated factor group. For

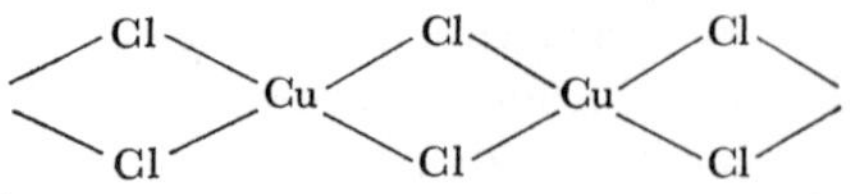

$CuCl_2$, the appropriate factor group is isomorphous with D_{2h}, so that we expect, to a first approximation (i.e. ignoring the long bonds), that there will be two infrared active CuCl stretching modes. Just two are found, at 328 and 275 cm^{-1}, respectively.[15] Whether there are absorption bands near 300 cm^{-1} corresponding to vibrations of the long 'bonds' is a vexed question. Cupric chloride dihydrate, $CuCl_2,2H_2O$, consists of *trans*, almost

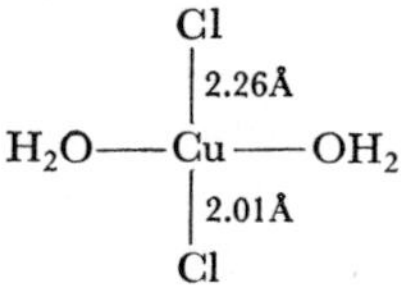

square units, packed in layers so that the chlorine atoms form 'long bonds' (2·95 Å) to the adjacent groups. For each unit we expect only a single infrared active Cu—Cl stretching mode; in fact[15] only one such band is observed at 299 cm^{-1}. Adams and Lock,[15] in discussing several other chain polymers of the general type CuX_2L_2, conclude that stretching of 'long bonds' does not give rise to characteristic absorptions in the usual range. This has been disputed by Campbell, Goldstein and Grzeskowiak,[17] who find for a range of unidentate and bidentate ligands, two bands in the normal region, and furthermore show that the two vibrations are X-sensitive; they explain the occurrence of only one band for $CuCl_2$, $2H_2O$ as possibly arising from a coincidence between the lower ν(Cu—Cl) band and the strong band at 242 cm^{-1} confidently assigned to the ($Cu—OH_2$) vibration. It would seem that a careful force constant analysis will be necessary to decide this point.

Square planar ions are formed by several metal atoms. $CuCl_4^{2-}$, for example, in the salt $[CH_3NH_3]_2[CuCl_4]$ adopts this habit, and the spectrum[15] shows the single (e_u) copper-chlorine stretching vibration at 284 cm^{-1}. All three expected lines have been observed[18] for $PtCl_4^{2-}$, $PdCl_4^{2-}$ and $PdBr_4^{2-}$. Magnus' green salt $[Pt(NH_3)_4]^{2+}[PtCl_4]^{2-}$ is quite interesting because the anticipated metal-metal interactions appear not to have any effect on the spectrum for the out-of-plane deformation mode, $\nu_2(a_{2u})$, which is identical in frequency and intensity with the corresponding mode in K_2PtCl_4. The site symmetries for the anions are quite high: in K_2PtCl_4 it is the same as that of the free ion, whilst in Magnus' green salt it is C_4, for which no removal of degeneracy is expected, and no splitting of ν_6 or ν_7 is observed. The spectra of the ion $Pd(NH_3)_4^{2+}$ observed[19] in the crystal

$Pd(NH_3)_4Cl_2$ are more complicated than would be expected from a vibrational analysis of the isolated ion. Strong coupling between the molecules in the unit cell is postulated to explain this. The mixed amino-halo-complexes of palladium can occur either as *cis* or *trans* stereo-isomers; complete assignments have been given[19] for $Pd(NH_3)_2Cl_2$, $Pd(ND_3)_2Cl_2$, $Pd(NH_3)_2Br_2$ and $Pd(NH_3)_2I_2$.

The tetrahalides of tellurium are particularly noteworthy from a stereochemical viewpoint.[20, 21] In the gas phase $TeCl_4$ has the same structure as SF_4, namely a trigonal bipyramid with one of the equatorial positions occupied by the lone pair. In solution in benzene it appears to have the same structure, for the spectrum shows five lines with the same appearance as the corresponding lines of SF_4, but shifted to lower frequencies. The spectrum of the solid is quite different, the strong band at 280 cm^{-1} disappearing and a new line showing up at 151 cm^{-1}. The Raman spectrum and electrical conductivity of the melt suggest that the structure in the solid is $TeCl_3^+Cl^-$. The infrared results support this formulation, the four expected modes for the assumed C_{3v} symmetry occurring at 363 (ν_1), 186 (ν_2), 353 (ν_3) and 51 (ν_4) cm^{-1}. Similar conclusions apply to $TeBr_4$ and TeI_4. More subtle analysis of the spectra indicates that there is some residual covalent interaction between the $TeCl_3^+$ cation and the Cl^- anion. Thus, the Raman bands of $TeCl_3^+$ shift to higher frequencies in going from $TeCl_4$ to $TeCl_3^+AlCl_4^-$, and the spectrum of $TeCl_4$ does contain a few additional, but very weak features, which cannot be explained as arising from $TeCl_3^+$. The structure is perhaps better written as $TeCl_3^{(1-\delta)+} \ldots Cl^{(1-\delta)-}$.

(*c*) *Miscellaneous molecules*

The heavier Group Vb trihalides AsI_3, SbI_3, $BiBr_3$ and BiI_3 have been studied by Manley and Williams.[22] The observed infrared spectra of crystalline samples show no obvious 'solid state' effect, and can be analyzed in terms of the C_{3v} symmetry of the isolated molecule. For AsI_3 six absorption bands are found at 226, 217, 201, 148, 102 and 74 cm^{-1}. A calculation of the expected frequencies for AsI_3, based on comparisons with $AsBr_3$, PBr_3 and PI_3, gives 219 (a_1), 209 (e), 90 (a_1) and 72 (e) cm^{-1} for the four expected fundamentals. From this, the bands at 226, 201, 102 and 74 cm^{-1} are identified as the fundamental frequencies. The frequency order $a_1 > e$ is confirmed by a force constant analysis, which shows that real force constants can be obtained only with the choice 226 (ν_1), 201 (ν_3), 102 (ν_2) and 74 (ν_4). The two extra bands in the spectrum are assigned to the first and second overtones of ν_4 expected to occur at 147 and 221 cm^{-1}.

Tungsten chloride pentafluoride WF_5Cl provides an example of a rare type of molecule—a volatile MX_5Y species. Infrared and Raman spectra have been reported by Adams and his colleagues[23] who have compared their results with those of Cross and his collaborators[24] for SF_5Cl. Eleven fundamental frequencies are expected distributed thus:

$$\Gamma(Q) = 4A_1 \text{ (i.r. and Raman)} + 2B_1 \text{ (Raman)} + B_2 \text{ (Raman)} + 4E \text{ (i.r. and Raman)}$$

The spectrum was assigned making use of Raman polarization data, PQR band contours in the gas-phase infrared spectrum and splitting of modes and the appearance of forbidden modes in the solid-state infrared spectrum. Adams *et al.*[23] number the fundamental modes for this molecule so that they correspond with similar modes in SF_5Cl although this violates the usual convention; ν_4, for example, principally involving stretching of the W—Cl bond, is not the lowest frequency a_1 mode. This particular mode is quite interesting since it has almost the same frequency as ν_4 for SF_5Cl. Taken together with the estimated rather short W—Cl bond length, the frequency is thought to indicate a π contribution to this bond.

Halogen-bridged complexes of the types X_2Y_6 and $X_2Y_4L_2$ have been investigated by Adams and his colleagues.[25] The spectra for the planar ions, such as $Pt_2Cl_6^{2-}$ and $Pd_2Cl_6^{2-}$ and molecules, such as Au_2Cl_6, can be assigned in terms of the D_{2h} symmetry of the bridged dimer. Force constants have been calculated and it is found that the stretching force constants for the terminal and the bridging X—Y bonds are rather similar. In Au_2Cl_6, as an example, these are 2·384 and 1·740 mdyne Å^{-1}, values that contrast with those for the non-planar molecule Al_2Cl_6 which are related by a factor of two. An explanation of this difference in terms of the greater ionicity of Al—Cl bonds, and of the smaller distortion involved in sharing an edge between two tetrahedra, has been suggested by Adams and Churchill.

Complete vibrational spectra of the halides and complex halides of cadmium have been reported by Davies and Long.[26] The tetrahedral CdX_4^{2-} species are isoelectronic with the tin tetrahalides SnX_4 and the indium tetrahalide ions InX_4^- studied by Woodward and his colleagues. The a_1 frequency of these species rises as the charge decreases, varying from 260 cm^{-1} for $CdCl_4^{2-}$ to 321 cm^{-1} for $InCl_4^-$ and to 367 cm^{-1} for $SnCl_4$. The symmetric stretching frequencies for the linear CdX_2 molecules are higher than those for the tetrahalide ions although for both modes only the halogen atoms move; from which we may conclude that the force constant and hence bond order are higher for CdX_2 than for CdX_4^{2-}. Vibrational spectra for the mercurous halides have been given by Goldstein,[27] who has shown that the previous assignment of ν_3 as 110 cm^{-1} is incorrect and that absorption at this frequency is due to the bending fundamental ν_5. Force constants calculated for Hg_2Cl_2 are given as: Hg—Cl = 1·135 and Hg—Hg = 2·02 mdyne Å^{-1}. Many other molecules containing metal-metal bonds have been investigated by means of far-infrared spectroscopy.[28] Contrary to earlier predictions,[29] it turns out that modes involving stretching of Sn—X bonds where X is a transition metal are intense in both Raman and far-infrared spectra. The range of frequency in which metal-metal stretching modes occur is 165–235 cm^{-1}.

One of the most fascinating molecules studied recently[30, 31] has been the tetrameric phosphonitrilic chloride $(PNCl_2)_4$. The molecular symmetry is different in the vapour, liquid and solid phases and, moreover, the habit adopted in the solid state depends on the method of sample manufacture and on such factors as working and ageing after preparation. This molecule has well been described as 'flexible'. The method of vibrational analysis used by Hisatsune[31] recognizes that D_{4h} is the highest symmetry the cyclic tetramer can have and suggests that the various structures adopted correspond to subgroups of D_{4h}. The infrared and Raman spectra suggest that the symmetry of the molecule in the liquid phase corresponds to D_{2d}. The representation of D_{2d} formed by the normal coordinates is

$$\Gamma(Q) = 6A_1 + 5A_2 + 5B_1 + 6B_2 + 10E$$

(the reduction given by Manley and Williams[30] is in error). A complete assignment on this basis for the liquid can be arrived at, the a_1 frequencies, for example, occurring at 538, 427, 402, 375, 333 and 295 cm^{-1} There are two distinct solid forms and these have been described as the S_4 and the C_i forms although there is some suggestion that the molecular symmetry in the C_i form is really C_{2h}. There are several differences in the two spectra, perhaps the most obvious of which is the presence of a band near 890 cm^{-1} for the S_4 form which is missing for the C_i form. The gas-phase symmetry is very likely D_{4h}. It would seem that for this flexible molecule ring-puckering is very easy, and in the condensed phases the ring does in fact pucker in various ways to lower the packing energy.

Carbon suboxide and carbon subsulphide are formally inorganic molecules even though the former is the anhydride of malonic acid. The C_3O_2 molecule has nearly always been regarded as linear O═C═C═C═O and this has been finally confirmed by the observation in the infrared spectrum of the missing fundamental ν_7 at 63 cm^{-1} in the gas phase and at 72 cm^{-1} in the liquid. Smith and Leroi[32] report both liquid and crystal spectra for C_3O_2, the latter being particularly interesting because of the clear indication from the lattice spectrum of a phase change at 115 K. Below this temperature four strong lattice bands at 24, 46, 84 and 109 cm^{-1} are evident but they all disappear above 115 K. A transition at this temperature was not noticed in the heat capacity measurements, thus highlighting the power of far-infrared 'lattice mode' spectroscopy to pinpoint subtle effects. Smith and Leroi suggest that the phase transition is merely a molecular reorientation with the crystal remaining in the orthorhombic class but losing the crystallographic centre of symmetry. Carbon subsulphide C_3S_2 is likewise shown to be linear but surprisingly the ν_7 fundamental occurs[33] at a higher frequency 94 cm^{-1} (gas) and is much more intense than the corresponding feature of C_3O_2. This is ascribed to the possibility of back-bonding to the sulphur atom, a phenomenon not possible with oxygen. The halogeno-cyanoacetylenes and halogenodiacetylenes also form linear molecules, the far-infrared spectra of which have been investigated by Christensen and his colleagues.[34] Again there is a low frequency bending mode which for Cl—C≡C—C≡N occurs at 129 cm^{-1} and which in the gas phase shows the typical PQR structure of a perpendicular band.

A series of pyrophosphates involving divalent cations has been investigated by Hezel and Ross.[35] The P—O—P angle of 134° in the pyrophosphate anion is quite large and the spectrum can be analyzed quite well in terms of D_{3h} symmetry. Departures from D_{3h} symmetry caused by both the non linear P—O—O framework and the rather low site symmetries in the crystal are, however, also evident from the spectra. A band varying in position between 340 and 270 cm^{-1}, which would arise from an e″ mode for D_{3h} symmetry is observed; and, in the region below 100 cm^{-1}, a series of bands occurs which are most probably explained as the overtones of a librational (a_2) oscillation of the PO_3 groups about their figure axes. For CdP_2O_7 the librational transitions found at 23, 34, 46, 52, 63, 73, 81, 86, 96 cm^{-1} can be assigned as the highly anharmonic overtones ($0 \rightarrow 2$, etc.) of a fundamental expected at 11·75 cm^{-1}. From the anharmonicity, a barrier height can be calculated, which for this particular salt is 733 cm^{-1}. The value varies with cation, ranging from 718 (Mn^{2+}) to 1380 cm^{-1} (Cu^{2+}).

Far-infrared spectra of very many complex compounds have been reported and analysed. Since it would be quite impossible to mention every one, a few representative examples are listed. *Cis* and *trans*-glycine chelates of Pd^{II}, Pt^{II} and Cu^{II} have been studied[36] and approximate force fields derived. The bond-stretching force constants indicate that the bond strengths are in the order $Pd^{II} > Pt^{II} > Cu^{II}$. The gallium trihalides, which normally exist as stable dimers, can also react with neutral donors, such as phosphines, to give stable products containing single GaX_3 moieties. The far-infrared spectra of these products show[37] that $H_3P \cdot GaCl_3$, for example, is a monomeric molecule with C_{3v} symmetry rather than $[Ga(PH_3)_2Cl_2]^+[GaCl_4]^-$. Complexes between phosphorus halides and boron halides are similar. Thus for PI_3BI_3, a P—B force constant of 1·0 mdyne $Å^{-1}$ has been deduced.[38] Complexes like $Cr(CO)_5PPh_3$ and similar molecules have been investigated[39] and the frequencies of metal-phosphorus stretching vibrations tentatively assigned to the 200–130 cm^{-1} region. Metal–sulphur stretching vibrations occur[40] over a wider range (298–170 cm^{-1}).

From the selection of far-infrared investigations discussed above, it will be seen that the availability of the new interferometric spectrometers has provided chemists with much improved technical resources with which to investigate the structures of inorganic materials. This, coupled with the advent of laser Raman spectrometers, will mean that the momentum of structural investigation, built up in the fifties and sixties by Woodward and others, will continue unabated into the seventies.

Acknowledgement

This work forms part of the research programme of the Division of Molecular Science, National Physical Laboratory.

References

1. Woodward, L. A., and Anderson, L. E., *J. Chem. Soc.*, 1957, 1284; Creighton, J. A., and Woodward, L. A., *Trans. Faraday Soc.*, 1962, **58**, 1077.
2. Woodward, L. A., and Creighton, J. A., *Spectrochim. Acta*, 1961, **17**, 594.
3. Adams, D. M., and Gebbie, H. A., *Spectrochim. Acta*, 1963, **19**, 925.
4. Adams, D. M., Gebbie, H. A., and Peacock, R. D., *Nature*, 1963, **199**, 278.
5. Adams, D. M., Chatt, J., Davidson, J. M., and Gerratt, J., *J. Chem. Soc.*, 1963, 2189.
6. Adams, D. M., and Morris, D. M., *J. Chem. Soc.*, *A*, 1967, 1666.
7. Adams, D. M., and Morris, D. M., *J. Chem. Soc.*, *A*, 1967, 1669.
8. Adams, D. M., and Morris, D. M., *J. Chem. Soc.*, *A*, 1967, 2067.
9. Adams, D. M., and Morris, D. M., *J. Chem. Soc.*, *A*, 1968, 694.
10. Greenwood, N. N., and Straughan, B. P., *J. Chem. Soc.*, *A*, 1966, 962.
11. Woodward, L. A., and Ware, M. J., *Spectrochim. Acta*, 1963, **19**, 775.
12. Butcher, F. K., Gerrard, W., Mooney, E. F., Rees, R. G., and Willis, H. A., *J. Organometall. Chem.*, 1963, **1**, 431.
13. Huggins, K. G., Parrett, F. W., and Patel, H. A., *J. Inorg. Nuclear Chem.*, 1969, **31**, 1209.
14. Bradbury, J., Forest, K. P., Nuttall, R. H., and Sharp, D. W. A., *Spectrochim. Acta*, 1967, **23A**, 2701.
15. Adams, D. M., and Lock, P. J., *J. Chem. Soc.*, *A*, 1967, 620.
16. Zbinden, R., 'Infra-red spectroscopy of high polymers', Academic Press, New York and London, 1964.

17. Campbell, M. J., Goldstein, M., and Grzeskowiak, R., *Chem. Comm.*, 1967, 778.
18. Adams, D. M., and Morris, D. M., *Nature*, 1965, **208**, 283.
19. Perry, C. H., Athans, D. P., and Young, E. F., *Spectrochim. Acta*, 1967, **23A**, 1137.
20. Greenwood, N. N., Straughan, B. P., and Wilson, Anne E., *J. Chem. Soc., A*, 1966, 1479.
21. Adams, D. M., and Lock, P. J., *J. Chem. Soc., A*, 1967, 145.
22. Manley, T. R., and Williams, D. A., *Spectrochim. Acta*, 1965, **21**, 1773.
23. Adams, D. M., Fraser, G. W., Morris, D. M., and Peacock, R. D., *J. Chem. Soc., A*, 1968, 1131.
24. Cross, L. H., Roberts, H. L., Goggin, P., and Woodward, L. A., *Trans. Faraday Soc.*, 1960, **56**, 945.
25. Adams, D. M., Chandler, P. J., and Churchill, R. G., *J. Chem. Soc., A*, 1967, 1272; Adams, D. M., and Churchill, R. G., *J. Chem. Soc., A*, 1968, 2141; Adams, D. M., and Chandler, P. J., *J. Chem. Soc., A*, 1969, 588; Adams, D. M., and Chandler, P. J., *Chem. Comm.*, 1966, 69.
26. Davies, J. E. D., and Long, D. A., *J. Chem. Soc., A*, 1968, 2054.
27. Goldstein, M., *Spectrochim. Acta*, 1966, **22**, 1389.
28. Carey, N. A. D., and Clark, H. C., *Chem. Comm.*, 1967, 292; Adams, D. M., Crosby, J. N., and Kemmitt, R. D. W., *J. Chem. Soc., A*, 1968, 3056.
29. Gager, H. M., Lewis, J., and Ware, M. J., *Chem. Comm.*, 1966, 616.
30. Manley, T. R., and Williams, D. A., *Spectrochim. Acta*, 1967, **23A**, 149.
31. Hisatsune, I. C., *Spectrochim. Acta*, 1969, **25A**, 301.
32. Smith, W. H., and Leroi, G. E., *J. Chem. Phys.*, 1966, **45**, 1767.
33. Smith, W. H., and Leroi, G. E., *J. Chem. Phys.*, 1966, **45**, 1778.
34. Christensen, D. H., Johnsen, I., Klaboe, P., and Kloster-Jensen, Else, *Spectrochim. Acta*, 1969, **25A**, 1569.
35. Hezel, A., and Ross, S. D., *Spectrochim. Acta*, 1968, **24A**, 131.
36. Walter, J. L., Hooper, C. S. C., and R. J., *Spectrochim. Acta*, 1969, **25A**, 647.
37. Balls, A., Greenwood, N. N., and Straughan, B. P., *J. Chem. Soc., A*, 1968, 753.
38. Chantry, G. W., Finch, A., Gates, P. N., and Steele, D., *J. Chem. Soc., A*, 1966, 896.
39. Chalmers, A. A., Lewis, J., and Whyman, R., *J. Chem. Soc., A*, 1967, 1817.
40. Adams, D. M., and Cornell, J. B., *J. Chem. Soc., A*, 1967, 884.

CHAPTER 5

Single-Crystal and High-Temperature Gas-Phase Raman Spectroscopy

I. R. BEATTIE

A. Introduction

The use of lasers as a source of the exciting radiation in commercial Raman spectrometers has led to the adoption of Raman spectroscopy as a routine physical technique in chemistry. However, two major developments which have arisen from this application of lasers have yet to be widely used by chemists. They are gas-phase Raman spectroscopy (notably at temperatures above ambient) and oriented single-crystal Raman spectroscopy. It is probable that the investigation of matrix-isolated species by Raman spectroscopy will also become of importance in the characterization of unstable species and high temperature vapours.

The Raman effect is a weak effect. In the passage of light through a transparent, non-absorbing, dust-free liquid, the ratio of primary ('Rayleigh') scattered light intensity to incident light intensity may be no more than 10^{-4} per centimetre of path length. The corresponding ratio of Raman to Rayleigh scattering, for a very strong Raman band, may be of the order of 10^{-2}. The bulk of the earlier literature using mercury arc sources is concerned with colourless, pure liquids. The reasons for this are not difficult to appreciate: (i) colourless, because the mercury-blue line (4358 Å) is strongly absorbed by (for example) even pale-yellow liquids†; (ii) pure, because the concentration of the molecular scattering species is at a

† However, work was carried out notably by Stammreich *et al.*[1] and Woodward[2] using sources other than the mercury arc.

maximum (frequently of the order of 10 M); (iii) liquids, because illumination of solids is difficult using an arc source, and further the scattering of the exciting line by powdered solids is considerable, necessitating the use of double monochromators of very high discrimination.

Lasers in Raman spectroscopy have the advantage that several excitation frequencies are normally available. The laser beam is highly collimated, plane polarized and coherent (although this last property is rarely utilized by chemists).† The study of coloured materials, including solids, thus becomes routine because the exciting radiation may be chosen to suit the material under examination, while the directional properties of the laser emission lead to ease of illumination even of opaque materials. In the case of gases, at atmospheric pressure the concentration of molecular species is only about 0·05 molar (less than one-hundredth that of many comparable liquids or solids), so that, to obtain spectra comparable to those from the same species in the liquid state, one requires a more intense light source, a more efficient sampling and monochromator assembly or increased sensitivity in the detector. In practice, the argon ion laser, which is capable of providing more than 1 watt of energy in one line (4880 Å) is a useful source. In the gas phase, fine structure on vibrational bands and studies of pure rotation bands can be of value in determining molecular parameters.[2a] However, because of the great accuracy attainable through the use of microwave techniques (applicable to molecules having a permanent dipole moment), only a restricted number of species is of particular interest for such pure rotation studies at high resolution. Subject to the requirements of volatility at a reasonably low temperature and of at least one small moment of inertia, the (restricted) number of inorganic compounds of interest includes PF_5, IF_7, XeF_6‡ and possible species of the type $H_nM—MH_n$. (Unfortunately even if PF_5, IF_7 and XeF_6 are not effectively spherical tops, the intensity of the rotational bands is likely to be extremely weak.) Although rotational spacings are frequently small (less than 1·0 cm^{-1} in the case of chlorine) it is possible to work close to the exciting line by making use of the strong polarization of the Rayleigh line compared to the bands of the rotational spectrum (which are depolarized). However, as a technique for studying molecular structures, high-resolution Raman spectroscopy of gases is of restricted interest, although recent results obtained using an interferometric technique offer considerable promise.[2b] By contrast, gas-phase measurements at high temperature exploiting vibrational Raman spectroscopy represent a physical technique which is likely to be of wide application and importance.

† It implies of course a very narrow linewidth.

‡ O_h, I_h and T_d species do not show pure rotation spectra but their vibration/rotation spectra are of importance.

The most elegant application of lasers in Raman spectroscopy is to the study of oriented single crystals, making use of the unique directional properties of laser radiation. In favourable circumstances it is possible to assign unambiguously the symmetry of the optically active modes of crystals to obtain the components of the derived polarizability tensor for each crystal mode. If the crystals contain 'discrete' molecules or ions, then by further assuming that the molecule or ion carries over unchanged into the crystal its derived polarizability tensor for each fundamental mode of vibration, it is possible in favourable cases to calculate, from single crystal measurements, the relative components of the derived polarizability tensor for each mode of vibration of the molecule or ion. In this way the depolarization ratios and relative intensities in the solution, melt or gas-phase Raman spectra *for the same molecular species* can be calculated.

In the subsequent sections of this discussion we shall deal in some detail with high-temperature gas-phase and single-crystal Raman spectroscopy. The treatment will be straightforward and aimed principally at the inorganic chemist, not the spectroscopist. Matrix-isolation methods involving Raman spectroscopy[3] have been omitted, not because they are unimportant, but because the techniques for working at concentrations around 1 % in the solid phase are likely in the immediate future to be available only to the specialist.

B. High-temperature gas-phase Raman spectroscopy

It is difficult to overemphasize the importance of high-temperature Raman spectroscopy in characterizing inorganic vapours, although as far as we are aware, the only work carried out at temperatures above 100°C prior to the work to be described here is that on HgX_2[4] and on P_4,[5] in both cases using arc excitation. Molecular weight data in the gas phase can frequently be obtained unambiguously from the direct application of the gas laws. It is thus possible to obtain the Raman spectra (including polarization data†) of rigorously defined molecular species in the isolated condition, in the complete absence of solvent bands and under isothermal conditions.‡ A classical, pre-laser example of such studies is the work by Selig and Claassen[5a] on vanadium pentafluoride. Frequently gas-phase studies represent the only way of studying a discrete molecular species, apart from matrix-isolation techniques.

With some commercially available Raman spectrometers it is possible

† The interpretation of polarization data and relative intensities requires care because different excitation, sample optics, sample nature, monochromator or detector can lead to different experimental results. This is particularly true of coloured materials.[5b]

‡ For coloured compounds heating of the gas in the (focused) 1 watt laser beam may be serious.

to obtain gas-phase Raman spectra at temperatures up to 1000°C with no modifications to the instrumentation. The work to be described here was carried out using a Spex 1401 monochromator and a Spectra Physics Model 140 argon ion laser. The experimental technique is elementary. The cell is a piece of silica or borosilicate tubing (depending on the temperature) 10 mm in diameter, about 5 cm long and normally rounded at one end. This is connected via a constriction to a vacuum system, degassed and an appropriate quantity of material introduced to give a pressure of about

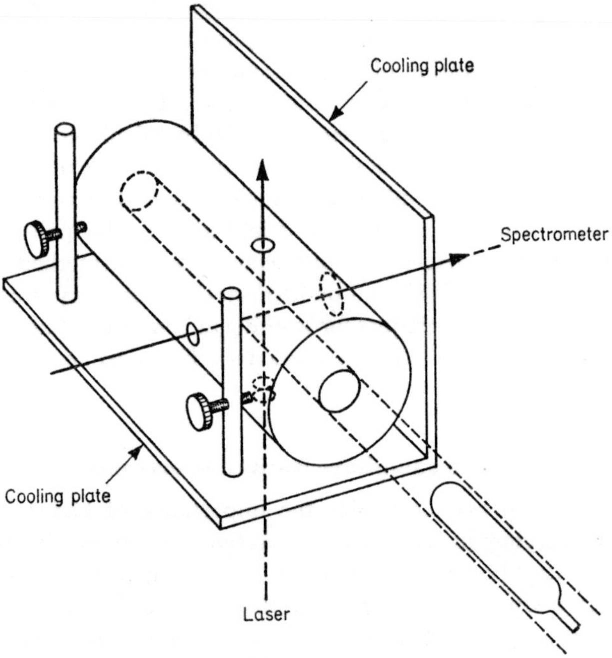

FIGURE 5.1. *Furnace and cooling plates for high-temperature, gas-phase Raman spectroscopy.*

one atmosphere of gas at the working temperature. The tube is then sealed at the constriction. The tubular furnace,[6] designed so that the cell is a reasonable sliding fit (see Figure 5.1) has four holes situated with their centres in a plane half-way down the furnace and perpendicular to the furnace axis. The two small holes in line allow for passage of the laser beam through the sample and furnace (normally as a single pass experiment). The two rather larger holes at 90° to this line allow for collection of the Raman radiation and for visual observation of the cell contents during a run. The spectrometer is to some extent protected from the furnace by hollow plates through which cold water is rapidly circulated.

When carrying out gas-phase studies it is essential to maximize the signal before each series of experiments. This is conveniently carried out by making use of the intense resonance fluorescence spectrum of iodine or the rotational spectrum of nitrogen. In either case it is useful to use a cell of the same internal and external diameter as that to be used in subsequent experiments. Such gas-phase studies have the following points in their favour:

(i) It is possible to study vapour-phase species in a closed, isothermal system at known pressure and temperature.

(ii) There are no interfering solvent bands.

(iii) It is possible to obtain polarization data on species which have been rigorously defined by ancillary PVT measurements.

(iv) Glass and silica are excellent materials for general resistance to chemicals.

(v) There is little problem due to fragility of optical materials or gaskets, nor is there any need to vary the window materials to suit different frequency ranges.

(vi) The cells are, in essence, sealed ampoules which can be taken directly from a vacuum line or preparation line to the spectrometer.

(vii) For Raman spectra because $B' \approx B''$ the individual Q lines ($\Delta J = 0$, $\Delta v = 1$) of totally symmetric modes of linear molecules, symmetric tops and spherical tops tend to coincide. Even for asymmetric tops a number of the lines of the Q branch are bunched together near the band origin. Because this is not true of P and R or O and S branches (deriving from $\Delta J = \mp 1$ or ∓ 2), frequently the only branch observed for a totally symmetric mode is the line-like Q branch. As this refers to $\Delta J = 0$ the problem of rotational broadening with increase of temperature is lessened for totally symmetric modes, although the problem of hot bands remains.

The interpretation of the (low resolution) gas-phase vibrational data is usually straightforward, provided that accompanying molecular weight studies are available, that several temperatures (and pressures) are examined, and that resonance fluorescence effects, if present, are recognized. A few examples will suffice to show the power and the likely applications of this technique.

Figure 5.2 shows the Raman spectrum of an approximately equimolar mixture of $HgCl_2$ and $HgBr_2$. The spectrum is relatively insensitive to changes of temperature.[7] Apart from the features due to $HgCl_2$ and $HgBr_2$, the spectrum shows new polarized bands at 385 and 253 cm^{-1}, which are due to stretching modes of HgBrCl. In the symmetrical, linear triatomic molecules HgX_2 the bending mode should be inactive. However, the bending mode of linear BrHgCl should be Raman active and depolarized. From previous work[8] it is expected to occur around 100 cm^{-1}. In our spectra there are possibly weak bands just below 100 cm^{-1}, but we were not able to obtain polarization data. These results contrast sharply with those[9] for gaseous $SnCl_2$ at 650°C (Figure 5.3). The presence of two strong polarized bands, one in the stretching region (352 cm^{-1}) and one in the bending region (120 cm^{-1}) points unambiguously to the presence of a bent triatomic species.

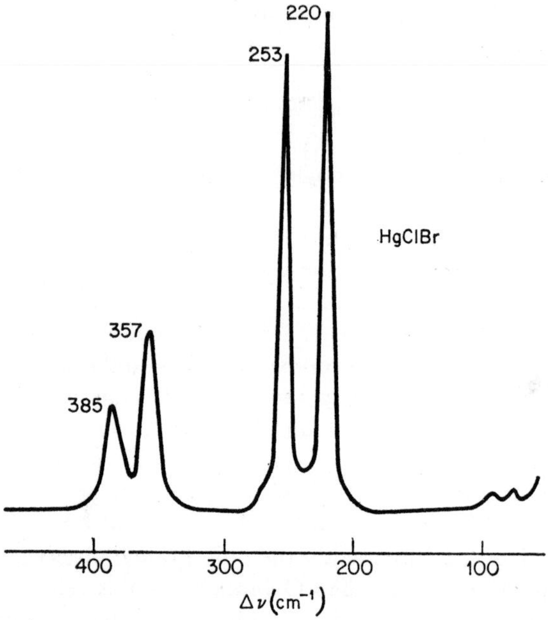

FIGURE 5.2. *The gas-phase Raman spectrum of an equimolar mixture of $HgCl_2$ and $HgBr_2$ at 475°C. (Reproduced with permission from Beattie and Horder[7].)*

A similar experiment carried out with gaseous $PbCl_2$ gave the results shown in Figure 5.4. It is immediately apparent that this is not a spectrum characteristic of the Raman effect for a triatomic molecule. It is, however, characteristic of a resonance fluorescence spectrum. In resonance fluorescence spectroscopy a molecule is excited by a photon and is thereby raised to a particular vibrational/rotational level of an excited *electronic* state. This level must be such that it lies at an energy precisely $h\nu$ (for the relevant laser line) above the appropriate ground state vibrational/rotational level. For a diatomic species with a $\Sigma - \Sigma$ transition in excitation, the selection rule $\Delta J = \pm 1$ will operate. (We assume molecules which undergo collisions are deactivated.)

The excited molecules can then decay back to the ground state emitting resonance fluorescence radiation. For the case chosen the selection rule $\Delta J = \pm 1$ is operative but any Δv is allowed. Thus we note that for a $\Sigma - \Sigma$

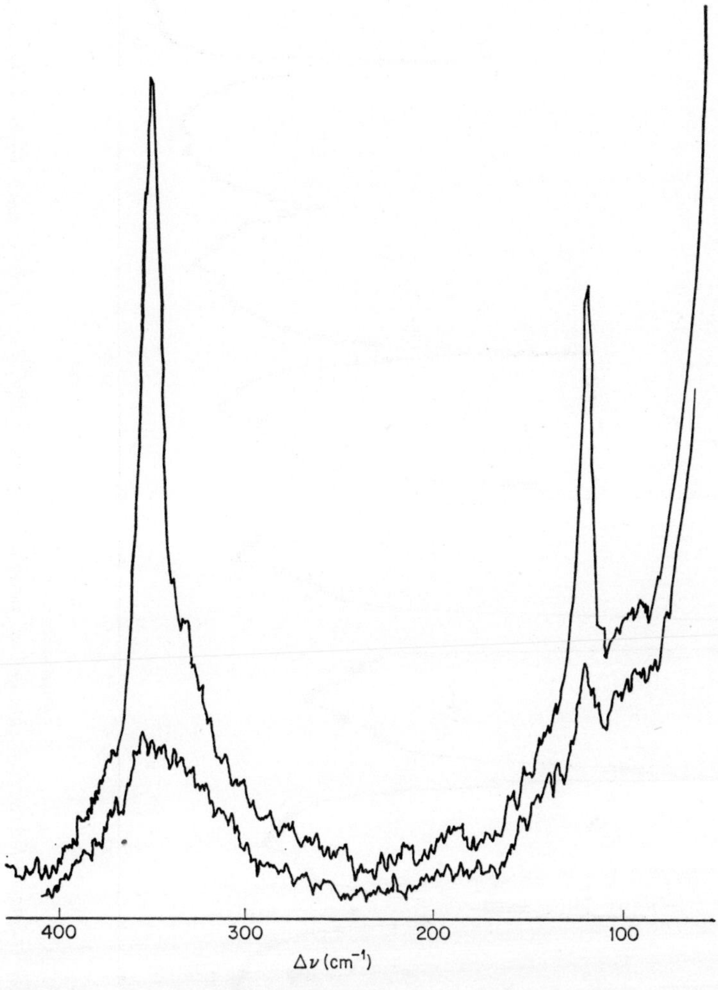

FIGURE 5.3. *The gas-phase Raman spectrum of $SnCl_2$ at 650°C. (Reproduced with permission from Beattie and Perry[9].)*

transition each band should be split to give a doublet (P and R branches corresponding to $\Delta J = \pm 1$, with separation $B(4J + 2)$ for a rigid rotor, where B is the rotational constant appropriate to the ground electronic state and J is the rotational quantum number applicable to the excited

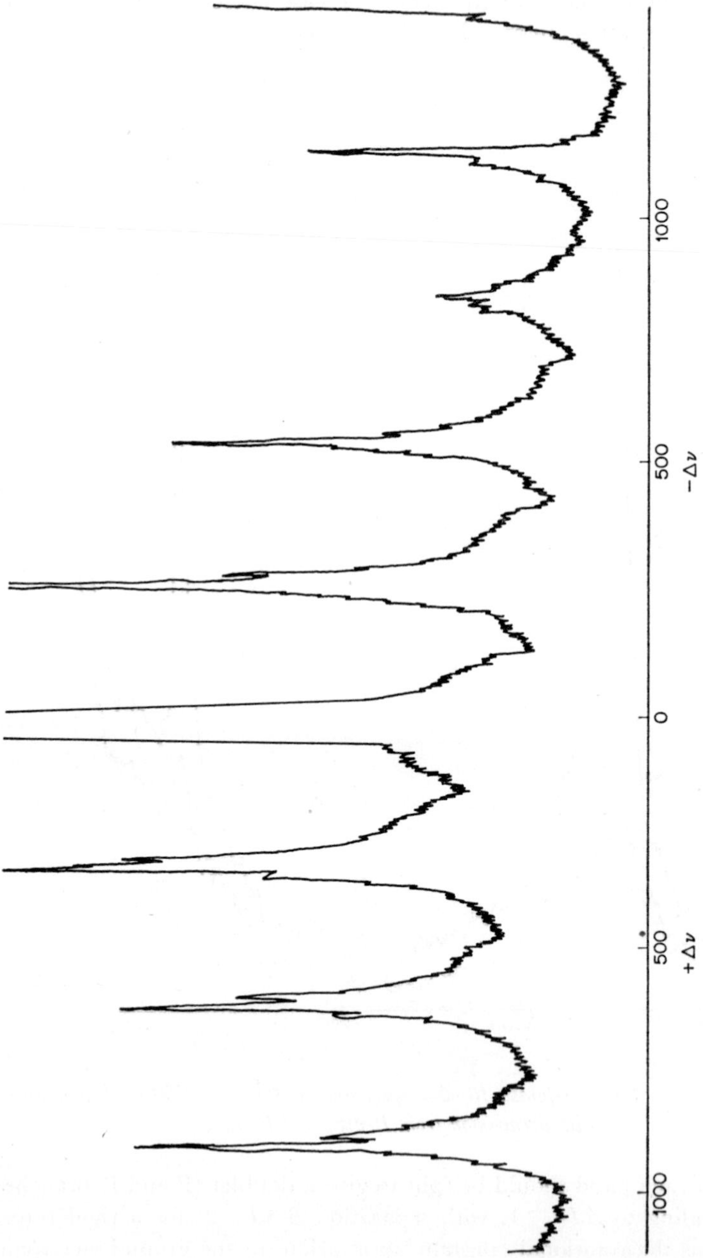

FIGURE 5.4. *The resonance fluorescence obtained on irradiating gaseous $PbCl_2$ at 1000°C with 4880 Å light. (Reproduced with permission from Beattie and Perry*[9]*.)*

state). Clearly, if it were possible to observe all lines on the anti-Stokes side, the last spacing would correspond to $v'' = 1 \rightarrow v'' = 0$, which is the Raman transition we are trying to observe.

The last two bands on the anti-Stokes (high frequency) side of the exciting line for Figure 5.4 occur at an energy separation of 298 cm^{-1}, while the anharmonicity, assuming a diatomic rigid rotor, is approximately 0·9 cm^{-1}. These values are strikingly close to those for PbCl in the $X^2\Pi_{1/2}$ state: $\omega_e - 2\omega_e x_e = 302$ cm^{-1}; $\omega_e x_e = 0{\cdot}9$ cm^{-1}. We thus assume that Figure 5.4 represents a resonance fluorescence spectrum of PbCl. Note the typical wing pattern of the spectrum, determined by the transition probability, so that it is not possible to be certain that the last band *observed* necessarily corresponds to $v'' = 0$. (The lines here are expected to be triplets due to the selection rule $\Delta J = 0, \pm 1$ for a $^2\Sigma - {}^2\Pi$ transition.)

In our experience this phenomenon of resonance fluorescence is the greatest single interfering factor in the Raman spectroscopy of gases.[10] As more powerful, tunable lasers become available, it will become correspondingly less troublesome. A resonance fluorescence spectrum may be recognized by all or some of the following features:

(i) The occurrence of a regularly repeating pattern of bands with an intensity distribution which is not that to be expected for combination or overtone bands. Note particularly the anti-Stokes components (although these are not an essential characteristic of resonance fluorescence spectra) and the wing shape of the group of bands, deriving from Franck-Condon factors.

(ii) The lines are effectively depolarized. This behaviour is frequently found for resonance fluorescence spectra of gases at lower pressures.† At high pressures or in solution the lines may become appreciably polarized.[10]

(iii) The process is highly efficient and frequently dominates any Raman effect also present. It is known that only a trace of PbCl is present in gaseous $PbCl_2$ at 1000°C.

(iv) Isotopic and rotational splittings are frequently observed where these would not normally be observable in the Raman effect. This arises from the $(4J + 2)$ multiples of B, which, for a rigid-rotor diatomic, is directly proportional to the reciprocal of the moment of inertia. Thus for $^{80}Se_2$, B is of the order of 0·09 cm^{-1} but J is of the order of 40. Thus, the observed splitting between P and R lines is 13·8 cm^{-1}. By contrast, for $^{78}Se_2$ a different pair of levels would be involved, leading to a new spectrum, with a different separation of P and R lines.

In an attempt to remove problems due to resonance fluorescence, the $PbCl_2$ spectrum was re-run in the presence of approximately five

† However for small values of J or for Q branches ($\Delta J = 0$) the bands may be appreciably polarized at low pressures in the gas phase.

atmospheres of chlorine. A true Raman spectrum was then obtained, comparable with that discussed above for $SnCl_2$ but of poorer quality.

The structure of crystalline inorganic compounds may be deduced from X-ray studies. In the gas phase, electron diffraction, microwave, infrared and Raman spectroscopy may be used to determine molecular structures. Raman and infrared spectroscopy have the great advantage that they may be applied readily to solids, melts, solutions and gases. Phosphorus pentoxide is an example where gas-phase Raman spectroscopy enables polarization measurements on the molecular species P_4O_{10} to be obtained[11] and confirms previous molecular weight and electron diffraction studies showing retention of the cage structure in the gas phase (see Figure 5.5). As phosphorus pentoxide is not to our knowledge appreciably soluble in any solvents with which it does not react and we could not obtain melt Raman spectra (and further, the properties of the melt suggest the presence of chain rather than cage species) it is difficult to see how else this polarized data could have been obtained. The three polarized fundamentals at 1440, 717 and 553 cm^{-1} may be regarded as deriving from 'breathing frequencies' of (a) four terminal oxygens (b) two strongly coupled modes derived from four phosphorus units and six bridging oxygens.

Tellurium tetrachloride represents a particularly interesting example. X-ray studies show the solid to have a tetrameric $TeCl_4$ unit in which the tellurium has distorted octahedral co-ordination, with three adjacent bridging chlorine atoms in each octahedron.[12] The vibrational spectrum has been interpreted in a variety of ways. In solution in benzene the molecular weight data are the subject of controversy. In solution in acetonitrile the compound has been separately reputed to behave as a 1:1 electrolyte and as a nonconductor.[13] The melt has the high conductivity characteristic of an ionic salt.[14] In the gas phase the compound is monomeric. Gas-phase Raman spectroscopy unambiguously identifies four totally symmetric fundamental modes of vibration of this monomer[15] (see Figure 5.6). The results show that the geometry of the compound is not a regular tetrahedron. The observation of two polarized 'stretching' modes and two polarized 'deformations' strongly suggests a C_{2v} structure based on a trigonal bipyramid with a vacant equatorial position. It is apparent that this structure is not present in the solid state. The spectrum of the melt is disappointing but suggests a closer structural analogy between melt and solid than between melt and gas. The 'acetonitrile solution spectrum' shown in Figure 5.6 is compounded of spectra run in acetonitrile and perdeuteroacetonitrile with solvent bands omitted. The results agree with the presence of a molecular species comparable with that found in the gas phase.

The most important aspect of gas-phase Raman spectroscopy at high temperatures is the ability to obtain the spectra of discrete (frequently

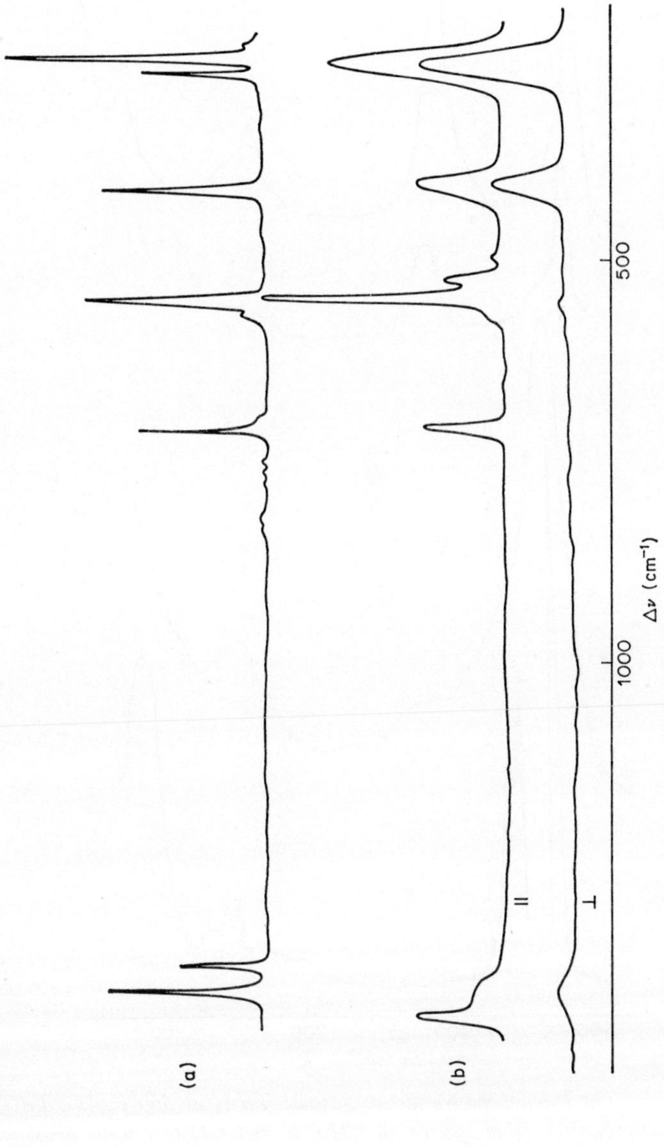

FIGURE 5.5 *The Raman spectrum of (a) powdered* P_4O_{10} *(b) gaseous* P_4O_{10} *at 750°C. (Reproduced with permission from Beattie, Livingston, Ozin and Reynolds*[11]*.)*

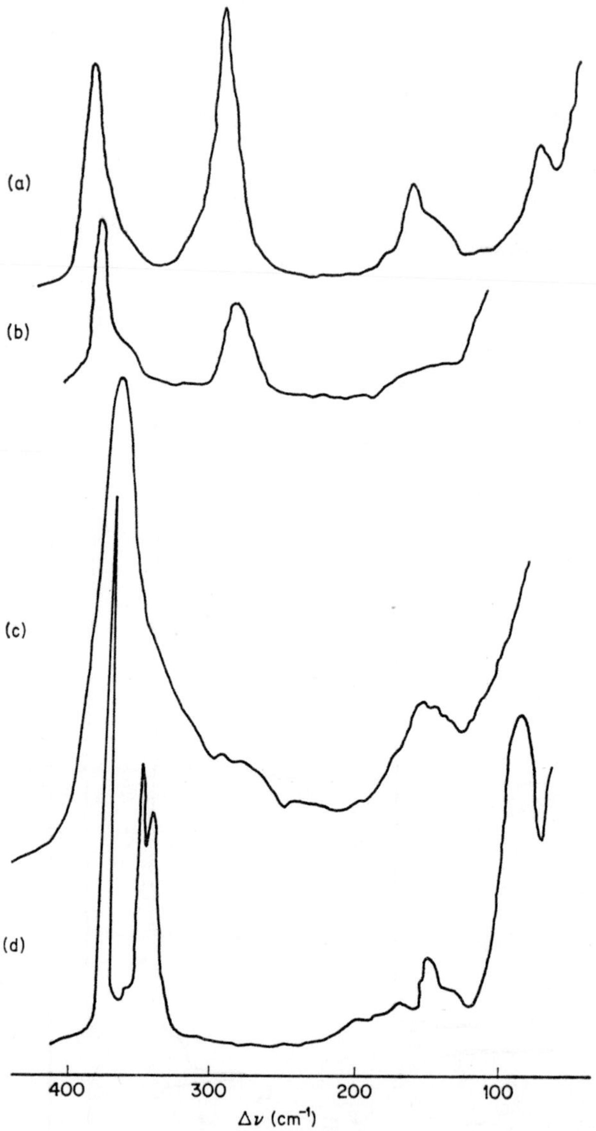

FIGURE 5.6. *The Raman spectra of (a) gaseous $TeCl_4$ at 350°C, (b) $TeCl_4$ in acetonitrile (synthesised from spectra in CH_3CN and CD_3CN with omission of solvent bands), (c) molten $TeCl_4$ (d) powdered $TeCl_4$. (Reproduced with permission from Beattie, Horder and Jones[15].)*

novel) species of known molecular weight. At Southampton the following have been studied:

(i) To characterize discrete monomers: MCl_5 (M = P, Sb, Nb, Ta, Mo); MBr_5 (M = Nb, Ta)[16]; BeX_2 (X = Br, I)[17]; ZnX_2 and HgX_2 (X = Cl, Br, I)[7]; Re_2O_7[18]; WCl_6[19]; P_4O_6; P_4O_{10}; As_4O_6[11]; $WOCl_4$; MoO_2Cl_2; $NbOCl_3$[20]; Fe_2Cl_6[21]; Re_3Cl_9[21]; $Fe(C_5H_5)_2$[22]; $TeCl_2$; $SnCl_2$; $GeCl_2$; $PbCl_2$; $PbBr_2$[9]; $TeCl_4$[15]; SeO_2[23].

(ii) To study equilibria in the gas phase: $M_4 \rightleftarrows 2M_2$ (M = P, As)[10]; $2MX_3 \rightleftarrows M_2X_6$ (M = Al, Ga, In; X = Cl, Br, I)[6]; $HgX_2 + HgY_2 \rightleftarrows 2HgXY$ (X, Y = Cl, Br, I)[7]; $GaBr_3/GaCl_3$ in equilibrium with $GaBr_2Cl$/$GaBrCl_2$[7]; the vapour from Ga_2Cl_4[7]; $SnCl_2 + SnBr_2 \rightleftarrows 2SnBrCl$.[9]

C. Single-crystal Raman spectroscopy

The most elegant application of lasers in Raman spectroscopy is to the study of oriented single crystals. Before the advent of lasers single-crystal research in Raman spectroscopy was largely carried out by Mathieu and his co-workers.[24] The application of laser excitation to such problems has been developed particularly by Porto and co-workers.[25] Single-crystal Raman spectroscopy is principally of interest to the chemist as a means of unambiguously assigning vibrational frequencies of molecules or ions to particular symmetry classes. This method is so powerful that no paper dealing with new vibrational assignments for crystalline materials can be considered complete without single-crystal data provided that:

(i) The powdered material yields a Raman spectrum of reasonable quality (note however that single-crystal spectra are frequently of much higher quality than powder spectra).

(ii) The X-ray structure is known and does not render single-crystal Raman studies of little value.

(iii) Single crystals of about 2 mm face or greater can be grown.

The chemist needs the unambiguous assignment of fundamental modes to develop adequate force fields to calculate vibrational spectra of related molecules. Single-crystal Raman spectroscopy is a useful adjunct to X-ray studies, particularly where different atoms of similar X-ray scattering efficiency occur in the same crystal, or where the position of light atoms (including hydrogen) is obscured in detail by the presence of heavy atoms. This approach is of interest in structures containing a heavy metal plus oxygen, for example, or where there may be unsuspected disorder in the crystal. For centrosymmetric crystals the prediction of Raman activity is given rigorously by group theory.

Gilson[26] has given an excellent account of problems which may be encountered in single-crystal Raman studies. As the present article is intended to act as an introduction to the subject, detailed discussion will be limited to crystals which are (i) 'non-absorbing' with respect to the incident light and to the Raman light (thereby problems of dichroism, rapid fluctuation of refractive index with wavelength and resonance effects are avoided), and (ii) centrosymmetric (thereby the group theoretical predictions of the Raman-active modes may be rigorously applied and the difficult problem of enantiomorphism is avoided).

When light falls on an isotropic solid, part is reflected and part is refracted (with polarization changes which are dependent upon the nature of the incident light and also upon the angle of incidence). In the absence of absorption bands the refracted ray is attenuated only by scattering processes. When light falls on an anisotropic medium such as a crystal, the refracted ray (in which we are interested for Raman studies) is normally split into two rays, one or both of which do not obey the normal laws of refraction. We shall not go into the details of extraordinary refraction but note here that such splitting of the *incident* light and of the *collected* Raman light must normally be avoided if meaningful results are to be obtained.

For a crystal the variation of the refractive index with direction in the crystal may be represented by a triaxial ellipsoid (commonly called the indicatrix). The axes of this ellipsoid coincide in direction with the crystallographic axes in orthorhombic symmetry. Thus, each crystallographic plane (100), (010), (001)—equivalent to yz, xz and xy planes—contains a principal (elliptic) section of the ellipsoid (or indicatrix) with the major and minor axes of the ellipses parallel to crystallographic x and y, x and z or y and z. For an orthorhombic crystal there are two sections of the indicatrix which yield a circle.[27] The directions perpendicular to the planes of these circles are optic axes. Thus orthorhombic crystals are bi-axial. The optic axes do *not* coincide with the axes of the indicatrix.

Examination of an orthorhombic crystal under a polarizing microscope will normally show two extinction directions. (If by coincidence the observation is an optic axis, the crystal will appear to be isotropic because the section of the indicatrix perpendicular to the direction of observation is circular.) These extinction directions define directions along which polarization characteristics of light passing through the medium are not materially affected by passage through the medium. They thus define directions suitable for the direction of the electric vector of the incident light or suitable for the direction of the analyser when collecting Raman light.

If the infrared or Raman spectrum of a solid of known structure is being studied, it is essential to carry out a factor-group analysis. We shall consider the orthorhombic crystal[28] $Na_2S_2O_6,2H_2O$, but for the sake of simplicity ignore the sodium ions and the water molecules. The space group is $P2_1/n\ 2_1/m\ 2_1/a$, or in the Schoenflies notation D_{2h}^{16}. The symbol D_{2h}^{16} represents the appropriate factor group (space group less translations) which is isomorphous with the corresponding point group D_{2h}. The difference between the factor group and the point group lies, in essence, in the fact that extended elements of symmetry, such as glide planes or screw axes, become more restricted elements—mirror planes and rotation axes. A factor-group analysis of a crystal is closely similar to the well known

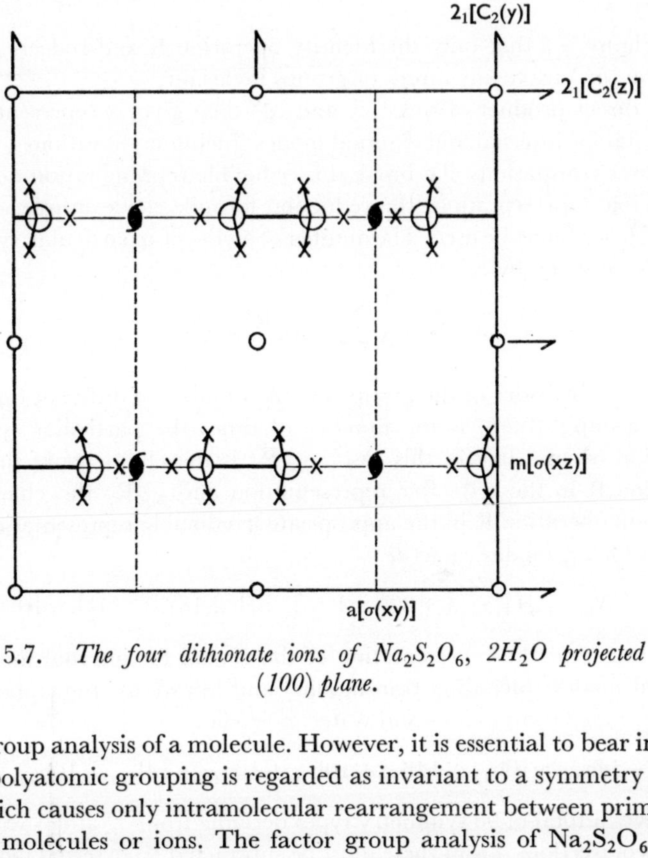

FIGURE 5.7. *The four dithionate ions of $Na_2S_2O_6$, $2H_2O$ projected on the (100) plane.*

point-group analysis of a molecule. However, it is essential to bear in mind that a polyatomic grouping is regarded as invariant to a symmetry operation which causes only intramolecular rearrangement between primitively related molecules or ions. The factor group analysis of $Na_2S_2O_6,2H_2O$ considering only the four dithionate ions in the primitive cell (see Figure 5.7) is given in Table 5.1. Rows (a) and (b) give the symmetry operations under the groups D_{2h} and D_{2h}^{16}, respectively; the effect of carrying out the appropriate symmetry operations is given in row (c) for the x, y and z axes, in row (d) for the individual atoms of the dithionate ions and in row (e) for the dithionate ions each regarded as a point. It is apparent

TABLE 5.1.[a] *The factor-group analysis of $Na_2S_2O_6$, $2H_2O$ under D_{2h}^{16}*

(a) Point group D_{2h}	E	$C_2(x)$	$C_2(y)$	$C_2(z)$	i	$\sigma(xy)$	$\sigma(xz)$	$\sigma(yz)$
(b) Factor group D_{2h}^{16}	E	2_1	2_1	2_1	i	a	m	n
(c) Cartesians x, y, z	3	−1	−1	−1	−3	1	1	1
(d) External + internal + librations	32	0	0	0	0	0	16	0
(e) External	4	0	0	0	0	0	4	0

[a] Beattie, Gall and Ozin.[28]

from Figure 5.7 that only the identity operation E and reflection in the xz plane (σ_{xz}) leave any atoms or groups invariant.

The direct product of rows (c) and (d) then gives a representation of the crystal for internal and external modes (including librations), together with three translations. To break this reducible representation down into irreducible representations the reduction formula, for example, of Mitra and Gielisse,[29] may be used. The number of modes of a particular symmetry type, N_k, is given by

$$N_k = \frac{1}{N} \sum_j h_j \chi_k(\text{R}) \chi_j(\text{R})$$

where N is the order of the group (i.e. the number of different operations in the group = 8), h_j is the number of times the particular symmetry operation occurs (one in this case), $\chi_k(\text{R})$ is the character of the group operation R in the reducible representation and $\chi_j(\text{R})$ the character of the group operation R in the appropriate irreducible representation.

Thus for b_{1g} modes we write

$$N_{b_{1g}} = \tfrac{1}{8}[1 \times (3 \times 32) \times 1 + 1 \times (1 \times 16) \times (-1)] = 10$$

We thus find the representation of the crystal for internal modes and external modes (including translations) and librations (remembering we have excluded sodium ions and water molecules)

$$\Gamma_{\text{cryst}} = 14a_{1g} + 10b_{1g} + 14b_{2g} + 10b_{3g} + 10a_{1u} + 14b_{1u} + 10b_{2u} + 14b_{3u}$$

This gives a total of 96 symmetry types deriving from 32 atoms with three degrees of freedom. From these must be subtracted (i) external modes and (ii) librations.

(i) External modes (including translations): These arise from movement of the $S_2O_6^{2-}$ species regarded as points. The reducible representation is obtained from the product of lines (c) and (e) of Table 5.1. Thus,

$$\Gamma_{\text{ext}} = 2a_{1g} + b_{1g} + 2b_{2g} + b_{3g} + a_{1u} + 2b_{1u} + b_{2u} + 2b_{3u}$$

These 12 modes deriving from four point ions contain the three translations $b_{1u} + b_{2u} + b_{3u}$ of the D_{2h} group, as shown in the conventional character table for this point group.

(ii) Librations: These derive from movements of the $S_2O_6^{2-}$ ions regarded as rigid species of symmetry D_{3d}. For such a rigid body there is clearly one a_{2g} libration (effectively a restricted rotation about the S—S link) and one e_g libration (about two axes perpendicular to the S—S link). Inspection of correlation tables from $D_{3d} \rightarrow D_{2h}$ shows:

$$D_{3d} \rightarrow (C_s) \rightarrow D_{2h}$$

$$a_{2g} \rightarrow (a'') \rightarrow b_{1g} + b_{3g} + a_{1u} + b_{2u}$$

$$e_g \rightarrow (a' + a'') \rightarrow a_{1g} + b_{1g} + b_{2g} + b_{3g} + a_{1u} + b_{1u} + b_{2u} + b_{3u}$$

so that

$$\Gamma_{lib} = a_{1g} + 2b_{1g} + b_{2g} + 2b_{3g} + 2a_{1u} + b_{1u} + 2b_{2u} + b_{3u}$$

This leaves

$$\Gamma_{internal} = 11a_{1g} + 7b_{1g} + 11b_{2g} + 7b_{3g}$$

for the Raman-active modes deriving from internal vibrations of the dithionate ion.

Conventional point-group analysis of the dithionate ion shows that

$$\Gamma_{mol} = 3a_{1g} + 3e_g + a_{1u} + 2a_{2u} + 3e_u$$

In theory the infrared-active modes of the free ion can couple leading to Raman-active components in the crystal. However, these bands in practice are likely to be weak so that as a first order approximation we may ignore them. We are thus interested in correlation of a_{1g} and e_g modes from the D_{3d} dithionate ion in the C_s site (one plane only) and in the D_{2h} crystal. For such correlations the correct orientation of the elements of symmetry applicable to the molecule and the character table must be maintained relative to those of the molecule in the site and in the crystal.

Correlating D_{3d} and the site symmetry (the plane of symmetry $\sigma(xz)$ being also the xz plane of the crystal), we find from correlation tables

$$D_{3d} \rightarrow (C_s) \rightarrow D_{2h}$$

$$a_{1g} \rightarrow (a') \rightarrow a_{1g} + b_{2g} + b_{1u} + b_{3u}$$

$$e_g \rightarrow (a'') \rightarrow b_{1g} + b_{3g} + a_{1u} + b_{2u}$$

(with an additional arrow from e_g to (a'))

Thus, in the Raman spectrum of $Na_2S_2O_6 \cdot 2H_2O$ we expect to find two components $(a_{1g} + b_{2g})$ for each a_{1g} mode of the D_{3d} ion and four components $(a_{1g} + b_{2g} + b_{1g} + b_{3g})$ for each e_g mode of the free ion.

In order that the single-crystal Raman data may be quantitatively interpreted in terms of the oriented gas-phase model it is necessary to carry out a transformation from molecular to crystallographic axes. We define molecular z as the S—S bond direction (note particularly that in Figure 5.7 molecular z is *not* parallel to crystal z) and for convenience make molecular x lie in the crystallographic σxz plane. Molecular y is then parallel to crystal y. The transformation from molecular axes to crystal axes brought about by a rotation of $\theta°$ around y is shown in Figure 5.8.

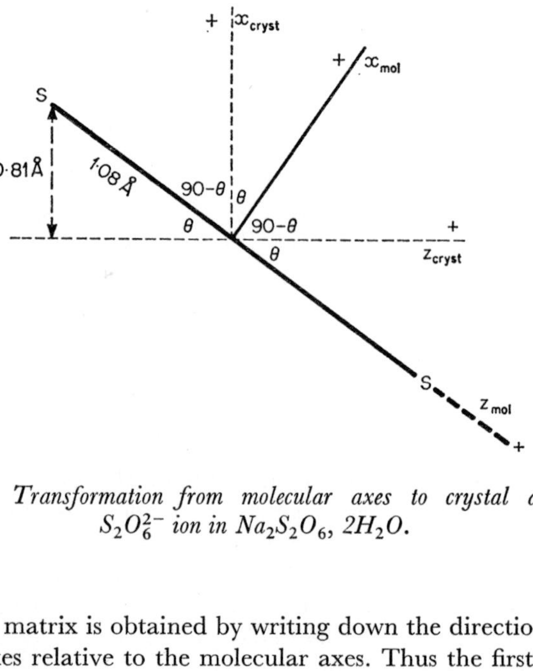

FIGURE 5.8. *Transformation from molecular axes to crystal axes for the* $S_2O_6^{2-}$ *ion in* $Na_2S_2O_6$, $2H_2O$.

The required matrix is obtained by writing down the direction cosines of the crystal axes relative to the molecular axes. Thus the first row of the matrix is formed by the cosines of the angles:

crystal x/molecular x   crystal x/molecular y   crystal x/molecular z

Maintaining the directions shown in the diagram we obtain

$$\cos\theta \quad 0 \quad \cos(90+\theta)$$

The whole matrix may then be written

$$\begin{pmatrix} x \\ y \\ z \end{pmatrix}_{\text{cryst}} = \begin{pmatrix} \cos\theta & 0 & -\sin\theta \\ 0 & 1 & 0 \\ \sin\theta & 0 & \cos\theta \end{pmatrix} \begin{pmatrix} x \\ y \\ z \end{pmatrix}_{\text{mol}}$$

which for $\theta = 49°$ becomes

$$T = \begin{pmatrix} 0{\cdot}656 & 0 & -0{\cdot}755 \\ 0 & 1 & 0 \\ 0{\cdot}755 & 0 & 0{\cdot}656 \end{pmatrix}^{\dagger}$$

We may now utilize this transformation matrix to calculate the Raman tensor components. For convenience we write α' as the derived polarizability components for the molecule (or ion) and R for the Raman (derived polarizability) tensor components of the crystal. Thus,

$$R_{\text{cryst}} = T\alpha'_{\text{mol}}\, T^t$$

where T^t is the transpose (equal to the inverse) of the transformation matrix. Thus, for a_{1g} modes of the $S_2O_6^{2-}$ ion under D_{3d} symmetry we obtain from character tables the symbols $\alpha_{xx} + \alpha_{yy}$, α_{zz}. This implies that $\alpha_{xx} = \alpha_{yy} \neq \alpha_{zz}$. In the matrix notation

$$\alpha'_{\text{mol}} = \begin{pmatrix} a & 0 & 0 \\ 0 & a & 0 \\ 0 & 0 & b \end{pmatrix}$$

where $\alpha'_{xx} = \alpha'_{yy} = a$ and $\alpha'_{zz} = b$.
Thus

$$R_{\text{cryst}} = \begin{pmatrix} 0{\cdot}656 & 0 & -0{\cdot}755 \\ 0 & 1 & 0 \\ 0{\cdot}755 & 0 & 0{\cdot}656 \end{pmatrix} \begin{pmatrix} a & 0 & 0 \\ 0 & a & 0 \\ 0 & 0 & b \end{pmatrix} \begin{pmatrix} 0{\cdot}656 & 0 & 0{\cdot}755 \\ 0 & 1 & 0 \\ -0{\cdot}755 & 0 & 0{\cdot}656 \end{pmatrix}$$

$$= \begin{pmatrix} 0{\cdot}43\,a + 0{\cdot}56\,b & 0 & 0{\cdot}5\,a - 0{\cdot}5\,b \\ & a & 0 \\ & & 0{\cdot}56\,a + 0{\cdot}43\,b \end{pmatrix}$$

The separation into a_g and b_{2g} correlation doublets under D_{2h}^{16} follows by taking diagonal and off-diagonal terms to form the respective matrices. The observed intensities are proportional to the squares of the components for the Raman tensors. Thus, for a_{1g} modes we predict zero intensity for (xy) and (yz) observations with a_{1g} components showing activity only in

† A useful check on the correctness of the matrix (assuming orthogonal coordinates) is that

$$\sum_j t_{ij}^2 = 1.$$

and that

$$\sum_k t_{ik}\,t_{jk} = 0.$$

The determinant of the total matrix should also be unity.

(xx), (yy), and (zz) and b_{2g} components showing activity only in (xz) measurements. Proceeding in this way we predict

D_{3d}	D_{2h}^{16}	(xx)	(yy)	(zz)	(xy)	(xz)	(yz)
a_{1g}	a_{1g}	✓	✓	✓	0	0	0
	b_{2g}	0	0	0	0	✓	0
e_g	b_{1g}	0	0	0	✓	0	0
	b_{3g}	0	0	0	0	0	✓

Note that it is not necessary to carry out the transformation for the four molecules in the unit cell. This is because these four molecules are symmetrically related. If there had been another set of four molecules in the unit cell, related to one another by the symmetry of the cell but not related to the other set of four molecules in the cell, then a further transformation would have been necessary. It is a simple and instructive exercise to carry out the transformation for each of the four molecules in the primitive cell. It is evident that only modes deriving from e_g fundamentals of the $S_2O_6^{2-}$ ion should show activity in (xy) or (yz) observations.

To carry out the orientation studies on large, well-developed and air-stable crystals, such as those of $Na_2S_2O_6,2H_2O$, is straightforward, and excellent spectra are to be expected. The orientation may be carried out with the aid of morphological examination (including studies under the polarizing microscope), of cleavage in conjunction with other behaviour, such as Raman spectroscopic data from the oriented crystal, and of X-ray studies of a chip taken in a known orientation from a large crystal. It is useful to check the alignment in the Raman spectrometer using a piece of polaroid to examine the beam emerging from the crystal and ensuring that the unwanted ray is at a minimum. For a fixed orientation of the crystal it is then possible to obtain four (two pairs of) measurements, of which two only should be identical. Thus, for the xy face of the crystal perpendicular to the propagation direction of the incident radiation, with y vertical and 180° collection (as is usual in the Cary 81 spectrometer), we may take z(yy)z, z(yx)z, z(xx)z and z(yx)z observations. (The nomenclature here gives in order, outside the brackets, the propagation direction of the incident ray and the collection direction. Inside the brackets the direction of the electric vector of the incident light is followed by that of the analyser.) The (xy) and (yx) measurements will normally be identical if the readings are ideal.

The results shown in Figure 5.9 refer to z(xx)z and z(yx)z measurements on $Na_2S_2O_6,2H_2O$. Inspection of the predicted activity immediately shows that the 1101 and 708 cm^{-1} bands arise from a_{1g} modes of $S_2O_6^{2-}$. The

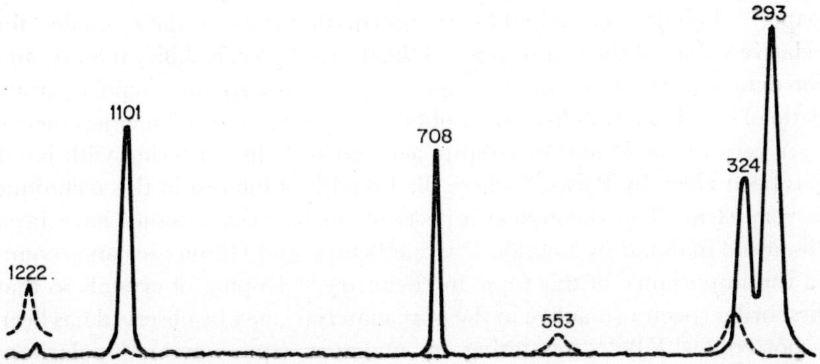

FIGURE 5.9. z(xx)z *(solid line) and* z(yx)z *(dashed line) measurements on a single crystal of* $Na_2S_2O_6$, $2H_2O$. *(Reproduced with permission from Beattie, Gall, and Ozin*[28].*)*

553 and 1222 cm^{-1} bands arise from e_g modes. Note the small shifts of the band at 1222 cm^{-1} corresponding to a_{1g}(xx) and b_{1g}(xy) components. The closeness of the 324 and 293 cm^{-1} bands renders assignment from this one observation open to question. As is shown by the more extensive measurements reproduced in Table 5.2, the band at 293 cm^{-1} derives from an a_{1g} mode of $S_2O_6^{2-}$. The 324 cm^{-1} band is an a_{1g}(xx) component and the 334 cm^{-1} band a b_{1g}(yx) component of an e_g multiplet. The original

TABLE 5.2. *Single-crystal Raman polarization data for the internal modes of* $Na_2S_2O_6$, $2H_2O$[a]

D_{3d}	D_{2h}^{16}	(cm^{-1})	z(xx)z	z(yy)z	x(zz)x	x(zx)z	z(yx)z	x(yz)x
	b_{1g}	1222·5	0	0	0	0	15	0
e_g	b_{3g}	1217	0	0	0	0	0	1·5
	a_g+b_{2g}	1213·5	3	10	19	2·2	0	0
a_g	a_g+b_{2g}	1100·5	68	92	85	10·7	1	7
a_g	a_g+b_{2g}	707·5	57·5	40	72	5·3	0·5	4
	b_{3g}	588	0	0	0	0	0	10·5
e_g	b_{1g}	554	0	0	0	0	6.5	0
	a_g+b_{2g}	543	0	17	9·5	2·7	0	0
	b_{1g}	334	0	0	0	0	11	0
	b_{3g}	330	0	0	0	0	0	23
e_g	b_{2g}	326	0	0	0	4·9	0	0
	a_g	324	51	4	34	0	0	0
	a_g	293	95	0·5	73	0	0	8
a_g	b_{2g}	291·5	0	0	0	89	0	0

[a] Peak heights measured on an arbitrary scale, 6328 Å excitation (Beattie, Gall and Ozin[28]).

paper[28] should be consulted for transformation data on the e_g modes, for relative values of the components of the derived polarizability tensors, and for details of the calculation of the relative intensity and depolarization ratio of solution bands from data obtained by single-crystal measurements.

A very readable and interesting account of light scattering with lasers has been given by Porto,[30] where the breadth of interest in this technique is very clear. The theoretical aspects of single-crystal studies have been discussed in detail by Loudon,[31] while Beattie and Gilson give an account of the importance of this topic in chemistry.[32] Doping of crystals so that first order spectra (inactive in the pure material) may be observed has been reported for KBr[33]; scattering by magnons (spin-waves),[34] polaritons (mixed phonon/photon excitation)[35] and F-centres[36] has also been reported. Single-crystal studies have been performed on the following compounds:

Oxides: ZnO[25, 35]; Al_2O_3[37]; TiO_2[38]; Cr_2O_3[39]; Fe_2O_3[39]; V_2O_5[40]; MoO_3[32].

Halides: MgF_2[38, 41]; $LaCl_3$[42]; BiF_5; $SbCl_4F$[43]; Ga_2Cl_6[32]; Hg_2Cl_2.[32]

More complex species: $CuCl_2,2H_2O$; $(CH_3NH_3)_2CuCl_4$; Cs_2CuCl_4[44]; $(Me_3N)_2InCl_3$[45]; *trans*-$[PtHCl(PEt_3)_2]$[46]; $Cs_3Tl_2Cl_9$[47]; K_2PtCl_6[32]; $NbOCl_3$[20].

The above list is not intended to be comprehensive, and is strongly biased towards compounds of chemical interest.

Acknowledgement

I am indebted to several colleagues at Southampton for helpful comments on this manuscript.

References

1. See for example, Stammreich, H., Kawai, K., and Tavares, Y., *Spectrochim. Acta*, 1959, **15**, 438.
2. See for example, Woodward, L. A., and Creighton, J. A., *Spectrochim. Acta*, 1961, **17**, 594.
2(a). See for example, Stoichieff, B. P., 'Advances in Spectroscopy', Volume I, ed. H. W. Thompson, Interscience, New York, 1959, p. 91.
2(b). Barrett, J. J., Second International Raman Conference, Oxford, 1970.
3. See for example, Beattie, I. R., and Collis, R. E., *J. Chem. Soc., A*, 1969, 2960.
4. Braune, H., and Engelbrecht, G., *Z. phys. Chem.* (Leipzig), 1932, **B19**, 303.
5. Venkateswaran, C. S., *Proc. Ind. Acad. Sci.*, 1935, **2A**, 260.
5(a). Selig, H., and Claassen, H. H., *J. Chem. Phys.*, 1965, **44**, 4039.

5(b). See for example, Beattie, I. R., and Gall, M. J., *Inorg. Nuclear Chem. Letters*, 1968, **4**, 677.
6. Beattie, I. R., and Horder, J. R., *J. Chem. Soc.*, *A*, 1969, 2655.
7. Beattie, I. R., and Horder, J. R., *J. Chem. Soc.*, *A*, 1970, 2433.
8. Janz, G. J., and James, D. W., *J. Chem. Phys.*, 1963, **38**, 902, and references therein.
9. Beattie, I. R., and Perry, R. O., *J. Chem. Soc.*, *A*, 1970, 2429, and references therein.
10. For a general discussion see Beattie, I. R., Ozin, G. A., and Perry, R. O., *J. Chem. Soc.*, *A*, 1970, 2071.
11. Beattie, I. R., Livingston, K. M. S., Ozin, G. A., and Reynolds, D. J., *J. Chem. Soc.*, *A*, 1970, 449.
12. Buss, B., and Krebs, B., *Angew. Chem.*, *Internat. Edn.*, 1970, **9**, 463.
13. Beattie, I. R., Jones, P. J., and Webster, M., *J. Chem. Soc.*, *A*, 1969, 218, and references therein.
14. See for example, Beattie, I. R., and Chudzynska, H., *J. Chem. Soc.*, *A*, 1967, 984.
15. Beattie, I. R., Horder, J. R., and Jones, P. J., *J. Chem. Soc.*, *A*, 1970, 329.
16. Beattie, I. R., and Ozin, G. A., *J. Chem. Soc.*, *A*, 1969, 1691.
17. Beattie, I. R., and Reynolds, D. J., unpublished work.
18. Beattie, I. R., and Ozin, G. A., *J. Chem. Soc.*, *A*, 1969, 2615.
19. Ozin, G. A., unpublished work.
20. Beattie, I. R., Livingston, K. M. S., Ozin, G. A., and Reynolds, D. J., *J. Chem. Soc.*, *A*, 1970, 1210.
21. Beattie, I. R., and Ozin, G. A., unpublished work.
22. Beattie, I. R., and Perry, R. O., unpublished work.
23. Beattie, I. R., and Cheetham, N., unpublished work.
24. See for example, Mathieu, J.-P., 'Spectres de vibration et symétrie des molecules et des cristaux', Hermann and Cie, Paris, 1945.
25. See for example, Damen, T. C., Porto, S. P. S., and Tell, B., *Phys. Rev.*, 1966, **142**, 570; Porto, S. P. S., Giordmaine, J. A., and Damen, T. C., *Phys. Rev.*, 1966, **147**, 608.
26. Gilson, T. R., and Hendra, P. J., 'Laser Raman Spectroscopy', Wiley, London, 1970.
27. Hartshorne, N. H., and Stuart, A., 'Crystals and the Polarizing Microscope', Arnold, London, 1960.
28. Beattie, I. R., Gall, M. J., and Ozin, G. A., *J. Chem. Soc.*, *A*, 1969, 1001, and references therein.
29. Mitra, S. S., and Gielisse, P. J., 'Progress in Infrared Spectroscopy', Plenum Press, New York, 1964.
30. Porto, S. P. S., *The Spex Speaker*, 1968, **13**, No. 2, p. 1.

31. Loudon, R., *Adv. Phys.*, 1964, **13**, 423 (errata 1965, *Adv. Phys.*, 1965, **14**, 621).
32. Beattie, I. R., and Gilson, T. R., *Proc. Roy. Soc.*, 1968, **A307**, 407.
33. Hurrell, J. P., Porto, S. P. S., Damen, T. C., and Mascarenhas, S., *Phys. Letters*, 1968, **A26**, 194.
34. Fleury, P. A., Porto, S. P. S., Cheesman, L. E., and Guggenheim, H. J., *Phys. Rev. Letters*, 1966, **17**, 84.
35. Porto, S. P. S., Tell, B., and Damen, T. C., *Phys. Rev. Letters*, 1966, **16**, 450.
36. Worlock, J. M., and Porto, S. P. S., *Phys. Rev. Letters*, 1965, **15**, 697.
37. Porto, S. P. S., and Krishnan, R. S., *J. Chem. Phys*, 1967, **47**, 1009.
38. Porto, S. P. S., Fleury, P. A., and Damen, T. C., *Phys. Rev.*, 1967, **154**, 522.
39. Beattie, I. R., and Gilson, T. R., *J. Chem. Soc.*, *A*, 1970, 980.
40. Beattie, I. R., and Gilson, T. R., *J. Chem. Soc.*, *A*, 1969, 2322.
41. Krishnan, R. S., and Russell, J. P., *Brit. J. Appl. Phys.*, 1966, **17**, 501.
42. Asawa, C. K., Satten, R. A., and Stafsudd, O. M., *Phys. Rev.*, 1968, **168**, 957.
43. Beattie, I. R., Livingston, K. M. S., Ozin, G. A., and Reynolds, D. J., *J. Chem. Soc.*, *A*, 1969, 958.
44. Beattie, I. R., Gilson, T. R., and Ozin, G. A., *J. Chem. Soc.*, *A*, 1969, 534.
45. Beattie, I. R., and Ozin, G. A., *J. Chem. Soc.*, *A*, 1969, 542.
46. Beattie, I. R., and Livingston, K. M. S., *J. Chem. Soc.*, *A*, 1969, 2201.
47. Beattie, I. R., Gilson, T. R., and Ozin, G. A., *J. Chem. Soc.*, *A*, 1968, 2765.

CHAPTER 6

High-Resolution Raman Spectroscopy of Gases

H. G. M. EDWARDS

A. Introduction

Studies of high-resolution Raman spectra are undertaken to obtain information about molecular structure and parameters. An analysis of rotational Raman spectra can give information on moments of inertia, internuclear bond distances and angles, molecular symmetry and nuclear statistics. In this article, the progress of high-resolution Raman spectroscopy will be traced and an account given of the impact that has been made by recent discoveries and advances in instrumentation.

B. Historical development

The Raman effect, which was discovered in 1928 by C. V. Raman[1] after a theoretical prediction by Smekal,[2] was rapidly applied to studies of the rotational spectra of gases. Ironically, the first rotational Raman spectrum reported[3] was obtained from liquid hydrogen, two rotational transitions being observed.

The rotational Raman spectra of gases were investigated first by Rasetti[4] and Wood[5] in 1929, and the moments of inertia of nitrogen, hydrogen, oxygen and hydrogen chloride were calculated. Rasetti[6] also obtained an ill-resolved rotational spectrum of nitric oxide. In 1931 Houston and Lewis[7] observed the rotational Raman spectrum of carbon dioxide as a series of equidistant lines. Given in Table 6.1 is a comprehensive list of all the molecules investigated in this early period up to 1935. The effects of pressure broadening, which resulted in unresolved rotational and rota-

TABLE 6.1. *Survey of early rotational Raman spectra*

Year	Reference	Observer	Molecule
1929	3	Maclennan *et al.*	H_2 (liquid)
	4	Rasetti	H_2, N_2, O_2
	5	Wood	HCl
	12	Dickinson *et al.*	NH_3
1930	6	Rasetti	NO
1931	7	Houston *et al.*	CO_2
	8	Bhagavantam	CO, NO
1932	13	Amaldi	CO
1933	14	Amaldi *et al.*	NH_3
	11	Houston *et al.*	CH_4, C_2H_2, C_2H_4, C_2H_6, NH_3
1935	15	Bhagavantam	H_2
	16	Teal *et al.*	H_2, HD, D_2

tion–vibration lines, were also noted by Bhagavantam[8] in the spectra of carbon monoxide and nitric oxide at pressures of twenty atmospheres or more.

The polarizability theory of Placzek[9] was developed and applied to rotational Raman scattering[10] and to vibrational band contours in gases under conditions of low resolution. The results of Houston and Lewis[11] in 1933 implied that the spherically symmetric methane molecule has no pure rotational Raman spectrum, in agreement with theory; the rotation–vibration lines of the C—H stretching vibration at $\Delta\nu = 3022$ cm^{-1} were photographed after a hundred hours' exposure, but the pure rotational lines were absent. Ammonia was shown to be a symmetric top, and the first rotational Raman investigations of hydrocarbons other than methane were also made.

In the years between 1935 (when Teal and MacWood[16] observed rotational Raman transitions in hydrogen and deuterium) and 1951 (when Andrychuk[17] obtained the rotational Raman spectrum of fluorine), no further papers appeared in the field of rotational Raman spectroscopy. These lean years seem quite inexplicable in view of such a promising start. However, the improvement of experimental techniques prepared the way for the revitalization of rotational Raman spectroscopy that resulted from the work of Stoicheff, Welsh and others in the 1950's. The period 1935 to 1951 was not completely 'inert' with regard to the investigation of the Raman spectra of gases generally, since Cabannes[18] and Nielsen[19] reported the vibrational Raman spectra of gases under conditions of low resolution.

In the 1950's the application of the Toronto-arc, a low-pressure mercury arc source, together with high-resolution grating spectrographs and multi-reflection cells, to the study of rotational and rotation–vibration Raman spectra, resulted in great success. Consistent advances in instrumentation, design of equipment and in techniques of detection gave access for the first time to the rotational Raman spectra of weakly scattering molecules. A comprehensive review of the rotational and rotation–vibration Raman work of this period has been given by Stoicheff.[20] An impression of the types of molecule investigated is given by reference to Table 6.2. In addition

TABLE 6.2. *Pure rotational Raman spectra*

Reference	Observer	Molecule	B_0 cm^{-1}	I_0, $\times 10^{-40}$ g cm^2.
17	Andrychuk	F_2	$0\cdot8828$	$31\cdot70_5$
31	Stoicheff	N_2	$1\cdot9897_3$	$14\cdot067_0$
32	Stoicheff	CO_2	$0\cdot3904_0$	$71\cdot69_3$
		CS_2	$0\cdot1091_0$	$256\cdot54$
33	Stoicheff	H_2	$59\cdot339_2$	$0\cdot4716_8$
		HD	$44\cdot667_8$	$0\cdot6266_0$
		D_2	$29\cdot910_5$	$0\cdot9357_6$
34	Weber	O_2	$1\cdot4378$	—
35	Stoicheff	C_2N_2	$0\cdot1575_2$	$177\cdot69$
36	Stoicheff	C_2H_2	$1\cdot1769_2$	$23\cdot78_2$
		diacetylene	$0\cdot1468_9$	$190\cdot54$
		MeCCMe	$0\cdot1122_0$	$249\cdot46$
37	Stoicheff	$ZnMe_2$	$0\cdot1347_8$	$207\cdot66$
		$CdMe_2$	$0\cdot1140_5$	$245\cdot41$
		$HgMe_2$	$0\cdot1162_0$	$240\cdot87$
38	Stoicheff	C_6H_6	$0\cdot1896_0$	$147\cdot59$
39	Romanko, Stoicheff	C_2H_4	$0\cdot10012$	—
40	Stoicheff	$HC{\equiv}CI$	$0\cdot1062_2$	$263\cdot50$

to the pure rotational Raman spectra, much work was done on rotation–vibration Raman spectra, principally by Welsh and his co-workers. The rotation–vibration Raman spectra of ethane,[41] ethylene,[42] acetylene,[43] methane[44] and cyclopropane[45] were thus recorded.

Two more recent landmarks were the use of a photoelectric recording Raman spectrometer to record rotation–vibration bands in gases by Yoshino and Bernstein,[21] and the application of the newly discovered laser source[22] to high-resolution studies of gases.[23] The development of the laser as a Raman source will be discussed in the following section of this article.

C. Experimental arrangement and advent of laser sources

Experimentally the investigation of the Raman spectra of gases has been fraught with difficulties because of the intrinsically low intensity of Raman scattering. When the rotational features are to be observed, under conditions of high resolution, then the following criteria must be satisfied:

(i) Fairly low pressures of gas must be used to minimize line broadening (although the Raman scattering is proportional to the concentration of scattering species).

(ii) Spectrographs (or spectrometers) of high resolving power are essential (but these are of low light-collecting power).

(iii) A light source of very high intensity must be used to excite the Raman spectra, in conjunction with efficient collecting optics.

(iv) The spectral line-width of the exciting line must be as small as possible.

(v) An efficient detection system, whether photographic or photo-electric, is required.

In the very early years, Raman sources consisted of batteries of discharge lamps, and the early means of detection was the photographic plate. Gas pressures were increased up to eighty atmospheres in an effort to observe weakly scattering species. Exposure times were commonly of the order 20–100 hours for pure rotational spectra, and an order of magnitude greater for the even weaker rotation–vibration spectra. The major advances in instrumentation and apparatus that were so successfully applied to the rotational Raman studies of the 'fifties' consisted of the following:

(a) The development of low-pressure, water-cooled, mercury arcs of low continuum and high intensity by Welsh *et al.*,[24] Stoicheff,[25] Claassen[26] and Woodward.[27]

(b) The use of multi-reflection Raman cells, whereby the Raman scattering collected could be increased by as much as 40-fold using concave mirrors.[25, 28]

(c) The construction of high-resolution spectrographs, such as that of Stoicheff with a 21-ft concave grating and the faster Littrow spectrograph at the University of Toronto.

The molecules listed in Table 6.2 were investigated with apparatus such as that described above. Few attempts were made to use discharge lamps other than those of the mercury-arc type, and this imposed a practical limit of resolution of about 0·2 cm^{-1}, the half-width of the natural mercury line. With the advent of the laser, its application to the field of Raman spectroscopy was quickly realized. The characteristics of lasers that distinguish them from other radiators are their very high powers, extreme directionality, polarization and small spectral line-width. The pulsed ruby laser, discovered by Maiman *et al.*[22] in 1960, was employed by Porto and Wood[29] to excite the Raman spectra of liquids. Koningstein[30] also demonstrated the superiority of the c.w. He—Ne laser as a Raman source.

An intriguing situation developed when the new laser excitation was applied to the studies of gases; the narrow line-width of the laser emission was favourable for high-resolution studies of rotational Raman spectra, but the very coherence and narrowness of the laser beam created new problems in the small quantities of gas that could be illuminated for observation of the excitation of the rotational Raman spectra. Thus, again, the field of high-resolution Raman spectroscopy was denied its opportunity for expansion, until cell-design met the requirements of the laser source. It was not until 1965 that the attempts at laser-excitation of rotational Raman spectra were successful, when Weber and Porto[23] used a He—Ne laser to produce the pure rotational spectrum of methylacetylene. Only very recently, then, have papers been published in which lasers have been used for the excitation of high-resolution spectra of gases.

With the discovery of more lasers operating in different regions of the visible spectrum, a great new advantage of the laser as a source for the excitation of Raman spectra becomes apparent. For the first time the Raman spectroscopist is now able to select critically an exciting line to suit his requirements, be these power or wavelength, in the visible region. Laser physicists are still a long way from constructing the perfect laser, which should have reasonably high powers in quite a few lines, be stable, and be reasonably inexpensive to maintain. At present, there is a large selection of lasers available commercially. However, although the intensity of scattering in the Raman effect is proportional to ν^4, where ν is the frequency of the exciting radiation, the selection of a laser line of high ν (i.e. in or near the ultraviolet) is not always advantageous 'chemically'; such radiation initiates decomposition in many molecules that would be of interest to the high-resolution spectroscopist. For this reason, a blue or green line represents a useful compromise; the argon-ion laser, with c.w. radiation at 4880 and 5145 Å, provides two such lines.

A critical comparison of excitation with a laser and with a mercury-arc source is outlined in Table 6.3. Although the scattering intensity relative to Hg 4358 Å is in all cases lower for laser-excitation, the two factors which in themselves advocate the use of lasers as Raman sources are:

(i) The very much smaller sample volume needed for laser-excitation. Thus, small quantities of expensive, isotopically-substituted molecules may be studied. By this means, the amount of information obtained from the rotational Raman spectrum is increased.

(ii) The smaller spectral line-width of a laser, relative to the mercury-arc

TABLE 6.3. *Comparison of lasers and arcs for the excitation of Raman spectra of gases*

	Laser			
	He—Ne	Ar^+	Pulsed ruby	Hg arc
Light source	Gas	Gas	Solid-state	Vapour
Exciting line(s), Å	6328	4880 5145	6923	4358 (2537, 5416)
Power in line				
(a) outside cavity	80 mW	$1\frac{1}{2}$ W	10–100 MW	
(b) inside cavity	1 W	50–70 W	—	2 W
No. of sources needed	1	1	1	Battery of 2–4 lamps and light furnace
Line-width, cm^{-1}	0·05	0·15	0·1*	0·20–0·25
Volume of sample irradiated	———	1–20 ml	———	1–5 l
Multi-reflection cell	———	Unnecessary	———	Necessary
Relative scattering intensity (due to ν^4 law)	0·23	0·64 0·52	0·14	1
Maximum moment of inertia, I_{max} resolvable with line-width given, $\times 10^{-40}$ g cm^2	1300	450	900	300

* Normal line-width of a ruby laser equipped with a Q-switch and a sapphire resonator; with a Fabry-Perot etalon for mode-selection line-widths as small as 0·005 cm^{-1} could be achieved.

Hg 4358 Å line, makes accessible the study of 'heavy' molecules, in which the rotational lines are very closely spaced. Where the laser line-width is the limiting factor for spectral resolution, a 'maximum' moment of inertia, I_{max}, may be calculated. For molecules having an $I_o > I_{max}$ rotational features will not be resolved using the laser-excitation in question. At the foot of Table 6.3 is given the approximate value of I_{max} for each particular method of excitation.

There are, however, factors other than the width of the exciting line which can affect the resolution of rotational features in the Raman spectrum. Broadening of Raman lines can arise from two sources: Doppler broadening and pressure broadening. Doppler broadening can be estimated from the expression:

$$\Delta\nu_{1/2}/\nu_0 = 7 \times 10^{-7}(T/M)^{1/2}$$

where $\Delta\nu_{1/2}$ is the width at half-height of a spectral line of frequency ν_0 (in wavenumbers), and the sample of molecular weight M is at absolute temperature T. For benzene or carbon disulphide the Doppler line-width is approximately 0·04 cm^{-1} at room temperature, and for dimethylacetylene (which has one of the closest line spacings yet investigated in the rotational Raman effect) it is 0·06 cm^{-1}.

Pressure broadening is important, especially in cases where the resolution of lines closer than 0·4 cm^{-1} apart is to be achieved[36]. The effects due to pressure broadening are rather difficult to estimate at lower pressures, about 0·1 atmosphere, where most of the high-resolution work with heavy molecules would be done. Several recent investigations of pressure broadening at high pressures have been made by Jammu[46] and others, and 'broadening coefficients' have been calculated. Stoicheff found that the broadening of the rotational lines of N_2 is linear with pressure up to 25 atmospheres, and the 'broadening coefficient', $\Delta\nu_{1/2}/p$, is 0·043 cm^{-1} $atmosphere^{-1}$. For isotropic scattering the broadening is less than for anisotropic scattering, and for the latter the broadening decreases with increasing rotational quantum number. A detailed impact theory of Raman scattering has been proposed by Fiutak and Van Kranendonk[47]; both elastic and inelastic collisions contribute additively to the line width for anisotropic scattering, but isotropic scattering is affected only by inelastic collisions. A more detailed exposition of the collision broadening of Raman lines is given in the next chapter.

A few general remarks should be made concerning the use of high-resolution grating spectrographs or spectrometers with laser excitation. The three parameters, resolving power, speed (i.e. light-collecting power) and signal intensity, should be considered. Spectrographs (or spectrometers) of sufficiently high resolving power to utilize the extremely narrow spectral line-width of the He—Ne laser could be constructed, but the optical arrangement would be such that to obtain the spectrum under conditions of high resolution would be a slow process. The luminosity of such a spectrograph would have been decreased to the extent that the signal would be perhaps only one-twentieth of its intensity under less stringent conditions of resolution. Other parameters that warrant consideration include laser stability, the stability of the gas sample under laser irradiation, and thermostatting problems for constant-temperature control over very long periods of time. Laser stability is not merely an academic problem of maintaining the desired power levels for long 'exposures'; it is also an economic one, since laser plasma-tubes have a finite lifetime. Rather special rooms are required, ideally, to minimize the effects of temperature ambience or fluctuation on the spectrometer and detector. Artificial 'line broadening', for example, on a photographic plate, can be caused by expansion or contraction of metal parts in the spectrograph. Temperature has some smaller effects on the detection system, too, and increase in grain size on the photographic plate or an alteration in the dark current of a sensitive photomultiplier can result from temperature ambience.

Resolution at the detector may, of course, be increased by using the grating (or gratings) in a higher order. Most modern Raman spectrometers incorporate double monochromators, being two-grating instruments with a very high discrimination against stray light intensity. The gratings are frequently used in higher orders to achieve the necessary resolution. However, signal intensity decreases markedly as one proceeds to the higher orders of diffraction of a grating, and selection of the required order is dictated by the blaze-angle of the grating. Further, a change from blue to red excitation can provide a useful increase in dispersion at the detector. Thus, Weber *et al.*[49] obtained the pure rotational Raman spectrum of methylacetylene with He—Ne-laser excitation in the ninth order of a diffraction grating, giving 2·0 cm^{-1} per mm at the photographic plate. This dispersion was 30 per cent greater than that achieved in the blue region (Hg 4358 Å), assisting materially towards the resultant quality of the rotational spectrum.

Apart from the effects of external parameters on detector efficiency, it is of interest to examine the detectors themselves. Basically, the detection system for high-resolution Raman spectroscopy may be photographic or photoelectric. With respect to the former, a photographic plate is selected which is sensitive at the wavelength of the laser line, or, what is more important, at that of the spectroscopic feature being investigated. The problem here is that the sensitivity of photographic plates falls off with increasing wavelength, and special 'sensitization' processes must be used to record spectra photographically in the red region of the spectrum. An account of the difficulties inherent in photographing Raman spectra in the red is givenby Stammreich *et al.*[48]. Although red-sensitive plates and film are available commercially (e.g. the Eastman Kodak 103 a-E and Ia-E series), for high-resolution purposes one must be careful to ensure that the resolving power (in line pairs mm^{-1}) has not been sacrificed for increased speed. Increased speed of the photographic plate can be effected in several ways, but a common result of most sensitizing procedures is an increase in grain size—and this is most undesirable for high-resolution work. Both Rao *et al.*[37] and Weber *et al.*[49] have reported on certain methods whereby the sensitivity of plates is increased without apparent loss of resolving power, the former author for the mercury blue region and the latter for He—Ne-laser excitation at 6328 Å.

In recent years Raman spectroscopists have been provided with a very sensitive detection system, namely, a photomultiplier coupled to pulse-counting equipment. A pulse-counting detection system for the photo-electric recording of Raman spectra was first reported by Porto.[50] With this system a very high signal-to-noise ratio is attainable by discrimination against random noise levels and by cooling the photomultiplier tube to reduce the dark current. Although red-sensitive photomultipliers are available, their sensitivity falls off as the wavelength increases towards the red. Thus, such a photomultiplier is, typically, some three times more sensitive at 4880 Å than it is at 6328 Å. So it is seen that, whichever method of recording is selected for high-resolution Raman spectroscopy, one is faced with perhaps a three-fold or greater decrease in sensitivity on going from the blue-green into the red-region of excitation. When one also considers that the ν^4 factor of Raman scattering intensity contributes a further decrease by a factor of three or so, for the same shift in spectral region, there is an understandable incentive to work in the blue or green regions of the spectrum.

Mention should also be made of the versatility of laser sources regarding the arrangement of the sample cell. The sample may be placed in a cell which can be irradiated either within the laser cavity itself or outside it. One of the laser cavity mirrors is normally totally-reflecting, and the laser beam emerges from the cavity via a mirror, whose transmission is usually some 3–5 per cent only of the light intensity within the cavity itself. Thus, Raman spectroscopists in their search for enhanced power of incident radiation have conducted experiments within the laser cavity. If the laser beam is focussed on the sample cell within the cavity, then increases in intensity of Raman scattering of 10–100-fold or more have been obtained relative to similar situations with the cell outside the cavity. However, factors such as the alignment and design of the cell within the cavity and the light-absorptive properties of the sample and impurities within it are of great importance for such intensity factors to be realized in practice. The presence of a substance which absorbs at the wavelength of laser excitation, however small that absorption may be, has a great effect on the 'lasing' characteristics. Likewise the cell windows should be designed so that reflection losses at the interfaces are reduced to a minimum, and should be set at the Brewster angle to preserve the linear polarization properties of the laser beam. The manipulation of a focussing lens within the cavity demands much skill. All such sources of potential loss of light intensity should be critically examined in the actual experimental assembly used, lest the 'gain' of the laser should be reduced to such an extent that 'lasing' ceases.

It is in connection with the polarization properties of the laser beam that a very useful application to rotational Raman spectroscopy becomes evident. The bane of rotational Raman spectroscopy has always been the overwhelming intensity of the Rayleigh line. In photographic work this can result in a severely over-exposed portion of the photographic plate, thereby masking much, if not all, of the region of interest. In the earlier experiments using discharge lamps some work was carried out to design suitable post-filters with the characteristics of a very sharp absorption band at the Rayleigh frequency and a 'clear' region on both the Stokes and anti-Stokes sides. Although this idea was fairly successful for helium discharge lamps, no suitable post-filter could be found for mercury lamps. The mercury discharge resonance line at 2537 Å has been used to excite Raman spectra, since the Rayleigh portion of the scattered light could be extensively absorbed by passage through mercury vapour prior to its being focused on the slit of the spectrograph. In experiments using polarized laser excitation a simple solution to the problem is to observe the Raman effect in a direction parallel to the polarization vector in the incident laser beam. The Rayleigh line is due to isotropic scattering, so that, if the scattered radiation is observed in a direction coinciding with the direction of vibration of the electric vector of the exciting laser radiation, the line will vanish to an extent which depends on its depolarization ratio (e.g. $\rho = 0{\cdot}01$ for the Rayleigh line in O_2 or N_2). Rotational Raman lines are depolarized with $\rho = 0{\cdot}75$ for linearly polarized exciting radiation, so some loss of intensity is incurred by viewing their depolarized component. However, the improved spectral quality gained by such a manoeuvre is normally considered well worthwhile.

D. The determination of structural parameters

The excitation and analysis of pure rotational and rotation–vibration Raman spectra of gases can yield information concerning rotational constants and centrifugal-distortion constants in the vibrational ground state and in various excited states. In addition, Coriolis coupling coefficients and the anharmonicity constants in the potential-energy function are accessible.

In the case of a non-polar molecule, i.e. a molecule which does not possess a permanent dipole moment, the pure rotational transitions are not normally observed in the infrared or microwave spectrum. Certain vibrational transitions are also forbidden in infrared absorption, e.g. the symmetric stretching frequency of a centro-symmetric molecule. Even when vibrational coincidences occur in the infrared and Raman spectra, while much information is available from infrared studies of rotation–vibration spectra, the less stringent selection rules of the Raman effect yield rotation–vibration bands which frequently give more information than their infrared counterparts: e.g. the rotation–vibration work on CH_4, and the calculation of the rotational constants for CH_3D from an analysis of the ν_4 Raman band.

Thus, the full power of high-resolution Raman spectroscopy becomes manifest in the study of homopolar molecules and molecules which possess a centre of symmetry. Even in the case of polar molecules, the pure rotational Raman spectrum can sometimes furnish evidence to support the results of a microwave investigation. For example, most observed microwave spectra involve only 'low-J' transitions, which yield highly accurate values of the rotational constants, but the determination of the ground state centrifugal-distortion constants (D) has been carried out in only a few cases. Thus, the rotational Raman spectrum involving some thirty or forty J-transitions is likely to yield more accurate D-values than the corresponding microwave spectrum, in which some six or so J-transitions are normally available for analysis.

Since the perturbation operator for the Raman effect involves the polarizability, and not the dipole moment, as in microwave absorption spectroscopy, significantly different selection rules arise in the two cases. These are summarized in Table 6.4, which gives the more important selection rules

TABLE 6.4. *Selection rules for pure rotational spectra*

Molecule	Microwave	Raman
Linear	$C_{\infty v}$ only $\Delta J = \pm 1$	$C_{\infty v}$ and $D_{\infty h}$ $\Delta J = \pm 2$
Symmetric top	C_{nv} only $\Delta K = 0$ $\Delta J = \pm 1$	All symmetries $\Delta K = 0$ $\Delta J = \pm 1, \pm 2$
Spherical top	Forbidden	Forbidden
Asymmetric top	$\Delta J = 0, \pm 1$	$\Delta J = 0, \pm 1, \pm 2$

for pure rotational transitions in the Raman and in the microwave spectrum. In certain cases the Raman selection rules allow those transitions present in the microwave plus further transitions, but the most important difference between the two sets of selection rules is, as discussed above, the presence of pure rotational Raman spectra for centro-symmetric molecules.

The virtues of high-resolution Raman spectroscopy have so far been extolled, but the technique is less accurate than either microwave or infrared methods. Microwave spectral absorptions in the region 12,000 to 30,000 MHz can be measured to an accuracy of 50 KHz, i.e. 1–5 p.p.m. The position of each rotational Raman line can normally be measured to within about 0·02 cm^{-1}, which is a full order of magnitude worse than the accuracy obtained in the infrared. Accordingly each Raman shift in a pure rotational Raman spectrum is measured with a precision much inferior to that obtained in the corresponding microwave spectrum. This is largely counteracted by the fact that the Raman spectrum consists of many lines (sometimes up to 100 or so), all of which are given by the same few rotational constants. A least-squares analysis is thus normally applied to give constants of an appreciable accuracy: e.g. from the rotational Raman spectrum of benzene and perdeuterobenzene Stoicheff[38] determined the following rotational constants for C_6H_6,

$$B_0 = 0{\cdot}18960 \pm 0{\cdot}00005 \text{ cm}^{-1}$$

$$D_J = (2{\cdot}2 \pm 1{\cdot}0) \times 10^{-8} \text{ cm}^{-1}$$

$$r_0(\text{C—C}) = 1{\cdot}397 \pm 0{\cdot}001 \text{ Å}$$

$$r_0(\text{C—H}) = 1{\cdot}084 \pm 0{\cdot}006 \text{ Å}$$

The main source of error lies not in the experimental accuracy but rather in the assumptions made in calculating the molecular geometry from the moments of inertia.

The uncertainties which arise in connection with the determination of the rotational constants, and therefore in the evaluation of the principal moments of inertia, may be described as follows:

(i) The small deformations due to rotation change the rotational constants and introduce small amounts of potential energy. In the determination of rotational constants from microwave spectroscopy approximations in the rotational Hamiltonian neglect the contribution due to centrifugal distortion.

(ii) The assumption that vibrational and rotational energy terms can be separated is not strictly correct. This is the explanation of the 'inertial defect' noted for planar molecules.

While the source of the first uncertainty can be eliminated, it is not yet possible to solve the second in all cases.

(a) *Isotopic substitution*

Except for diatomic molecules the moments of inertia of one isotopic species are not sufficient to determine the molecular dimensions. However, sufficient information for the complete determination of a molecular structure becomes available if different isotopically substituted species can be studied. The only common reference structure for these different species is the 'equilibrium structure'. Hence, a correct structure calculation can be performed only if we are able to determine *all* the necessary vibrational contributions to the observed, effective rotational constants. Unfortunately, at present our ability in this respect is limited, and only the smallest molecules can be handled correctly.

For example, the case of benzene may be cited. Stoicheff[38] determined the B_0 value from the rotational Raman spectrum. The spectrum of benzene-d_6 was also obtained, enabling both the C—C and the C—H distances to be calculated. The corresponding internuclear distances are rigorously identical in both molecules *only* if the molecules are in their equilibrium configurations, i.e. the bond distances are then r_e values. The determination of B_e, and hence r_e, for benzene is a formidable undertaking; the values of some twenty different constants representing the change in B-value for the different vibrational levels must be known. It should be noted that Stoicheff estimates the error in the assumption of r_0 as the 'equilibrium' distance, instead of r_e, as being within 0·005 Å. As would be expected, the moment of inertia is not as sensitive a measure of the C—H distance as might have been desired (see above).

The main advantages of isotopic substitution are as follows.

(i) The coordinates of the atoms can be obtained individually when isotopic species are available.

(ii) The coordinate uncertainties are independent of mass, so that hydrogen atoms can be located with the same accuracy as a heavier atom.

The main disadvantages are:

(i) Internuclear distances of atoms very close to the axis of rotation cannot be accurately determined, since their squares make a small contribution to the moment of inertia.

(ii) Some atoms have no suitable isotopic variants, e.g. ^{19}F.

(*b*) *Comparison with electron-diffraction methods*

Internuclear distances, as measured by gas-phase electron diffraction, are average values summed over the intramolecular thermal vibrations. Root-mean-square vibrational amplitudes of ordinary bonded or non-bonded atomic distances lie in the range 0·03–0·2 Å. These values are an order of magnitude greater than the uncertainty in current experimental measurements, so a careful consideration of vibrational effects is necessary. For a rigid linear molecule a non-bonded distance obtained from electron-diffraction results should be easily calculated from the bonded distances. Because of the bending vibrations, however, an apparent decrease in the non-bonded distances is observed (the Bastiansen-Morino shrinkage effect).

The internuclear distances obtained from electron diffraction and from spectroscopy (Raman, infrared and microwave) can be compared in terms of an 'average structure'. The small corrections for vibrational effects in electron diffraction necessary for this comparison are based on the quadratic force field. Therefore, gas-phase electron diffraction, combined with a careful vibrational analysis, provides a powerful technique for the determination of molecular structures. The experimental error in the best spectroscopic and electron-diffraction structure determinations is about 0·001 Å or less. Thus, the significance of experimental structural parameters in terms of this precision depends on the asymmetries of the potential functions and upon the method of averaging over the molecular rotation–vibration states (which is appreciably different for electron-diffraction and spectroscopic methods). The problem is relatively simple for diatomic molecules, but sufficiently complicated for polyatomic molecules that in only a few cases have true equilibrium bond lengths been determined by spectroscopic methods. In few electron-diffraction investigations have the results been reduced to equilibrium parameters.

A critical experimental comparison of the diffraction and spectroscopic methods has been made,[51] in which the different types of averaging over the molecular motions are taken into account. A very interesting comparison is made between the diffraction and spectroscopic results for CH_4 and CD_4, and, after suitable corrections for vibrational effects, the equilibrium bond distances, r_e, derived from both methods are seen to be in very good agreement (Table 6.5).

TABLE 6.5. *Comparison of diffraction and spectroscopic bond parameters*

Electron diffraction			Spectroscopy		
C—H distance	CH_4	CD_4	C—H distance	CH_4	CD_4
$r_{average}$ (experimental), Å	$1{\cdot}106_8$	$1{\cdot}102_7$	r_0 (experimental), Å	$1{\cdot}094_0$	$1{\cdot}092_3$
r_e (calc.), Å	$1{\cdot}084_7$	$1{\cdot}086_3$	r_e (calc.), Å	$1{\cdot}085_0$	$1{\cdot}085_6$

E. Theory and examples of rotational Raman spectra

The theory of rotational Raman spectra is well documented in the classic works of Herzberg[52] and Stoicheff[20] and will not be reproduced in full here. Instead the elements of essential theory will be outlined and illustrated by suitable examples of rotational Raman spectra discussed below.

It should be noted, however, that rotational eigenfunctions have important symmetry properties which make them positive or negative with respect to an inversion of all the particles at the origin of the molecular system. For molecules possessing a centre of symmetry the eigenfunctions are symmetric or antisymmetric with respect to an exchange of the identical nuclei. The statistical weights of the symmetric and antisymmetric levels are different and depend on the spin and statistics of the equivalent nuclei. In linear molecules of point group $D_{\infty h}$, if the spins of all the nuclei not at the centre of symmetry are zero, the antisymmetric levels are missing. If only one pair of equivalent nuclei have non-zero spin ($I_n \neq 0$), the ratio of statistical weights of the symmetric and antisymmetric levels is $(I_n+1)/I_n$ if the nuclei follow Bose-Einstein statistics, and $I_n/(I_n+1)$ if they follow Fermi-Dirac statistics.

Rotational selection rules have been derived by Placzek and Teller;[10] for pure rotational spectra the selection rules of totally symmetric vibrations apply. Molecules of cubic symmetry (e.g. CH_4, SF_6) do not exhibit a pure rotational Raman spectrum. Rotational branch intensities of Raman features have been calculated,[10] and deductions made concerning the rotational structure of vibrational Raman bands.[20]

Selected rotational Raman spectra will now be discussed.

(a) *Nitrogen*

For diatomic molecules the rotational energy levels of the ground state are given by,

$$E/hc = F_0(J) = B_0 J(J+1) - D_0 J^2(J+1)^2 + \ldots$$

where E is the rotational energy (in ergs) and $F_0(J)$ is the rotational term value (in cm^{-1}). B_0, the rotational constant for the lowest vibrational level, is $h/8\pi^2 cI_0$, where I_0 is the moment of inertia for this level. J is the rotational quantum number, and D_0 is the centrifugal distortion constant (which is very small compared to B_0). For a vibrating diatomic molecule B_0 should be replaced by B_v, where,

$$B_v = B_e - \alpha(v + \tfrac{1}{2})$$

B_e being the rotational constant for the equilibrium configuration and α the constant which determines the dependence of B on the vibration.

The occurrence of a pure rotational Raman spectrum for linear, non-polar molecules is due to the fact that the polarizability changes during a rotation of the molecule about an axis perpendicular to its internuclear axis. The selection rules are $\Delta J = 0, \pm 2$. The rule $\Delta J = 0$ gives the Rayleigh line and defines the centre of the rotational spectrum; the Stokes and anti-Stokes rotational lines (S-branches) are given by $\Delta J = +2$.

The displacements of the rotational lines from the Rayleigh line are given by

$$|\Delta\nu| = (4B_0 - 6D_0)(J + \tfrac{3}{2}) - 8D_0(J + \tfrac{3}{2})^3$$

A series of very nearly equidistant lines on either side of the Rayleigh line is obtained with a line spacing of approximately $4B_0$, though the separation of the first rotational line from the Rayleigh line is $6B_0$. Precise values of the constants B_0 and D_0 are obtained graphically by plotting $|\Delta\nu|/(J+\frac{3}{2})$ against $(J+\frac{3}{2})^2$, the intercept giving $(4B_0-6D_0)$, and the slope $8D_0$. The $\Delta\nu$ values are obtained by measuring the distance between corresponding Stokes and anti-Stokes lines and dividing by two; this obviates the necessity of determining the exact centre of the intense Rayleigh line.

The pure rotational Raman spectrum of N_2 is illustrated in Figure 6.1. To obtain such a spectrum the gas was contained at a pressure of 10 atmospheres in a specially constructed cell placed within the cavity of an r.f. excited Ar^+ laser. The slit-width was 50 μ at 4880 Å. The strong, polarized Rayleigh scattering was reduced by observing in the direction of the depolarized component.

It is seen that, in contrast to vibrational Raman spectra, almost as many anti-Stokes as Stokes lines are observed. This is readily explained by the large number of molecules which are thermally excited into the various rotational energy levels at room temperature, whereas only the lowest vibrational levels are populated to any extent. A striking feature of the spectrum is the intensity alternation of 2:1 for even:odd lines, which is characteristic of a Bose gas with a ground electronic state of Σ_g^+ and a nuclear spin of 1. The line spacing is approximately 8 cm^{-1}, and the effect of the high pressure, which results in a broadening of the rotational lines giving the imperfectly resolved spectrum, is clearly seen.

The most accurate values of the rotational constants for the ground state of nitrogen are obtained[31] from its Raman spectrum:

$$B_0 = 1{\cdot}9897_3 \pm 0{\cdot}0003\ \text{cm}^{-1}$$

$$I_0 = (14{\cdot}067_0 \pm 0{\cdot}002) \times 10^{-40}\ \text{g cm}^2$$

$$r_0 = 1{\cdot}1000_6 \pm 0{\cdot}0001\ \text{Å}$$

$$D_0 = (6{\cdot}1 \pm 0{\cdot}5) \times 10^{-6}\ \text{cm}^{-1}$$

This analysis was based on the assumption that the observed rotational lines arise from transitions in the vibrationless ground state. There is a vibrational state at 2330 cm^{-1} above the ground state, the population of which is small at room temperature. In a heavier molecule such as benzene, however, there is a degenerate vibrational state which is 404 cm^{-1} above the ground state, and transitions among the rotational levels of this vibrational state may contribute to the intensity of the observed rotational lines. Since the rotational constants for this upper vibrational state are probably only slightly different from those in the ground state, the two rotational spectra will be almost superimposed. Dependent on the values of the rotational constants and the thermal population in the upper vibrational state, the observed rotational lines may be broadened and shifted slightly relative to

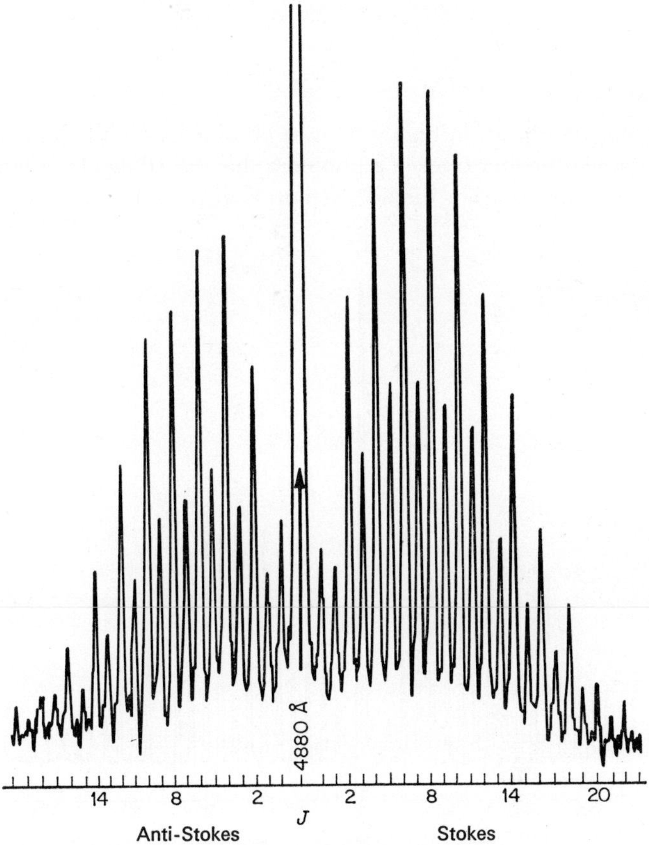

FIGURE 6.1. *The pure rotational Raman spectrum of* N_2.

their ground-state situation. Thus, any difference in the rotational constants in both states will lead to an apparent B value which is not exactly B_0, but a weighted average of B_0 and B_1.

In order to calculate B_e, and hence the equilibrium internuclear distance r_e, α_e must be known; this can be evaluated from a plot of B_v versus $(v + \frac{1}{2})$. For nitrogen, in the absence of sufficient data concerning B_v, an accurate value of α_e is not available. An approximate value of α_e was obtained from electronic data and the following constants for the equilibrium configuration were calculated:[31]

$$B_e = 1{\cdot}9987_4 \pm 0{\cdot}0003_5 \text{ cm}^{-1}$$

$$I_e = (14{\cdot}003_5 \pm 0{\cdot}002_6) \times 10^{-40} \text{ g cm}^2$$

$$r_e = 1{\cdot}0975_8 \pm 0{\cdot}0001 \text{ Å}$$

(*b*) *Carbon dioxide*

The spectrum shown in Figure 6.2 was obtained by W. J. Jones *et al.*;[53] it is a microphotometer trace of a photographic recording of the rotational Raman spectrum of carbon dioxide, at a pressure of 1 atmosphere contained

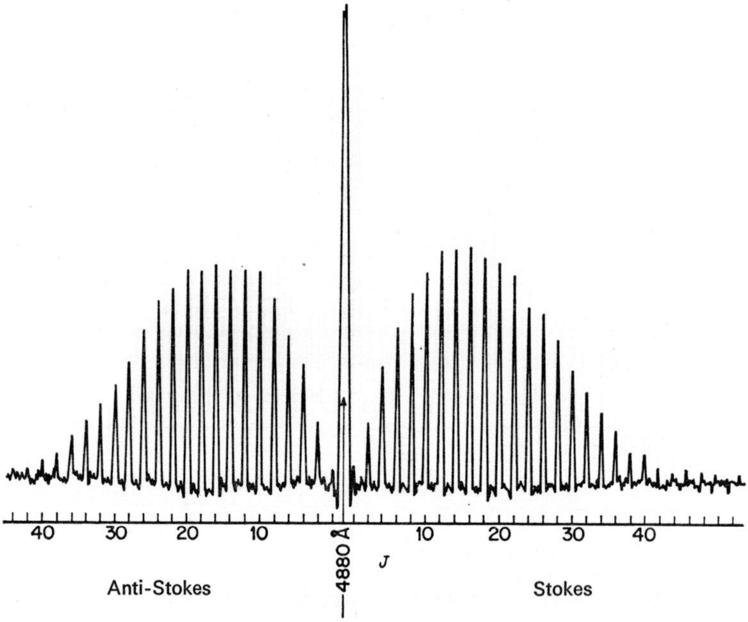

FIGURE 6.2. *The pure rotational Raman spectrum of* CO_2.

in a Brewster-windowed cell within the cavity of an argon-ion laser operating at 4880 Å. The spectrum was resolved in the second order of a 1200 lines mm^{-1} Bausch and Lomb grating spectrograph at a reciprocal linear dispersion of 12·4 cm^{-1} per mm, with an exposure time of about an hour. The line separation is approximately 3 cm^{-1}.

The CO_2 molecule has a centre of symmetry (point group $D_{\infty h}$) and ^{16}O has a nuclear spin of zero. It therefore obeys Bose-Einstein statistics and alternate rotational levels are missing. Since the ground electronic state is $^1\Sigma_g^+$ it is the antisymmetric rotational levels which are 'missing' and only lines of even J-value are observed.

The expression for the displacement of the rotational lines from the Rayleigh line is,

$$|\Delta\nu| = (4B_0 - 6D_0)(J + \tfrac{3}{2}) - 8D_0(J + \tfrac{3}{2})^3$$

The first rotational line is $6B_0$ from the Rayleigh line, and thereafter the line-spacing is approximately $8B_0$. The rotational constants were evaluated in the usual way[32] giving,

$$B_0 = 0{\cdot}3904_0 \text{ cm}^{-1}$$

$$I_0 = 71{\cdot}69_3 \times 10^{-40} \text{ g cm}^2$$

$$r_0 = 1{\cdot}162 \text{ Å}$$

(c) *Oxygen*

In oxygen alternate rotational levels are likewise missing because of the zero nuclear spin of the ^{16}O atom. Only lines of odd K-value are present since the ground electronic state is $^3\Sigma_g^-$ (because the ground state has electronic angular momentum, the quantum number K, or N, is used to designate the rotational levels). The line-spacing is $8B_0$, as in the case of carbon dioxide, but the first line is $10B_0$ from the exciting line. The rotational line separation is $\sim 11{\cdot}5$ cm^{-1}.

Oxygen was first investigated by Rasetti,[4] and later by Weber,[34] who obtained the following rotational constants:

$$B_0 = 1{\cdot}4378 \text{ cm}^{-1}$$

$$B_e = 1{\cdot}4457 \text{ cm}^{-1}$$

$$D_0 = 5{\cdot}6 \times 10^{-6} \text{ cm}^{-1}$$

During a further investigation of the rotational Raman spectrum of oxygen in 1966, Jammu *et al.*[46] noted the presence of weak satellites on either side of the $K = 1$ Stokes and anti-Stokes lines. At higher K-values the satellites diminish rapidly in intensity and are no longer observable at $K = 7$. The satellites arise from transitions between the J-sublevels of the rotational states. For each K-value there are three substates corresponding to the three values of the total angular momentum, $J = K$, $K \pm 1$, and the rotational Raman selection rule $\Delta K = 2$, $\Delta J = 0, 1, 2$ applies, i.e. each K-line is a triplet.

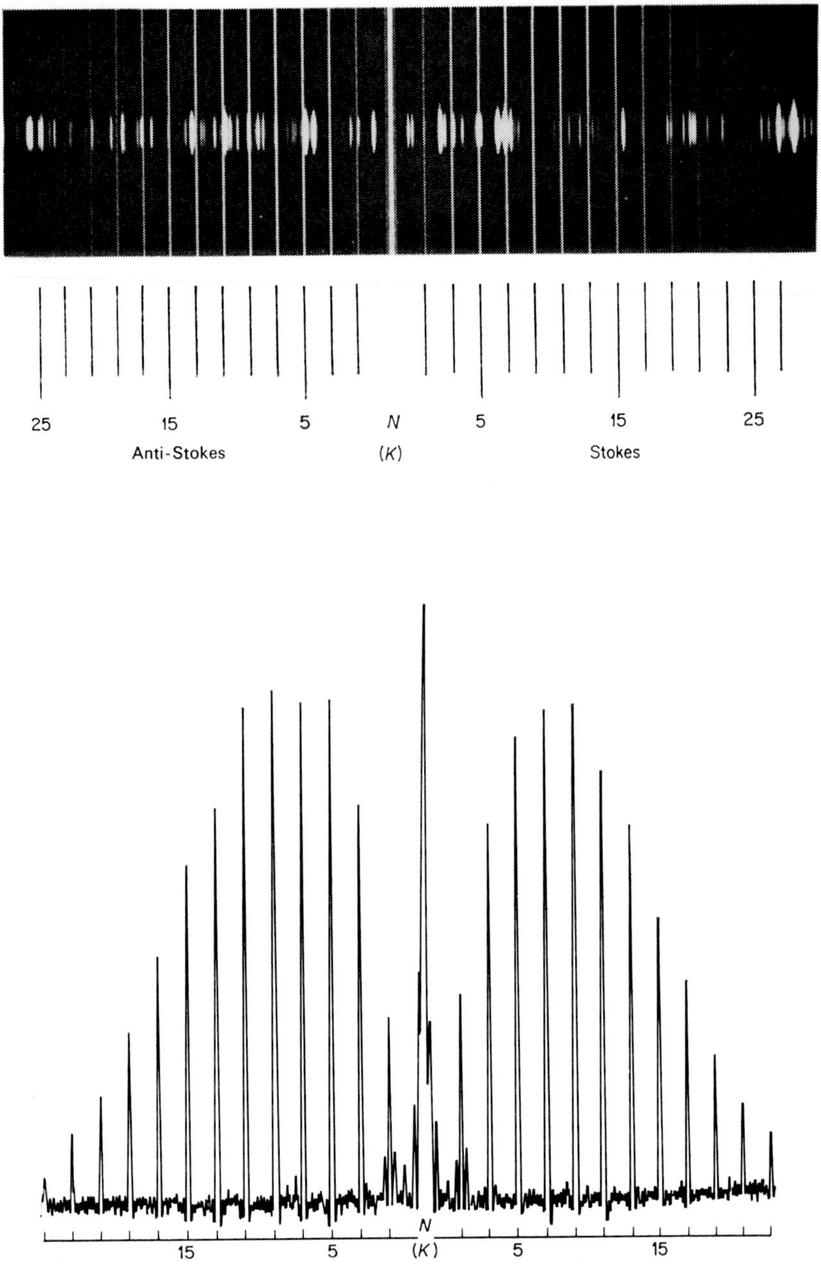

FIGURE 6.3. *Pure rotational Raman spectrum of* O_2*: (a) plate; (b) microphotometer trace.*

Figure 6.3 shows a recent rotational spectrum of oxygen obtained under conditions of high resolution by Jones and Shotton.[54] The oxygen was at a pressure of one atmosphere in a flow cell, which was mounted inside the cavity of an argon-ion laser giving 8 watts power at 4880 Å. The slit-width was 0·3 cm^{-1} and the exposure time two hours. The triplet structure of the K-levels (designated N-levels in the figure) is clearly seen for low K-values; also two transitions are observed very close to the laser line (about 1·98 cm^{-1} distant), which correspond to a change in the J-sublevel for no change in rotational angular momentum.

(*d*) *Fluorine*

Fluorine provides a good example of the power of rotational Raman spectroscopy. The molecule does not exhibit a spectrum in the infrared, nor in the microwave, and the absorption spectrum in the ultraviolet and visible regions is continuous. Thus, no information was previously available concerning the rotational constants of F_2 in the ground state.

The Raman spectrum of fluorine under conditions of high resolution was obtained by Andrychuk[17] in 1951. The line separation is $4B_0$, about 3·5 cm^{-1}, and the intensity alternation for successive rotational lines 3:1. This is consistent with a spin of $\frac{1}{2}$ for the ^{19}F nucleus. From the observation that the odd-J transitions are the strong features, and assuming that the ground state of fluorine is $^1\Sigma_g^+$, it was concluded that the ^{19}F nuclei follow Fermi statistics.

The most recent determination of the rotational constants of F_2 yields the following data:[55]

$$B_0 = 0{\cdot}8841 \pm 0{\cdot}0006\ \text{cm}^{-1}$$

$$D_0 = 3{\cdot}47 \times 10^{-6}\ \text{cm}^{-1}$$

$$r_0 = 1{\cdot}4168 \pm 0{\cdot}0005\ \text{Å}$$

(*e*) *Chloroform*

The microwave spectrum of chloroform has been investigated by several workers and the value of B_0 for the ground state in several isotopic species has been calculated. The hyperfine structure has been resolved by Wolfe[56] and by Long *et al.*,[57] who determined the centrifugal distortion constants D_J and D_{JK}. It was suggested that D_J is negative and J-dependent.

A negative value for D_J is incomprehensible,[58] since it implies that the chloroform molecule is subjected to a centripetal contraction, rather than a centrifugal expansion, when rotating about an axis perpendicular to the top axis. A rotational Raman study was undertaken by Weber,[59] principally to determine the sign of D_J. Some 84 J-transitions were observed on the Stokes side of the exciting line and the following result was deduced,

$$D_J = (6{\cdot}2 \pm 2) \times 10^{-8}\ \text{cm}^{-1}.$$

in good agreement with the theoretical value of $4{\cdot}80 \times 10^{-8}$ cm^{-1}.

(*f*) *Nitric oxide*

Rasetti[6] excited the rotational Raman spectrum of NO in 1930, using the mercury resonance line at 2537 Å. The rotational wings were only partially resolved, but an approximate value for B_0 was obtained. The pure rotational Raman spectrum of nitric oxide has been recorded under conditions of high resolution by Jones and Shotton at Cambridge.[60] The spectra, as illustrated in Figure 6.4, were obtained from vacuum-purified

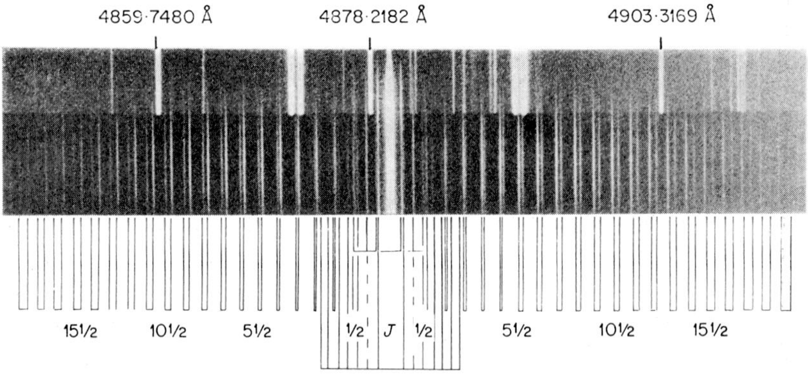

FIGURE 6.4. *The rotational Raman spectrum of NO.*

samples of NO contained in a multi-reflection cell, which was placed inside the cavity of an r.f.-excited argon-ion laser, operating at 4880 Å with a power of 10 watts. The Rayleigh scattering was minimized by observing along the electric vector of the incident radiation. The exposure time was four hours on Ilford Zenith Astronomical plates, with a gas pressure of 600 mm Hg.

The ground electronic state of NO is $^2\Pi$. Spin-orbit interaction splits this into $^2\Pi_{1/2}$ and $^2\Pi_{3/2}$ substates which have different components of the total angular momentum, J, along the internuclear axis. Two series of lines, constituting the R- and S-branches, would be expected for each substate.

The more intense components of the observed S-branch arise from the $^2\Pi_{1/2}$ state, and the less intense components (at a higher wavenumber shift) from the $^2\Pi_{3/2}$ state. The splitting between corresponding S-lines increases with J as a consequence of the slightly different rotational constants in the two substates. The B_0 and D_0 values, calculated using a mean $\Delta\nu$ displacement, are as follows:

$$B_0 = 1{\cdot}6961_4 \pm 0{\cdot}0001 \text{ cm}^{-1}$$

$$D_0 = (5{\cdot}46 \pm 0{\cdot}20) \times 10^{-6} \text{ cm}^{-1}$$

which compare favourably with the values obtained from a high-resolution infrared study,[61]

$$B_0 = 1{\cdot}6960_9 \text{ cm}^{-1}$$

$$D_0 = 5{\cdot}47 \times 10^{-6} \text{ cm}^{-1}$$

A series of R-branch lines, extending some 40 cm^{-1} on each side of the exciting line, are less readily identified. The line intensity decreases with increasing J-value, in agreement with theoretical prediction.

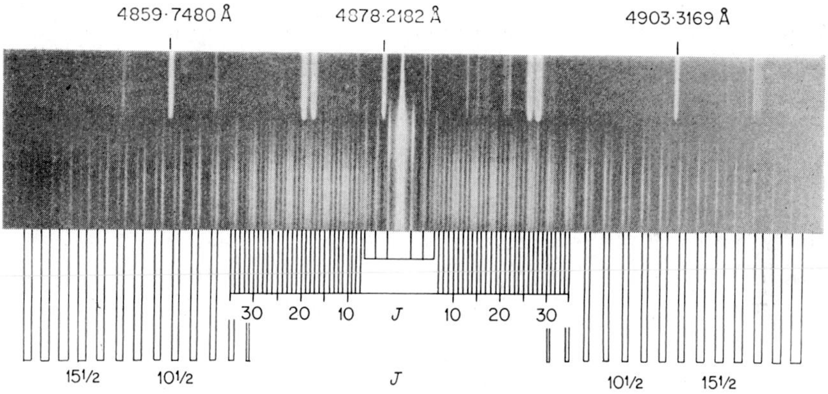

FIGURE 6.5. *The rotational Raman spectrum of NO containing ~1 per cent N_2O impurity.*

A weak feature in the spectrum between $J = 16\frac{1}{2}$ and $J = 17\frac{1}{2}$ of the $^2\Pi_{1/2}$ S-branch has been assigned to the $^2\Pi_{3/2} - {}^2\Pi_{1/2}$ transition, corresponding to a 120 cm^{-1} shift from the exciting line [Q-branch (depolarized) of the electronic transition].

The sample purity was found to be crucial for the observation of Raman scattering. Traces of NO_2, commonly present in nitric oxide, absorb at 4880 Å and thereby diminish the laser power remarkably, especially in intra-cavity work. Traces of nitrous oxide proved troublesome for another reason: Figure 6.5 shows the rotational Raman spectrum of a nitric oxide sample that was contaminated with <1 per cent nitrous oxide. The intensity of Raman scattering from the impurity is so great that its rotational spectrum is clearly observable. Thereby the low-J value region is complicated, and the unwary may even be misled into concluding that the impurity spectrum is an integral part of the rotational spectrum of NO, perhaps even an R-branch!

(*g*) *Cyanogen*

Cyanogen provides an example of a molecule, the rotational Raman spectrum of which could not be completely analysed owing to a deficiency of experimental data. Stoicheff[35] obtained the rotational Raman spectrum of $(C^{14}N)_2$ in 1954, the frequency shifts of the Raman lines relative to the exciting line being given by:

$$|\Delta\nu| = (4B_0 - 6D_0)(J + \tfrac{3}{2}) - 8D_0(J + \tfrac{3}{2})^3$$

The calculated rotational constants were,

$$B_0 = 0{\cdot}15752 \pm 0{\cdot}00015 \text{ cm}^{-1}$$

$$D_0 = (4 \pm 2) \times 10^{-8} \text{ cm}^{-1}$$

To calculate the C—C *and* C—N bond distances it is necessary to have two B_0-values for isotopically substituted cyanogens, e.g. cyanogen-^{13}C or cyanogen-^{15}N. Both ^{13}C and ^{15}N isotopes are quite expensive, and the rather large sample volumes required by Stoicheff tended to preclude the investigation of isotopic compounds other than deuterium analogues. With the coming of laser excitation and the consequent small sampling volumes, such investigations are now within the reach of the high-resolution Raman spectroscopist.

Thus, at Cambridge[62] an investigation into $(C^{14}N)_2$ and $(C^{15}N)_2$ has been carried out. The rotational Raman spectra obtained are shown in Figures 6.6 and 6.7. The spectra were photographed in the third order of a diffraction grating (linear dispersion 4 cm^{-1} per mm) with a slit-width of 0·3 cm^{-1}. Exposure times were about four hours at pressure of $\sim\frac{1}{3}$ atmosphere cyanogen in the intra-cavity cell. The line separation is about 0·45 cm^{-1} and the alternation in line intensity corresponds to that expected from theory, being 2:1 for $(C^{14}N)_2$ and 3:1 for $(C^{15}N)_2$.

A complexity arises in the rotational Raman spectrum of cyanogen because the first excited vibrational state is only 230 cm^{-1} above the ground state. Since some two-fifths of the molecules are in this vibrational state at room temperature, the contribution from these molecules to the rotational spectrum observed for the ground state cannot be neglected. Calculations are presently being made to evaluate the rotational spectra in the light of this complication.

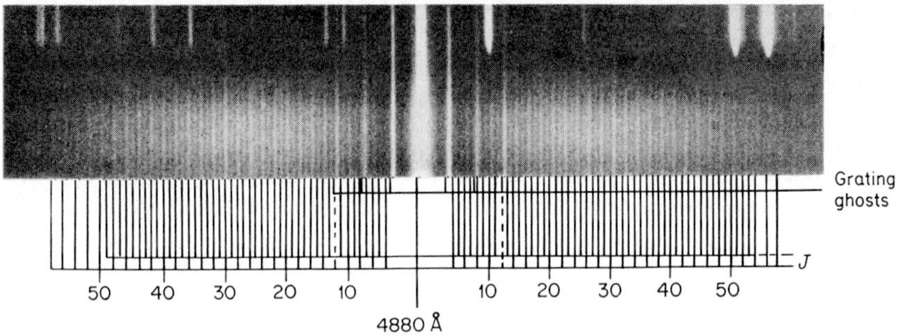

FIGURE 6.6. *The rotational Raman spectrum of* $(C^{14}N)_2$.

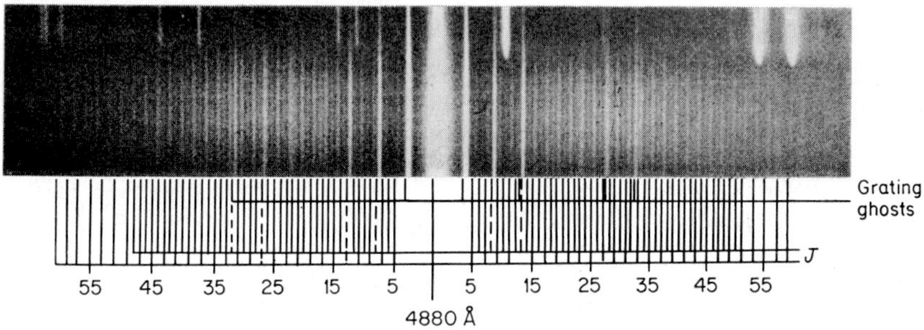

FIGURE 6.7. *The rotational Raman spectrum of* $(C^{15}N)_2$.

F. Future perspectives

It has been seen that the application of lasers to rotational Raman spectroscopy has resulted in an improvement in sample handling and in a greater versatility in the choice of an exciting line. In the near future further advances in laser physics should yield more high-powered lasers and novel exciting lines. For example, a c.w. neodymium-doped YAG (yttrium aluminium garnet) laser, which operates at 1·06 μ, is being developed to give powers in the range of 500 watts. By passing the laser beam through a non-linear frequency-doubling crystal, c.w. radiation at 5300 Å would be obtained with an efficiency of about 3 per cent, giving some 15 watts at 5300 Å. In the ultraviolet, a low powered He—Cd laser has been developed, working at 3250 Å.

Advances of a similar kind are being made in the field of pulsed lasers. High powers (100 MW) are now easily available with Q-switched ruby lasers. With such high powers in short periods of time, the pulsed laser (with a narrow spectral line-width of $<0{\cdot}1$ cm^{-1}) could well be applied to high-resolution gas work. There are also applications in the field of rapid-scanning Raman spectroscopy. Further, with high-energy lasers the spectra of excited molecular states or of short-lived free radicals could be investigated.

The use of laser excitation to observe rotation–vibration spectra has yet to be fully exploited. However, values of B_e, the rotational constant at the potential energy minimum, could be determined, in principle, from measurements of the rotational constants B_v in the excited vibrational states v.

Analysis of rotational Raman spectra can yield three experimental quantities that can be used for the calculation of force constants, namely, vibration frequencies, centrifugal distortion constants and Coriolis coupling coefficients. For a symmetric-top molecule, by combination of Raman and infrared data from the same degenerate band it is feasible that the rotational constants A and B, and the Coriolis coupling coefficient ξ can be calculated.

Finally, if the conditions of resolution are not very high, or if the molecule concerned is relatively 'heavy', the individual rotational transitions will not be observed, but rather a band profile or contour. As in infrared spectra, measurements of band shape and branch separation could yield some useful information on molecular structure and assist the assignment of vibrational bands, for example by the calculation of approximate moments of inertia.

Acknowledgement

The author would like to acknowledge the great help rendered by Dr. W. Jeremy Jones and Dr. K. C. Shotton of the University of Cambridge in making available the spectra reproduced in this article.

References

1. Raman, C. V., and Krishnan, U., *Nature*, 1928, **121**, 501; Raman, C. V., *Indian J. Phys.*, 1928, **2**, 387.
2. Smekal, A., *Naturwiss.*, 1923, **11**, 873.
3. Maclennan, J. C., and McLeod, J. H., *Nature*, 1929, **123**, 1960.
4. Rasetti, F., *Proc. Acad. Nat. Sci.*, 1929, **15**, 515; Rasetti, F., *Phys. Rev.*, 1929, **34**, 367.
5. Wood, R. W., *Nature*, 1929, **123**, 166; *Phil. Mag.*, 1929, **7**, 744.
6. Rasetti, F., *Z. Physik*, 1930, **66**, 646.
7. Houston, W. V., and Lewis, C. M., *Proc. Acad. Nat, Sci.*, 1931, **17**, 229.
8. Bhagavantam, S., *Indian J. Phys.*, 1931, **6**, 319; *Phys. Rev.*, 1932, **42**, 437.
9. Placzek, G., *Handbuch der Radiologie*, 1934, **6**, 205 (Leipzig, Akad. Verlag).

10. Placzek, G., and Teller, E., *Z. Physik*, 1933, **81**, 209.
11. Houston, W. V., and Lewis, C. M., *Phys. Rev.*, 1933, **44**, 903.
12. Dickinson, R. G., Dillon, R. T., and Rasetti, F., *Phys. Rev.*, 1929, **34**, 582.
13. Amaldi, E., *Z. Physik*, 1932, **79**, 492.
14. Amaldi, E., and Placzek, G., *Z. Physik*, 1933, **81**, 259.
15. Bhagavantam, S., *Proc. Indian Acad. Sci.*, 1935, **A2**, 303.
16. Teal, G. K., and MacWood, G. E., *J. Chem. Phys.*, 1935, **3**, 760.
17. Andrychuk, D., *Canad. J. Phys.*, 1951, **29**, 151.
18. Cabannes, J., and Rousset, A., *J. Phys. Radium*, 1940, **1**, 155 and ff.
19. Nielsen, J. R., and Ward, N. E., *J. Chem. Phys.*, 1942, **10**, 81.
20. Stoicheff, B. P., 'Advances in Spectroscopy', Vol. 1 (ed. H. W. Thompson), Interscience, New York, 1959, p. 91.
21. Yoshino, T., and Bernstein, H. J., *J. Mol. Spectroscopy*, 1958, **2**, 213.
22. Maiman, T. H., Hoskins, R. H., D'Haenes, I. J., Asawa, C. K., and Evtuhor, V., *Phys. Rev.*, 1961, **123**, 1145.
23. Weber, A., and Porto, S. P. S., *J. Opt. Soc. Amer.*, 1965, **55**, 1033.
24. Welsh, H. L., Crawford, M. F., Thomas, T. R., and Love, G. R., *Canad. J. Phys.*, 1952, **30**, 577.
25. Stoicheff, B. P., *Canad. J. Phys.*, 1954, **32**, 330.
26. Claassen, H. H., and Nielsen, J. R., *J. Opt. Soc. Amer.*, 1953, **43**, 352.
27. Woodward, L. A., and Waters, D. N. J., *J. Sci. Instr.*, 1957, **34**, 222.
28. Welsh, H. L., Cumming, C., and Stansbury, E. J., *J. Opt. Soc. Amer.*, 1951, **41**, 712; Welsh, H. L., Stansbury, E. J., Romanko, J., and Feldman, T. J., *J. Opt. Soc. Amer.*, 1955, **45**, 338.
29. Porto, S. P. S., and Wood, D. L., *J. Opt. Soc. Amer.*, 1962, **52**, 251.
30. Koningstein, J. A., and Smith, R. G., *J. Opt. Soc. Amer.*, 1964, **54**, 1061.
31. Stoicheff, B. P., *Canad. J. Phys.*, 1954, **32**, 630.
32. Stoicheff, B. P., 'Symposium on Molecular Structure and Spectroscopy', Columbus, Ohio, 1955, 1956; *Canad. J. Phys.*, 1958, **36**, 218.
33. Stoicheff, B. P., *Canad. J. Phys.*, 1957, **35**, 730.
34. Weber, A., and McGinnis, E. A., *J. Mol. Spectroscopy*, 1960, **4**, 195.
35. Møller, C. K., and Stoicheff, B. P., *Canad. J. Phys.*, 1954, **32**, 635.
36. Callomon, J. H., and Stoicheff, B. P., *Canad. J. Phys.*, 1957, **35**, 373.
37. Rao, K. S., Stoicheff, B. P., and Turner, R., *Canad. J. Phys.*, 1960, **38**, 1516.
38. Stoicheff, B. P., *Canad. J. Phys.*, 1954, **32**, 339.
39. Romanko, J., Feldman, T., Stansbury, E. J., and McKellar, A., *Canad. J. Phys.*, 1954, **32**, 735; Dowling, J. M., and Stoicheff, B. P., *Canad. J. Phys.*, 1959, **37**, 703.
40. Jones, W. J., Stoicheff, B. P., and Tyler, J. K., *Canad. J. Phys.*, 1963, **41**, 2098.

41. Romanko, J., Feldman, T., and Welsh, H. L., *Canad. J. Phys.*, 1955, **33**, 588; Shaw, D. E., and Welsh, H. L., *ibid.*, 1967, **45**, 3823; Lepard, D. W., Shaw, D. E., and Welsh, H. L., *ibid.*, 1966, **44**, 2353.
42. Feldman, T., Romanko, J., and Welsh, H. L., *Canad. J. Phys.*, 1956, **34**, 737.
43. Feldman, T., Shepherd, G. G., and Welsh, H. L., *Canad. J. Phys.*, 1956, **34**, 1425.
44. Stoicheff, B. P., Cumming, C., St John, G. E., and Welsh, H. L., *J. Chem. Phys.*, 1952, **20**, 498; Feldman, T., Romanko, J., and Welsh, H. L., *Canad. J. Phys.*, 1955, **33**, 138; Herranz, J., and Stoicheff, B. P., *J. Mol. Spectroscopy*, 1963, **10**, 448.
45. Mathai, P. M., Shepherd, G. G., and Welsh, H. L., *Canad. J. Phys.*, 1956, **34**, 1448; Jones, W. J., and Stoicheff, B. P., *Canad. J. Phys.*, 1964, **42**, 2259.
46. Jammu, K. S., St John, G. E., and Welsh, H. L., *Canad. J. Phys.*, 1966, **44**, 797.
47. Fiutak, J., and Van Kranendonk, J., *Canad. J. Phys.*, 1962, **40**, 9; 1963, **41**, 21.
48. Stammreich, H., *Spectrochim. Acta*, 1956, **8**, 41; Stammreich, H., Forneris, R., and Tavares, Y., *ibid.*, 1961, **17**, 1173.
49. Weber, A., Porto, S. P. S., Cheesman, L. E., and Barrett, J. J., *J. Opt. Soc. Amer.*, 1967, **57**, 19.
50. Porto, S. P. S., *Bull. Amer. Phys., Soc.*, 1966, **11**, 79.
51. Bartell, L. S., Kuchitsu, K., and de Neui, R. J., *J. Chem. Phys.*, 1961, **35**, 211.
52. Herzberg, G., 'Infrared and Raman Spectra of Polyatomic Molecules', Van Nostrand, Princeton, 1945.
53. Jones, W. J., Shotton, K. C., and Carpenter, J., personal communication.
54. Jones, W. J., and Shotton, K. C., personal communication.
55. Claassen, H. H., Selig, H., and Shamir, J., *Appl. Spectroscopy*, 1969, **23**, 8.
56. Wolfe, P. N., *J. Chem. Phys.*, 1956, **25**, 276.
57. Long, M. W., Williams, Q., and Weatherly, T. L., *J. Chem. Phys.*, 1960, **33**, 508.
58. Aliev, M. R., Subbotin, S. I., and Tyulin, V. I., *Optics and Spectroscopy*, 1968, **24**, 47.
59. Weber, A., *J. Mol. Spectroscopy*, 1964, **14**, 53.
60. Jones, W. J., and Shotton, K. C., *Canad. J. Phys.*, 1970, **48**, 632.
61. Keck, D. B., and Howes, C. D., *J. Mol. Spectroscopy*, 1968, **26**, 163.
62. Edwards, H. G. M., Jones, W. J., and Shotton, K. C., unpublished work.

CHAPTER 7

Intermolecular Force Effects in the Raman Spectra of Gases

C. G. GRAY AND H. L. WELSH

A. Introduction

The study of Raman spectra (or indeed spectra of any type) has in principle two uses. One is the determination of the energy levels of the system under study; for molecules in their ground electronic state, this requires ideally the investigation of Raman and infrared spectra of gases at low pressures where interactions are effectively absent. The other is associated with the effects of intermolecular forces and molecular motions on the spectrum. This field has great importance *per se,* since the spectral effects of molecular interactions can lead to a better understanding of the nature and strengths of the forces involved. Also, since many structural investigations have to be carried out with substances in the condensed phases, the effects of interactions on the molecular parameters should be known so that they can be eliminated if necessary.

One of the outstanding problems in present-day science is the structure and molecular dynamics of liquids. One is confronted here with a complicated case of many-body interactions, and it is clear that many different lines of approach may have to be used in solving the problem. Some clarification of the behaviour of molecules in such systems can be gained by studying the spectrum of some molecular species, whose parameters in the free state are accurately known, as a function of increasing density in the gaseous state, preferably both above and below the critical point. A significant amount of experimental data is being accumulated in this field and, when it is fully interpreted, should give important information on many phases of molecular interactions and molecular dynamics.

Raman spectroscopy of gases is in principle perhaps better suited than infrared spectroscopy for the study of the forces between molecules in their ground electronic states. There are two different types of Raman scattering, i.e., isotropic (polarized) and anisotropic (depolarized) scattering, which behave differently in several ways with respect to molecular interactions. Moreover, the advent of lasers which facilitate the excitation of Raman spectra under controlled conditions should make such studies much more profitable. Although Brillouin and Rayleigh scattering will not be treated here, important advances in the investigation of molecular interactions are also being made using these types of spectra.

The sequence of events as the pressure of a gas is increased is, as far as the Raman spectrum is concerned, roughly as follows. At low pressures the rotational and rotation–vibration Raman lines have their natural width, usually with some Doppler broadening superimposed. As the pressure increases the widths of the lines increase because of collision damping. When the broadening of the isotropic scattering (the Q branch components of a totally symmetric vibrational band) leads to overlapping of the separate transitions, the phenomenon of frequency degeneracy begins and the broadening ceases; at higher densities the Q branch shows collisional narrowing. The rotational wings of the vibrational lines and the rotational spectrum itself also show frequency degeneracy, but at higher densities, since the line spacing is greater than that of the Q branch. This effect, often called rotational hindering, can be discussed theoretically most easily in terms of band correlation functions and band moments. Accompanying these changes in line or band shapes there are also changes in the mean frequencies of the Raman transitions. All of these effects are quite complicated since they are produced by various kinds of isotropic and anisotropic† interactions (overlap, dispersion, dipole–dipole, quadrupole–quadruopole, etc.) which have different ranges and are active in varying degrees for the different effects.

In the following sections these intermolecular force effects on the Raman spectra of simple molecular gases will be reviewed. As the 'simplest' of the simple gases, hydrogen will assume a predominant position in the discussions.

† The words *isotropic* and *anisotropic* are used throughout this chapter in two different senses: to designate the light scattering originating in the scalar and traceless-symmetric parts, respectively, of the molecular polarizability tensor (Placzek[1]); and to designate non-dependence or dependence, respectively, of the interaction of a pair of molecules on their orientations.

B. The effect of pressure on line and band shapes

The starting point in modern theories of spectral line and band shapes is the correlation function expression for the spectral shape function $I(\omega)$:

$$I(\omega) = \int_{-\infty}^{+\infty} e^{-i\omega t} C(t)\, dt \tag{7.1}$$

For Raman scattering ω is the shift of the scattered frequency from the exciting frequency, and $C(t)$ is the time correlation function for the appropriate component of the molecular polarizability tensor, α_{ij},

$$C(t) = \langle \alpha_{ij}(0)\alpha_{ij}(t) \rangle \tag{7.2}$$

where the brackets indicate an equilibrium ensemble average. Various formulations of the theory—quantum mechanical, semiclassical, and classical—are found in the literature; detailed discussions of equations (7.1) and (7.2) are given, for example, by Fiutak and Van Kranendonk,[2] Cowley,[3] Gordon,[4], and Gray.[5]

The Raman spectrum and the correlation function for gases consist of two non-interfering components: isotropic $(j=0)$† scattering originating in the isotropic part of the polarizability $\alpha_{ij}^{(0)}$, and anisotropic $(j=2)$ scattering originating in the anisotropic part $\alpha_{ij}^{(2)}$; the antisymmetric part of the polarizability $\alpha_{ij}^{(1)}$ is ordinarily zero. The anisotropic part of $C(t)$ can be expressed in the invariant form,

$$\langle \boldsymbol{\alpha}^{(2)}(0) : \boldsymbol{\alpha}^{(2)}(t) \rangle \tag{7·3}$$

where the double dot expresses a full contraction of the tensor indices. If perturbations of the vibrational motion are neglected, a purely rotational correlation function can be separated off from equation (7.3) if the molecule has sufficient symmetry; for example, for a diatomic molecule, the rotational part of (7.3) can be expressed as

$$\langle \sum_m Y_{2m}(\boldsymbol{u}_0)^\dagger Y_{2m}(\boldsymbol{u}_t) \rangle \tag{7.4}$$

where $\boldsymbol{u}_t$ is a unit vector along the symmetry axis. In the classical limit equation (7.4) becomes

$$\langle P_2(\cos\theta_t) \rangle \tag{7.5}$$

where $\cos\theta_t = \boldsymbol{u}_0 \cdot \boldsymbol{u}_t$, and $P_2(x)$ is the second Legendre polynomial. If vibrational perturbations are neglected, the correlation function and the spectral shape depend directly on the anisotropic intermolecular forces and only indirectly on the isotropic forces, the latter mainly determining the paths followed by the molecules during collisions.

† Here and in what follows, a quantity corresponding to $j=0$ transforms under rotations like a spherical harmonic Y_0, etc.

(*a*). *The impact theory of line broadening*

A correct discussion of the shapes of non-overlapping lines of infrared bands was given by Anderson[6] working within the framework of the so-called classical-path impact theory. In classical-path theories one treats the motions of the centres of mass of the molecules classically and the internal motions quantum mechanically. In impact theories it is assumed that: (a) the collisions are of sufficiently short duration, compared to the time between collisions, that the direct spectral effect of the radiation processes during collisions can be neglected, and (b) successive collisions of the radiating molecule are uncorrelated. In Anderson's theory the effects of both inelastic collisions (lifetime limiting broadening) and elastic collisions (phase shifts and reorientation effects) are taken into account. The theory has been extended to overlapping lines by Baranger[7] and Kolb and Griem.[8]

Fiutak and Van Kranendonk[2] have extended the Anderson theory to Raman scattering and obtained a general expression for the shape of an isolated line. For nonresonant scattering the shape reduces to the Lorentzian form as for absorption lines, but the widths and shifts are in general different for isotropic ($j=0$) and anisotropic ($j=2$) scattering, and different from infrared dipole ($j=1$) absorption lines.† If vibrational perturbations are neglected, corresponding lines of the rotation and the rotation–vibration spectrum have the same shape.

The impact theory is restricted by the basic assumptions to low densities and to intermolecular forces of sufficiently short range. For non-overlapping lines the part of the correlation function corresponding to the line $j_i \rightarrow j_f$ is an exponentially damped harmonic function, and the corresponding line shape is Lorentzian. However, the exponential behaviour of the correlation function is not valid for times which are too short, i.e., less than the average duration of a collision τ_d. The Lorentzian shape is therefore inapplicable for frequencies displaced from the line centre by amounts large compared with τ_d^{-1}.

The half-width and shift of the line $j_i \rightarrow j_f$ are given, respectively, by the real and imaginary parts of the 'optical collision frequency' $n\bar{v}\sigma_{if}(j)$, where n is the number of perturbing molecules per unit volume, $\bar{v}$ is the average relative velocity of a radiating and a perturbing molecule, and $\sigma_{if} = \sigma'_{if}(j) + i\sigma''_{if}(j)$ is the optical cross-section for the transition $j_i \rightarrow j_f$ and for radiation of character j. The optical cross-section is an integral over the impact parameter b,

$$\sigma_{if}(j) = 2\pi \int_0^\infty \sigma_{if}(j;b)b \, db \tag{7.6}$$

† However, a quadrupole spectrum (e.g., the rotation–vibration quadrupole spectrum of hydrogen), since it arises from a $j=2$ radiative element, is broadened in just the same way as the anisotropic Raman spectrum.

where $\sigma_{if}(j;b)$, the differential optical cross-section, is given by

$$\sigma_{if}(j;b) = 1 - p_f \Sigma C(j_i j j_f; m_i m m_f) C(j_i j j_f; m'_i m m'_f) \times \langle j_i m'_i | S(b) | j_i m_i \rangle \langle j_f m'_f | S(b) | j_f m_f \rangle^* \quad (7.7)$$

In equation (7.7), $p_f = (2j_f + 1)^{-1}$, C is a Clebsch-Gordan coefficient, and the summation is over all the m's. $S(b)$ is the classical-path collision operator and is given by

$$S(b) = T \exp\left\{ \frac{-i}{\hbar} \int_{-\infty}^{+\infty} V_I(t)\, dt \right\} \quad (7.8)$$

where T is the Dyson time-ordering operator and $V_I(t)$ is the 'interaction-picture' time-dependent interaction for a collision. Equation (7.7) is valid for monatomic perturbers (e.g. broadening of N_2 lines by Ar), but can be easily generalized for perturbers whose internal states can change during collisions (e.g. broadening of N_2 lines by N_2 or CO).

To compute the optical cross-section given by equation (7.7) for a given anisotropic interaction and a given collision, approximations must be introduced to evaluate the S matrix. The most useful scheme to date has been that of Anderson,[6] in which the Born approximation for $S(b)$ is used for the weak (distant) collisions, and extrapolation methods based on physical arguments are used for small impact parameters.

To proceed further, the nature of the anisotropic intermolecular force must be specified, at least as far as its angular dependence is concerned. We denote by $V_\Lambda(\Lambda \equiv l_1 l_2 l)$ the interaction involving spherical harmonics of order l_1 l_2 and l in the orientations of the radiating molecule, the perturbing molecule, and the intermolecular axis, respectively. In multipole interactions we have $l = l_1 + l_2$, but this relation does not necessarily hold for induction, dispersion and overlap interactions[9, 10]. For simplicity we consider axially symmetric molecules, with the multipolar interaction energy denoted by $V_\lambda(\lambda \equiv l_1 l_2)$, so that V_{11} is the dipole–dipole, V_{12} the dipole–quadrupole interaction, etc. V_λ is given by

$$V_\lambda = 4\pi\epsilon_\lambda(R) \sum C(l_1 l_2 l; m_1 m_2 m) Y_{l_1 m_1}(\boldsymbol{\Omega}_1) Y_{l_2 m_2}(\boldsymbol{\Omega}_2) Y^*_{lm}(\boldsymbol{\Omega}) \quad (7.9)$$

where $l = l_1 + l_2$, and

$$\epsilon_\lambda(R) = \frac{(-1)^{l_2}}{2l+1} \left\{ \frac{4\pi(2l+1)!}{(2l_1+1)!(2l_2+1)!} \right\}^{1/2} (Q_{l_1} Q_{l_2}/R^{l+1}) \quad (7.10)$$

The orientation $\boldsymbol{\Omega}_i \equiv \theta_i\ \phi_i$ of the symmetry axis of molecule i and the orientation $\boldsymbol{\Omega}$ of the intermolecular axis $\boldsymbol{R}$, chosen as pointing from molecule 1 to molecule 2, refer to an arbitrary space-fixed frame. Q_l is the scalar multipole moment of order l for an axially symmetric charge distribution,

$$Q_l = \int \rho(r,\theta) r^l P_l(\cos\theta)\, d\tau \tag{7.11}$$

The form (7.9) for V_λ exhibits explicitly the rotational invariance of the interaction and the time dependence; $\boldsymbol{R}_t \equiv (R_t, \boldsymbol{\Omega}_t)$ is a definite function of the time in the classical-path theory, and is determined by the isotropic part of the potential.

Spherical harmonic expressions for the other anisotropic interactions can also be written down. We remark that if two interactions of types Λ and Λ' are simultaneously present, the optical cross-sections of second order in the interaction are additive if and only if $\Lambda \neq \Lambda'$. Thus, for example, the differential cross-sections add for the various London dispersion interactions, which correspond to $\Lambda = 000$, 022, 202, 220, 222 and 224, but the 224 part interferes with the quadrupole–quadrupole interaction, which also corresponds to $\Lambda = 224$.

Details of calculations of the optical cross-sections are given in refs. 5, 9, and 11. The results are expressed in terms of the elastic and inelastic collision cross-sections for the levels j_i and j_f. These in turn depend on two dimensionless parameters characterizing the collisions: (*i*) a 'strength' parameter, $\gamma = \epsilon_\Lambda(b)\tau/\hbar$, where $\epsilon_\Lambda(b)$ is the strength of the anisotropic interaction, and $\tau = b/v$ is of the order of the duration of the collision; (*ii*) a 'slowness' parameter, $x = \omega\tau$, which is the ratio of the duration of the collision to the Bohr period, ω^{-1}, where ω is the frequency corresponding to the change in the total rotational energy induced by the collision. Collisions in which the changes in rotational energies of the radiating and perturbing molecules exactly compensate each other correspond to $\omega = 0$, and are termed resonant collisions. The inelastic cross-sections for nonresonant collisions decrease quite rapidly with increasing ω.

An interesting feature of the line widths is their J dependence (where $J \equiv j_i$). At large J the widths generally tend to decrease with increasing J, since both the elastic and the inelastic cross-sections tend to zero for large J. The elastic collision effects (phase shifts and reorientations) tend to zero because the anisotropic interaction tends to average out for large J. The inelastic cross-sections tend to zero because the non-resonant collisions give the main contribution at large J; a radiating molecule in a high J state is likely to encounter a perturber in a much lower J state.

In the next two sections the experimental data on Raman line broadening will be reviewed and compared with the theoretical predictions. It is convenient to discuss anisotropic and isotropic scattering separately.

(*b*). *Line broadening in anisotropic scattering*

Early experimental studies of Raman line broadening were essentially qualitative, and the first quantitative results were obtained by Stoicheff[12] and St John.[13] The latter showed that for several simple gases the widths of rotational lines vary linearly with the pressure in the range 10–40 atm and have a pronounced dependence on J. In later investigations Mikhailov,[14] Bazhulin and Lazarev[15, 16] measured the broadening of rotational lines and some vibrational lines of simple molecules; their results have been summarized by Bazhulin.[17] Although these authors found that the broadening is proportional to the pressure, no J dependence of the broadening coefficient was noted, apparently because their results were confined to a narrow range of J values near the maximum of the rotational band. More recently Pinter[18] confirmed the J dependence predicted by the theory but quoted half-widths much larger than those calculated by Van Kranendonk.[11]

A study at higher spectral dispersion (11·2 cm^{-1}/mm) was carried out by Jammu *et al.*[19] for the rotational lines of the nonpolar gases O_2, N_2, and CO_2, and the weakly polar gas CO at room temperature; the O_2 and N_2 lines were also studied with helium and argon as perturbing gases. As examples of the results obtained, the self-broadening coefficient ($\Delta\sigma/p$ in cm^{-1}/atm†) is shown as a function of J for nitrogen and carbon dioxide in Figure 7.1. The broadening coefficient shows a monotonic decrease with increasing J with some irregularities which cannot be ascribed entirely to experimental error.

Figure 7.1 shows also theoretical broadening coefficients for nitrogen and carbon dioxide calculated by Gray and Van Kranendonk[9] assuming quadrupolar and anisotropic dispersion interactions. In view of the approximations made in the calculations the overall agreement of the experimental and theoretical values is satisfactory. It should be noted that the calculations do not involve any adjustable parameters: all molecular constants used in the calculations were obtained from sources other than the broadening experiments. The main error in the theoretical J dependence probably results from the neglect of the anisotropic overlap forces and the use of the classical path formalism for the inelastic collisions.

The experimental values for the $S_0(0)$ and $S_0(1)$ lines of hydrogen shown in Figure 7.1 were obtained by Cooper *et al.*[20] in the experiment on Dicke narrowing discussed below. The theoretical values calculated by Van Kranendonk assuming only quadrupolar interaction are in excellent agreement. The smallness of the coefficients is a consequence of the large separation of the few levels populated at room temperature; the inelastic

† $\Delta\sigma$ is the *half* width at half intensity.

collisions must be mainly exactly resonant. On the other hand, for N_2 and CO_2 with much smaller level-spacings nonresonant inelastic collisions can make an important contribution to the shortening of the lifetime of the levels and thus to the line broadening.

The theoretical broadening coefficients for hydrogen chloride, shown in Figure 7.1, were calculated by Gray[5] with dipole–dipole, dipole–quadrupole, quadrupole–dipole, and quadrupole–quadrupole interactions taken into account. There are unfortunately no accurate experimental data

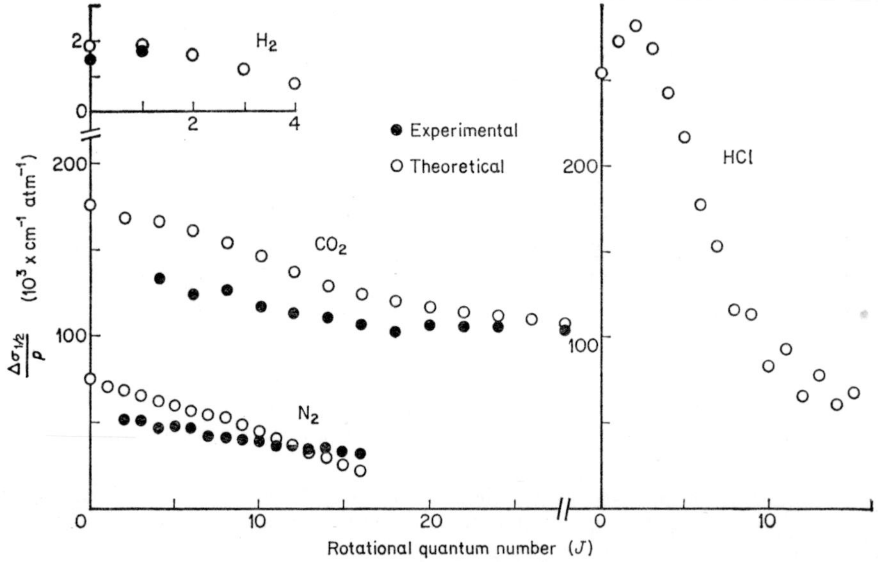

FIGURE 7.1. *Experimental and theoretical broadening coefficients for the rotational Raman lines of some simple molecules at room temperature.*

available for comparison, but some results of Marcoux[21] show order-of-magnitude agreement with the calculations. The strong dipolar interactions give rise to somewhat larger broadening coefficients than for the simple nonpolar molecules. The coefficients show a distinctive J dependence, with a maximum in the region of $J \approx 3$, the most highly populated level at room temperature, and an alternation at high J values. The maximum occurs because the dipole–dipole interaction, which is important for small J values, shows pronounced resonance; a radiating molecule with $J \approx 3$, collides most often with a molecule in the correct level for a resonant collision to occur. For larger J the dipole–quadrupole interaction gives the main contribution. The alternation at high J values is due to the selection rules, $\Delta J_1 = \pm 1$, $\Delta J_2 = \pm 2$, associated with this interaction.

The broadening coefficients for anisotropic Raman scattering are generally of the same order of magnitude as those found in infrared and microwave spectra for the same molecular type. It is clear that, in order to utilize in high-resolution Raman spectroscopy of gases the very narrow exciting lines now available in laser sources, low gas pressures must be used, especially for polar and the heavier nonpolar molecules. This will create some difficulties in fine structure studies since Raman scattering, unlike infrared absorption, is a second-order effect. Nevertheless, the use of lasers in place of high-intensity mercury lamps for excitation should lead to a significant improvement, since the resolution obtained with mercury lamp excitation is effectively limited by the width of the cores of the Hg lines and not by pressure broadening.

(*c*). *Pressure broadening and narrowing in isotropic scattering*

The Q branch of the fundamental band of a totally symmetric vibration usually shows a high degree of polarization; by far the main contribution to the intensity of the branch therefore arises from isotropic ($j = 0$) Raman scattering. The broadening of individual isotropically scattered lines can be studied easily only in the case of hydrogen, for which the lines of the fundamental Q branch are well separated by the rotation–vibration interaction. The widths of the first four Q(J) lines were found by Varghese[22] to vary linearly with the gas density in the pressure range 30 to 600 atm at room temperature; the broadening coefficients ($\Delta\nu_{1/2}$†) are given in Table 7.1. Also given in the table are the theoretical coefficients calculated

TABLE 7.1. *Pressure broadening coefficients of the Q lines of hydrogen*

Line	Frequency, cm^{-1}	Broadening coefficient, $10^3 \times cm^{-1}$/amagat: Experimental	Broadening coefficient, $10^3 \times cm^{-1}$/amagat: Theoretical
Q(0)	4161·1	2·32 ± 0·04	2·2
Q(1)	4155·2	1·40 ± 0·03	0·6
Q(2)	4143·5	2·53 ± 0·03	2·6
Q(3)	4125·9	3·66 ± 0·04	3·6

by Van Kranendonk[11] for quadrupole–quadrupole interaction alone. The widths of the Q(J) lines are significantly smaller than those quoted above for the rotational S(J) lines, since only inelastic collisions contribute to the width of isotropically scattered lines. Physically, this is due to the fact that

† $\Delta\nu_{1/2}$ is the *full* width at half intensity.

the isotropic part of the polarizability is unaffected by changes in the molecular orientation.

For three of the lines the agreement of the experimental and theoretical coefficients in Table 7.1 is excellent; the discrepancy for the Q(1) line may be due to some broadening of the vibrational transition which was not taken into account in the calculation. If it is assumed that the vibrational broadening is mainly a coupling effect proportional to the relative population of the initial state (0·65 for the $J = 1$ state of normal hydrogen at 300 K), then the effect would be small for the other lines and the agreement for these lines would not be destroyed.

The low value of the coefficient for Q(1) as compared with the other lines is an interesting example of the effect of resonance in broadening collisions. Although the $J = 1$ state is by far the most highly populated, inelastic collisions with exact resonance, e.g., (1,3) → (3,1) are very rare. In contrast, the resonance factor in the broadening formula for the Q(0) line has the value 1 for the collisions (0,2) → (2,0), 0·75 for (0,3) → (2,1), and 0·01 for (0,4) → (2,2).

Width measurements for the Q(1) line have also been made by Foltz *et al.*[23] in d.c. field-induced absorption (an induction phenomenon like the Raman effect), and by Lallemand and Simova[24] in stimulated Raman scattering; their values of $\Delta\nu_{1/2}/\rho$, $1{\cdot}77 \times 10^{-3}$ and $2{\cdot}1 \times 10^{-3}$ cm^{-1}/amagat, respectively, are somewhat larger than the spontaneous Raman measurement in Table 7.1. Lallemand and Simova have noted that the width of the Q(1) line, calculated for room temperature by the Van Kranendonk theory,[11] can be given in the form

$$\Delta\nu_{1/2}/\rho = [2{\cdot}48(n_2/n) + 4{\cdot}5(n_3/n) + 3{\cdot}6(n_4/n)] \times 10^{-2}\ \text{cm}^{-1}/\text{amagat} \tag{7.12}$$

where (n_J/n) is the fractional population of the state J, and that this equation is not in agreement with their width measurements with different ortho-para ratios. We suggest that a vibrational broadening term, $\simeq 14(n_1/n) \times 10^{-2}$, implying the vibrational coupling effect discussed above, can explain their results.

For heavier molecules (e.g. N_2) the rotation–vibration interaction is so small that the Q(J) lines are resolvable only at quite low pressures. With increasing pressure the lines quickly overlap and only the width of the whole Q branch can be measured. It might be expected that further increases in pressure would increase the width of the Q branch, but the experimental data show a quite different result. Mikhailov[25] found that for oxygen and nitrogen the width remains constant in the pressure range 15 to 126 atm. More extended measurements by May *et al.*[26] show that the width actually decreases with increasing density after overlapping is complete. Figure 7.2 shows the change in the profile of the Q branch of the fundamental band of nitrogen as the density is increased from 44 to 359 amagat, the latter density corresponding to about 600 atm.; a marked narrowing of the band is evident. In Figure 7.3 the half-width of the Q branch corrected for the finite spectrographic slit is plotted as a function of the density; the figure also shows the similar behaviour of the Q branch of the fundamental band of carbon monoxide, the highest density here corresponding to a pressure of 3000 atm.

The explanation of the collisional narrowing of overlapping lines is briefly the following. The lines overlap when the collision rate, i.e., the

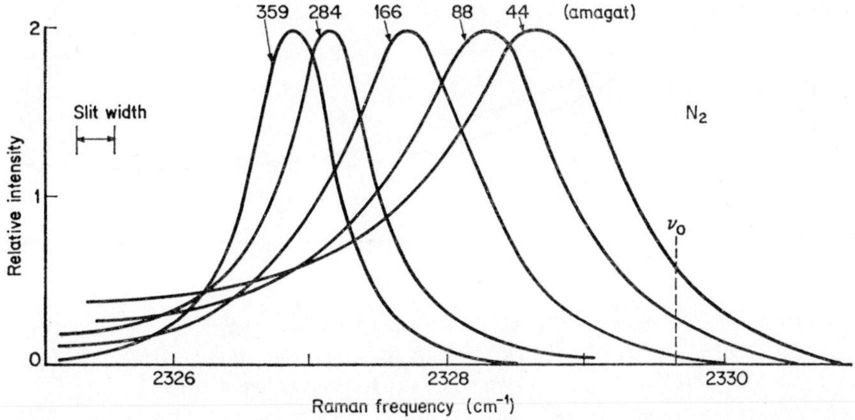

FIGURE 7.2. *Pressure narrowing of the Q branch of the fundamental Raman band of nitrogen at various gas densities at room temperature. The profiles are normalized to the same peak intensity.* ν_0 *is the calculated band-origin.* (*May, A. D., Varghese, G., and Stryland, J. C., personal communication.*)

rate at which a molecule changes its radiative frequency, becomes of the order of the spacing between the lines. In this situation inelastic collisions can change the state of the molecule without changing the radiative frequency; this has been termed by Fiutak and Van Kranendonk[2] the region of frequency degeneracy. When the density is further increased until the collision frequency is large compared with the line spacing, overall narrowing of the band occurs.

The theoretical calculation of the shape of a band of overlapping lines is complicated (cf. for example, Baranger,[7] Kolb and Griem[8]); in addition to the usual difficulties one must solve a secular problem of dimension equal to the number of overlapping lines. The resultant shape is no longer a superposition of unperturbed Lorentzian shapes, and interference effects are important. Detailed calculations of the collision narrowing of Raman Q branches have been made by Alekseev and Sobelman.[27] At high densities the profile of the band acquires a Lorentzian shape with a half-width which varies as the reciprocal of the density. The profiles of the Q branch of nitrogen calculated by Alekseev and Sobelman bear a general resemblance to the experimental profiles in Figure 7.2; however, the calculations were carried out in the impact approximation and may not apply

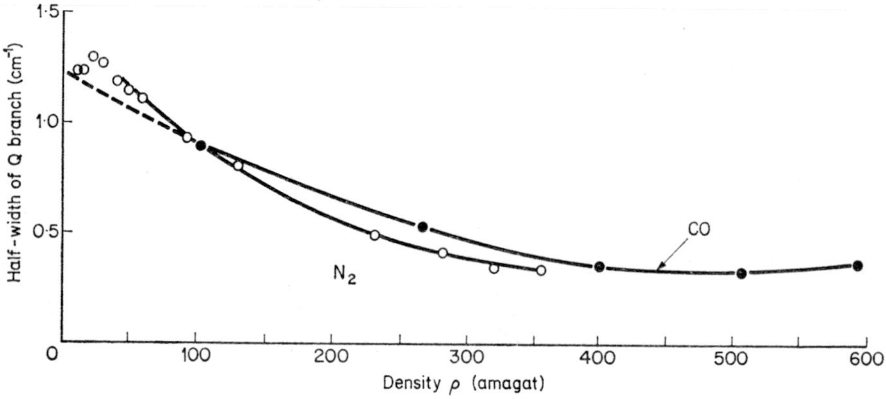

FIGURE 7.3. *The half-width of the Raman Q branch of nitrogen and of carbon monoxide as a function of the gas density at room temperature. (May, A. D., Varghese, G., and Stryland, J. C., personal communication.)*

strictly at high densities. Nevertheless, the curves of Figure 7.2 are not inconsistent with a relation of the form:

$$\Delta\nu_{1/2} = a/\rho + b\rho \qquad (7.13)$$

where $b\rho$ can be construed as collision broadening of the vibrational transition which was not taken into account in the calculation of the collision narrowing term a/ρ.

Collision narrowing of overlapping lines is a general spectral effect and has also been detected in magnetic resonance[28, 29] and in infrared absorption.[30, 31] However, it has a somewhat special significance in Raman scattering, in that for the Q branch of a totally symmetric Raman band it begins at relatively low densities and, at liquid densities, can become quite extreme for nonpolar molecules. The narrowness of such bands has of course been used for the assignment of totally symmetric frequencies in the Raman spectra of liquids. The extreme narrowness which can be attained in some cases has only recently been made evident with laser excitation and interferometer measurement giving a resolving power of the order 5×10.[5]

Thus, Clements and Stoicheff[32] have measured half-widths of 0·067 and 0·117 cm^{-1} for the fundamental Q branches in liquid nitrogen and oxygen, respectively. Such very small line widths have significance in the excitation of stimulated Raman spectra.

(d). Dicke narrowing of the Raman lines of hydrogen

Under conditions for which collision broadening is negligible, a spectral line emitted, absorbed or scattered by a gas whose molecules have a Maxwellian velocity distribution is expected to have a Gaussian profile of definite width due to the Doppler effect. However, it has been shown by Dicke[33] that as the pressure is raised collisions can lead to a narrowing of the Doppler profile provided the ordinary collision broadening is not too large. The physical origin of the effect is rapid frequency switching, as in the collisional narrowing of overlapping lines. In the limit when the mean free path of the molecules is small compared with the wavelength of the light the line acquires a Lorentzian profile with a half-width given by $\Delta\nu_{1/2} = 4\pi D/\lambda^2$, where D is the self-diffusion coefficient of the gas. Since $D = D_0/\rho$, where D_0 is the diffusion coefficient at STP and ρ is the density, the half-width of the line should vary inversely as the density.

It is probable that Dicke narrowing in optical transitions can be easily observed only for hydrogen for which, as pointed out above, the impact broadening is very small. The effect was detected in the infrared quadrupole spectrum of hydrogen by Rank and Wiggins.[34] From stimulated Raman scattering in compressed hydrogen Lallemand *et al.*[35] showed that the line width of the Q(1) line decreases up to about 10 atm. because of Dicke narrowing, but at higher pressures collisional broadening becomes predominant. A careful interferometric study of the widths of the rotational S(0) and S(1) lines in the spontaneous Raman scattering excited by a He—Ne laser was carried out by Cooper *et al.*[36] Figure 7.4 shows a plot of their results for the density range 2 to 40 amagat at room temperature. At the lowest density the width is effectively the Doppler width; as the pressure rises the width decreases rapidly to a minimum at about 10 amagat. At higher pressures the widths increase and can be used to calculate the broadening coefficients given above for these lines.

(e). Effect of intermolecular forces on Raman band shapes

Since the line spacing in the wings of rotational–vibrational Raman bands is greater than in the Q branch, overlapping of the pressure-broadened rotational lines occurs at higher densities. The change of band shape which then ensues is indicative of hindering of the rotational motion by the anisotropic intermolecular forces. The fact that in some simple liquids (O_2, N_2, CH_4) the intensity distribution in the rotational wings of fundamental bands could be approximately fitted by overlapping Lorentzian curves for the individual lines was taken as evidence that molecular rotation is almost free in these liquids.[37] However, even in these cases an anomalously

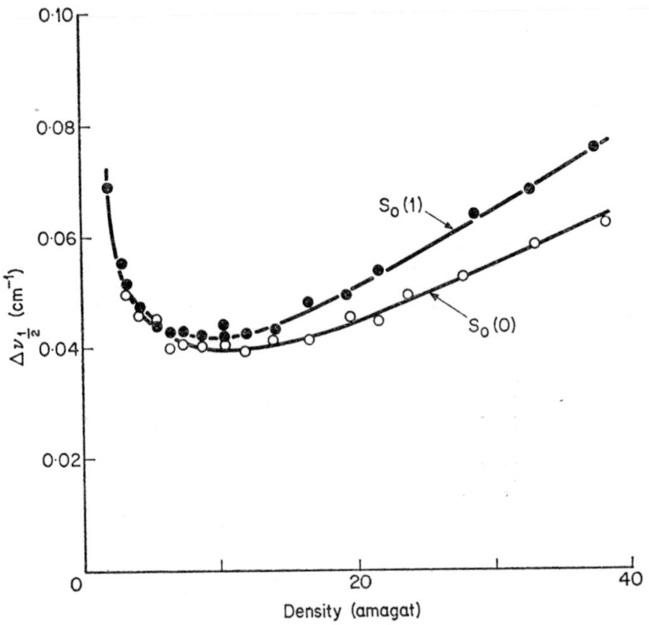

FIGURE 7.4. *Dicke narrowing and collision broadening in the rotational Raman spectrum of hydrogen. The observed half-widths include a small instrumental contribution.* (*Cooper, V. G., May, A. D., Hara, E. H., and Knaap, H. E. P.,* Canad. J. Phys., *1968*, **46**, *2019.*)

high intensity was observed in the centre of the band, and this shows that a type of 'density narrowing' of the band is occurring. Also, it is of course well-known that for more complex molecules even degenerate vibrational bands often appear as moderately sharp lines in the liquid state.

The change in shape of a Raman or infrared band due to rotational relaxation is best discussed in terms of the correlation function introduced in Equation (7.1). The calculation of rotational correlation functions is a problem of current theoretical interest (cf. Steele,[38] Shimuzu,[39] Gordon,[40]

Sears,[41] Fixman and Rider[42]); so far, the discussions have been based mainly on various stochastic models.

Experimental correlation functions are obtained from the experimental band shapes by Fourier inversion of Equation (7.1):

$$\hat{C}(t) = \int_{-\infty}^{+\infty} e^{i\omega t}\, \hat{I}(\omega)\, d\omega \tag{7.14}$$

$\hat{I}(\omega)$ is obtained from the experimental anisotropic band shape by dividing by $(\omega + \omega_0)^4$, where ω_0 is the exciting frequency, and then normalizing to unit area; this normalizes $\hat{C}(t)$ so that $\hat{C}(0) = 1$. Up to now, anisotropic Raman bands of gases have not been analyzed in this way, but some bands of liquids have been discussed (Gordon,[4] Blumenfeld[43]). It is therefore convenient to illustrate the procedures with an example taken from a recent study of some infrared bands by Armstrong *et al.*[31]

Figure 7.5 shows the doubly degenerate ν_3 band of methane perturbed by nitrogen at various densities at room temperature; the redistribution of the intensity with increasing perturber density is evident. The corresponding correlation functions are given in Figure 7.6, along with the classical theoretical correlation function for a free rotator.[38] Only the real part of $\hat{C}(t)$ is shown; the imaginary part is small. It is seen that the experimental correlation functions follow the free rotator function for times $\lesssim 1{\cdot}2 \times 10^{-13}$ s; after this initial period the correlation functions decay to zero, the rate of decay increasing with increasing density.

Gordon[44] has pointed out that rotation hindering can be studied in a simple way by calculating the lower order moments of the band intensity distribution, which are the derivatives of $\hat{C}(t)$ at the origin. The nth moment is defined by

$$M(n) = \int \omega^n\, \hat{I}(\omega)\, d\omega \tag{7.15}$$

If vibrational perturbations and quantum effects are neglected, the mean square torque $\langle \boldsymbol{\tau}^2 \rangle$ acting on a spherical top molecule can be expressed in terms of the second and fourth moments as follows:

$$M(4) - (5/2)M(2)^2 = (2/3I^2)\langle \boldsymbol{\tau}^2 \rangle \tag{7.16}$$

where I is the moment of inertia. In Figure 7.7 are shown the values of $\langle \boldsymbol{\tau}^2 \rangle$ for the methane molecule perturbed by nitrogen, as calculated from band profiles such as those shown in Figure 7.6. Within the rather large experimental error, $\langle \boldsymbol{\tau}^2 \rangle$ is linear in the density of the perturber.

Theoretical calculations of the mean square torque in a low density gas have been carried out for multipolar, induction, dispersion, and overlap

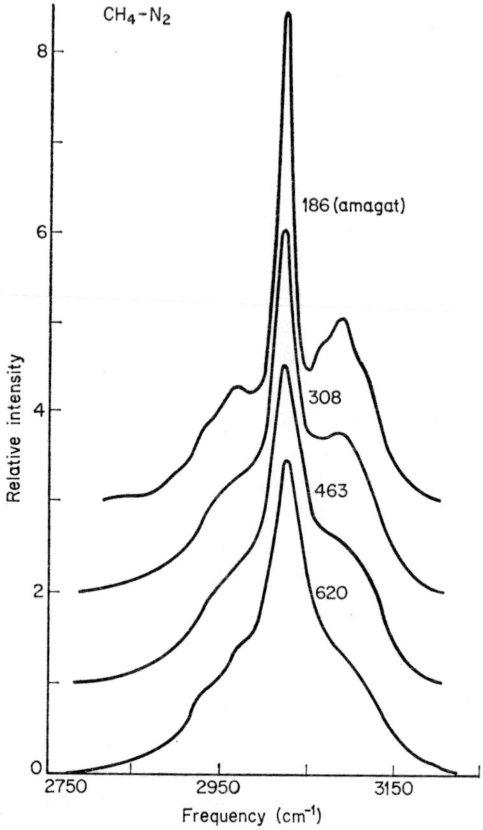

FIGURE 7.5. *The effect of various densities of nitrogen on the shape of the* ν_3 *fundamental infrared absorption band of methane. The profiles have been normalized to equal areas, and the relative intensity scales shifted for clarity of presentation. (Reproduced with the permission of Armstrong, R. L., Blumenfeld, S. M., and Gray, C. G.,* Canad. J. Phys., *1968*, **46**, *1331.)*

interactions by Armstrong *et al.*[31] and Gray.[45] For example, for the multipole interaction of type λ, $\langle \boldsymbol{\tau}^2 \rangle$ is given by

$$\langle \boldsymbol{\tau}^2 \rangle = l_1(l_1+1)n\sigma^3(A_\lambda Q_{l_1} Q_{l_2}/R^{l+1})^2 I_{2l+2}(T) \tag{7.17}$$

where n is the number of perturbing molecules per unit volume, σ is one of the Lennard-Jones parameters, and A_λ is a numerical constant,

$$A_\lambda = (-1)^{l_2}\left[\frac{(2l)!}{(2l_1+1)!(2l_2+1)!}\right]^{1/2} \tag{7.18}$$

The $I_n(T)$ terms are dimensionless temperature-dependent integrals,

$$I_n(T) = \int e^{-\beta V_0}(\sigma/R)^n \, dR/\sigma^3 \tag{7.19}$$

V_0 being the Lennard-Jones potential; these integrals have been evaluated and tabulated by Buckingham and Pople.[46] The theoretical value of the

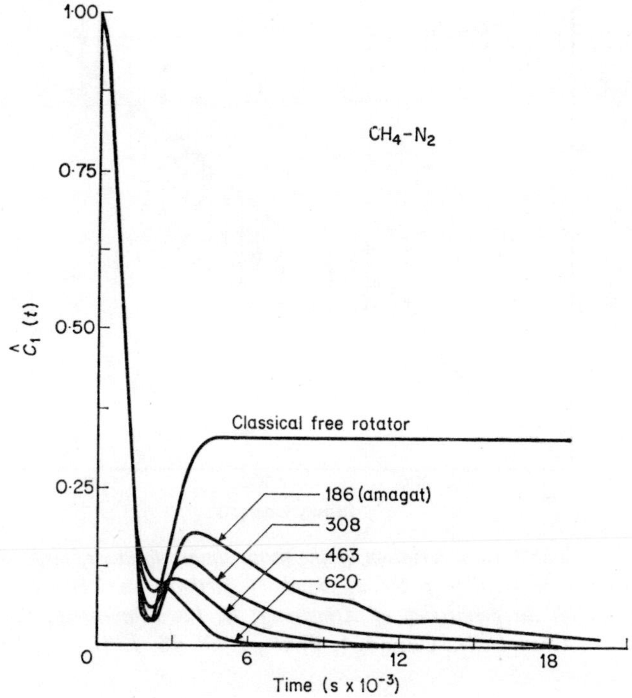

FIGURE 7.6. *The effect of various densities of nitrogen on the rotational correlation function of methane as calculated from the profiles of the ν_3 band in Figure 7.5. (Reproduced with the permission of Armstrong, R. L., Blumenfeld, S. M., and Gray, C. G.,* Canad. J. Phys., *1968*, **46**, *1331.)*

torque for CH_4—N_2 due to the octupole–quadrupole interaction is about half the experimental value; the remainder is probably due to the higher-order multipole interactions and to the overlap interaction.

The investigation of infrared and Raman band shapes in compressed gases is clearly a promising field of study of intermolecular forces. Raman spectra are particularly suitable for study since the $\Delta J = \pm 2$ selection rule gives a wider band than in infrared absorption. The anisotropic part of totally symmetric Raman bands can be easily studied with laser excitation by removing the intense isotropic component by a suitable arrangement of polarization filters.

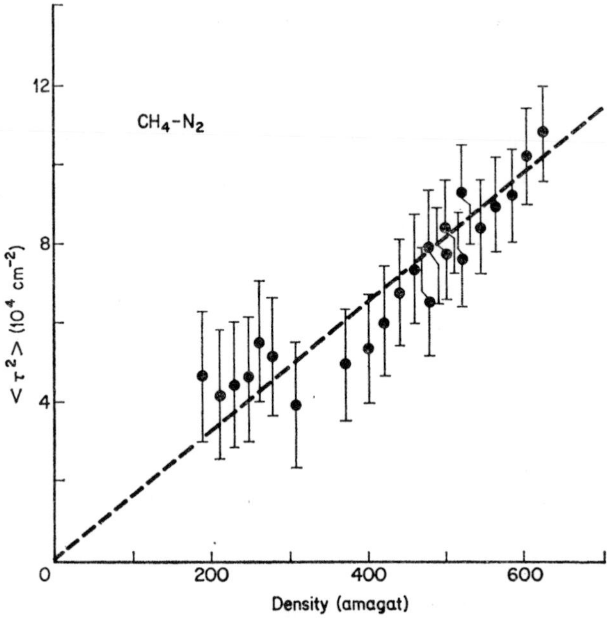

FIGURE 7.7. *The density variation of the mean square torque as deduced from a moment analysis of profiles of the* ν_3 *band of methane in* CH_4-N_2 *mixtures.* (*Reproduced with the permission of Armstrong, R. L., Blumenfeld, S. M., and Gray, C. G.*, Canad. J. Phys., *1968*, **46**, *1331.*)

C. Frequency shifts due to isotropic intermolecular forces

In a practical sense the perturbation of the vibrational frequencies of molecules is perhaps the most important of the spectral effects of intermolecular forces. There have therefore been many attempts to calculate the vibrational frequency change in going from the free molecule to a condensed system by relating it to the properties of the molecule and to the macroscopic properties of the condensed medium, e.g. the Kirkwood-Bauer-Magat formula and the detailed discussion of the so-called solvent shift by Buckingham.[47] The problem is however very complex since it depends in principle not only on the kinds of intermolecular forces operating in a particular case but also on the molecular configurational distribution function. The simplest cases to treat are undoubtedly the low-pressure gas and the pure crystal. If lattice vibrations are neglected the molecular distribution function in the crystal is known but inclusion of lattice vibration effects is usually necessary and complicates the problem. Thus, the structures of the infrared and Raman spectra of solid parahydrogen have been analyzed in detail by Van Kranendonk and co-workers (cf. the review by Van Kranendonk and Karl[48]). At the other end of the scale the investigation of the vibrational shifts in low pressure gases, where the low density pair distribution function can be assumed to hold, has attractive possibilities.

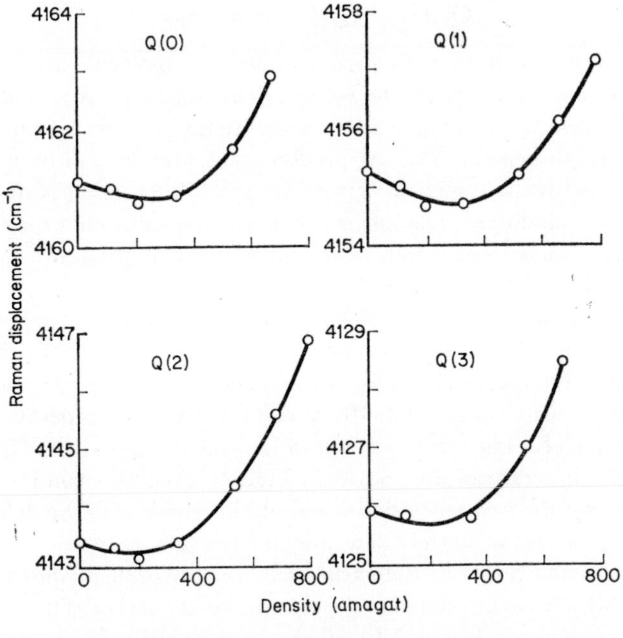

FIGURE 7.8. *Frequencies of the fundamental vibrational Raman lines of hydrogen as a function of the gas density for pressures up to 2000 atm.* (*Reproduced with the permission of May, A. D., Degen, V., Stryland, J. C., and Welsh, H. L.,* Canad. J. Phys., *1961,* **39**, *1769.*)

The Raman spectrum of compressed hydrogen is particularly suited for such studies since the interactions of hydrogen molecules are reasonably well understood. In addition, although the vibrational frequency perturbations are not large they can be studied with good precision even at high pressures since, as has been noted above, the pressure broadening of the Q lines is quite small. A brief review of the experimental results and their interpretation is given below.

The Q branch of the fundamental band was studied by May *et al.*[49] at densities up to 800 amagat (~2000 atm.) at room temperature and, in more detail, up to 400 amagat at 300 and 85 K. Since the widths of the Q lines did not exceed ~1·4 cm^{-1}, at the highest densities, the frequency shifts could be measured to about 0·01 cm^{-1} with a high resolution spectrograph. The behaviour of the Q(*J*) frequencies over the density range up to 800 amagat at room temperature is shown in Figure 7.8. The Raman shifts at first decrease somewhat with increasing density and then increase rapidly at the higher densities. The data can be expressed in a 'virial type' expansion in powers of the density:

$$[Q(J)]_\rho = Q(J) + a_J \rho + b_J \rho^2 \tag{7.20}$$

where Q(*J*) is the Raman frequency of the free molecule and a_J and b_J are *J*-dependent and temperature-dependent constants. It is evident that the negative linear term, $a_J\rho$, is associated with the pressure region in which the attractive dispersion forces are predominant, and the positive quadratic term, $b_J\rho^2$, with the region in which the main interaction is due to the repulsive overlap forces. The linear coefficient is easier to interpret and we shall confine our attention to it in the following; the quadratic coefficient has been discussed by May and Poll.[50]

The behaviour of a_J has been delineated in more detail by lower pressure experiments by May *et al.*[49] and, more recently, by Looi.[51] In the latter investigation the frequency shifts in hydrogen and deuterium were measured in the density range 10 to 100 amagat at various temperatures from 315 down to 85 K. The measurements of Looi for the Q(0) line of hydrogen at various temperatures are shown in Figure 7.9 as an example. In this pressure range the frequency shift is essentially linear with the density and values of a_J can be accurately determined from such graphs.

It was already noted in the earlier experiments that a_J shows a *J* dependence which can be related to the occupational number of the rotational state; this is shown in Figure 7.10, where the data of Looi for a_J are plotted against the fractional population, n_J/n. It is evident that, for a given temperature, a_J has the form

$$a_J = a_i + a_c(n_J/n) \tag{7.21}$$

where a_i and a_c are constants independent of *J*. The value of a_i varies from $-2{\cdot}0 \times 10^{-3}$ to $-8{\cdot}6 \times 10^{-3}$ cm^{-1}/amagat for the temperature range 315 to 85 K; in the following discussion a_i will be interpreted in terms of the vibrational frequency perturbation arising from the isotropic dispersion and overlap forces. On the other hand, a_c is practically temperature-independent with a mean value of $-1{\cdot}7 \times 10^{-3}$ cm^{-1}/amagat; it can be shown that a_c arises mainly from vibrational coupling through the dispersion forces.

A calculation of the perturbation of the vibrational frequency of a nonpolar diatomic molecule by isotropic Van der Waals forces can be made with the impact theory (Gray[5]), but we shall outline here the treatment of May *et al.*[49] based on the so-called statistical theory. The molecule is assumed to be an anharmonic oscillator with an internuclear potential of the form

$$u = f(r - r_e)^2 + g(r - r_e)^3 + j(r - r_e)^4 \qquad (7.22)$$

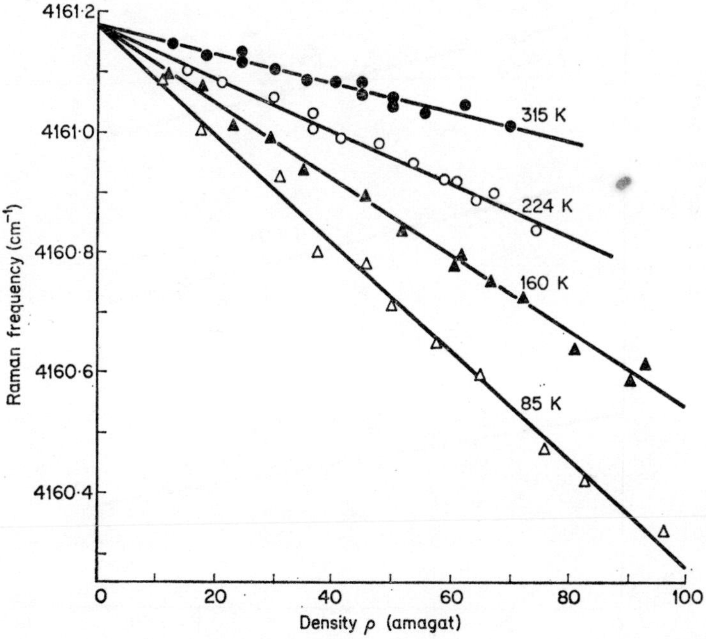

FIGURE 7.9. *Density dependence of the Q (0) Raman frequency of hydrogen as a function of the gas density at various temperatures.* (*Looi, E. C., and Stryland, J. C., personal communication.*)

The isotropic intermolecular potential of a pair of molecules 1 and 2 distant R_{12} apart is taken as

$$V_{12}(R_{12}) = V_{12}^{e} + V_1'(r_1 - r_e) + (\tfrac{1}{2})V_1''(r_1 - r_e)^2 + \ldots$$
$$+ V_{12}''(r_1 - r_e)(r_2 - r_e) + \ldots + V_2''(r_2 - r_e)^2 + \ldots \qquad (7.23)$$

where V_1' is written for $\partial V_{12}/\partial r_1$, etc. The perturbation of the frequency of a Q transition of molecule 1 by the second molecule is found to be of the form

$$\Delta Q = \Delta\omega_e - 2\Delta x_e\,\omega_e = C_1 V_1' + C_2 V_1'' \qquad (7.24)$$

where C_1 and C_2 can be calculated from the known constants of the free

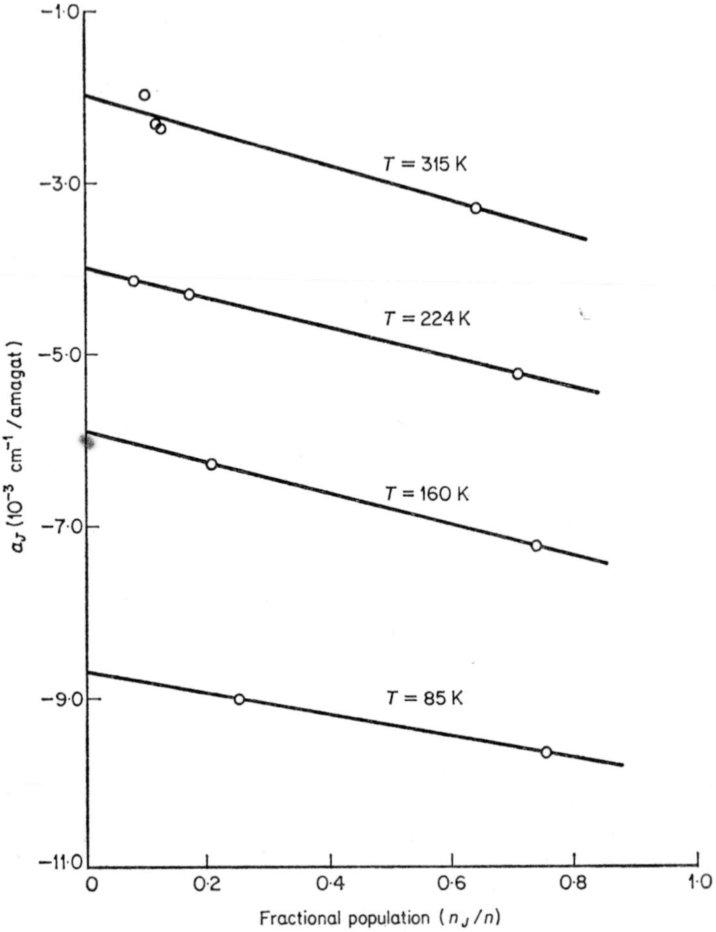

FIGURE 7.10. *Variation of the linear shift coefficient, a_J, with the fractional population of the rotational state, n_J/n, for the vibrational Raman lines of hydrogen at various temperatures. (Looi, E. C., and Stryland, J. C., personal communication.)*

H_2 molecule. If the intermolecular potential is expressed in the Lennard-Jones (6, 12) form,

$$V = \frac{A}{R_{12}^{12}} - \frac{B}{R_{12}^{6}} \tag{7.25}$$

we obtain

$$\Delta Q = \frac{C_1 A_1' + C_2 A_1''}{R_{12}^{12}} - \frac{C_1 B_1' + C_2 B_1''}{R_{12}^{6}}$$

$$= \frac{K_{\text{rep}}}{R_{12}^{12}} - \frac{K_{\text{att}}}{R_{12}^{6}} \tag{7.26}$$

Thus, ΔQ can be expressed as the difference of two terms, one due to the repulsive overlap forces and the other due to the attractive dispersion forces, and characterized by the parameters, K_{rep} and K_{att}, respectively.

The mean value of the shift is obtained by summing the ΔQ over all the perturbing molecules, multiplying the result by the molecular distribution function, and integrating over all configurations. This amounts to weighting equation (7.26) with the pair distribution function, $g(R_{12})$, and integrating over R_{12}. The result is that the linear shift coefficient a_i can be expressed as:

$$a_i = K_{rep} I_{12} - K_{att} I_6 \tag{7.27}$$

where I_{12} and I_6 are temperature-dependent integrals as in equation (7.19) above.

The parameters, K_{rep} and K_{att}, can be obtained from experimental values of a_i for two different temperatures, and the extensive data of Looi are consistent with the values, $K_{rep} = (3{\cdot}58 \pm 0{\cdot}11) \times 10^{-90}$ cm^{11} and $K_{att} = (3{\cdot}83 \pm 0{\cdot}08) \times 10^{-45}$ cm^5.

Thus, the repulsive part of the linear shift, $a_i(\text{rep}) = K_{rep} I_{12}$, is positive, and the attractive part, $a_i(\text{att}) = -K_{att} I_6$, is negative. The values of $a_i(\text{rep})$ and $a_i(\text{att})$ for four different temperatures are given in Table 7.2.

TABLE 7.2. *The linear shift coefficient*, $a_i = a_i\ (\text{rep}) + a_i\ (\text{att})$

Temperature K	a_i 10^{-3} cm^{-1}/amagat	a_i (rep) 10^{-3} cm^{-1}/amagat	a_i (att) 10^{-3} cm^{-1}/amagat
315	−2·0	21·8	−23·8
224	−4·0	19·0	−23·0
160	−5·9	17·1	−23·0
85	−8·6	15·2	−23·8

As might be expected, $a_i(\text{rep})$ is strongly temperature-dependent and $a_i(\text{att})$ is practically constant in the range 83 to 315 K. The smallness of a_i at 315 K arises from the near cancellation of the repulsive and attractive contributions. At a somewhat higher temperature the cancellation would be exact, and at still higher temperatures the repulsive term would be predominant, producing an overall 'blue' shift. Since with modern methods experiments can be carried out at quite high temperatures, the Raman vibrational shifts in hydrogen might be used to explore the form of the repulsive part of the intermolecular potential, the study of which is difficult by other means.

The parameters, K_{att} and K_{rep}, obtained from the analysis of the compressed gas data, can be used to evaluate the frequency shift due to the isotropic interactions in solid hydrogen. For this purpose equation (7.26) can be summed over the crystal lattice of hydrogen as follows:

$$\Delta Q_i = \sum_{\text{lattice}} \left\{ \frac{K_{rep}}{R_{12}^{12}} - \frac{K_{att}}{R_{12}^{6}} \right\}$$

$$= \frac{12{\cdot}13}{d^{12}} K_{rep} - \frac{14{\cdot}45}{d^{6}} K_{att} \qquad (7.28)$$

where the lattice sums are appropriate for hexagonal close packing and the lattice constant, d, is 3·75 Å. The calculation gives

$$\Delta Q_i = 5{\cdot}9 - 20{\cdot}5 = -14{\cdot}6 \text{ cm}^{-1} \qquad (7.29)$$

The experiments of Soots, Allin, and Welsh[52] on the Raman spectra of samples of solid hydrogen with varying ortho-para ratios give

$$\begin{aligned} \Delta Q(0) &= -8{\cdot}7 - 2{\cdot}65C_p \\ \Delta Q(1) &= -9{\cdot}9 - 2{\cdot}65C_o \end{aligned} \qquad (7.30)$$

where the terms in the fractional para or ortho concentrations, C_p and C_o, represent coupling shifts. The isotropic shifts, −8·7 and −9·9 cm^{-1}, are somewhat less negative than that calculated in equation (7.29) from the gas data. The discrepancy undoubtedly arises from the use of equation (7.28) which assumes that the molecules are fixed on the lattice sites, and thus neglects their zero-point motions. The result is that the role of the attractive forces is weighted too heavily, and too great a negative shift is obtained. The effect of the zero-point motion is difficult to calculate accurately; however, the considerations of May and Poll[50] indicate that this explanation for the discrepancy is a plausible one.

Acknowledgements

The authors are indebted to Professor J. C. Stryland, Professor A. D. May, Dr. G. Varghese, and Dr. E. C. Looi for permission to quote some unpublished data, and to the Canadian Journal of Physics for permission to reproduce Figures 7.4, 7.5, 7.6, 7.7, and 7.8.

References

1. Placzek, G., *Marx Handbuch der Radiologie*, 1934, **6**, 209.
2. Fiutak, J., and Van Kranendonk, J., *Canad. J. Phys.*, 1962, **40**, 1085; 1963, **41**, 21.
3. Cowley, R. A., *Proc. Phys. Soc., London*, 1964, **84**, 281.
4. Gordon, R. G., *J. Chem. Phys.*, 1965, **42**, 3658; 'Correlation Functions for Molecular Motions', in 'Advances in Magnetic Resonance' (ed. J. S. Waugh), 1968, Academic Press, New York, Vol. 3, p. 1.

5. Gray, C. G., *Ph.D. Thesis*, University of Toronto, 1967.
6. Anderson, P. W., *Phys. Rev.*, 1949, **76**, 647.
7. Baranger, M., *Phys. Rev.*, 1958, **111**, 481, 494; **112**, 855.
8. Kolb, A. C., and Griem, H. R., *Phys. Rev.*, 1958, **111**, 514.
9. Gray, C. G., and Van Kranendonk, J., *Canad. J. Phys.*, 1966, **44**, 2411.
10. Gray, C. G., *Canad. J. Phys.*, 1968, **46**, 135.
11. Van Kranendonk, J., *Canad. J. Phys.*, 1963, **41**, 433.
12. Stoicheff, B. P., *Ph.D. Thesis*, University of Toronto, 1950.
13. St John, G. E., *Ph.D. Thesis*, University of Toronto, 1952.
14. Mikhailov, G. V., *Sov. Phys.-JETP* 1959, **9**, 974; 1960, **10**, 1114.
15. Bazhulin, P. A., and Lazarev, Yu. A., *Optics and Spectroscopy*, 1960, **8**, 106.
16. Lazarev, Yu. A., *Optics and Spectroscopy*, 1962, **13**, 373.
17. Bazhulin, P. A., *Sov. Phys.-Uspekhi*, 1962, **5**, 661.
18. Pinter, F., *Optics and Spectroscopy*, 1964, **17**, 428.
19. Jammu, K. S., St John, G. E., and Welsh, H. L., *Canad. J. Phys.*, 1966, **44**, 797.
20. Cooper, V. G., May, A. D., Hara, E. H., and Knaap, H. F. P., *Canad. J. Phys.*, 1968, **46**, 2019.
21. Marcoux, J., *Ph.D. Thesis*, University of Toronto, 1956.
22. Varghese, G., *Ph.D. Thesis*, University of Toronto, 1967.
23. Foltz, J., Rank, D. H., and Wiggins, T. A., *J. Mol. Spectroscopy*, 1966, **21**, 203.
24. Lallemand, P., and Simova, P., *J. Mol. Spectroscopy*, 1968, **26**, 262.
25. Mikhailov, G. V., *Trudy. Fiz. Inst. Akad. Nauk*, 1964, Pt. XXVII, 50.
26. May, A. D., Varghese, G., and Stryland, J. C., private communication.
27. Alekseev, V. A., and Sobelman, I. I., *Acta Phys. Polon.*, 1968, **34**, 579; *IEEE J. Quant. Elect*, 1968, QE-4, 654.
28. Bloembergen, N., Purcell, E. M., and Pound, R. V., *Phys. Rev.*, 1948, **73**, 679.
29. Anderson, P. W., *J. Phys. Soc. Japan*, 1954, **9**, 316.
30. Vu, H., Atwood, M., and Vodar, B., *J. Chem. Phys.*, 1963, **38**, 2671.
31. Armstrong, R. L., Blumenfeld, S. M., and Gray, C. G., *Canad. J. Phys.*, 1968, **46**, 1331.
32. Clements, W. R. L., and Stoicheff, B. P., *Appl. Phys. Letters*, 1968, **12**, 246.
33. Dicke, R. H., *Phys. Rev.*, 1953, **89**, 472.
34. Rank, D. H., and Wiggins, T. A., *J. Chem. Phys.*, 1963, **39**, 1348.
35. Lallemand, P., Simova, P., and Bret, G., *Phys. Rev. Letters*, 1966, **19**, 1239.
36. Cooper, V. G., May, A. D., Hara, E. H., and Knaap, H. F. P., *Canad. J. Phys.*, 1968, **46**, 2019.

37. Crawford, M. F., Welsh, H. L., and Harrold, J. H., *Canad. J. Phys.*, 1952, **30**, 81.
38. Steele, W. A., *J. Chem. Phys.*, 1963, **38**, 2411; *Phys. Rev.*, 1969, **184**, 172.
39. Shimuzu, H., *J. Chem. Phys.*, 1965, **43**, 2453.
40. Gordon, R. G., *J. Chem. Phys.*, 1966, **44**, 1830.
41. Sears, V. F., *Canad. J. Phys.*, 1967, **45**, 237.
42. Fixman, M., and Rider, K., *J. Chem. Phys.*, 1969, **51**, 2425.
43. Blumenfeld, M., in 'Molecular Dynamics and Structure of Solids' (eds. Carter and Rush, Nat. Bur. Stand., Special Publication 301, 1969), p. 441.
44. Gordon, R. G., *J. Chem. Phys.*, 1963, **39**, 2788; 1964, **40**, 1973; **41**, 1819.
45. Gray, C. G., *J. Chem. Phys.*, 1969, **50**, 549.
46. Buckingham, A. D., and Pople, J. A., *Trans. Faraday Soc.*, 1955, **51**, 1173.
47. Buckingham, A. D., *Proc. Roy. Soc., London*, 1958, **A248**, 169; 1960, **A255**, 32; *Trans. Faraday Soc.*, 1960, **56**, 753.
48. Van Kranendonk, J., and Karl, G., *Rev. Mod. Phys.*, 1968, **40**, 531.
49. May, A. D., Degen, V., Stryland, J. C., and Welsh, H. L., *Canad. J. Phys.*, 1961, **39**, 1769; Varghese, G., *Canad. J. Phys.*, 1964, **42**, 1058.
50. May, A. D., and Poll, J. D., *Canad. J. Phys.*, 1963, **43**, 1836.
51. Looi, E. C., *Ph.D. Thesis*, University of Toronto, 1970.
52. Soots, V., Allin, E. J., and Welsh, H. L., *Canad. J. Phys.*, 1965, **43**, 1985.

CHAPTER 8

Raman Spectra of Alkali Halides

R. S. KRISHNAN

A. Introduction

Raman spectroscopy is a powerful tool for the study of the structure of matter and the physicochemical problems of the crystalline states of matter. However, in the case of cubic crystals, because of the symmetry properties, most and in some cases all the fundamental modes of vibration are inactive in the Raman effect. But they can manifest themselves as overtones and combinations in the second order Raman effect. Being a comparatively feeble phenomenon the second order Raman effect is more difficult to record and to interpret. In spite of these limitations, considerable progress has been made in the study of the Raman spectra of cubic single crystals. In this chapter, it is proposed to give a historical review of studies of alkali halide crystals with special reference to the interpretation of experimental Raman data on the basis of lattice dynamics.

B. Historical background

Using the powerful 2537 Å wavelength mercury resonance radiation for excitation, Rasetti[1] was the first to record the Raman spectrum of a rock salt crystal. He commented on the peculiar structure of the spectrum which consisted of a few lines superposed over a continuum and was different from the normal spectra of single crystals like calcite, quartz, etc. In the photograph reproduced in the paper by Fermi and Rasetti[2] the discrete character of the Raman spectrum of rock salt was quite evident. They attributed the observed spectrum to a second order spectrum as its high frequency limit was roughly twice the known Reststrahlen frequency of rock salt. Because of the non-availability of good single crystals of other alkali halides and the lack of a proper source of exciting radiation, no further work was done during the period from 1931 to 1946.

In 1946 interest in the subject of the Raman spectra of simple cubic crystals was revived by a new atomistic theory of lattice dynamics proposed by Raman.[3] He put forward his theory in order to stress the importance of the small but finite number of lines that are actually observed in the Raman and infrared spectra of simple crystals. On the basis of his theory the possible modes of vibrations of the NaCl and CsCl types of lattices were fully worked out. None of these are allowed to appear in the first order Raman effect, but there was a clear possibility that their overtones and combinations might appear as second order in strongly exposed Raman spectra. I interpreted with some success[4] the frequency shifts of the lines or intensity maxima in the observed Raman spectrum of rock salt which was reproduced in the paper by Fermi and Rasetti,[2] (this being the only available data at that time) as overtones and combinations of some of the fundamental modes of the rock salt lattice as envisaged in the Raman theory. In order to test the predictions of the Raman theory, I developed the Rasetti technique of using the powerful 2537 Å radiation as a source and succeeded in recording the second order spectra of diamond,[5] rock salt[6] and other alkali halides.

The publication of my experimental results on diamond and rock salt prompted Born and Bradburn[7] to apply the Born–von Karman theory[8] to the Raman effect in cubic crystals. Many of the observed features of the spectrum of rock salt were not satisfactorily explained by them. Further experimental work on the Raman spectra of alkali halides using the 2537 Å excitation was also carried out by Menzies and Skinner in England and by Stekhanov and his co-workers in Russia. Studies on the temperature dependence of the intensity distribution and polarization were also carried out.

The dynamics of the alkali halide structures was the subject of a number of treatments using various models for the forces and the dispersion data (frequency versus wave vector) obtained from neutron scattering experiments. Using the van Hove singularities[9] corresponding to certain critical points in the density of states curve and introducing appropriate selection rules for Raman scattering for the rock salt structure worked out by Burstein, Johnson and Loudon,[10] the frequency shifts corresponding to the intensity maxima of peaks or lines in the second order Raman spectra of alkali halides were accounted for satisfactorily. Since the intensity distribution in a second order Raman spectrum is very much dependent on the exact nature of the phonon spectra, various attempts have been made to evaluate them on the Born theory employing different lattice dynamical models including the shell model; many papers have appeared recently on this subject. The availability of laser excitation for Raman studies in recent years has considerably extended the scope of these investigations.

C. Experimental technique

The powerful 2537 Å wavelength radiation of mercury was found to be highly suitable for exciting the second order spectra of cubic crystals.[1, 5] Its use has many advantages. Firstly, this line has a high intrinsic intensity. Secondly it is a resonance line and can be reabsorbed by mercury vapour. A cell containing a small quantity of mercury placed between the crystal and the spectrograph will completely absorb the exciting radiation and thereby give a clear background on which weak low-frequency Raman lines can be easily recorded. Thirdly the enormously increased scattering power arising from the λ^{-4} law enables one to use small specimens of the crystal, shorter exposure times and a low-dispersion quartz spectrograph to record the spectra. In order to get the maximum output intensity of the 2537 Å radiation it is necessary to use a water-cooled quartz mercury arc in which the discharge is squeezed against the wall of the tube to reduce to the minimum the self absorption of the radiation inside the arc itself.

Since its advent, the laser has been found to be an ideal source for Raman studies. The light from the laser is highly monochromatic, plane polarized, highly directional and very intense. These characteristics make it very easy to measure accurately the polarization properties of the Raman spectra for different crystal orientations and different states of polarization of the exciting radiation. From such measurements the group theoretical symmetries of the Raman active phonons are deduced unambiguously. The laser beam may be focused to a very small spot with very high power density of about 10^3 W cm^{-2}. The laser source makes low temperature measurements easy. This is very important for cubic crystals as the analysis of the second order Raman spectrum is easier if there are no difference bands; these have negligible intensity at very low temperatures.

D. Theoretical considerations

In order to understand the nature of the Raman spectra (first and second order) of crystals, it is necessary to be able to evaluate the vibrations of the lattice and their interaction with electromagnetic radiation. In this section the development of the various theories of lattice dynamics is reviewed.

(a) Born–von Karman theory

The equations for the natural frequencies of vibration of a cubic binary (two atoms per unit cell) ionic crystal were given by Born and von Karman in 1912.[8] The numerical calculations involved in deducing from this the actual frequencies of a particular crystal are very complicated and tedious even when approximations are introduced, and therefore the theory remained dormant for a considerable time. Only with the availability of modern electronic computers, have calculations been carried out on a large number of cubic crystals.

The theory of Born and von Karman regards a crystal as an interacting mechanical system of pN particles where p is the number of particles in the unit cell and N is the number of unit cells in the lattice. The radiation with which the crystal interacts is regarded as being enclosed in a large box, the dimensions of the crystal being very small compared with the box, yet very large compared with the size of the unit cell. Thus the crystal is considered as a large molecule and the Dirac theory of the particle-radiation interaction as given by Placzek is applicable. In order to determine the frequency spectrum, the classical equations of motion for the lattice are set up; the solutions of the independent normal vibrations are plane waves. If the length of the parallelepiped of N cells along any one direction is $2na$ where n is the number of cells in that direction and $2a$ is the lattice constant, then all wavelengths $2na/f$ where f is any integer, are permissible along that direction. These are practically infinite in number and are in the nature of a wave motion in a continuum. The next step in Born's theory is to identify these waves with the $3Np$ normal modes of oscillations which the parallelepiped should exhibit. This is done by assuming a correspondence between the finite number of dynamical oscillations and the infinite number of possibilities in which the wave motion can take place in the medium, and equating them by setting a limit to the wavelength. In effect, the vibration spectrum of a crystal with p nonequivalent atoms in the unit cell has $3p$ branches of continuously varying frequencies out of which ($3p$-3) constitute the optical branches and the remaining 3 acoustic branches. The ($3p$-3) frequencies in the limit $q \to 0$ are normally the ones that are recorded in the first order Raman or infrared spectrum. In the case of alkali halides the degenerate optical mode frequency at $q = 0$ is forbidden to appear in Raman scattering and therefore it is only the overtone which one can expect to record. In the second order spectrum, two phonons are either created or destroyed, the conservation of energy and momentum requiring

$$\omega_{\mathrm{i}} = \omega_{\mathrm{s}} \pm \omega_j \pm \omega_{j'} \tag{8.1}$$

$$\boldsymbol{k}_{\mathrm{i}} = \boldsymbol{k}_{\mathrm{s}} \pm \boldsymbol{q}_j \pm \boldsymbol{q}_{j'} + N\boldsymbol{K} \tag{8.2}$$

where ω_i, ω_s, ω_j, $\omega_{j'}$ denote the frequencies of the incident photon, the scattered photon, and those of the phonons from the branches j and j', $\boldsymbol{k}$ and $\boldsymbol{q}$ are the corresponding wave vectors of the photons and phonons and $\boldsymbol{K}$ is the reciprocal lattice vector. Since photon wave vectors are negligible compared to Brillouin zone dimensions, wave vector conservation for second order scattering requires effectively that the wave vectors of the two phonons, $\boldsymbol{q}_j$ and $\boldsymbol{q}_{j'}$, should be equal and opposite. Therefore the second order Raman spectrum of an alkali halide should be of the nature of a continuum consisting of 21 two phonon branches arising from the original six single phonon branches and extending from the exciting line corresponding to zero frequency shift up to a frequency shift corresponding to the maximum frequency of the lattice vibrations. The intensity distribution in this continuum is proportional to a weighted density of lattice states in which two phonons of equal and opposite wave vector are present. The weighting arises from the frequency and wave vector dependence of the interactions involved in the scattering process.

(*b*) *Methods of solution*

The solution of the Born–von Karman equation was attempted by Kellermann[11] by considering simplifications and a numerical computation of frequencies for a number of special values of the phase differences and thus building up the frequency distribution function for the NaCl lattice. Kellermann's treatment gives us a knowledge of the local values of the distribution function; but to get a fairly good idea of the shape of the curve, one must calculate for an immense number of points in the Brillouin zone. This is extremely tedious. Another method of solution was attempted by Montroll[12] by reducing the problem to that of the moment problem in statistics. He used the matrix theorem that the sum of the nth powers of the characteristic values of a matrix is equal to the trace of the nth power of that matrix. $g(\omega)$ could be stated analytically as a function of the lattice constants of the crystal. Montroll's treatment gives immediately a general idea of the trend of the curve, but local values are unreliable unless one considers a large number of orders. Houston[13] attempted to solve the Born–von Karman equation by calculating the distribution function along selected directions in the reciprocal lattice and interpolating between them. The secular equation for the vibration frequencies of a crystal lattice can be reduced to an equation of order $3p$ in the squares of the frequencies where p is the number of different atoms per unit cell. The solution gives the vibration frequencies in terms of the propagation vectors of the standing waves. These vectors will fill the space in a Brillouin zone of the reciprocal lattice. The solution gives the frequency at any point in the Brillouin zone as a function of the coordinates of the point. The process could be repeated for all points, but the work involved is very great. In order to reduce the labour of numerical evaluation, Houston calculated the distribution of frequencies within narrow cones along a number of specific directions in the crystal and by summing up he obtained the distribution function. He also made use of the symmetry of the crystal and thereby reduced the labour in the case of cubic crystals by a factor of 48. Although Houston's approximation is less laborious to apply, one often gets spurious peaks after the first maximum in the distribution function.

(*c*) *Raman's theory of lattice dynamics*

The experimental observation of a finite number of discrete lines in the Raman spectra of crystals led Raman to postulate a new theory of lattice dynamics. In a series of papers Raman[6, 14] put forward an atomistic theory of the vibrations of crystal lattices. According to him the vibration spectrum of the crystal corresponding to the optical branch is discrete in character unlike the quasi-continuous frequency distribution obtained from the Born dynamics. Unlike the Born theory, the Raman theory does not treat the lattice as a large mechanical system but focuses its attention on a 'super cell' having eight times the volume of the ordinary Bravais cell. In other words, while the theory of Born assumes that the vibration pattern repeats itself in a unit consisting of $N = n \times n \times n$ cells where n is a very large number, the Raman theory considers n to be equal to 2. The basic postulate of Raman's theory is the definition of the normal mode as stated by Lord Rayleigh, i.e., 'all particles in a normal vibration have at any instant the same or opposite phase of vibration'. In a normal mode the atoms in the adjacent cells must vibrate either in the same or opposite phase. Therefore if the ratio of the amplitudes of the atoms in adjacent cells along the three principal directions are α, β, γ, then these can take only the values $+1$ or -1. Also the values of α, β, γ, must be the same for all the non-equivalent atoms in the unit cell. If the two nearest equivalent atoms have the same phase, the third must also have the same phase and the phases of the motion are pictured as +++.... If they have opposite phases, the motion must be like +−+−.... These two possibilities arise for translation along one of the primitive axes. The two other axes are each associated with two such cases and thus the total number of discrete possibilities is 8. If there are p atoms in the unit cell, there are $3p$ equations for the modes for each case. Therefore excluding the three translations, there are only $24p$-3 modes out of which ($3p$-3) modes are invariant to lattice translations, while the $21p$ modes have opposite phases at least along one of the three axes and are therefore forbidden to appear in the Raman effect as fundamentals. Quantitative rules governing the intensities of these modes and their activity in infrared and Raman processes have been given by Raman.[15] The $3p$-3 modes could give rise to a first order Raman spectrum if allowed by the selection rules. In the second order Raman spectra, overtones and combinations of $21p$ normal modes are expected in addition to those of the $3p$-3 modes. Raman has given a physical picture of the $21p$ modes in any cubic crystal. In the case of the rock salt lattice, for example, the $21p$ modes are the oscillations relative to each other of the alternate planes of equivalent atoms in the crystals, the planes being the octahedral and the cubic. Since there are four sets of octahedral and three sets of cubic planes these oscillations acquire four-fold and three-fold degeneracies respectively. A further double degeneracy occurs for the transverse modes and the total number of discrete vibrations is reduced further. Excluding the translations, the total number of modes with distinct frequencies is 9 and 11 for NaCl and CsCl lattices respectively. A normal coordinate analysis of these modes is available from the planar force constant methods of Ramanathan[16] for the different cubic lattices.

On the Raman theory, one can expect 21 overtones and combinations to appear crowded together over a small region of frequency shifts in the second order Raman spectra of alkali halides. My collaborators and I[17] at Bangalore have explained many of the observed features and especially the frequency shifts of the prominent peaks or lines in the second order Raman spectra of rock salt, KBr, KI, CsBr, NaI, etc., as overtones and allowed combinations of some of the ($24p$-3) Raman modes. In most of these cases, the frequencies of the fundamental Raman modes have been evaluated under simplifying assumptions.

(*d*) *Frequency distribution function*

According to the Born–von Karman theory, the frequencies of the normal modes (ω) are related to their wave vectors ($\boldsymbol{q}$) by the dispersion relation

$$\omega = \omega_j(\boldsymbol{q}) \tag{8.3}$$

where the index j signifies a particular branch of the multi-valued function. This function will manifest itself directly in the wavelength shift of X-rays scattered by one phonon processes. Because of its extremely small relative magnitude, this shift is not readily accessible to measurement and hence the function $\omega_j(\boldsymbol{q})$ has to be inferred from the measurements of the scattered intensity. Hence there will be some uncertainty in its estimation.

Placzek and Van Hove[18] showed that in the case of slow neutron scattering, the relative change in energy or wavelength was such that the difficulty of direct measurement of the dispersion relation was considerably reduced.

The energy distribution of neutrons incoherently scattered by a one phonon process is directly connected with the frequency distribution function of the crystal and hence each observed group of scattered neutrons will yield a pair of values of ω and $\mathbf{q}$ belonging to the dispersion relation which can thus be constructed by repeated observation. This was done experimentally for the first time for germanium by Brockhouse and Iyengar,[19] thereby establishing the validity of the Born–von Karman theory of the vibrations of a crystal. By comparing the experimentally observed dispersion curves with those to be expected theoretically from various lattice dynamical models, it is possible to check the validity of any particular model for a crystal.

(*e*) *Lattice dynamical models*

Since second-order Raman spectra are dependent on the exactness of the phonon spectra, it is desirable to have the dynamics of crystal lattices worked out on a realistic model. As indicated earlier, inelastic neutron scattering gave the possibility of observing the dispersion relations with better precision. The data obtained therefrom can be used to check the validity of the theoretical lattice dynamical models and the assumptions made therein. The first model to be used was the rigid ion model proposed by Kellermann.[11] This was shown by Woods *et al.*[20] not to give a satisfactory description of the phonon dispersion curves of alkali halides. Different lattice dynamical models have been employed to explain the phonon spectra of alkali halides. The models differ from each other primarily in the type and extent of interactions between the ions. For example, in an ionic crystal like an alkali halide one has to consider the attractive Coulomb long-range interaction between the ions. Because of its long range the second-order coupling coefficients coming from such a potential were found to converge very slowly and Born and Thomson[21] used the well-known theta-function transformation of Ewald[22] to convert the sum into two rapidly converging sums, one in the crystal space and the other in the reciprocal space. One has also to consider the short-range overlap interaction between the ions which is repulsive and responsible for the stability of the lattice, counterbalancing the attractive Coulomb forces. Different workers introduce this repulsion into the dynamical calculations in different ways, either from well-established potential functions such as the Born–Mayer potential for alkali halides or through the use of parameters which in turn can be determined from experimentally determined quantities such as the elastic constants. Thus, the lattice dynamical models differ primarily in this part of the calculation. During the last few years many refinements have been made by incorporating special features, as for example, the field and distortion polarizabilities.

Detailed work on the vibration spectrum of sodium chloride was carried out by Kellermann.[11] He used the rigid-ion model in which the ions do not get deformed during the lattice vibration and carry the potential field along with them. He treated the ions as point charges and considered the electrostatic interaction between the ions and the short range interaction between the nearest neighbours, by assuming a central potential function whose derivatives were expressed in terms of two parameters which he determined from the compressibility and equilibrium condition. He worked out the dispersion curves for the various symmetry directions. But his model failed to explain satisfactorily even the variation of specific heat with temperature as well as the dielectric properties of alkali halides. Since the ions were treated as point charges without any polarizability, the high frequency dielectric constant of the material in this model should have a value of unity whereas for NaCl, for example, it is 2·25.

Lyddane and Herzfeld[23] calculated the dispersion curves taking into account the polarizability of ions, but they found some unstable vibrational modes for which q is close to the Brillouin zone boundary in the [100] direction. Later Hardy[24] traced the origin of the unstable modes as due to the inappropriate values of the polarizability used by Lyddane and Herzfeld. He showed that the unstable modes can be removed by using the more refined values of Tessman *et al.*[25] But neither theory gave the correct ratio of ω_{LO} and ω_{TO} for $q = 0$, satisfying the famous LST relation.

This difficulty led to the introduction of distortion polarizability by Szigeti[26] and later to the deformation dipole (DD) model of Karo and Hardy.[27] Szigeti has shown that the dielectric constants can be fitted only when we also assume the ions to carry an effective charge e^* which is less than e, in addition to including the electronic polarization mechanism associated with the overlap repulsion. Then, in an undistorted lattice the charge distribution about a given ion has cubic symmetry for alkali halides; but when the lattice gets perturbed by the plane wave, the charge in the overlap regions gets redistributed and this redistribution produces a new contribution to the dipole moment in addition to the polarization due to the effective electrostatic field. This had not been considered either in the rigid ion model of Kellermann or in the later refined calculation of Lyddane and Herzfeld.

Hardy incorporated this contribution by assuming that the deformation dipole varies as a short range potential of the Born–Mayer type. He assumed that it is present only for the negative ion and is situated at the centre of the ion. The parameters representing this contribution were evaluated from Szigeti's second relation using the concept of virtual work. Later Hardy and Karo[28] extended these models to incorporate second neighbour interaction as well and the deformation dipole was assumed to be situated not necessarily at the centre, but at a point on the bond, the distance of which from a given ion was estimated by fitting it to some measured quantities, preferably to the phonon frequencies obtained from neutron scattering experiments.

To account for the observed dispersion curves in NaI, Woods *et al.*[20] indicated that a more satisfactory model was provided by the shell model first proposed by Cochran.[29] Cochran extended the idea of Dick and Overhausser[30] and developed further a more mechanical model known as the shell model where the distortion polarizability is introduced in a more straightforward manner. Both the deformation dipole model and the shell model are based on the principle that the perturbed charge distribution of the lattice can be expressed in terms of the multipole expansion. In both the models the expansion was stopped after the dipole terms and only the dipole–dipole interactions are considered.

Cochran's shell model, however, utilizes a mechanical analogy and thus presents a concrete picture of the physical situation. The physics of the mechanical polarization suggested by Szigeti is presented in a simpler way in the shell model. In alkali halides for example, we have two ions which are nearest neighbours. Their electron clouds overlap and as a consequence the ions repel one another If the ions are deformed by an electric field their electron clouds distort and this distortion leads to an alteration in the repulsion. Conversely, the short range repulsion leads to polarization. Cochran explains this in his mechanical model as follows. Each ion has been replaced by a shell of charge Ye and a core with charge Xe. The net charge of the ion is $(X + Y)e = Ze$. Each shell is connected to its own core by an isotropic spring with restoring force K. This spring constant accounts for the polarizability of the free ion which is given by Ye^2/K. In addition there are springs which connect the two shells. The core of one ion is connected to the core of the other ion by a spring of constant D and the core of one ion is connected to the shell of the other by a spring of constant F. Thus in the shell model we have two cores and two shells instead of the two rigid ions of the alkali halides. Thus there are separate equations for the core and the shell and hence the dynamical matrix will be of dimension double that of the rigid ion model.

For crystals in which the energies of the electronic transitions are much larger than the energies of the lattice vibrations, as is the case for the alkali halides, the well-known adiabatic, harmonic and electrostatic approximations to the equation of motion are adequate. When the ions are displaced, the forces between them are calculated from the change in the potential energy of the crystal. This change arises in part from the distortions of the electric moments produced on the ions. The shell model is a first approximation in which only the dipole moments are included and these are placed at the centres of the ions. Six degrees of freedom, three from the displacement and three from the dipole moment are now associated with each ion in the crystal and, in general, there will be coupling between all of them. The forces are calculated by dividing the interactions into (a) long range Coulomb forces between point dipoles and (b) short range repulsive forces between the neighbouring ions. The long range forces are calculated using the dimensionless coefficients introduced by Kellermann. The equations of motion are expressed in terms of the quantities which are more physically meaningful than the core shell constant K and shell charges Y. These quantities are the electrical polarizabilities of the ions.

In their preliminary calculations, Woods *et al.*[20] took into consideration the short range repulsive forces between the first nearest neighbours only. Later Dolling *et al.*[31] extended the shell model calculations of the dynamics of the alkali halide lattices by taking into account short range forces between both first and second nearest neighbour atoms in the crystal, the polarizabilities of both ions and the possibility that the ionic charge might be less than one electronic charge. The arbitrary parameters of the shell model were obtained by means of a least squares fit to the measured dispersion relations for the lattice vibrations, the dielectric constant and the elastic constants. By using the adiabatic approximations with the mass of the shell set equal to zero, one can express the shell displacements through the core displacements and then the dimension of the dynamical matrix reduces to that of the rigid ion model.

The shell model has been applied to explain the dispersion curves of NaI,[20] KBr,[20] KI,[31] NaF,[32] LiF,[33] NaCl,[34] etc., and the results have been satisfactory. Although the simple version of the shell model is sufficient in many cases, in order to get a better fit with the experimental values the simple model has been further refined by taking into consideration the polarizability of both the ions, more distant neighbours, an effective charge on the ion and so on. Martson and Dick[35] have employed an exchange charge model for the alkali halides with results less satisfactory than the simple shell model.

More recently Schroder[36] has extended the shell model by introducing two more degrees of freedom arising from the compressibility of the electronic shells due to intrinsic interaction. This is now known as the 'breathing shell model'. The dispersion curves worked out on this model are in better agreement with the experimental results. The same model has been successfully employed for explaining the observed dispersion relations for LiF[36] and NaCl.[37] The extension of the breathing shell model calculations to other ionic crystals is yet to be carried out.

Because of its formal approach through the use of disposable parameters, Cochran's shell model has been applied successfully to other types of cubic lattices. The chief objection to the shell model is its parametric approach with a large number of parameters to be fitted. Some of the values obtained for the parameters are unusual and cannot be explained physically. However the overall validity of the shell model and its success in explaining the phonon dispersion curves and many of the consequent properties of the crystal, including the phase transitions as in the ferroelectrics, has now been well established. Cowley[38] has given a quantum mechanical justification of the shell model by deriving Cochran's equation of motion independently using a quantum mechanical approach. The formalism of Tolpygo[39] is based on the same formalism as that of Cochran, although his approach may appear to be slightly different.

E. Raman spectra—Theoretical considerations

On the semi-classical radiation theory, the intensity of Raman scattering from a crystal is proportional to $|M|^2$ and inversely proportional to the fourth power of the wavelength of the scattered light, where M is the electric moment set up in the crystal by the electric vector Re $(\boldsymbol{E}^{-i\omega t})$ of the incident beam. If $\alpha_{\rho\sigma}$ is the polarizability tensor associated with the electrons in the crystal, then

$$M_\rho = \Sigma \alpha_{\rho\sigma} E_\sigma \tag{8.4}$$

The notation used here is taken from Loudon's paper.[40] $\alpha_{\rho\sigma}$ is given by the expression[18]

$$\alpha_{\rho\sigma} = \alpha^0_{\rho\sigma} + \Sigma_\mu \alpha_{\rho\sigma,\mu} r_\mu + \Sigma_{\mu,\nu} \alpha_{\rho\sigma,\mu\nu} r_\mu r_\nu + 0(r^3) \tag{8.5}$$

where

$$\alpha_{\rho\sigma,\mu} = \left(\frac{\partial \alpha_{\rho\sigma}}{\partial r_\mu}\right)_{r=0}, \qquad \alpha_{\rho\sigma,\mu\nu} = \left(\frac{\partial^2 \alpha_{\rho\sigma}}{\partial r_\mu \partial r_\nu}\right)_{r=0} \tag{8.6}$$

and r is the relative displacement amplitude of two nuclei. The term linear in r gives rise to the first order Raman scattering, the quadratic term gives rise to second order Raman scattering and so on.

First order Raman scattering does not occur for crystals with NaCl and CsCl structures. In these crystals each atom is at a centre of symmetry and consequently the fundamental TO and LO modes do not produce a linear change in the electronic polarizability. These crystals do exhibit a well defined second order spectra with a number of prominent peaks or lines.[5, 6]

(a) *Born–Bradburn method*

The first attempt to apply the Born-von Karman theory to the Raman effect in crystals was by Born and Bradburn[7] who made use of the calculations of Kellermann[11] to explain the observed spectrum of rock salt.[6] Born and Bradburn expanded the electronic polarizability tensor up to the terms proportional to the products of the amplitudes of two nuclear displacements and calculated the scattering intensity. To simplify the problem, only the products of the nuclear displacements of nearest neighbours were taken to contribute to the second order term in the electronic polarizability expansion and the dominant contribution to the density of states arises from the region of the Brillouin zone close to the centre of the hexagonal faces. As a result of this approximation each pair of phonon branches makes a contribution to the second order continuum intensity at the frequency shift 2ω equal to the combined density of states of the two branches at frequency ω multiplied by a proportionality factor which is constant for a given pair of branches. The matrix elements which give the dependence of the polarizability on the lattice vibrations were not known. The matrix element for one wave vector was therefore calculated and assumed to be constant for all the other modes. For NaCl the thirty-six narrow maxima representing the density maxima of the combination in pairs of the six frequency branches (3 optical and 3 acoustic) of the exact theory had to be reduced to sixteen and the same selection rules at the point $L(\frac{1}{2}\frac{1}{2}\frac{1}{2})$ in the reduced zone assumed to hold good for all points. Using the information extracted in this way, Born and Bradburn were able to explain some of the features of the second order spectrum of rock salt. They had to make an arbitrary choice of the coupling constants and the intensity factors in the second order scattering in an attempt to fit the experimental results. They succeeded in getting an intensity distribution which showed only four broad maxima, whereas the recorded spectrum exhibited no less than 9 peaks including an extremely sharp line at a frequency shift of 235 cm^{-1}. At best, Born and Bradburn's treatment could yield only the general form of the observed second order Raman effect in rock salt.

The Born–Bradburn method was improved upon by taking into account the selection rules at each point. The second order Raman intensity distributions for KBr and NaI were worked out by Cowley[41] with the help of the vibration spectra calculated, on the basis of the shell model, from the dispersion relations obtained from neutron scattering experiments. But the extreme sharpness of the Raman line at 235 cm^{-1} in NaCl, at 126 cm^{-1} in KBr and at 89 cm^{-1} in KI could not be accounted for with these theories. Krishnamurthy[42] has carried out a calculation of the intensity distribution by the Born–Bradburn method from the second order spectra of KBr, NaI and CsI and found reasonable agreement with the experimental data when the coupling constant is evaluated from some of the experimental peaks of the spectrum.

The disadvantage of the Born–Bradburn approach is that the polarizability components are unknown and have to be arbitrarily chosen to give a better fit to the experimental intensity profiles. Therefore in recent years the structure in the second order spectra has been interpreted from the two phonon density of states with appropriate selection rules, since accurate and detailed data on the phonon density of states could be obtained from realistic models whose validity can be checked with the experimental phonon dispersion relation.

(b) *Critical point approach*

With the quasi-continuous frequency distribution function as envisaged in the Born theory, it became difficult to account for the fine details of the second order Raman spectra of cubic crystals. It was also recognized that, independent of the specific scattering mechanism which gives rise to a frequency dependence of the interaction, the second order Raman spectra may be expected to exhibit a structure arising from structure in the combined density of states of the pairs of vibration modes. Hardy and Smith[43] suggested that a more complete interpretation of the absorption spectra of cubic crystals, using a critical point analysis, can lead to experimental values of phonon frequencies at specific symmetry points in the Brillouin zone. Johnson and Loudon[44] pointed out that a similar approach to the Raman effect would be more fruitful in interpreting the observed spectra on the basis of the Born lattice dynamics.

A critical point on a phonon branch is defined as a point where every component of $\nabla_q \omega(q)$ is either zero, or changes sign discontinuously. When all the components are zero the critical point is analytic; when one or more components change sign discontinuously it is singular. The importance of critical points arises from the circumstance that a plot of phonon density of states against frequency may exhibit a discontinuity of slope at a critical point. This property of critical points was first investigated by Van Hove.[9] His work was later extended by Phillips[45] who set up a method for determining the positions and types of critical points. Using the results of these authors in conjunction with the lattice dynamical calculations and neutron scattering results, Johnson and Loudon[44] have determined the positions and types of critical points in silicon and germanium. In second order Raman spectra taken at ordinary temperatures one is mainly concerned with the summation states, in which two phonons of equal and opposite wave vector are created. The relevant two phonon density of states is obtained from two phonon dispersion curves constructed by adding together all pairs of phonon branches at each wavevector q.

In a cubic crystal with 2 atoms per unit cell ($p = 2$) there are 6 single phonon branches and 21 two phonon branches and the critical points of each of the latter must obey the same theorems as the critical points on single phonon branches. The structure of the observed Raman spectrum is influenced by the frequency dependence of the interactions involved, as well as the frequency dependence of the two phonon density of states. The final details are governed by the space group selection rules for multiphonon processes which were indicated by Birman[46] and applied to crystals of diamond structure. With the use of modern computing facilities one can work out the phonon distribution for any lattice on a fine mesh of q points and get a contour of the two phonon density of states with all the fine details.

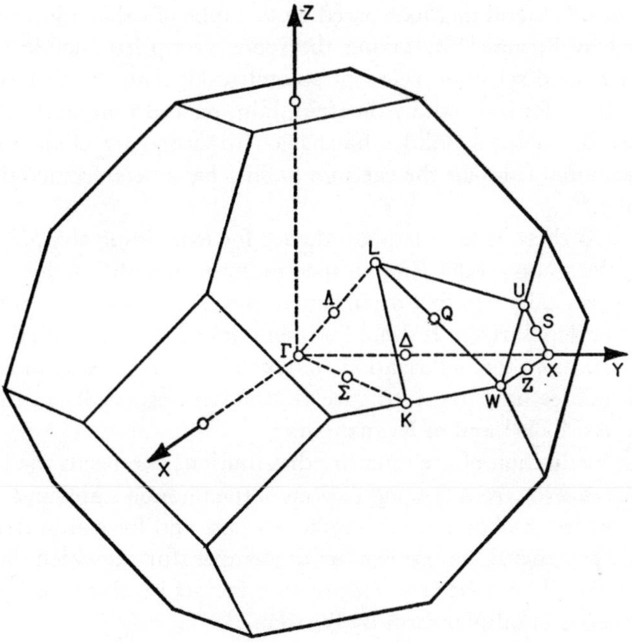

FIGURE 8.1. *The Brillouin zone of face centred cubic lattice showing the critical points.*

Using this procedure, Karo *et al.*[47] compared the combined density of states curve with two phonon dispersion curves of caesium halides and brought out the correspondence between the critical points and the peak positions in the second order spectra.

Most of the critical points in the phonon spectra occur at points of high symmetry in the Brillouin zone such as Γ, L, X and M in the case of a f.c.c. lattice (see Figure 8.1). The remaining critical points can occur on lines or planes of symmetry or at general positions in the zone. There are distinct selection rules, however, for the different wave vector points, to find which phonon pairs are allowed to contribute to the second order Raman scattering. The selection rules can be worked out by using group theoretical arguments. As an example let us state the condition for the combination of two phonons at $q = 0$ to appear in Raman effect. The Raman transition is allowed if the direct product of the irreducible representations of the two phonons contains irreducible representations in common with the polarizability tensor. For overtone states the Kronecker square of the irreducible representation of the particular state must be taken to determine the selection rules. In principle the same procedure can be used for a general wave vector. General methods based on this type of calculation have been developed by Birman[46] by taking the space group irreducible representations. Detailed selection rules for second order Raman scattering have been tabulated for rock salt by Burstein, Johnson and Loudon.[10] The selection rules for caesium halides have been discussed by Ganesan *et al.*[48] Some additional rules for the caesium halides have been framed by Merlit and Poulet.[49]

At present there is no adequate theory for explaining the polarization effects in the second order Raman spectra, even though Loudon[50] has discussed theoretically whether a particular combination can be completely polarized or depolarized. It should be mentioned here that with the advent of laser excitation, the polarization characteristics of the Raman scattering in alkali halides have received special attention, especially in the recent works of Worlock[51] and of Krauzman.[52]

For the evaluation of the intensity distributions one needs the temperature factors arising from the populations of the phonon states and these are different for the Stokes and anti-Stokes spectra and for combinations and overtones. In general they depend on the temperature at which the spectra are evaluated. The necessary temperature factors involved in these situations are given in tabular form by Burstein.[53]

(*c*) *Combined density of states approach*

The observed structure in the second order Raman spectra of alkali halides arises mainly from the structure present in the two phonon combined density of states curves. Cochran and Johnson[54] have successfully exploited this concept to explain the two phonon infrared spectrum of GaAs, even with the use of the same constant for the phonon–phonon coupling for all the q values, showing thereby that the positions of the peaks are primarily governed by the density of states maxima rather than by any selection rules. Another characteristic of the Raman lines, namely, the line shape can also be roughly guessed from symmetry considerations of the critical points. Polarization characteristics can also be explained in a general way as was done by Loudon. The crucial characteristic turns out to be the relative intensity distribution in the spectrum and this can be explained only if one introduces the Raman polarizability tensor in the combined density of states picture with the corresponding temperature factors for the Stokes, the anti-Stokes and difference bands. For a rigorous treatment, one should in principle resort to the fourth order perturbation theory to obtain the second order polarizability tensor. Although the treatment is an extension of Loudon's[55] theory of the first order scattering, it would be formidably complex and cannot be carried out at present for want of precise knowledge of electron–photon and electron–phonon matrix elements for all the q points. A formal but simplified treatment for the combined density of states taking into consideration the temperature factors also has been given by Johnson and Sennett.[56] Calculations on these lines have been made by Karo and Hardy for NaCl[57] and for caesium halides[47] using the phonon spectrum derived from the deformation dipole model. Krishnamurthy and Haridasan have done similar calculations on CsBr[58] using the phonon spectrum given by the rigid ion model and on CsCl[59] using the shell model spectrum. The relative intensity distribution in the Raman spectrum of NaF was estimated by Hardy *et al.*[60] using the deformation dipole model. The polarizability tensor components necessary for the evaluation were picked out from the scattering geometry and their Fourier transforms were expressed in terms of certain parameters. On the assumption that the polarizability tensor is affected by the nearest neighbour interactions alone, the number of parameters is reduced using symmetry properties. Since the spectrum is found to be completely polarized the P_{YZ} component of the polarizability is taken as zero. This enables the disposable parameters to be reduced to three, namely, a, c and d_2. For the relative intensity distribution these can be replaced by two parameters, namely the ratios c/a and d_2/a. The form of the spectra was calculated for the limiting cases (1) $c = d_2 = 0$ and (2) $a = d_2$, $c/a = -2\rho/r_0$ where ρ is the Born–Mayer constant. The results obtained by Karo and others are interesting in many respects. The inclusion of the polarizability tensor results in a marked suppression of subsidiary structures in the combined density of states curve especially in the low frequency region. These structures do not show up in scattering.

Taking into consideration the second neighbour central forces between fluorine ions in the calculation for a 'DDNNN model', Hardy and Karo[61] have presented a comprehensive theoretical account of the second order Raman spectra of the alkali fluoride sequence of crystals, NaF, KF, RbF and CsF. In the case of NaF, the theoretical results are found to be in excellent agreement with the experimental intensity distribution in the second order spectra. Similar comparisons could not be made for the other fluorides for want of experimental data.

The work of Hardy and Karo, referred to above, clearly indicates that in future more realistic calculations can be undertaken for other alkali halides and similar crystal structures, as and when more reliable and precise experimental results become available on which to base or check the calculations.

(*d*) *Relation between Born dynamics and Raman dynamics*

From a critical examination of the Raman dynamics in terms of the dynamics of Born, I[62] was able to establish that the concept of the phonons from critical points in the density of states curve and their enhanced activity in optical processes was nothing but the analogue of Raman's super-cell vibrations, and that the Raman dynamics afford a physical and easy way of explaining the observed Raman and infrared absorption spectra. Let us consider the implications of this statement. The critical points L, X, M, W and also those along the Σ, Δ and Λ directions in cubic lattices are along the symmetry directions and hence the atomic vibrations correspond to the movements of planes of atoms in phase and hence the equations of motion for this case can be written down easily by considering the linear chain of atoms. Further, the critical points L, X and M are situated at the extremities of the dispersion curves along the [111], [100] and [110] directions and, therefore, involve a phase difference of π at least along one of the three axes. In particular an examination of the lattice vibration eigenvectors shows clearly that for the critical points L for NaCl and R for CsCl, the six branches of the phonon spectrum are such that in three of them one set of planes of atoms oscillates, while the others are at rest and in the next three the other set of atoms oscillates with the first set of planes of atoms at rest, the direction of oscillations for the NaCl structure being the [111], $[11\bar{2}]$ and $[1\bar{1}0]$ while that for the CsCl structure is arbitrary. Since there are four equivalent L points and three X points for the NaCl structure and one R, three M and three X points for the CsCl structure the lattice branches from these points are degenerate. A further double degeneracy is introduced for the transverse phonons. These critical points, together with the triply degenerate Brillouin zone centre Γ, give a total of $(24p\text{-}3)$ modes which are identical with the super-cell modes envisaged on the Raman theory for NaCl and CsCl structures. Thus it is seen that the Raman theory postulated in 1943[3] affords an extremely practicable and simple method for explaining the second order or two phonon scattering processes and has now received full justification from the critical point analysis of Birman, Johnson and Loudon.

To explain some of the minor details of the observed spectrum, other critical points situated along the symmetry directions W, Σ, Δ, Λ, etc., which correspond to higher order super-cell modes in the Raman model, have to be taken into account. In the case of diamond-like crystals, at the zone boundary X(100), the LO and LA modes become accidentally degenerate in the Born theory. This is also confirmed in the Raman theory. I[5] worked out the selection rules for the second order Raman spectrum of diamond using Bhagavantam's character table for the super-cell of diamond. These selection rules are in agreement with those of Smith[63] based on lattice dynamical eigenvectors, etc. Another point of interest is the invalidity of the rule of mutual exclusion as shown by Venkatarayudu[64] for the diamond super-cell. The same conclusion has been arrived at by Birman[46] for second order processes from the critical point approach.

F. Experimental results

(*a*) *Introduction*

The term 'alkali halide' is applied to compounds of the alkali elements Li, Na, K, Rb and Cs with the halogen elements F, Cl, Br and I. They are twenty in number and from the crystallographic point of view they fall into two categories: (1) those which crystallize in the face centred cubic system which is known as the NaCl or rock salt structure, and (2) those which crystallize in the body centred cubic system which is known as the CsCl structure. The three caesium salts, CsCl, CsBr and CsI come into the second category, while all the other alkali halides come into the first category. All of them except LiF and NaF are freely soluble in water. Most of them can be crystallized from aqueous solutions. LiBr, LiI, KF and CsCl are highly deliquescent. The alkali halides are generally grown in the form of single crystals from the melt using the Kyropoulous method.

Using the 2537 Å mercury-arc excitation, Raman data have been collected for a large number of alkali halides, viz., LiF,[65] LiCl,[66] NaCl,[1, 6, 67, 68, 69] NaBr,[69, 70] NaI,[17] KCl,[52, 69, 71] KBr,[70, 72, 73] KI,[17, 70] RbCl,[74] RbBr,[69] RbI,[75] CsF,[74] CsCl,[74] CsBr[76, 77] and CsI.[74, 78] Polarization studies were made for NaCl, NaBr, KCl, KBr and RbBr by Menzies and Skinner,[69] for KI and CsBr by Narayanan[76] and for CsBr by Stekhanov and Korolkov.[77] The variation of the intensity of the second order spectra with temperature of LiCl,[66] NaCl,[67, 79] KI,[80] RbCl[74] and CsBr[77] have been investigated by Stekhanov's group over the temperature range 90 to 700 K.

Using helium-neon laser radiation, Worlock[51] analysed the polarization characteristics and temperature dependence at 300 and 90 K of the Raman spectra of NaCl, CsBr and KCl to find the irreducible representations of the singularities. Similar investigations were carried out recently by Hardy, *et al.*[60] on the Raman spectrum of NaF using argon ion laser radiation. Recently Krauzman[52, 81] has made a detailed analysis of the polarization characteristics and the frequency shifts of the numerous maxima appearing in the second order spectra of NaCl, KBr, KI and RbI employing the argon laser radiation. The spectra were recorded at crystal temperatures of 300 and 90 K.

(*b*) *Results*

The main results are summarized below. The frequency shifts of the prominent lines or peaks in the Raman spectra of alkali halides reported by the various authors are collected in Table 8.1. Although in most cases only the second order Raman spectrum has been recorded, in a few cases the recorded spectrum and the observed peaks in the region of high frequency shifts refer to the third order. Krauzman[52] has reported many more peaks in NaCl, KBr and KI than those given by others. These are not included in Table 8.1. Raman effect data are not available for LiBr, LiI, KF and RbF as they have not been obtained in the form of single crystals because of their highly deliquescent nature.

A comparative study of the spectra of alkali halides leads to the following conclusions:

(i) In every case the Raman spectrum is very complex with a continuous background extending over a wide range of frequency shifts not necessarily starting from the exciting line. The spectra exhibit numerous sharp maxima. These are clearly visible in the typical spectrograms and the microphotometer records reproduced in Figures 8.2 and 8.3 respectively.

(ii) The highly individual character of the vibration spectrum of each halide is reflected in its Raman spectrum which shows marked variations in intensity distribution from halide to halide.

(iii) Qualitatively, one can classify the spectra of the alkali halides into three groups.

Group I. Those halides, viz., NaF, KCl and RbBr, for which the mass ratio of the cation and anion is nearly one, exhibit weak continuous spectra with a small number of less prominent peaks.

Group II. CsCl, CsBr and CsI exhibit intense spectra with peaks evenly distributed over the entire region. The peaks stand out very prominently from the background.

Group III. All the other alkali halides belonging to the NaCl type except those coming under Group I exhibit fairly intense spectra with a large number of prominent and sharp peaks which include one very sharp and intense line. The frequency shift of this intense line for the various alkali halides is shown in bold letters in Table 8.1.

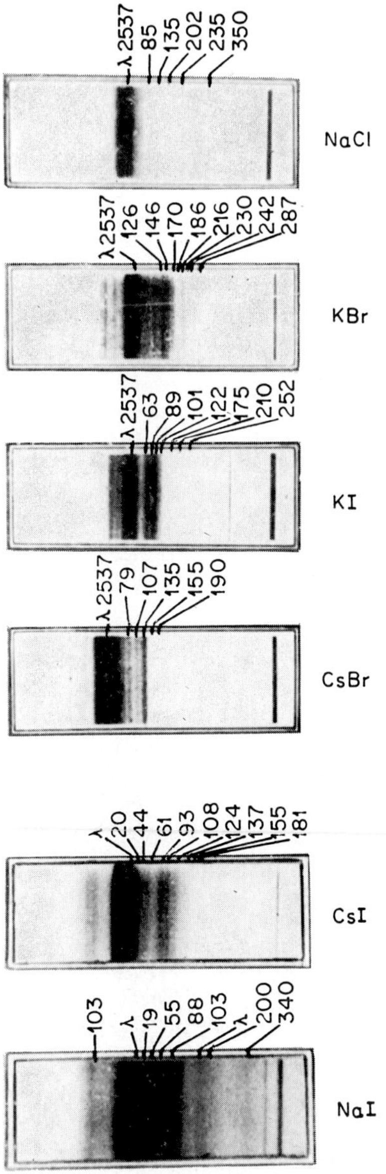

FIGURE 8.2. *Raman spectra of alkali halides obtained with 2537 Å excitation. The frequency shifts of the Raman lines are marked in wave numbers. These spectra were photographed with a Hilger Medium Quartz Spectrograph.*

TABLE 8.1. *Raman spectra of alkali halides*

Crystal	LiF	LiCl	NaF	NaCl	NaBr	NaI	KCl	KBr	KI	RbCl	RbBr	RbI	CsF	CsCl	CsBr	CsI
Mass-ratio	0·37	0·2	0·82	0·65	0·29	0·18	0·91	0·48	0·31	0·42	0·94	0·69	0·14	0·26	0·6	0·95
	(a)	(b)	(c)	(d)	(e)	(f)	(g)	(h)	(f)	(i)	(j)	(k)	(i)	(i)	(l)	(i)
Frequency shifts (in cm^{-1})	570	49	245	85	31	19	121	46	63	35	50	55	51	42	25	20
		86														
	676	116	404	135	64	42	177	61	**89**	48	80	76	65	61	40	28
	715	128	421	140	116	58	205	76	101	56		**82**	83	81	54	44
	764	147	440	162	**152**	88	213	86	102	75		115	107	112	75	52
	814	159	503	184	181	**103**	242	116	175	85		156	116	158	105	61
	892	168	524	199	254	120	293	**125**	210	112		160	152	170	125	67
		207	577	202		132	331	146	252	132		171	170	212	134	94
		227		220		200	349	170		155			206	310	163	107
				235		255		186		**200**			225	353	176	113
		274		258		310–370		216		230			275	366		122
		292		270				230		240			298			128
		298		276				242		250			320			135
		307		286				287		280			358			160
		337		300						315			390			167
		357		314						326			402			180

375	320	357	184
405	326	364	200
432	343	432	
444	350		
453			
472			
498			
522			
540			
558			
618			

References:

(a) Krishnan, R. S., and Narayanan, P. S., unpublished work.
(b) Stekhanov, A. I., Korolkov, A. P., and Eliashberg, M. P., *Soviet Phys. Solid State*, 1962, **4**, 945.
(c) Hardy, J. R., Karo, A. M., Morrison, I. W., Sennett, C. T., and Russell, J. P., *Phys. Rev.*, 1969, **179**, 837.
(d) Krishnan, R. S., *Proc. Roy. Soc.*, 1946, **A187**, 188; *Proc. Indian Acad. Sci.*, 1947, **A26**, 419, 432.
(e) Stekhanov, A. I., and Petrova, M. L., *Soviet Phys. JETP*, 1949, **19**, 1108.
(f) Krishnan, R. S., and Narayanan, P. S., *Jour. Indian Inst. Sci.*, 1957, **39**, 85.
(g) Stekhanov, A. I., *Soviet Phys. JETP*, 1950, **20**, 330.
(h) Krishnan, R. S., and Narayanan, P. S., *Proc. Indian Acad. Sci.*, 1948, **28**, 296.
(i) Stekhanov, A. I., and Korolkov, A. P., *Soviet Phys. Solid State*, 1966, **8**, 734.
(j) Menzies, A. C., and Skinner, J., *J. Phys. Radium*, 1949, **9**, 93.
(k) Krishnamurthy, N., *Proc. Phys. Soc.*, 1965, **85**, 1025.
(l) Stekhanov, A. I., and Korolkov, A. P., *Soviet Phys. Solid State*, 1963, **4**, 2311.

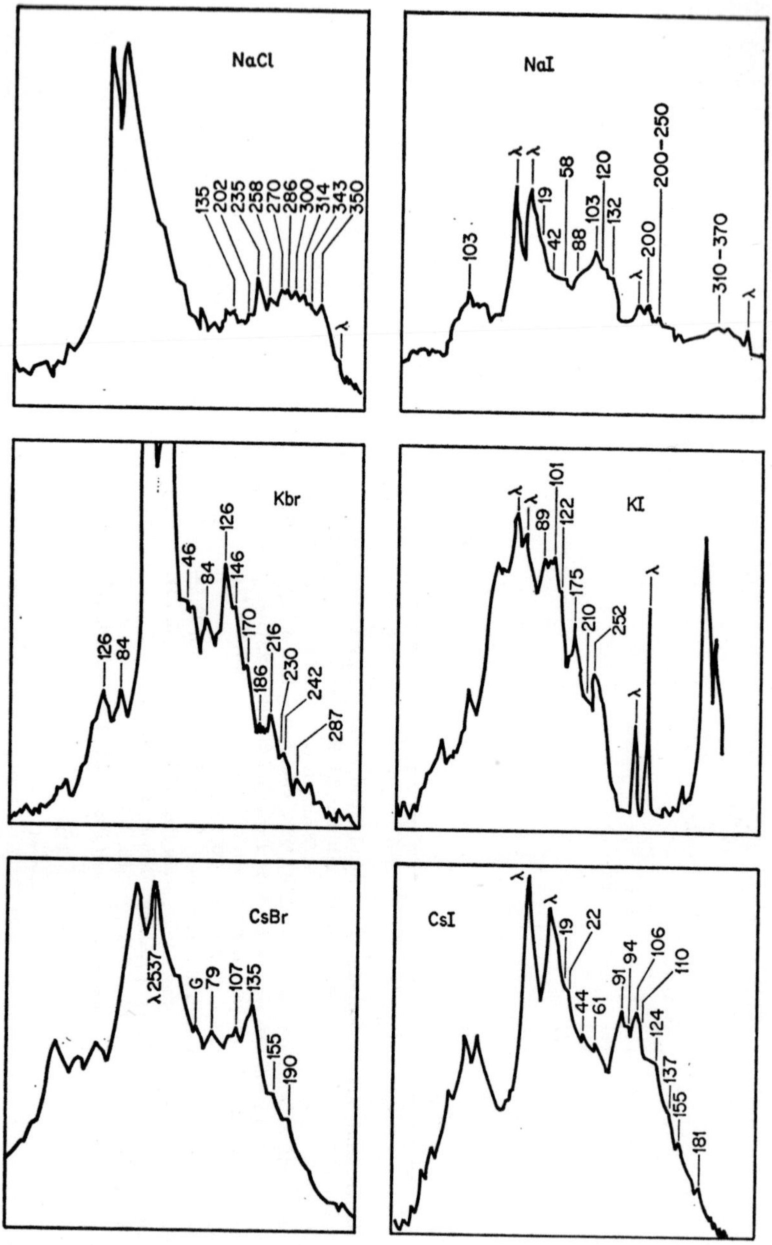

FIGURE 8.3. *Microphotometer records of the Raman spectra of alkali halides.*

(iv) The intensity of Raman scattering increases as one goes from the fluoride to the iodide showing that the electron shells of halogen atoms play a principal role in the scattering intensity as they possess large polarizability. The increased intensity of the spectra of caesium halides has been attributed to the scattering from the alkali ions also, as the ionic polarizability of caesium is close in magnitude to that of chlorine and is only 1·4 times smaller than that of bromine.

(v) In the case of NaI which has the lowest mass ratio between the cation and anion among the alkali halides studied so far, the peaks in the second order spectra are bunched together to form four groups of which the second one from the exciting line contains the more intense peaks. In the case of CsI where the mass ratio is 0·95 the intense peaks at 20, 94 and 108 cm^{-1} are closely placed doublets.

(vi) The spectra exhibited by alkali halides belonging to Group I are completely polarized. Laser studies[51, 52] have indicated that the major part of the intensity distribution in the second order spectra in these crystals belongs to the irreducible representation or the symmetry type A and only a negligible part comes under symmetry type E. The F symmetry type spectrum in these crystals exhibits one single strong line corresponding to the combination (LO + TO). In the case of the halides belonging to Group II, the peaks are all depolarized. The entire second order spectrum appears to be a mixture of A, E and F symmetry types. In the case of the halides coming under Group III the entire spectrum is polarized except the intense sharp line which is depolarized. Accurate polarization studies[51, 52] have revealed that the major part of the second order spectrum in these crystals belongs to symmetry type A, while the intense sharp line belongs to symmetry type F.

(c) *Discussion*

(i) *Lithium halides.* Experimental data on the dispersion relation from neutron scattering are available only for LiF.[33] Shell model calculations of Dolling *et al.*[33] give better agreement with the observed data for dispersion relations than the rigid ion model or DD (deformation dipole) model. The experimental data on the Raman spectrum of LiF are explained on the combined density of states (CDS) scheme without taking into consideration the selection rules.[33, 65] I have explained the data on LiCl on the basis of the critical point analysis (CPA).[62] No data are available for LiBr and LiI.

(ii) *Sodium halides.* Dispersion relations are available from neutron scattering studies for NaF,[32] NaCl[82] and NaI.[20] DD model calculations of Karo and Hardy[28] are found to be satisfactory to explain the observed dispersion data of NaF and fairly satisfactory for NaCl and NaI. In the case of NaCl, a breathing shell model calculation[37] gave a good fit with experimental data.

The Raman data for NaF have been satisfactorily explained by Hardy and others[60] on the CDS scheme taking polarizability also into consideration. The spectrum of NaCl is accounted for fairly satisfactorily[28] on the CDS scheme using the DD model. I have explained the peaks in the Raman spectra of NaCl, NaBr and NaI on the basis of CPA.[62] No calculations have been made so far on the CDS scheme.

(iii) *Potassium halides.* Dispersion relations have been determined from neutron scattering experiments for KBr[83] and KI[31] and shell model calculations have been found to be satisfactory to explain the observed results. The frequency shifts of the peaks in the Raman spectra of KBr and KI have been accounted for by myself[62] and Krauzman[81] on the basis of CPA. A similar analysis has been made by Karo and Hardy for KCl and by Krisnamurthy for KBr.[42] So far no CDS analysis of the Raman spectra of potassium halides has been carried out.

(iv) *Rubidium halides.* No experimental data on dispersion relations of rubidium halides are available. The phonon branches for RbCl[59] and for RbI[75] have been worked out in the framework of the shell model. The frequencies calculated at critical points have been used to assign the second order Raman spectra of these crystals. No CDS calculations have been made for rubidium halides.

(v) *Caesium halides.* For caesium halides no neutron scattering data are available. However, the phonon branches for the CsCl, CsBr and CsI have been worked out on the shell model[59] and the observed peaks in their Raman spectra have been accounted for on the basis of the phonon frequencies calculated at critical points. In the case of CsCl only the combined density of states for the two-phonon Raman spectrum has been worked out and the peaks in the CDS curve agree well with the Raman frequencies. I have also explained these on the basis of CPA.[62] Karo *et al.*[47] have calculated the intensity distribution in the second order Raman spectra of CsCl, CsBr and CsI from the CDS curves derived by allowing for the polarizability and overlap deformation of the ions in the rigid ion model. Their theoretical curves are found to be in reasonably good agreement with the experimental Raman data even though no allowance has been made for the wave vector dependence of the Raman polarizability tensor. Hardy and Karo[84] have theoretically calculated the second order Raman spectrum of CsF based on the deformation dipole model with short range repulsive interactions between second neighbour negative ions on the assumption that the Raman polarizability tensor is determined by the configurations of the first neighbour ions. The frequencies corresponding to the maxima are in agreement with the observed frequency shifts in the Raman effect.[74]

The principal phonon frequencies of 16 alkali halides having the NaCl type structure and 3 halides having the CsCl type structure obtained from neutron scattering experiments, etc., are given in Tables 8.2 and 8.3. In the last two columns of each of these Tables are given the source of the data and the reference to the literature respectively. The corresponding Raman super-cell modes are also indicated in the second row.

Finally, the Lyddane–Sachs–Teller (LST) rule,[85] derived from purely rigorous macroscopic phenomenological considerations without appealing to the nature of the interatomic forces in the crystal, gives us an indication of the extent of the second order spectra in the alkali halides. According to this rule, the second order spectra should terminate at a frequency corresponding to the overtone of the maximum frequency of the lattice vibrations which is the limiting longitudinal optical frequency ν_0 ,[Γ(LO)] for $k = 0$. This frequency can be obtained from neutron scattering experiments. It can also be calculated from the limiting transverse optical frequency ν_1, [Γ(TO)] and the static (ϵ_0) and high frequency (ϵ_∞) dielectric constants using the LST relation

$$\frac{\nu_0}{\nu_1} = \sqrt{\left(\frac{\epsilon_0}{\epsilon_\infty}\right)} \tag{8.7}$$

This limiting longitudinal optical mode calculated with the help of the infrared active transverse optical mode is found to agree well with the experimentally observed value.

G. Mixed crystals and crystals doped with impurities or defects

(*a*) *Introduction*

As has been pointed out earlier, the characteristic lattice vibration spectrum of alkali halide crystals is continuous with a very complicated frequency distribution. Any disturbance in the ideal periodicity of the lattice by the introduction of impurities or defects will lead to changes in the spectrum of the vibrations propagated through the crystal. This question has been discussed from the theoretical point of view in the article by Wallis and Maradudin.[86] In the case of crystals of the alkali halide type where normally a first order Raman spectrum does not appear, because of the crystal symmetry, the impurity or defect of the crystalline lattice may destroy the crystal symmetry to such an extent as to make the first order spectrum active in the Raman effect in some cases. This will be in the form of a continuum and not restricted to limiting frequencies of modes at $k = 0$. In other crystals the impurity may produce localized vibrations which show up in scattering. Direct evidence for the occurrence of localized vibrations of a

TABLE 8.2. *Principal phonon frequencies of NaCl type alkali halides*

Mode	Γ LO	Γ TO	L LO	L TO	L LA	L TA	X LO	X TO	X LA	X TA		
Crystal	ν_0	ν_1	ν_2	ν_3	ν_4	ν_5	ν_6	ν_7	ν_8	ν_9	Source	Reference
LiF	660	305	630	300	387	206	456	345	350	256	NSE	a
LiCl	580	275	477	218	165	106	324	265	149	100	DDMC	b
LiBr	325	159										
NaF	414	239	313	206	300	160	284	266	266	146	NSE	c
NaCl	264	162	226	140	173	118	182·5	174	142·5	87·5	NSE	d
NaBr	204	139	212	114	93	71	186	159	80	51	DDMC	b
NaI	170	120	130	126	62	41	170	116	77	53	NSE	e
KF	326	190	274	186	146	97	177	141	138	88	DDMC	b
KCl	214	142									DXC	f
KBr	163	113	144	96·5	92	70	133·5	118	73	42	NSE	d
KI	139	101	125	93	64·5	51·5	108	102·5	51	31	NSE	g
RbF	286	156										
RbCl	182	115	138	102	100	66	113	105	90	43	CPA	h
RbBr	127	88										
RbI	103	75	97	64	68	42	78	82	36	20	CPA	i
CsF	300†	115	259	213	109	55	102	85	95	21	CPA	h

NSE = data collected from neutron scattering experiments. They are reliable.

DDMC = data obtained from a comparison of the observed second order Raman spectrum and the two phonon spectrum calculated on the deformation dipole model.

CPA = data obtained from a critical point analysis of the Raman spectrum in the shell model.

† The LO frequency is estimated from the extent of the second order spectrum.

References:

(a) Dolling, G., Smith, H. G., Nicklaw, R. M., Vijayaraghavan, P. R., and Wilkinson, M. K., *Phys. Rev.*, 1965, **168**, 970.
(b) Karo, A. M., and Hardy, J. R., *Phys. Rev.*, 1963, **129**, 2024; Hardy, J. R., and Karo, A. M., *Proc. Lattice Dyn. Conf.*, Pergamon Press, 1965, 195.
(c) Buyers, W., *Phys. Rev.*, 1967, **153**, 923.
(d) Cowley, R. A., Woods, A. D. B., Brockhouse, B. N., and Cochran, W., *Phys. Rev.*, 1963, **131**, 1030.
(e) Woods, A. D. B., Cochran, W., and Brockhouse, B. N., *Phys. Rev.*, 1960, **119**, 980; Woods, A. D. B., Brockhouse, B. N., Cowley, R. A., and Cochran, W., *Phys. Rev.*, 1963, **131**, 1025.
(f) Lyddane, R. M., Sachs, R. G., and Teller, E., *Phys. Rev.*, 1941, **59**, 673.
(g) Dolling, G., Cowley, R. A., Schiltenhelm, C., and Thornson, I. M., *Phys. Rev.*, 1966, **147**, 577.
(h) Haridasan, T. M., and Krishnamurthy, N., *Indian Jour. Pure and Appl. Phys.*, 1968, **6**, 407.
(i) Krishnamurthy, N., *Proc. Phys. Soc.*, 1965, **85**, 1025.

TABLE 8.3. *Phonon frequencies of CsCl type alkali halides*

Mode	Γ LO	Γ TO	R LO	R LA	M LO	M TO	M LA	M TA	X LO	X TO	X LA	X TA		
Crystal	ν_0	ν_1	ν_2	ν_3	ν_4	ν_5	ν_6	ν_7	ν_8	ν_9	ν_{10}	ν_{11}	Source	Reference
CsCl	165	99	124	67	115	63	91	35	152	96	87	40	CPA	a
CsBr	112	73	79	63	82	61	47	40	92	62	81	41	CPA	a
CsI	85	62	58	54	69	45	39	38	75	45	62	39	CPA	a

Reference
(a) Haridasan, T. M., and Krishnamurthy, N., *Indian Jour. Pure and Appl. Phys.*, 1968, **6**, 407.

U centre in an alkali halide was obtained by Schaefer[87] in infrared absorption. If the concentration of impurity is high as in the case of mixed crystals, the lattice vibration of the host crystal will be very much altered and large changes will be observed in its second order Raman spectrum.

(*b*) *Mixed crystals*

Stekhanov and Eliashberg[88] were the first to investigate the Raman spectra of mixed crystals of KCl and KBr using 2537 Å excitation and reported the appearance of a first order spectrum in the low frequency region connected with the defects in the crystalline lattice. When the concentration of KCl was progressively increased, the prominent peaks of the KBr spectrum gradually levelled off. Using laser excitation, Hurrel *et al.*[89] recorded a first order Raman scattering mainly of A_{1g} symmetry, in addition to the usual second order spectrum, in a mixed crystal of KBr containing 8 per cent molar concentration of KCl, reflecting the appropriate projected one-phonon density of states for pure KBr.

(*c*) *Doped crystals*

The Raman spectrum of KCl containing Li^+ ions as impurities was recorded by Stekhanov and Eliashberg.[90] At 77 K they recorded a single narrow line at 208 cm^{-1} superposed over the continuous second order Raman spectrum. This line was broadened and shifted to 198 cm^{-1} at room temperature; since it appeared well within the allowed band frequencies of the pure KCl lattice, it corresponded to a resonant mode. Some of these observations were not confirmed by Kaiser and Mockel.[91] Stekhanov and Eliashberg[92] investigated the Raman spectra of KCl containing Li^+, Br^-, I^-, Na^+, Cs^+ and Rb^+ as impurities. The spectrum of $(KCl + I^-)$ was very blurred. The scattering spectra of crystals doped with either Na^+, Cs^+, Rb^+ consisted of a series of discrete bands on a continuous weak background. The observed bands were attributed to resonant modes. No first order spectrum of KCl was present. The theory of the impurity-induced Raman scattering by the F_{1g} modes in KCl containing U centres has been worked out by Xuam Xinh *et al.*[93] and by Gurevich *et al.*[94] According to Gurevich *et al.* the Raman spectrum of the doped crystal should display two peaks near the frequency limits of optical phonons of the pure crystal. The results obtained with a Li doped KCl crystal[90] were in qualitative agreement with the above conclusion. Recently Leigh and Szigeti[95] derived expressions for the first order Raman scattering arising from the electrostatic field of the impurity in NaCl and CsCl type lattices and showed that peaks would be expected in the Raman scattering at frequencies corresponding to both Γ(LO) and Γ(TO) modes.

Fenner and Klein[96] have made Raman scattering measurements on hydroxyl doped alkali halide (NaCl, KCl, KBr) crystals using the 4880 Å argon laser radiation. Besides the Raman line corresponding to the O—H stretching mode, a perpendicularly polarized Raman line corresponding to the librational mode of OH^- ion in KBr:OH^- and KCl:OH^- has been recorded at 4 K. A broad Raman line at 50 to 60 cm^{-1} with parallel polarization is also observed in the spectra of hydroxyl doped NaCl, KBr and KCl.

Callender and Persham[97] have investigated the Raman spectra of NaCl, KCl and KBr crystals doped with CN^-, OH^- or NO_2^- at about 10 K. At this temperature the second order spectra of the host crystals are of very low intensity. Besides the Raman lines due to the internal vibrations of the impurity ions, low frequency modes due to their rotational and translational degrees of freedom and also induced first order Raman spectrum of the host have been recorded in the low frequency region.

(*d*) *Crystals with* F *centres*

A study of the Raman spectra of F centres in alkali halide is of interest because they provide the most direct information about the phonons responsible for the line width of F centres and also about the perturbation of the vibrational modes of the host lattice by such F centres. Raman scattering by F centres in additively coloured and γ-irradiated NaCl and KCl has been investigated by Worlock and Porto[98] using laser excitation. In spite of the low concentration of F centres, a relatively large Raman cross-section is obtained through a resonant enhancement arising from the fact that the exciting radiation is near the F absorption band. However, strong and sharp localized modes were not observed. In the case of NaCl, they reported the appearance of polarized bands at 175 and 350 cm^{-1} and depolarized bands at 115 and 235 cm^{-1}, while in KCl a broad polarized double band extending from 40 to 200 cm^{-1} with a peak at 80 cm^{-1} was recorded. The observed peaks have been ascribed to impurity-induced first order Raman scattering. These experimental results are found to be in fairly good agreement with the calculations of the first order Raman spectra of F centres in NaCl and KCl carried out by Benedek and Nardelli.[99] An accurate analysis of the observed Raman spectra combined with a good knowledge of the vibrational states of crystals with F centres would give us a deeper insight into the electron-phonon interaction at colour centres.

Buchenauer *et al.*[100] investigated the Raman spectra of F centres in KF, NaBr and RbF using argon laser radiation for excitation at liquid helium temperature and analysed the Raman active modes into the different symmetry types. The Raman spectra of F centres vary considerably for different host lattices. In KF the three broad peaks observed are due to the first order Raman spectrum. The frequency shifts of the maxima are influenced by the F centres and are higher than the three extremes in the theoretical KF phonon density of states. In the case of NaBr which has a frequency gap in the region from 105 to 126 cm^{-1}, besides the perturbed first order Raman spectrum of NaBr, a new strong resonance peak at about 136 cm^{-1} belonging to A_{1g} symmetry is observed. In fact the resonance peak is a doublet with a separation of 11 cm^{-1}. In the case of RbF, besides the perturbed first order Raman spectrum of pure RbF, two peaks of moderate intensity are observed; one near 170 cm^{-1} with E_g symmetry and another near 125 cm^{-1} with A_{1g} symmetry. In a theoretical paper Benedek and Mulazzi[101] have discussed the conditions under which the second order spectra of the imperfect lattice display the peculiarities of the projected densities for the perturbed two phonon state. Their theoretical results have been used to interpret the F-centre-induced Raman spectra of NaBr reported by Buchenauer *et al.*[100] with some success.

H. Future trends

In conclusion, one can summarize the present position of our knowledge concerning the second order Raman spectra of alkali halides from both the theoretical and experimental standpoints, point out the gaps in our knowledge and suggest work for the future.

There are four important parameters concerning the Raman spectrum of any crystal, the precise determination of which is essential for a proper understanding and theoretical interpretation of the scattering process. These are

(i) frequency shifts of the peaks or lines,
(ii) the line shape or contour of each peak,
(iii) polarization characteristics of the peaks and their symmetry types, and
(iv) relative intensities of the peaks or the intensity versus frequency shift distribution curve over the entire Raman spectrum.

Information concerning all these four parameters are now available only for the two alkali halides which have been studied with laser excitation. In most cases the investigations were carried out using mercury 2537 Å excitation and only quantitative information concerning the positions of peaks is available; information concerning the remaining three parameters is of a very qualitative nature. Microphotometer records give only a rough estimate of the intensity distribution. With the advent of lasers and direct recording of intensities, precise information on the four parameters could be collected for all the alkali halides.

If any lattice dynamical theory is to be accepted as correct, it should be capable of giving quantitative predictions using an appropriate model and the results should compare well with the experimental values of the four Raman parameters mentioned above. Let us examine the present position on the theoretical side.

In the case of alkali halides, the observed dispersion curves have been satisfactorily accounted for by the lattice dynamical calculations made on the basis of the shell model of Cochran and the deformation dipole model of Hardy and Karo.

The frequency shifts of the peaks in the Raman spectra of many alkali halides have been explained on the basis of Raman super-cell modes or the critical point phonons obtained from neutron scattering experiments or more effectively from the combined density of states. Qualitative information on the line shapes can also be had from symmetry arguments on the critical points. Using group theoretical arguments Loudon has shown how in a rock salt crystal any particular peak corresponding to the combination of the two critical point phonons can be completely polarized or depolarized. Other than this, no theory, which would help in the unambiguous assignment of the Raman peaks, has been developed so far for predicting the polarization characteristics of the Raman lines of alkali halides. The combined density of states approach of Hardy and Karo taking into consideration the selection rules and temperature factors could explain only the observed positions of peaks in the Raman spectrum and not the actual intensity-versus-frequency-shift distribution curve. For a satisfactory explanation of the intensity distribution one has to bring into the theory the mechanism of Raman scattering involving the polarizability. One such attempt was made by Hardy *et al.*[60, 61] for NaF and CsF, when they expanded the polarizability tensors and formulated their Fourier transforms. The number of parameters entering the polarizability tensor, even with the assumption that only nearest neighbours affect the polarizability, becomes quite large. Fortunately in the case of NaF the number of parameters could be reduced from a knowledge of the precise scattering geometry and also from a knowledge of the observed polarization characters of the spectrum from laser excitation. Even then there were two disposable parameters and a judicious estimation of these parameters was mainly responsible for the satisfactory explanation of the observed intensity distribution. Thus the present method of attack lacks uniqueness in its approach; the parameters have to be chosen on an *ad hoc* basis with some physical reasoning. A similar limited approach is possible in the case of other alkali halides only if precise information is available concerning the polarization characteristics and intensity distribution based on an equally precise scattering geometry. With laser excitation, such measurements are expected to be made for all the alkali halides in the near future.

Loudon has given a satisfactory theory for the first order phonon scattering in crystals. One has to extend this theory to two phonon processes by following a fourth order perturbation theory. This procedure is formidably complex at the moment since it demands a knowledge of the electron-photon and electron-phonon matrix elements for all the wave vectors. In principle, one can obtain such data from band structure studies; but they are too meagre at present to be useful.

References

1. Rasetti, F., *Nature*, 1931, **127**, 626.
2. Fermi, E., and Rasetti, F., *Zeits. f. Phys.*, 1941, **71**, 689.
3. Raman, C. V., *Proc. Indian Acad. Sci.*, 1941, **A13**, 1; **A14**, 459; 1943, **A18**, 237.
4. Krishnan, R. S., *Proc. Indian Acad. Sci.*, 1943, **A18**, 298.
5. Krishnan, R. S., *Proc. Indian Acad. Sci.*, 1944, **A19**, 216.
6. Krishnan, R. S., *Proc. Roy. Soc.*, 1946, **A187**, 188; *Proc. Indian Acad. Sci.*, 1947, **A26**, 419, 432.
7. Born, M., and Bradburn, M., *Proc. Roy. Soc.*, 1947, **A188**, 161.
8. Born, M., and Von Karman, T., *Phys. Zeit.*, 1912, **13**, 297.
9. Van Hove, L., *Phys. Rev.*, 1953, **89**, 1189.
10. Burstein, E., Johnson, F. A., and Loudon, R., *Phys. Rev.*, 1965, **139**, 1239.
11. Kellermann, E. W., *Phil. Trans. Roy. Soc.*, 1940, **238**, 513.
12. Montroll, E. W., *J. Chem. Phys.*, 1942, **10**, 218; 1943, **11**, 481.
13. Houston, W. V., *Rev. Mod. Phys.*, 1948, **20**, 161.
14. Raman, C. V., *Proc. Indian Acad. Sci.*, 1947, **A26**, 339–398; 1951, **A34**, 61, 141.
15. Raman, C. V., *Proc. Indian Acad. Sci.*, 1961, **A54**, 253, 281.
16. Ramanathan, K. G., *Proc. Indian Acad. Sci.*, 1947, **A26**, 481.
17. Krishnan, R. S., and Narayanan, P. S., *Jour. Indian Inst. Sci.*, 1957, **39**, 85; Krishnan, R. S., and Krishnamurthy, N., *Zeit. Phys.*, 1963, **175**, 440; Krishnamurthy, N., Ph.D. Thesis, I.I.Sc., 1964.
18. Placzek, G., and Van Hove, L., *Phys. Rev.*, 1954, **93**, 1207.
19. Brockhouse, B. N., and Iyengar, P. K., *Phys. Rev.*, 1957, **108**, 894; 1958, **111**, 747.
20. Woods, A. D. B., Cochran, W., and Brockhouse, B. N., *Phys. Rev.*, 1960, **119**, 980; Woods, A. D. B., Brockhouse, B. N., Cowley, R. A., and Cochran, W., *Phys. Rev.*, 1963, **131**, 1025.
21. Born, M., and Thomson, J. H. C., *Proc. Roy. Soc.*, 1934, **A147**, 594.
22. Ewald, P. P., *Ann. Phys.*, 1921, **64**, 253.

23. Lyddane, R. H., and Herzfeld, K. F., *Phys. Rev.*, 1938, **54**, 846.
24. Hardy, J. R., *Phil. Mag.*, 1959, **4**, 1278.
25. Tessman, J. R., Khan, A. H., and Shockley, W., *Phys. Rev.*, 1953, **92**, 890.
26. Szigeti, B., *Proc. Roy. Soc.*, 1950, **A204**, 51.
27. Karo, A. M., and Hardy, J. R., *Phil. Mag.*, 1960, **5**, 859.
28. Karo, A. M., and Hardy, J. R., *Phys. Rev.*, 1963, **129**, 2024; Hardy, J. R., and Karo, A. M., 'Proc. Lattice Dyn. Conf.', Pergamon Press, 1965, p. 195.
29. Cochran, W., *Phil. Mag.*, 1959, **4**, 1082.
30. Dick, B. G., and Overhausser, A. W., *Phys. Rev.*, 1958, **112**, 90.
31. Dolling, G., Cowley, R. A., Schiltenhelm, C., and Thornson, I. M., *Phys. Rev.*, 1966, **147**, 577.
32. Buyers, W., *Phys. Rev.*, 1967, **153**, 923.
33. Dolling, G., Smith, H. G., Nicklaw, R. M., Vijayaraghavan, P. R., and Wilkinson, M. K., *Phys. Rev.*, 1968, **168**, 970.
34. Lagu, M., and Dayal, B., *Indian J. Pure Appl. Phys.*, 1968, **6**, 670.
35. Martson, R. L., and Dick, B. G., *Solid State Comm.*, 1967, **5**, 731.
36. Schroder, U., *Solid State Comm.*, 1967, **4**, 347.
37. Nusslein, V., and Schroder, U., *Physics Status Solidi*, 1967, **21**, 309.
38. Cowley, R. A., *Proc. Roy. Soc., London*, 1962, **A268**, 109.
39. Tolpygo, K., *Sov. Phys. Solid State*, 1961, **2**, 2367.
40. Loudon, R., *Adv. Phys.*, 1964, **13**, 423.
41. Cowley, R. A., *Proc. Phys. Soc.*, 1964, **A84**, 281.
42. Krishnamurthy, N., *Indian J. Pure Appl. Phys.*, 1964, **2**, 150.
43. Hardy, J. R., and Smith, S. D., *Phil. Mag.*, 1961, **6**, 1163.
44. Johnson, J. A., and Loudon, R., *Proc. Roy. Soc.*, 1954, **A281**, 274.
45. Phillips, J. C., *Phys. Rev.*, 1956, **104**, 1263.
46. Birman, T. L., *Phys. Rev.*, 1962, **127**, 1093; 1963, **131**, 1489.
47. Karo, A. M., Hardy, J. R., and Morrison, I., *J. de Phys.*, 1965, **26**, 668.
48. Ganesan, S., Burstein, E., Karo, A. M., and Hardy, J. M., *J. de Phys.*, 1965, **26**, 639.
49. Merlit, F., and Poulet, M. H., *C.R. Acad. Sci.*, 1967, **B265**, 326.
50. Loudon, R., *Phys. Rev.*, 1965, **137**, 1784.
51. Worlock, J. M., unpublished work, paper presented at 1966 Edinburgh Conference.
52. Krauzman, M., 'Light Scattering Spectra of Solids', Springer-Verlag, New York, 1969, p. 109.
53. Burstein, E., 'Phonons and Phonon Interactions' (ed. T. A. Bak), 1964, p. 276.
54. Cochran, W., and Johnson, F. A., 'Proc. Int. Conf. Semiconductors', Exeter, 1962, p. 498.

55. Loudon, R., *Proc. Roy. Soc.*, 1963, **A275**, 218.
56. Johnson, F. A., and Sennett, C. T., 'Proc. Int. Conf. Semiconductors', Kyoto, 1966, p. 53.
57. Karo, A. M., and Hardy, J. R., *Phys. Rev.*, 1966, **141**, 696.
58. Krishnamurthy, N., and Haridasan, T. M., *Indian J. Pure Appl. Phys.*, 1966, **4**, 337.
59. Haridasan, T. M., and Krishnamurthy, N., *Indian J. Pure Appl. Phys.*, 1968, **6**, 407.
60. Hardy, J. R., Karo, A. M., Morrison, I. W., Sennett, C. T., and Russell, J. P., *Phys. Rev.*, 1969, **179**, 837.
61. Hardy, J. R., and Karo, A. M., 'Light Scattering Spectra of Solids', Springer-Verlag, 1969, p. 99.
62. Krishnan, R. S., *Indian J. Pure Appl. Phys.*, 1965, **3**, 424.
63. Smith, M. L., *Phil. Trans. Roy. Soc., London*, 1948, **A241**, 105.
64. Venkatarayudu, T., *J. Chem. Phys.*, 1954, **22**, 1219.
65. Krishnan, R. S., and Narayanan, P. S., unpublished work.
66. Stekhanov, A. I., Korolkov, A. P., and Eliashberg, M. P., *Soviet Phys. Solid State*, 1962, **4**, 945.
67. Gross, E. F., and Stekhanov, A. I., *Nature*, 1947, **159**, 474; **160**, 568.
68. Welsh, H. L., Crawford, M. F., and Staple, W. J., *Nature*, 1949, **164**, 737.
69. Menzies, A. C., and Skinner, J., *J. Phys. Radium*, 1949, **9**, 93.
70. Stekhanov, A. I., and Petrova, M. L., *Soviet Phys.*, JETP, 1949, **19**, 1108.
71. Stekhanov, A. I., *Soviet Phys.*, JETP, 1950, **20**, 330.
72. Krishnan, R. S., and Narayanan, P. S., *Proc. Indian Acad. Sci.*, 1948, **28**, 296; *Nature*, 1949, **163**, 570.
73. Stekhanov, A. I., and Eliashberg, M. B., *Soviet Phys. Solid State*, 1961, **2**, 2096.
74. Stekhanov, A. I., and Korolkov, A. P., *Soviet Phys. Solid State*, 1966, **8**, 734.
75. Krishnamurthy, N., *Proc. Phys. Soc.*, 1965, **85**, 1025.
76. Narayanan, P. S., *Proc. Indian Acad. Sci.*, 1955, **A42**, 303.
77. Stekhanov, A. I., and Korolkov, A. P., *Soviet Phys. Solid State*, 1963, **4**, 2311.
78. Krishnamurthy, N., and Krishnan, R. S., *Indian J. Pure Appl. Phys.*, 1963, **1**, 239.
79. Maksimova, T. I., Stekhanov, A. I., and Chisler, E. V., *Soviet Phys. Solid State*, 1965, **7**, 1515.
80. Stekhanov, A. I., Gabrichidze, Z. A., and Eliashberg, M. B., *Soviet Phys. Solid State*, 1961, **3**, 964.

81. Krauzman, M., *Compt. rend. Acad. Sci., Paris*, 1967, **265B**, 689, 1029; 1968, **266B**, 186.
82. Raunio, G., Almgvist, L., and Stedman, R., *Phys. Rev.*, 1969, **178**, 1496.
83. Cowley, R. A., Woods, A. D. B., Brockhouse, B. N., and Cochran, W., *Phys. Rev.*, 1963, **131**, 1030.
84. Hardy, J. R., and Karo, A. M., *Phys. Rev.*, 1968, **168**, 1054.
85. Lyddane, R. M., Sachs, R. G., and Teller, E., *Phys. Rev.*, 1941, **59**, 673.
86. Wallis, R., and Maradudin, A., *Progr. Theoret. Phys.*, 1960, **24**, 1349.
87. Schaefer, C., *J. Phys. Chem. Solid.*, 1959, **12**, 233.
88. Stekhanov, A. I., and Eliashberg, M. P., *Optics Spect.*, 1961, **10**, 174; *Soviet Phys. Solid State*, 1961, **2**, 2096.
89. Hurrel, H. P., Porto, S. P. S., Damen, T. C., and Mascarenas, S., *Phys. Letters*, 1968, **26A**, 194.
90. Stekhanov, A. I., and Eliashberg, M. P., *Soviet Phys. Solid State*, 1964, **5**, 2185.
91. Kaiser, R., and Mockel, P., *Phys. Letters*, 1967, **25A**, 749.
92. Stekhanov, A. I., and Eliashberg, M. P., *Soviet Phys. Solid State*, 1965, **6**, 2718; Stekhanov, A. I., and Maksimova, T. I., *Soviet Phys. Solid State*, 1966, **8**, 737.
93. Xuam Xinh, N., Maradudin, A. A., and Coldwell Horsfall, R. A., *J. de Phys.*, 1965, **26**, 717.
94. Gurevich, L. E., Ipatova, I. P., and Klochikhin, A. A., *Soviet Phys. Solid State*, 1967, **8**, 2608.
95. Leigh, R. S., and Szigeti, B., 'Light Scattering Spectra of Solids', Springer-Verlag, 1969, p. 477.
96. Fenner, W. R., and Klein, M. V., 'Light Scattering Spectra of Solids', Springer-Verlag, 1969, p. 497.
97. Callendar, R. H., and Persham, P. S., 'Light Scattering Spectra of Solids', Springer-Verlag, 1969, p. 505.
98. Worlock, J. M., and Porto, S. P. S., *Phys. Rev. Letters*, 1965, **15**, 697.
99. Benedek, G., and Nardelli, G. F., *Phys. Rev.*, 1967, **154**, 872.
100. Buchenauer, C. J., Fitchen, D. B., and Page, J. B., 'Light Scattering Spectra of Solids', Springer-Verlag, 1969, p. 521.
101. Benedek, G., and Mulazzi, E., 'Light Scattering Spectra of Solids', Springer-Verlag, 1969, p. 531.

CHAPTER 9

Atomic Interactions in Molecular, Covalent, Ionic and Metallic Crystals

SAN-ICHIRO MIZUSHIMA AND
ISAO ICHISHIMA

A. Introduction

The structure of solids is related to the structure of the component atoms. Different types of crystals may be classified as molecular, covalent, ionic and metallic crystals. The atomic interactions in some of them are fairly well understood, but we do not consider that a satisfactory general systematic treatment has so far been developed.

Such being the case, we have considered, as a first step, the relation between crystal structure and cohesive energy for some simple substances. Such relations have been found to afford important clues to understanding the general problem of atomic interactions in solids. In this article we will endeavour to discuss this problem and set our findings in perspective.

B. Molecular crystals

Since structural chemistry started with the study of free molecules, let us first discuss some aspects of molecular crystals.

The development of quantum mechanics enables us to treat the atomic interactions in very simple molecules like H_2 on a purely theoretical basis. However, for more complicated molecules the interatomic forces are obtained by empirical or semi-empirical methods. For example, spectroscopists make assumptions about the molecular potential functions and determine the force constants from observed frequencies.[1]

Infrared and Raman vibrational spectra of many crystals consisting of molecules held together by van der Waals forces have many features in common with those of the corresponding free molecules, but in addition we find splitting of molecular vibrational frequencies and the appearance of lattice frequencies in crystal spectra.

The differences in the spectra between the crystalline and free states may be given a simple explanation. Let us take benzene as an example. The lattice frequencies (in cm^{-1}) of the crystal have been found to be $\nu_1 = 69 - 0{\cdot}100\ T$, $\nu_2 = 89 - 0{\cdot}083\ T$, $\nu_3 = 112 - 0{\cdot}127\ T$, and $\nu_4 = 141 - 0{\cdot}128\ T$, where T denotes the absolute temperature.[2] The molecular entropy of crystalline benzene can be calculated from these frequency values and is found to be in agreement with the thermal entropy. This shows that the temperature dependence of the frequencies is due to the anharmonicity in the lattice vibrations.[3]

Space does not allow us to consider the interatomic forces in the benzene molecule itself; these have been treated by many authors. We will, however, discuss the calculation of the intermolecular forces in a benzene crystal. The crystal structure is quite complicated, but we can consider that it is the interactions between two non-bonded hydrogen atoms (each of them belonging to different benzene molecules) which play the most important part. There are five different pairs of non-bonded hydrogen atoms, if we only take into account the non-bonded distances less than 3 Å: $r_1 = 2{\cdot}628$ Å, $r_2 = 2{\cdot}698$ Å, $r_3 = 2{\cdot}764$ Å, $r_4 = 2{\cdot}767$ Å and $r_5 = 2{\cdot}826$ Å. The corresponding force constants in mdyne/Å are estimated as 0·016, 0·012, 0·010, 0·0096 and 0·0080. If the lattice frequencies are calculated for each symmetry species, it can be shown that they are in good agreement with those observed[4] (see Table 9.1). The splitting of molecular vibrations can be calculated on the same basis and the calculated values are also in good agreement with those observed as shown in Table 9.1.

Therefore, we can conclude that the intermolecular forces in the benzene crystal arise mainly from the non-bonded interatomic forces between hydrogen atoms situated close to one another. This situation also holds for the crystals of many other hydrocarbons.[5]

It is usually the case that in non-polar molecular crystals including that of benzene, atoms in the same molecule are held by strong intramolecular (valence) forces, but molecules in the crystals are held together by much weaker forces. In consequence essentially the same spectra are observed both in the free and crystalline states, so far as the intramolecular vibrations are concerned.

However, in the case of molecules with an axis or axes of internal rotation

TABLE 9.1. *Lattice frequencies and the splitting of molecular vibration frequencies in crystalline benzene*[4]

Lattice frequencies at 270 K, cm^{-1}			Frequency splitting of molecular vibrations, cm^{-1}		
Symmetry species	ν_{obs}	ν_{calc}	Molecular vibrational frequencies	$\Delta\nu_{calc}$	$\Delta\nu_{obs}$
A_g	35	28			
	63	57			
	—	79			
			707	27	23
B_{1g}	35	33			
	—	70			
	105	102			
			410	15	12
B_{2g}	63	56			
	69	69			
	—	76			
			987	16	14
B_{3g}	—	60			
	69	72			
	105	100			

the situation is quite different. We find many more vibrational lines in the free state than in the crystalline state, in so far as all the minimum energy forms of internal rotation are not identical with each other.[1] This is due to the fact that the energy of internal rotation is not much greater than the intermolecular potential energy. For example, in 1,2-dichloroethane, $ClH_2C—CH_2Cl$, the frequency of torsional vibration has the order of magnitude of 100 cm^{-1}, which is not much greater than the lattice frequencies of the crystal.[6]

Therefore, it is no wonder that the crystal consists of only the trans form (with symmetry C_{2h}) and the liquid and gas consist of both trans and gauche forms (the latter differing from the former in the internal rotation angle by about 120° and having[1] the symmetry of C_2). For this reason we observe a smaller number of intramolecular vibrational lines in the crystalline state than in the liquid or gas. There are of course additional lines due to lattice vibrations[7] in the solid.

It may have been better to take $CH_3.CH_2—CH_2.CH_3$ as an example of internal rotation, because intermolecular forces in the crystal can also be considered to arise only from non-bonded hydrogen interactions as in the crystal of benzene. However, since 1,2-dichloroethane has been studied so fully it has been cited here as an example of the problem of internal rotation. In this crystal there are non-bonded H...Cl interactions in addition to non-bonded H...H interactions, but the former are not much greater than the latter and in consequence the situation remains essentially like that in an unsubstituted hydrocarbon.

If the molecule contains an atom or group which forms a hydrogen bond, the intermolecular force becomes much stronger, although it still is considerably less than the usual covalent forces.[8]

C. Covalent and ionic crystals

It is well known that in diamond the tetrahedrally disposed bonds of carbon atoms run through the whole crystal. A single perfect crystal is thus one molecule and we can, therefore, use the term covalent crystal.

It follows that the forces holding together atoms are the same as those of corresponding hydrocarbon molecules. This can be proved in various ways but we will deal here only with evidence from vibrational spectroscopy.

Group theory shows that the diamond lattice has only one principal normal mode of vibration which is triply degenerate and Raman active. The frequency ν has been calculated as[9]

$$4\pi^2 c^2 \nu^2 = \frac{1}{m}\left[\frac{8}{3}K + \frac{64}{3}H\right] \tag{9.1}$$

where c is the light velocity, m the mass of the carbon atom and K and H are the stretching and bending force constants. Using values of force constants analogous to those of the appropriate hydrocarbon molecules the calculated frequency, 1334 cm^{-1}, is found to be in good agreement with the observed value 1332 cm^{-1}.[10]

One might consider why the optically active vibration of the diamond crystal can be expressed as simply as in equation (9.1) which contains only K and H but not any constant corresponding, for example, to the non-bonded C...C interactions. The above normal vibration calculation shows that constants other than K and H have in fact a negligible effect on this vibration.

This is also the case for ionic crystals consisting of positive and negative ions with noble gas configurations. The frequency ν of the infrared active (triply-degenerate) vibrations of a KCl crystal is calculated as[9]

$$4\pi^2 c^2 \nu^2 = 2K\left[\frac{1}{m_1} + \frac{1}{m_2}\right] \tag{9.2}$$

where K is the stretching force constant between the neighbouring ions, m_1 the mass of K^+ and m_2 that of Cl^-. From the observed frequency, 141 cm^{-1}, the force constant K is calculated[11] as 0·11 mdyne/Å.

We may explain the situation as follows. At equilibrium the attractive forces and the repulsive forces are balanced. The former arise from the well-understood Coulomb terms and the latter from the repulsive potential which falls off very rapidly with increasing interatomic distance. Therefore, for a vibration of small amplitude around the equilibrium position (which gives rise to an observable spectral frequency), we would expect that the repulsive force, changing rapidly with distance, will play the most important part.

However, in the calculation of the cohesive energy, the Coulomb terms which are effective even at larger distances become important. This is the reason why we introduce the Madelung constant to calculate the cohesive energy of ionic crystals. To summarize, for ionic crystals of the NaCl type, the repulsive force plays the more important part in the optically active vibration whereas the Coulomb force is more important in the cohesive energy.

D. Types of atomic aggregate and the Periodic Table

So far we have treated those problems concerning interatomic forces, the nature of which is fairly well understood. However, if we want to discuss much more generally the atomic interactions in various types of atomic aggregates, we find that these have not been explained in a systematic way even for crystals involving atoms of only one kind. Let our next main object be to consider this problem, taking into account the individual nature of atoms as far as possible.[12] We consider that the crystal structure and the cohesive energy which is the energy required for breaking up a crystal into free atoms, will throw considerable light upon the understanding of the nature of atomic interaction in crystals of elementary substances.

In order to give a general idea of what we have in mind, let us first consider how the cohesive energies of crystals of elements vary.[13] These energies are presented in Fig. 9.1. The metallic elements Groups IA, IIA ... VIII, IB, IIB, are on the left and the non-metallic elements with some additional metallic ones, Groups IIIB, IVB ... VIIB, 0, are on the right. We shall refer to the elements on the left and right as the first and second categories in the following.

The number of elements belonging to the first category is greater than that in the second, but the crystal structures[14] exhibited by the first category are, except for a few elements, rather simple (f.c.c., h.c.p. and b.c.c.) whereas those of the second show a great variety and in some cases the structures are considerably complicated. The individual character of the atoms seems to be reflected more clearly in the nature of atomic aggregates in the second category than in the first.

It is evident that the difference in the nature of atomic aggregates arises from the different behaviour of outer electrons.

In free atoms these electrons are attracted by the positive nuclear charge, shielded partially by other electrons, and the attractive force becomes stronger with increasing atomic number. Furthermore, since there is little shielding by the electrons in the same shell the valence electrons tend to be more localized in the elements of the second category,

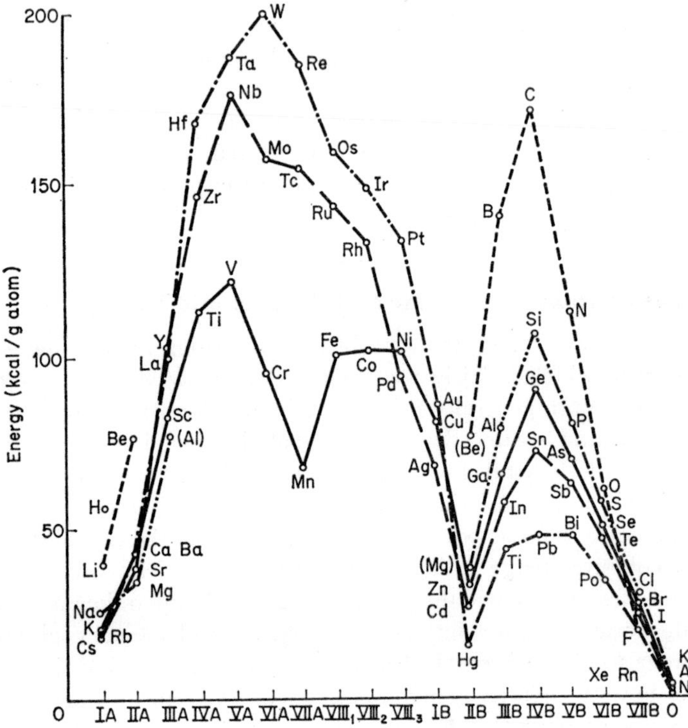

FIGURE 9.1. *Cohesive energy of crystals.*[13]

particularly for those elements shown on the right of Figure 9.1. This is one difference in the nature of atoms between the first and second categories. Another difference is that in the second category the outer electrons are s- and p-electrons, whereas in the first category they are s-electrons. Also in atoms of the transition elements the d-electrons, which lie just inside and have a principal quantum number less by one than the outermost s-electron, have to be considered.

In the second category, there are two s-electrons in the free state and so in the formation of atomic aggregates p-electrons with directional character will play an important part. Furthermore, there may be hybrid bonds formed by s- and p-electrons, the directional character being probably more pronounced. Therefore, the atoms of the second category of elements tend to form directional binding in the atomic aggregates, usually forming covalent bonds. For this reason the individual character of atoms in the second category of elements is generally reflected in the nature of atomic aggregates, each element exhibiting its own crystal structure, as mentioned above.

Such is not the case for the elements of the first category. Here s-electrons with non-directional character play an essential part in the formation of atomic aggregates and, therefore, the crystal structure is to be explained in the first place in terms of packing. Valence electrons move through the crystal and have been successfully treated collectively by the band theory.

However, the fact that the cohesive energy shows a maximum value in the middle of the first category of elements is evidence for the contribution of d-electrons to the atomic aggregation. The d-electrons in the free atom tend to be localized and to show directional character in contrast to the outer s-electrons. Therefore, the behaviour of these d-electrons of unclosed d-shells may be in a sense analogous to that of the p-electrons of the elements belonging to the second category.

The transition metals are a major part of the metallic elements of the first category. Since they are characterized by d-electrons, the specific behaviour of these electrons should be taken into account in order to understand the nature of the crystal structures of the first category of elements, even if they seem to be apparently simple.

Although we have divided the elements essentially into the two categories the nature of elementary substances changes gradually in the periodic table, and the second category contains some semi-conductors, semi-metals and metals. The substances in this category have been studied from various approaches.

In what follows we shall discuss the first category of elements in more detail. In this way we hope to present a general systematic view of elementary substances of both the first and the second categories.

E. Basic nature of metallic crystals

In order to describe the basic nature of metallic crystals, let us first consider those which consist of free electrons and ionic cores with closed shells. These electrons spread over the crystal and as is well known, the cores generally tend to be held together as closely as possible until the repulsive force becomes sufficiently strong. This is the usual explanation of why the IB elements (Cu, Ag and Au) exhibit the closest-packed structure (f.c.c.).

However, crystals of the alkali metals (Group IA) exhibit the b.c.c. structure, although they consist of nearly free electrons and closed shells. Before going into the explanation of this, we should note that transition metals of Group VA (V, Nb, Ta) and Group VIA (Cr, Mo, W) exhibit the b.c.c. structure and that almost all the elements which undergo transformation exhibit the b.c.c. structure at high temperatures. Li and Na which exhibit the b.c.c. structure at room temperature, transform to the closest-packed structure (h.c.p.) at low temperatures. (Li exhibits the f.c.c. structure by cold working.[15]) Since there is a great difference in hardness and density between Group IA (alkali metals) and Group VA and Group VIA elements, there should be a difference in atomic interactions between these two groups, although all of them exhibit the same crystal structure (b.c.c.).

As we have seen above, in some cases the thermal energy plays an important part in stabilizing the b.c.c. structure, alkali metals being considered to belong to this category. It is mainly the motion of the nuclei that is affected by the change of temperature. In the case of the electrons only a small portion of them near the Fermi level are affected, since electrons obey Fermi statistics and the Fermi energy is much higher than kT. We shall therefore treat the above problem, taking into account mainly the motion of the nuclei. The free energy F of the crystal is related to the enthalpy H, entropy S and absolute temperature T as:

$$F = H - TS, \tag{9.3}$$

In typically metallic crystals, the motion of electrons can be represented by a plane wave except for a certain volume surrounding an ionic core; since the electrons fill half of the Brillouin zone, the Fermi surface is not affected by the zone boundaries. Under such conditions, the b.c.c. structure having more room will to a certain extent have an energy higher than the closest-packed structures. On the other hand, if the core has more room to move, the randomness will be more pronounced, resulting in an increase of entropy S. Furthermore, the free energy F is related to the entropy term TS as shown in equation (9.3), and, therefore, H will become less important than TS above a certain temperature and the b.c.c. structure will be stabilized.[16]

The alkali metals (IA) have values of atomic volume and thermal expansion coefficient much higher than those of any other metals; furthermore, as shown in Fig. 9.1, their cohesive energies are very low. Therefore, the entropy term TS becomes very effective in decreasing the free energy F of equation (9.3). This is the most important factor in understanding the stability of the b.c.c. structure of alkali metals at room temperature.

We consider that in many cases in which the high temperature form exhibits the b.c.c. structure, the entropy relation explained above should always be taken into account, so far as the transformation of the structure is concerned.

F. Atomic aggregates with closed d-shells

Let us discuss further the difference between IA and IB elements. In addition to what we have already described, they differ markedly from each other in chemical properties, the former being very active, while the latter are inactive (noble metals).

IA or alkali metals have cores of the rare gas configuration (s^2p^6) with energies much lower than those of outer s-electrons. The shielding of the nuclear charge by these core electrons is effective, so that the s-valence electrons can move away from the core quite easily. Thus alkali metals can be considered to be a prototype of a metallic crystal consisting of cores and nearly free electrons. Evidently s-valence electrons can penetrate into the cores and the shielding will be less effective as the number of core electrons becomes larger, but we may treat the deviation from the Coulomb field by introducing a semi-empirical term (quantum defect) at least for alkali metals.[17]

On the other hand in IB or noble metals each of the closed cores has ten d-electrons with less effective screening of nuclear charge. In this case the exchange interaction between s-electrons and cores is strong, so it is desirable to explain the behaviour of these metals on the basis of this interaction. This is one of the reasons why the cohesive energies of IB metals are higher than those of IA metals.[18] Therefore, for IB metals, the enthalpy H of equation (9.3) will be more important than the entropy term TS, and they cannot exhibit the b.c.c. structure shown by IA metals (alkali metals).

The atoms of IIA elements have cores of electronic configurations which are the same as those of IA elements, the shielding of the nuclear charges by these core electrons being effective. Therefore, the energy of a p-level is not much higher than that of an s-level, and consequently in the atomic aggregates s-bands overlap p-bands to a considerable extent. This means that the two electrons per atom do not much increase the kinetic energy when atoms are held together to form aggregates On the other hand the charge of the core of a IIA element is higher than that of a IA element. Consequently, it is no wonder that the cohesive energy of IIA elements is higher than that of IA elements.

This is not the case for B-group elements each atom of which has two outer s-electrons in the free state. The cohesive energy of a IIB crystal is lower than that of IB and corresponds to the minimum between the two energy maxima shown in Fig. 9.1. In IIB elements the screening for p-electrons by closed d-shells is more effective than for s-electrons, since p-electrons cannot get so close to the nucleus as s-electrons. For this reason p-bands lie much higher than s-bands, resulting in less overlapping to decrease the cohesive energy.

As shown above, the existence of the closed d-shells is favourable to the atomic aggregation of IB-elements, but unfavourable to that of IIB elements. Therefore, the cohesive energies of IB elements are higher than those of IIB elements, whereas those of IA elements are lower than those of IIA elements.

Generally speaking, d-states form much narrower bands and where s- and d-bands are overlapped, they can mix with each other to a considerable extent, forming separated bands, if the symmetry allows. If both of the bands are completely filled with electrons, the energy will not be changed by mixing. If they are not filled, this will result in a decrease of energy. Therefore, the cohesive energy of IB elements becomes higher than that of IA elements by an amount more than that expected from the explanation given before.[19]

In conclusion, d-shells affect the nature of atomic aggregates considerably, even in the case of closed shells.

G. Crystal structure and cohesive energy of transition metals

The transition elements which represent the major part of the metallic elements are characterized by high cohesive energies, which change considerably with atomic number. These energies are considered to be due to the behaviour of the incomplete d-shell which plays a much more important part than a filled d-shell.

It is to be noted that the crystal structure of the transition elements shows the following regular change except for Mn, Fe and Co in the first long period.

IIIA	IVA	VA	VIA	VIIA	VIII_1	VIII_2	VIII_3	(IB)
h	h	b	b	h	h	f	f	f

Here, h, b and f denote, respectively, the h.c.p., b.c.c., and f.c.c. structures. It is also important to note that this regular change of crystal structure is closely related to the change of cohesive energy shown in Fig. 9.1 and that the energy increases in the order of f.c.c. → h.c.p. → b.c.c.

Let us first discuss VIII_3 elements (Ni, Pd, Pt) which lie next to the IB elements described in the previous sections. Their atoms have almost closed d-shells, the energies of which are not much different from those of the outermost s-state. For example, the electronic configuration of the ground state of Pt is $5d^9 6s^1$ and is similar to that of Au ($5d^{10} 6s^1$): in other words we have only one outermost s-electron in both of them even in the free state. This may be the cause of the similarity in the crystalline state and so be the reason why many properties of these two metals are similar.

Thus we can understand why VIII_3 elements exhibit the f.c.c. structure found for IB elements. However, since the d-core is not completely filled, the cohesive energy is higher than that of IB elements, and other properties (such as magnetic and electric) are also different. However, the basic nature of atomic aggregates of VIII_3 metals can be explained on the basis of closest packing.

The VA and VIA metals with the b.c.c. structure show the highest cohesive energies in each long period of the periodic table; and tungsten has the highest cohesive energy of all the elements. Therefore, the structure cannot be considered in the same way as the b.c.c. structure of alkali metals in which the entropy term plays an important part. That we have different kinds of b.c.c. structure for metallic elements can also be seen from the transformation of iron: α(b.c.c.) $\rightarrow$ γ(f.c.c.) $\rightarrow$ δ(b.c.c.). The b.c.c. structure of δ-Fe can be understood in principle by the effect of the entropy term, because this structure is stable above the transition point, below which the crystal exhibits the closest-packed structure. As the temperature is lowered still further, we have another transition point below which the crystal expands in volume and again exhibits the b.c.c. structure. This is the room temperature form of iron which, therefore, cannot be explained by the entropy term as in the case of alkali metals. Consequently, it is probable that the nature of atomic aggregates of α-Fe is in a sense similar to that of VA and VIA elements.

In conclusion the b.c.c. structure of this kind has a high value of the enthalpy H, although we have more room per atom as compared with the closest-packed structure. This cannot be explained in terms of the nature of typical metals described so far.

H. Contribution of d-electrons to the atomic aggregation of transition elements

An s-orbital is spherically symmetrical, whereas d-orbitals with two nodal surfaces are directional. These properties are related to the nature of atomic aggregates. Evidently the delocalized s-electrons characterize the metallic state and in this respect the transition metals are not exceptions. So the structure is to be treated in principle on the basis of packing. Even the b.c.c. structure is more closely packed than a structure like the diamond lattice.

In transition metals outermost s-electrons move almost freely, but d-electrons lying inside tend to be localized and hence the directional character of d-electrons should be taken into account.

In a free atom the d-level is five-fold degenerate and the corresponding five wave functions are expressed as: d_{xy}, d_{yz}, d_{zx} and d_{z^2} and $d_{x^2-y^2}$, the latter two extending along the directions of the rectangular coordinate axes. They may be expressed as $xyf(r)$, $yzf(r)$, $zxf(r)$, $(1/2\sqrt{3})(3z^2-r^2)f(r)$ and $\frac{1}{2}(x^2-y^2)f(r)$, where $f(r)$ denotes the radial distribution function. A linear combination of d_{xy}, d_{yz}, and d_{zx} forms the wave functions extended along the body diagonal axes of a body-centred cube.

Such wave functions will overlap with each other effectively, and we can have a high electron density between neighbouring atoms. Evidently lattice periodicity results in the different ways in which wave functions overlap to form bands. We may consider that in addition to the contribution of delocalized s-electrons to the atomic cohesion, we also have contributions from d-electrons which show a tendency to have a higher density between neighbouring atoms in this structure. Such a situation cannot be expected in the closest-packed structure.

If, therefore, we have a sufficient number of d-electrons, they stabilize the atomic aggregates more effectively in the b.c.c. structure than the closest-packed structure, even though the former is a less densely packed one.

Such a situation will be realized when there are three or more electrons in the d-level of the free atom and this level is almost as low in energy as the s-level. The d-levels of IA and IIA elements are certainly higher in energy than their s-levels, but the d-level comes down close to the s-level in VA elements with three d-electrons in the free atom. Therefore, we have good reason to believe that the d-electrons of VA and VIA elements contribute most effectively to the formation of the b.c.c. structure. Furthermore, since in such elements we have the largest number of electrons available for atomic aggregation, the cohesive energy will be at a maximum (see Figure 9.1).

The electronic configuration in the ground state of a free atom of Group IIIA is s^2d^1; more energy is required for excitation to s^1d^2. Consequently, the d-electrons cannot contribute to the b.c.c. structure as in VA and VIA elements. In other words, the less densely packed structure (b.c.c.) becomes more unstable than the closest-packed one. However, because of the unclosed d-shell without spherical symmetry, the crystal will tend to exhibit an anisotropic closest-packed structure: i.e. the h.c.p. structure. The wave function d_{z^2} is extended along the z-axis and the states in which this axis coincide with the c-axis of the h.c.p. structure will contribute most effectively to the atomic aggregation. Among these states, stable ones will correspond to those in which the density of electron clouds between atomic layers of the h.c.p. structure is high. This results in a decrease of the interatomic distance along the c-axis, so that the axial ratio c/a becomes less than the value of 1·633 corresponding to that of hexagonal closest packing. In IVA elements in which there is an additional d-electron, its contribution to atomic aggregates will be less anisotropic and the structure will approach the closest-packed hexagonal one. Evidently the change in the number of d-electrons affects the cohesive energy and so IIIA and IVA elements have values of this energy lying between that of IIA elements and those of VA and VIA elements.

If the number of d-electrons is larger than that of a half shell, the effective number of the electrons contributing to the atomic aggregation will be reduced. At the same time the effective nuclear charge becomes stronger with atomic number and the d-electrons tend to be localized. Therefore, the cores of VIII_3 elements behave like closed shells and these elements exhibit the f.c.c. structure as explained before. The same structure, exhibited by VIII_2 elements, can be understood in a similar way.

Thus in VIII_3 and VIII_2 elements the contribution of d-electrons to atomic aggregation is not so important, whereas in VA and VIA elements, as already indicated, the contribution is substantial. In VIIA and VIII_1 elements lying between the above two groups of elements in the Periodic Table, the contribution of d-electrons is intermediate. Therefore, in VIIA and VIII_1 elements the contribution of d-electrons is neither great enough to lead to the b.c.c. structure as in VA, and VIA elements, nor so insignificant as to lead to the f.c.c. structure of VIII_2 and VIII_3 elements. It is understandable that these elements (VIIA, VIII_1) exhibit an anisotropic closest-packed structure or the h.c.p. structure in accordance with what we have discussed about IIIA elements. It is also seen that VIIA and VIII_1 elements have values of cohesive energy lying between those of VA, VIA and VIII_2, VIII_3 elements.

I. Crystal structure and cohesive energy of lanthanides

We have seen above that the atoms of the transition elements are characterized by the inner d-electrons, the energies of which are not much different from those of the outermost s-electrons.

Inside the 5d-shell we have the 4f-shell which is successively filled with electrons increasing consecutively from 1 to 14. We have, therefore, a group of elements called lanthanides. The f-electrons lying inside are sufficiently localized that atomic interactions made directly through f-electrons are negligible. Thus the elements have properties similar to each other and it is generally difficult to separate one from the other by chemical methods. We shall now discuss how the nature of such atoms is reflected in the crystal structure and cohesive energy.

Except for Eu and Yb, these elements exhibit the h.c.p. structure or hexagonal structures essentially similar to h.c.p.[20] In any case they generally exhibit the closest-packed structures without cubic symmetry. The cohesive energy of La, Ce, Gd, Tb and Lu is close to the energy of IIIA elements. The energy decreases gradually as we proceed from La to Eu and from Gd to Yb (Figure 9.2),[13] and the minimum energy at Eu and Yb is almost equal to that of Ba belonging to IIA.

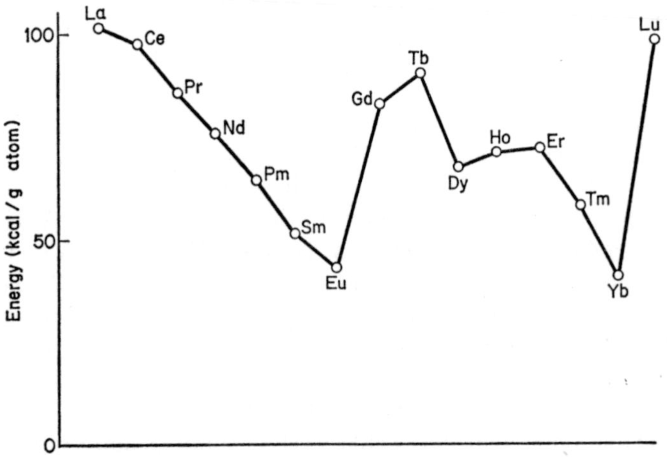

FIGURE 9.2. *Cohesive energy of the lanthanides.*[13]

Spectroscopic investigations generally show that the atoms of lanthanides in the free state have no d-electron except for those of La, Ce, Gd, Tb and Lu, each of which have one d-electron. The electronic configuration of the atoms of Ce, Gd, Tb and Lu is the same as that of IIIA elements except for the existence of the 4f-electrons which do not directly contribute to atomic aggregation. Thus we can understand that the cohesive energy of La, Ce, Gd, Tb and Lu is almost the same as that of Sc and Y. On the other hand the free atoms of Eu and Yb have, respectively, half closed and completely closed f-shells and no d-electron. These f-shells are so stable that no electron can be transferred from the f-state to the d-state to contribute to atomic aggregation in the crystal. In this way we can understand why the cohesive energy of Eu and Yb is as low as that of Ba. The energy difference between the d-levels and the f-levels of lanthanides is generally small in the free state and so the contribution of d-electrons to the atomic aggregates will gradually decrease in the order of La, Ce, ..., Eu, and Gd, Tb, ..., Yb. This will be the reason why the cohesive energy also decreases gradually in the same order as shown in Figure 9.2.

What we have explained above about the cohesive energy is closely related to the crystal structure. In Eu and Yb, the atoms of which have outer s-electrons and stable inner shells in the free state, the crystal structure is to be considered mainly in terms of packing of atoms. Thus we can understand why Eu exhibits the b.c.c. structure like Ba, and Yb the f.c.c. or b.c.c. structure. As mentioned above, in other lanthanide elements d-electrons are considered to contribute to atomic aggregation rather as in the IIIA elements. Thus the explanation of why the crystals exhibit the h.c.p. or certain hexagonal structures (i.e. closest-packed structures without cubic symmetry) is primarily the same as that given previously for IIIA elements.

J. The nature of 3d-electrons and the transformation of Fe

We have discussed the crystal structure and cohesive energy in relation to the nature of atomic aggregation. Let us now discuss Mn and Fe, both of which show exceptional behaviour among transition metals. They belong to VIIA and VIII_1 metals, which except for Mn and Fe exhibit the h.c.p. structure in accordance with the explanation given above. Mn transforms as $\alpha \rightarrow \beta \rightarrow \gamma \rightarrow \delta$ and Fe as $\alpha \rightarrow \gamma \rightarrow \delta$. The crystals of α- and β-Mn exhibit structures with many atoms in the unit cell, but the γ-modification exhibits the f.c.c. structure and the δ-modification the b.c.c. structure in both Mn and Fe.[14] For α- and β-Mn we have to introduce another factor or factors to understand the exceptional behaviour, but other modifications of Mn and Fe can be understood on the basis of the systematic theory developed here.

It is important to consider the role of the 3d-electrons. From the anomalous behaviour of the cohesive energy of the metallic elements in the first long period shown in Figure 9.1, we see that 3d-electrons have a character different from 4d- and 5d-electrons. Probably the main difference arises because the 3d radial distribution function does not have a nodal surface. This will tend to emphasize the localization of 3d-electrons and hence affect the nature of the atomic interactions in the following way.

As already explained, the b.c.c. structure of VA and VIA elements is understood essentially by the overlapping of wave functions extended body-diagonally. In the case of 3d-electrons this tendency is more pronounced and the contribution to atomic aggregation from this cause will be greater than for 4d- and 5d-electrons. Hence Fe exhibits the b.c.c. structure, notwithstanding that the corresponding VIII_1 elements of the second and third long periods exhibit the h.c.p. structure.

Such being the case, the b.c.c. structure of α-Fe is not so stable as that of VA and VIA elements. At higher temperatures the thermal motion will easily change the condition mentioned above and affect the overlapping of the orbitals unfavourably. So it will become difficult for the atomic aggregate to stay in this condition and the crystal will exhibit the closest-packed structure which is usually associated with the metals. The transformation of Fe from α(b.c.c.) to γ(f.c.c.) can be explained in this way.

The further transformation from γ(f.c.c.) to δ(b.c.c.) which takes place at still higher temperature can be understood in the first instance as arising from the contribution of the entropy term as explained. However, it is to be noted that the electron density between neighbouring atoms will be higher for the b.c.c. structure. Consequently the b.c.c. structure will be stabilized by this effect in addition to that of the entropy term. Thus we may understand that some of the observed properties of δ-Fe fall on the extrapolated curve of α-Fe when we plot the values of such properties against temperature, although there are discontinuous changes of properties at both $\alpha \rightarrow \gamma$ and $\gamma \rightarrow \delta$ transition points.

The role of the 3d-electrons mentioned above can also be seen from the transformation of Co which at higher temperatures exhibits the f.c.c. structure like other VIII_2 elements (Rh, Ir), but at lower temperatures usually shows both the h.c.p. and f.c.c. structures. This anomalous behaviour of Co can be considered to arise from the directional character of 3d-electrons. Thus Co tends to exhibit an anisotropic closest-packed structure (h.c.p.) characteristic of VIIA and VIII_1 elements, but this tendency is not strong so that at lower temperature this metal behaves as described above.

K. Conclusion

In this chapter an explanation of the nature of atomic interaction in crystals has been given. Our primary object has been to give an overall systematic view of atomic aggregates; the individual sections have not been treated in great detail because of the limit on space.

Generally speaking, non-metallic substances are characterized by localized electrons, although delocalized electrons, such as π-electrons in some organic compounds or electrons in conduction bands in semiconductors, play an important role. On the other hand metallic crystals are characterized by delocalized conduction electrons, but in some cases localization has to be taken into account. A treatment of crystal structures and cohesive energies such as described here is a first step towards the formation of a wider systematic view of the nature of atomic aggregation.

References

1. Mizushima, S., 'Structure of Molecules and Internal Rotation', 1954, Academic Press, New York; 'Raman Effect', Handbuch der Physik, **XXVI**, Springer-Verlag, Berlin.
2. Ichishima, I., and Mizushima, S., *J. Chem. Phys.*, 1950, **18**, 1686.
3. Ichishima, I., and Mizushima, S., *J. Chem. Phys.*, 1951, **19**, 388.
4. Harada, I., and Shimanouchi, T., *J. Chem. Phys.*, 1966, **44**, 2016.
5. Tasumi, M., and Shimanouchi, T., *J. Chem. Phys.*, 1965, **43**, 1245.
6. Ichishima, I., Kamiyama, H., Shimanouchi, T., and Mizushima, S., *J. Chem. Phys.*, 1958, **29**, 1190.
7. Mizushima, S., and Morino, Y., *Proc. Indian Acad. Sci.*, 1938, **8**, 315; Ichishima, I., and Mizushima, S., *J. Chem. Phys.*, 1950, **18**, 1420.
8. Mizushima, S., *Adv. Protein Chem.*, 1954, **9**, 299.
9. Shimanouchi, T., Tsuboi, M., and Miyazawa, T., *J. Chem. Phys.*, 1961, **35**, 1597.
10. Referred to by Lox, M., and Burstein, E., *Phys. Rev.*, 1955, **97**, 39.
11. Hiraishi, J., and Shimanouchi, T., *Spectrochim. Acta.*, 1966, **22**, 1483.
12. Mizushima, S., and Ichishima, I., *Proc. Japan Acad.*, 1966, **42**, 783 and 789.
13. See for example, Gschneidner, K. A., Jr., *Solid State Physics*, 1964, **16**, 275 (ed. F. Seitz and D. Turnbull), Academic Press.
14. See for example, Wyckoff, R. W. G., 'Crystal Structures', 2nd ed., Vol. 1, 1960, Interscience.
15. Barrett, C. S., *Phys. Rev.*, 1947, **72**, 245; *Acta Cryst.*, 1956, **9**, 671.
16. Zener, C., 'Elasticity and Inelasticity of Metals', 1948, The University of Chicago Press.
17. Brooks, H., *Phys. Rev.*, 1953, **91**, 1027.
18. Fuchs, K., *Proc. Roy. Soc.*, 1935, **A151**, 585.
19. See for example, Mueller, F. M., *Phys. Rev.*, 1967, **153**, 659.
20. See for example, Spedding, F. H., and Daane, A. H., 'The Rare Earths', 1961, Wiley.

CHAPTER 10

Structural Aspects of the Mössbauer Effect

J. F. DUNCAN

A. Introduction

In 1958 Mössbauer[1] reported the results of an experiment in which he was trying to observe nuclear resonance at low temperature. Resonant absorption of γ-radiation can only be expected if the line-width of the incident radiation overlaps with that of the absorption line. According to previous work this would occur most readily at high temperature. At low temperature, therefore, loss of the resonance would be expected. In fact, Mössbauer observed the resonance to increase. The explanation of this result depends on a phenomenon which had not previously been appreciated, namely that the recoil energy loss when a gamma-ray is emitted is about 10^{-3} eV, enough to throw the emitting line (about 10^{-8} eV line-width) out of resonance. Recoil-loss induced by the outgoing gamma-ray may be reduced to about 10^{-8} eV by incorporating the emitter in a crystalline lattice so that emitting and absorbing lines overlap sufficiently for resonance to be obtained. This was the basis of the Mössbauer discovery. In this article illustrations are given of how the technique may be used to investigate certain aspects of structural chemistry. Attention will be primarily confined to ^{57}Fe and ^{119}Sn on which most of the work so far has been done.

B. Method

A conventional Mössbauer assembly consists of an emitter, an absorber and a detector. If absorption occurs in the absorber then the intensity determined in the detector is less than it would be if the absorber were not present. In the case of ^{57}Fe the emitter is usually a source of ^{57}Co which decays to ^{57}Fe with a half-life of 260 days. The ^{57}Fe is formed in an excited state and decays via a 14·4 keV γ-ray from its first excited state with a lifetime (τ) of about 10^{-7} s. The half-life is important because by the Heisenberg uncertainty principle the line-width at half height ($\Gamma_{1/2}$) is determined by the relation

$$\Delta\tau\Delta\Gamma_{1/2} = h/2\pi$$

If τ is small then Γ is broad, and *vice versa*. To determine chemical effects the line-width must be of the same order of magnitude as the shifts expected. The γ-ray from the 14·4 keV level is that on which Mössbauer studies of iron compounds are done. Natural iron contains 2 per cent of ^{57}Fe which means that absorption from the ground to the first excited state in the absorber can occur, provided that the decay energy of the source and excitation level of the emitter are identical within a line-width, i.e. within 10^{-8} eV. In practice this is not the case, for chemical reasons related to the following three parameters.

(i) There is a shift, δ, of the centroid of the resonance spectrum determined by the electron density at the Mössbauer nucleus, viz.

$$\delta = A[\psi_s^2 - \psi_a^2] \tag{10.1}$$

where s and a refer to the source and absorber and A is a calculable constant. The largest contribution to ψ is due to s-electrons which have the largest density at the nucleus, determined for iron, however, by the screening that results from chemical interaction through the 3d shell.

(ii) Quadrupole interaction can occur, the energy of which is

$$\Delta E_Q = \frac{e^2 qQ(3I_z - I[I+1])}{4I(2I-1)} \tag{10.2}$$

In this equation *eq* is the electric field gradient, *eQ* the nuclear quadrupole moment and I the nuclear spin with a component I_z in the z-direction. Since eQ, I and I_z are constant for a given nucleus, ΔE_Q is a measure of the electric field gradient at the nucleus, $[d^2V/dx^2]$, and of the asymmetry of the electrical environment in which the Mössbauer nucleus finds itself.

(iii) The nuclear hyperfine magnetic field (H) is determined by the Hamiltonian

$$\mathscr{H} = -g_N \beta_N HI \tag{10.3}$$

where g_N and β_N are the nuclear g factor and Bohr magneton, respectively. There are also various chemical effects, including electron relaxation which may determine whether the resonance is observed as predicted by equation (10.3).

Since there are these changes in energy as a result of chemical interactions, resonance will not normally occur in a stationary system because the line-width is less than the shift in resonance energy between the absorber and the emitter. To produce resonance it is necessary to change the energy of one of them by the Doppler effect. If the source moves with respect to the absorber with a velocity v, a change in energy occurs equal to

$$\Delta E_\gamma = E_\gamma v/c \tag{10.4}$$

where E_γ is the nuclear γ-ray energy (14·4 keV for ^{57}Fe), v is the Doppler velocity and c is the velocity of light. For ^{57}Fe this corresponds to a change in energy of $4{\cdot}8 \times 10^{-8}$ eV, or 0·011 cal per mole, per cm per second Doppler velocity. A Mössbauer spectrum thus consists of a plot of γ-ray intensity as a function of relative velocity of source towards absorber; typical examples of spectra are illustrated in Fig. 10.1. It is not possible in a short chapter of this kind to describe in detail the various experimental techniques used for studying the Mössbauer effect, which may be found elsewhere.[2, 3, 4] Either electronic or mechanical methods may be used for inducing the Doppler effect. In practice, electronic methods are more convenient, but not always more accurate.

C. Applications in chemistry

(a) *General features*

Chemical applications of the Mössbauer effect depend on evaluating the parameters described in Section B from experimental data. The isomer shift, being related to the s-electron density at the nucleus, can be studied in relation to chemical bonding and the valence state of the constituent atoms. Such applications are numerous, for example, in studying the chemistry of solid systems and in analysis. The quadrupole interaction energy allows the symmetry of the electron cloud about the Mössbauer atom to be determined, and changes in site symmetry with changing conditions to be studied. By contrast, the hyperfine splitting of the magnetic resonances is significant in relation to the following: (i) ferromagnetism and related phenomena, (ii) the interaction between the s-electrons and the nucleus (by application of strong external fields), (iii) the interaction of orbital electron spins with the nucleus (by application of weak magnetic fields), and (iv) the internal fields generated by the orbital and spin vectors of atoms in cases where the relaxation times are long.

Apart from the structural problems, such as those arising in coordination chemistry, and in oxide systems of the spinel and amphibole type, these parameters may usefully be investigated in studying surfaces as well as

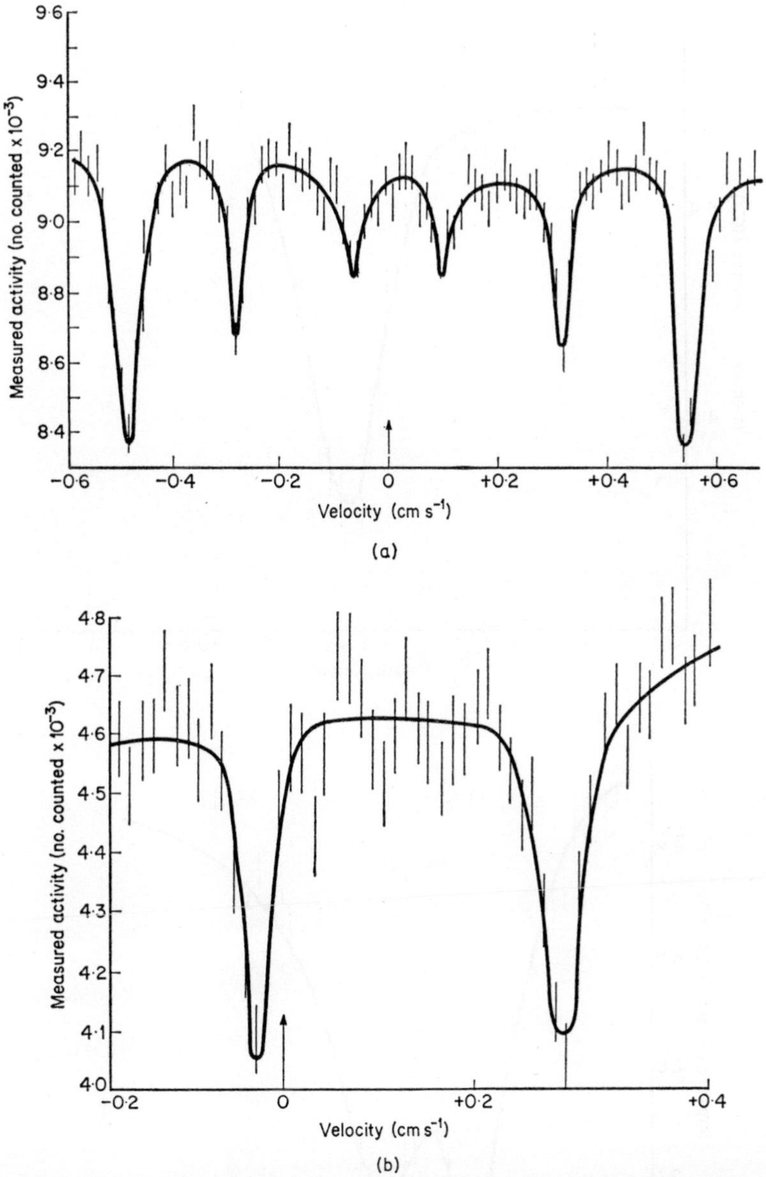

FIGURE 10.1. *Typical Mössbauer spectra: (a) magnetic splitting obtained for a copper-backed $^{57}Co/^{57}Fe$ source in a natural iron absorber; (b) quadrupole splitting (Fe^{2+}); (c) single resonance (Fe^{3+} and Fe^{II});*† *(d) quadrupole splitting (Fe^{III}). (Reproduced with permission from Duncan, J. F., and Cook, G. B., 'Isotopes in Chemistry', Oxford, 1968, pp. 183–184.)*

† Roman numeral superscripts are used to indicate the oxidation state of an element as distinct from the charge state of a free ion.

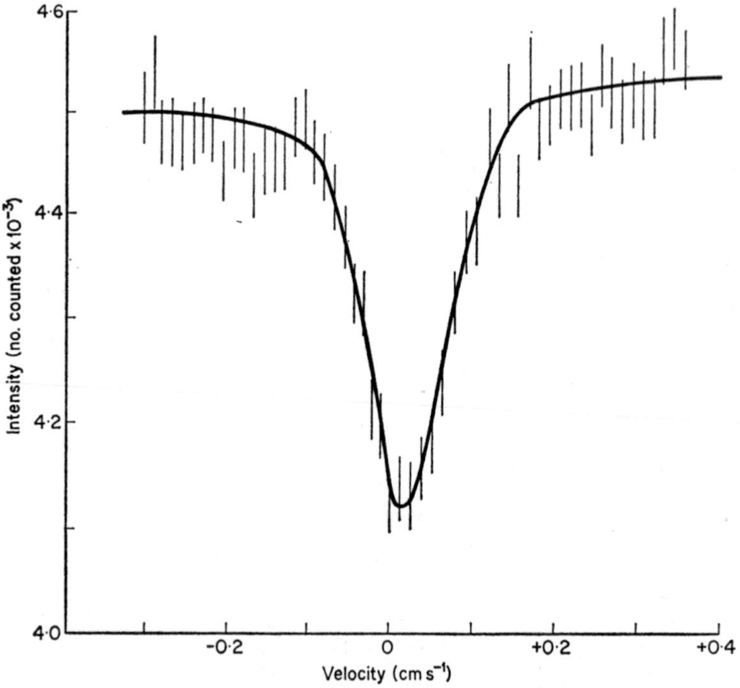

(c)

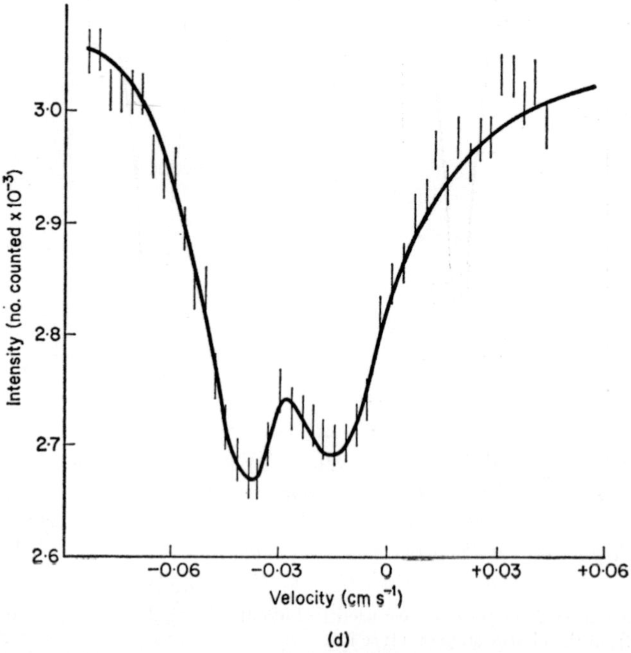

(d)

FIGURE 10.1

some liquids; measurements have also been applied to kinetic studies of chemical reactions. In addition interference of the scattered Mössbauer radiation has formed the basis of structural investigations similar to those conventional with X-rays.

(b) Valence states

The ability to distinguish between different valence states by Mössbauer spectroscopy arises because each valence state has a characteristic value of the isomer shift δ. In many cases it is possible not merely to determine quite clearly the valence state of the Mössbauer nucleus, but even to use Mössbauer spectroscopy as an analytical tool to determine the proportion of the element in its several valence states. Typical isomer shifts for some Mössbauer nuclei are summarized in Table 10.1.

TABLE 10.1 *Some typical values of isomer shift, δ*

Nuclide	Valence state	δ, cm sec^{-1}	Standard
^{57}Fe	Fe^{2+}	0·10 to 0·15	Natural iron
	Fe^{3+}	0·02 to 0·08	
	Fe^{II}	−0·04 to +0·04	
	Fe^{III}	−0·02 to +0·01	
^{119}Sn	Sn^{2+}	+0·25 to +0·5	SnO
	Sn^{IV}	−0·1 to +0·25	
^{129}I	I^{V}	+0·10 to +0·30	ZnTe source
	I^{I}	+0·11 to +0·19	
	I^{0}	Almost zero	
	I^{VII}	−0·2 to −0·47	
	I^{-}	−0·035 to −0·055	
^{151}Eu	Eu^{II}	−0·5 to −1·50	Sn source
	Eu^{III}	Zero to +0·14	

Numerous examples of the simple identification of valency of an element or of changes in valency during chemical reaction could be quoted. Thus, the valence state of iron atoms in iron silicide structures has been studied over the temperature range 4–1000 K. Two lines were observed, the separation of which was highly temperature dependent, showing them to be quadrupole split (see below). There was no evidence of more than one site for the iron atoms in the structure, and the isomer shift was characteristic of that expected for high-spin iron(II).

Distinction between different sites in a chemical compound may be illustrated by the complex iron cyanides. A variety of compounds of different composition (but similar fundamental structure) may be obtained by precipitation of the complex iron cyanides when solutions of either iron(II) or iron(III) chloride are mixed with either potassium ferrocyanide or potassium ferricyanide. These precipitates, of which Berlin green, Turnbull's Blue and Prussian Blue are the best known, contain a variable amount of alkali metal ions held interstitially, and iron atoms in crystallographically equivalent sites (see reference 4). However, the iron atoms are not in chemically identical positions, since if the cation is radioactive before the precipitate is formed, it does not exchange with the complex anion. The dilemma posed by these apparently contradictory facts has now been resolved by Mössbauer spectroscopy. No matter how the precipitate is prepared, its Mössbauer spectrum degenerates into one which implies that the iron cation is in the iron(III) state, and that the anion is almost identical with the ferrocyanide ion.

(c) *Symmetry of electron environment*

Site symmetry is normally investigated by determining the quadrupole interaction energy. However, it is not always easy to distinguish clearly between two resonances corresponding to Mössbauer atoms in different valence states, on the one hand, and a single quadrupole resonance on the other. An example is provided by the ethylenediaminetetraacetate salt of iron(III). This gives two resonances. Since one would expect the salt to be a hexa-coordinated compound of high-spin Fe^{3+}, these two resonances are probably due to the presence of a distorted structure, which would result in a large quadrupole energy. However, with such a complex structure there is always the possibility either that there is more than one type of site present, or that the product is one which contains iron atoms in two valence states.

If the electronic state of the Mössbauer atom is not spherically symmetric, whether it is ionic (as in Fe^{2+}, ${}^5T_2$) or covalent (as in $Fe^{III}(CN)_6^{3-}$, ${}^2T_2$), then strong quadrupole interaction is obtained. But even if the electronic wave function is completely spherically symmetric e.g. ${}^1A_1$ for $Fe^{II}(CN)_6^{4-}$ or ${}^6A_1$ for Fe^{3+}, distortion may arise from the introduction of asymmetry through adjacent ions. This causes an asymmetric electron occupancy of the orbitals, which may influence and distort the s-electron orbitals. In addition, substitution of a symmetric covalent compound with a hetero-atom or group will distort the electron wave function of the central (Mössbauer) atom and give quadrupolar splitting.

The spectrum of the nitroprusside ion ($Fe^{II}(CN)_5NO^{2-}$) contains two distinct resonances with $\Delta E_Q = 0{\cdot}1712 \pm 0{\cdot}0004$ cm s^{-1}. The isomer shift, δ, is not greatly different from that obtained with sodium hexacyanoferrate and, relative to natural iron, is close to zero velocity. However, the very large quadrupole interaction arises from the great asymmetry introduced when one of the six cyanide ligands of $Fe^{II}(CN)_6^{4-}$ is replaced by the NO^+ group. One of the commoner types of distortion is axially along the z-direction, for example in $Fe(py)_4X_2$ (py = pyridine; X = uninegative anion). In such cases, the value of the quadrupole interaction can be used to indicate the degree to which X donates electrons to the Mössbauer nucleus. By this means the order Cl < Br < OCN < SCN < I has been established. Furthermore, quantitative correlation with the observed magnetic moments for these high-spin iron(II) compounds can be obtained by assuming that the distortion is due to axial elongation in the X—Fe—X directions. The greater the spin multiplicity, the greater the electric field gradient caused by neighbouring ions and groups. Hence we expect, as is observed, that the ΔE_Q value should be for $Fe^{2+} > Fe^{3+}$, for $Fe^{2+} > Fe^{II}$ and for $Fe^{3+} > Fe^{III}$. If we now plot ΔE_Q against δ for ^{57}Fe we obtain the correlation diagram shown in Fig. 10.2, in which the iron atoms of given valence state occur in specific areas.

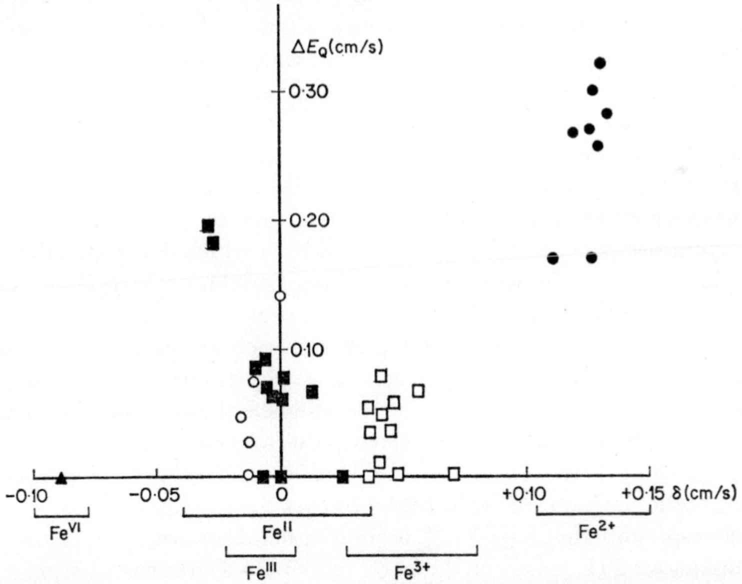

FIGURE 10.2. *Correlation diagram showing relation between ΔE_Q and δ for iron compounds,* ● Fe^{2+}; □ Fe^{3+}; ■ Fe^{II}; ○ Fe^{III}; ▲ Fe^{VI}. (*Reproduced with permission from Brady, P. R., Duncan J. F., and Mok, K. F.,* Proc. Roy. Soc., *1965*, **A287**, *350*.)

From Figure 10.2 it is possible to determine rapidly the valence state, the spin type and the degree of electronic asymmetry in any iron compound for which δ and ΔE_Q are known. This information, related directly to electronic structure, can in turn enable stereochemical structures to be inferred in many cases. A simple example of this is the identification of geometric isomers for which the ratios of ΔE_Q values can be shown to be 2·0, the value for the *trans* compound being the larger.

(*d*) *Magnetic interactions*

The study of nuclear hyperfine fields, both in the presence and in the absence of applied external fields, is one of the areas in Mössbauer spectroscopy where important advances of structural significance are to be expected in the next few years. Already this area has yielded quite unexpected information about the nature of the internal structure of the atom, and especially the realization that very large magnetic fields are to be found at the nucleus as a result of the orbital motions of the electrons. A study of these fields leads to further applications of chemical significance.

If the atom containing the Mössbauer nucleus is isolated from all interactions with other electron spins, then the relaxation time of the electrons may be so long that hyperfine structure is observed. This is normally the case with high-spin iron(III) compounds when the temperature is sufficiently low. There are two possible effects. Firstly, there may be line broadening or splitting; secondly, there may be a change in the isomer shift δ. One may normally assume that the precession time of the nucleus is extremely fast: the effects observed depend on the duration time of the Mössbauer event relative to the relaxation time of the atomic electrons. Temperature-dependent hyperfine interactions which occur with the very simple compound ferric ammonium alum, $Fe(NH_4)(SO_4)_2 12H_2O$, have been perhaps the most widely studied. The line-width obtained at room temperature with this compound is about twice the natural line-width, but when the material is cooled from room temperature to 20 K, magnetic splitting becomes evident, even in the absence of any applied magnetic field. The spectrum has been explained as due to crystal field splitting of the 5/2, 3/2, and 1/2 electronic states, but there is considerable overlap between some of the magnetically split peaks, so that the resolution of the various levels, especially that with $S = \frac{1}{2}$, is not clearcut. However, by studying the compound at 20 K, where all the levels are populated, the energy separation of the microstates has been shown to be about 0·08 cm^{-1}. This compound is magnetically so simple, and yet gives such surprising complexity in its Mössbauer spectrum that it has been extensively studied with externally applied fields both at room and at very low temperatures.

Hyperfine magnetic effects are observed in many other cases with high-spin iron(III) when the temperature is low, or the ion is in an isolated environment. For example, in glasses, when small quantities of the separated isotope ^{57}Fe are introduced, magnetic hyperfine spectra are observed corresponding to magnetic fields of the order of 500 kgauss at the nucleus.

Magnetic hyperfine spectra are obtained only when the orbital motion of the unpaired electron about the Mössbauer nucleus is restricted in some way. In the above examples the restriction of the electron motion of individual atoms is imposed by lowering the temperature, which has the effect of increasing the relaxation time of the atomic electrons. In almost all other cases the restriction is imposed by magnetic ordering, i.e. by coupling the spins of the Mössbauer atom with some other part of the system. Coupling may be with the spins of neighbouring atoms, as in ferromagnetic material; or with externally applied fields; or by inducing some degree of ordering (but not necessarily ferromagnetism) into paramagnetic material by a variety of devices. Lowering the temperature usually facilitates the production of hyperfine spectra, since, apart from the change in relaxation time, coupling is thus made easier.

In a ferromagnetic system, the spins of the atomic electrons are aligned parallel. At absolute zero of temperature the alignment is complete, but as the temperature is increased, and thermal agitation causes the alignment to become more and more random, the degree of magnetization decreases, in accordance with the well known Brillouin–Weiss law. It becomes zero at the characteristic Curie temperature. We therefore expect a system in which the Mössbauer nuclei exhibit hyperfine interactions due to ferromagnetic ordering to display a similar behaviour, and to be a function of temperature of similar form. There are many examples of this type of behaviour. It has been shown, for instance, that the hyperfine field observed at the Mössbauer nucleus in the case of ^{57}Fe in natural iron follows the saturation magnetization very closely in its temperature dependence. Similarly the magnetic field calculated from the nuclear hyperfine spectrum of atoms in the Fe_I site have been observed to follow closely a Curie plot for the ferromagnetic Fe_3P, using the Weiss molecular field model with a Curie temperature of 716 K and $J = 1$.[6] An important feature of this correspondence should be noted. The Curie dependence of the bulk magnetic moment refers to the alignment of the spins of the electrons in the unfilled orbitals of the atom. On the other hand, the nuclear hyperfine field is primarily due to the interactions of the s electrons with the nucleus. We could not, at first sight, have expected any correspondence between these two quantities, although it seems often to have been observed. The fact that such a correspondence is observed implies that the spin alignment caused by application of the field in the outer atomic electrons also induces a corresponding alignment in the inner s shells. This result, therefore, affords direct evidence that the inner shells of atoms are spin-polarized by effects in the outer shells, as required by the Fermi contact interaction. A consequence of structural importance is that it is possible to determine the atomic magnetic moments, and therefore the electron spin states of two atoms of different kind (e.g. two transition metals). Conventional determination of the magnetic moment by the Gouy or Faraday method can be used to measure the total spin contribution of the two elements together; and if one of them is a Mössbauer atom, determination of the nuclear hyperfine field enables the spin contribution of that element to be estimated. There are limitations to this technique because the hyperfine field is not always observable and no case in which the method has been used for diagnostic structural purposes has yet been reported. Nevertheless, it remains an exciting new possibility of structural analysis.

Perhaps a simpler case to consider is that of iron in alloy systems, which have been extensively studied. Thus, in iron–silicon alloys, the eight nearest-neighbour atoms of this body-centred system are those which determine the hyperfine field. As with many other alloys, the introduction of impurities into the system slowly reduces the field, although in some cases, such as FeV, the field of the impurity atom influences atoms well beyond the nearest neighbours. Alloys of iron with palladium and platinum show very high magnetization effects, much larger than can be due merely to the 3d electrons of iron. However, investigation of the hyperfine field in very dilute alloys of these elements indicates fields at the nuclei of about 300 kgauss, corresponding to a moment of about 2 Bohr magneton residing on each iron atom.

In magnetic oxides, such as spinels, and in many non-magnetic systems, the valence state of an atom has been widely studied by the Mössbauer effect. Investigation of the symmetry properties of iron sites in relatively ionic compounds has been made for FeF_6^{3-}, $FeCl_3$ and $FeBr_3$, all of which contain octahedrally coordinated, high-spin iron(III), and exhibit no quadrupole interactions. The isomer shift generally increases with the electronegativity of the ligand. A significant quadrupole interaction ($\Delta E_Q = 0{\cdot}055$ cm s^{-1}) was observed for $[(H_2O)FeF_5]^{2-}$ but not for the corresponding chloride complex at room temperature, suggesting that in the latter the iron atom is in a much more nearly octahedral electronic environment. By contrast, $FeCl_3, 6H_2O$ gives an unusually large $\Delta E_Q = 0{\cdot}10$ cm s^{-1}, although the isomer shift is characteristic of the high-spin Fe^{3+} ion. This suggests that a low-symmetry structure is present, which is different from the regular octahedral structure of the corresponding aluminium compound. One possible explanation is that some of the chloride ions are very close, if not directly bonded, to the iron atoms. Changes in symmetry can also occur as the composition is changed from that of the pure iron compound to that of the pure aluminium compound by progressive substitution of one ion by the other. This phenomenon is found also with iron(III) ammonium alum, which has been investigated by both e.s.r. and Mössbauer methods.[7]

An example in the field of coordination chemistry illustrates the unique information which can be obtained by the technique. The example concerns the magnetically anomalous structures of iron(II) complexes containing 2-(2-pyridylamino)-4-(2-pyridyl)-thiazole (papth), a tridentate ligand, and a variety of anions.[8] The magnetic moments of these compounds depend on the anion, varying from 5 to 1 Bohr magnetons. Two distinct classes of compounds are obtained: the first obey the Curie–Weiss law, with magnetic moments corresponding to Fe^{2+}, are normally yellow and include ClO_4^-, I^-, $C_2O_4^{2-}$ and SO_4^{2-} as anions; the second are magnetically anomalous, are normally brown in colour and include the thiocyanate ($\mu = 4{\cdot}5$ B.M. at room temperature), another sulphate (3·4 B.M.), the nitrate (2·4 B.M.), the chloride (1·9 B.M.) and the bromide (1·3 B.M.). The moments of these compounds vary from that characteristic of spin-paired Fe^{II} at low temperatures, to that characteristic of high-spin Fe^{II}. This behaviour shows quite clearly that the ligand field strength is very close to the point of crossover from high-spin to low-spin states. The question has been asked whether the anomalous behaviour could be attributed to (thermally dependent) equilibrium between the two spin types, or whether it would be better to regard the wave function in this region as made up from contributions from the wave functions of both the low-spin and the high-spin type. If the latter explanation holds, then quite unusual behaviour might be expected. The answer to this question in some measure is determined by the time scale of the experimental assembly which is used to investigate the phenomenon. Measurement of magnetic moments could hardly be expected to distinguish between these two viewpoints unless the time necessary to establish equilibrium between the different electron states is relatively long. If the time necessary for equilibration were greater than about 10^{-6} second, the Mössbauer spectrum would reveal the presence of the high-spin and the low-spin species, together in any particular specimen. On the other hand, if the time scale were less than 10^{-10} second, then the Mössbauer spectrum would reveal a time-averaged electronic environment (if it revealed anything at all), and the two theoretical viewpoints would become experimentally indistinguishable. Experimentally, it is found that the magnetically normal compounds have $\Delta E_Q = 0{\cdot}03$ cm s^{-1} and $\delta = 0{\cdot}1$ cm s^{-1}, as expected for high-spin Fe^{2+}. The anomalous compounds show much smaller values of ΔE_Q and

δ at low temperatures. Typical is Fe(papth)$_2$(SCN)$_2$, H_2O. Although its magnetic data show it to be high spin at room temperature, its Mössbauer spectrum is weak and diffuse, indicating that the electronic transitions are very rapid, with a time-scale of less than 10^{-7} s. Other compounds, such as Fe(papth)$_2$$SO_4$, 5$H_2O$, show two sharp absorptions, corresponding to the low-spin form, superimposed on a diffuse background of the other form. The energy difference between the two spin states is similar to thermal energies, since it is influenced by temperatures such that the spin-paired form is favoured by low temperatures.

(e) Single crystal studies

More detailed information about the environment of the Mössbauer atom can be obtained using single crystals rather than polycrystalline material. The simplest technique is to study the variation in the relative intensities of the quadrupole arms as a function of the angle between the direction of the γ-rays and the symmetry axis, possibly simultaneously applying an external magnetic field. Associated with such a technique, however, one may also investigate the stereochemical arrangements of the chemical bonds by studying the anisotropy of recoil. Such work is not confined to pure compounds. Trace Mössbauer atoms may be introduced into single crystals of other materials in order to investigate, for example, the formation and structure of defects and problems of site symmetry. The intensities of the two quadrupole arms, I_1 and I_2, may be shown theoretically to be in the ratio

$$\frac{I_1}{I_2} = \frac{3(1+\cos^2\theta)}{5-3\cos^2\theta} \tag{10.5}$$

where θ is the angle between the direction of the incoming γ-ray and the symmetry axis. The experimental method is therefore to measure I_1 and I_2 and plot I_1/I_2 as a function of θ, known from crystallographic methods. It is then relatively simple to determine the direction of the electric field gradient. Also, if q is positive, then I_1 refers to the quadrupole arm at the more negative velocities, and *vice versa*; this allows the sign of q to be determined. Remarkably few single crystals have been studied intensively. One of the most popular has been sodium nitroprusside. A difficulty of investigating this compound is that there are four orientations of the nitroprusside ion in one crystallographic plane, which means that the results do not display features related to a single orientation and are therefore more difficult to interpret. Nevertheless it is possible to show that q is negative for this system, implying that electrons are donated to the Fe^{2+} ion less readily by NO^+ than by CN^-.

If the bond energies holding a Mössbauer atom in its chemical environment are not isotropic, we should expect that the total intensity of the resonance (including both arms of any quadrupole interaction) would be dependent on the direction of observation. As an integral part of any investigation of quadrupole interaction intensities, therefore, it is also possible to obtain evidence for bond anisotropy from the line intensity determined as a function of angle. Very little work of this kind has been done because careful experimentation is necessary to eliminate the effects of variable Compton scattering as a function of thickness. For this reason, most work has involved a direct comparison between the intensities measured in two different directions. Thus Housley and Nussbaum[9] report an investigation of ^{57}Fe in a zinc host lattice, for which a quadrupole interaction was observed. The ratio of the line intensities was 3:5 in a direction perpendicular to the c-axis and 2:1 in the parallel direction, indicating a positive electric field gradient. In addition, the total fractional intensities of recoilless γ-rays were, respectively, 0·64 and 0·41. Assuming that the mean-square-amplitude of vibration, $\langle x^2 \rangle$, can be accurately calculated, using lattice dynamic theory, from its phonon dispersion curve measured by neutron scattering, one may calculate that $\langle x_{\parallel}^2/x_{\perp}^2 \rangle$ has a value of about 2·8. This may be compared with the experimental value obtained from the Mössbauer results of greater than 2. One may therefore conclude that the mean-square-amplitude of vibration of the iron atom in the direction parallel to the c-axis is more than twice as large as in the perpendicular direction.

Further structural information may be obtained by application of magnetic fields to single crystals. The most extensive study hitherto is that of ferric ammonium alum, with the iron atoms sometimes replaced by aluminium atoms so that the molar concentration of iron is only a few per cent. Even at room temperature the applied field can cause line broadening to a degree dependent on the orientation of the field relative to the crystal axes and to the γ-ray. At low temperature the broadening has been found to split in a complicated fashion giving 11 lines. This has been shown to be due to magnetic hyperfine splitting of the electron spin states $\pm\frac{5}{2}$, $\pm\frac{3}{2}$, and $\pm\frac{1}{2}$. However, if single crystals of the pure compound are studied at temperatures in the region of the absolute zero with magnetic fields of the order of 24 kgauss oriented parallel to the γ-ray and perpendicular to the absorber, a 4-line spectrum is obtained, the relative intensities of the lines being dependent on the conditions. The theory for interpreting these effects has not yet been adequately worked out, but it is clear that microstates with very small energy separations are being identified, the energies of which are affected by very small structural displacements.

(*f*) *Diffraction phenomena from single crystals*

The use of the Mössbauer effect in conjunction with diffraction phenomena arises because the wavelengths of the Mössbauer radiation are very frequently of the same order as those of X-rays, and like X-rays they can be diffracted by solid crystal lattices. The possibility of using this effect for structural studies and for other types of investigation has been developed in detail by Black *et al.*[10] There are two ways in which such diffraction can take place. The first is identical with what would occur if X-rays of the same energy impinged on the crystal—that is, the scattering is due to interference with the electrons of the atom. The second can occur as a result of the Mössbauer event itself.

From theory it appears that Mössbauer scattering is observable both for ^{57}Fe and ^{119}Sn, and possibly for ^{161}Dy and ^{169}Tm. In the case of ^{57}Fe, it has been observed with $K_4[Fe(CN)_6]$, $3H_2O$, for which the isomer shift is small enough for resonance to be obtained at zero velocity. In such a case it is possible from one single experiment to obtain structural information about the positions of all the atoms in the lattice (from Rayleigh scattering), and also about the Mössbauer atoms only from the scattering obtained at resonance. In addition, interference of the two types results in asymmetry of the velocity spectrum from which information about the relative phases can be obtained. The Mössbauer and Rayleigh scattering may also be distinguished by the temperature dependence of the resonances. In the former case, the intensity is given by the product of the fractions of the recoilless γ-rays for absorption and emission; in the case of X-ray scattering it is simply the well known Debye-Waller factor which controls the intensity directly. These features make the use of the Mössbauer effect for structural investigations a real possibility, but as yet the conventional anomalous dispersion X-ray technique appears to be more advantageous, except in the cases of compounds containing Mössbauer atoms of the same type in the different types of site. Only then is it worthwhile accepting the much lower detection sensitivity obtainable with the Mössbauer effect compared with X-rays. No clearcut demonstration of this application has yet been reported.

(*g*) *Investigations of reaction rates*

All the structural topics so far discussed have been concerned with static systems. Kinetic investigations can, of course, also be made. An important feature of Mössbauer methods is that changes in bonding of the Mössbauer atom in the same compound can be investigated, and the rates quantitatively determined. However, a limitation is that only those atoms bound sufficiently strongly in their environments to give a substantial fraction of recoilless interactions are detected, and quantitative work is not possible if the positions of the lines overlap sufficiently to prevent adequate resolution. There are three other possible causes of error. Firstly, errors due to different absorber thicknesses must be eliminated. Usually this is most easily achieved by using absorber thicknesses close to the maximum of the intensity-thickness absorption curve of the Mössbauer line. Secondly, reference to peak heights as a measure of the amount of material in a particular environment should be used only if one is sure that the line width is constant throughout the series of experiments. In other cases it is more satisfactory to use the integrated line area intensity. Thirdly, it is very important that any estimates of the amount of material present should depend on an absolute calibration, and should not be merely a relative estimate. The reason for this is that the Mössbauer atom will be visible in the spectrum only if the recoilless fraction is adequately large. Some of the Mössbauer atoms may therefore not be detected at all, so that complete conversion of the material from one form to another may not be obtained, even though there is no change in the spectrum with time. The usual way of making an absolute calibration is by means of synthetic mixtures of the reactant and product, and of any intermediate which is detected for comparison with the spectra actually obtained in the course of the reaction. These and other problems associated with Mössbauer studies of reaction kinetics are discussed elsewhere.[11]

Quantitative work on reaction kinetics has been done, mainly in Wellington, New Zealand, where the following studies have been made:

(i) the mechanism of formation of the aluminosilicate mullite ($3Al_2O_3.2SiO_2$) from kaolinite ($Al_2O_3.2SiO_2.2H_2O$);

(ii) the mechanism of the oxidation of ilmenite ($FeTiO_3$) and the rate at which TiO_2 and Fe_2O_3 separate and subsequently react to form *pseudo*-brookite Fe_2TiO_5 and other products;

(iii) the mechanism of the formation of the spinel phase $ZnFe_2O_4$ from Fe_2O_3 and ZnO.

In all these investigations the reaction has been studied by several different instrumental methods, namely, X-ray diffraction, infrared spectroscopy, and Mössbauer absorption. Detailed information about the structure of the transition state has emerged by comparing the results of the several different methods. Here we shall give further details about the last of these investigations; these illustrate most clearly the conclusions to be drawn, though similar conclusions emerge in the other investigations.

Work with X-ray and other techniques has shown[11] that the reaction $ZnO + Fe_2O_3 = ZnFe_2O_4$ proceeds by vapour transport of ZnO on to the surface of Fe_2O_3, followed by diffusion of zinc and iron cations throughout the oxide lattice. According to X-ray methods the observed reflections used for following the reaction are primarily due to metal atoms situated between the oxygen atoms in the oxygen layers. The Mössbauer effect, however, studies the iron atoms in lattice positions which are sufficiently 'stiff' to give a reasonable fraction of recoilless γ-rays. The type of spectra

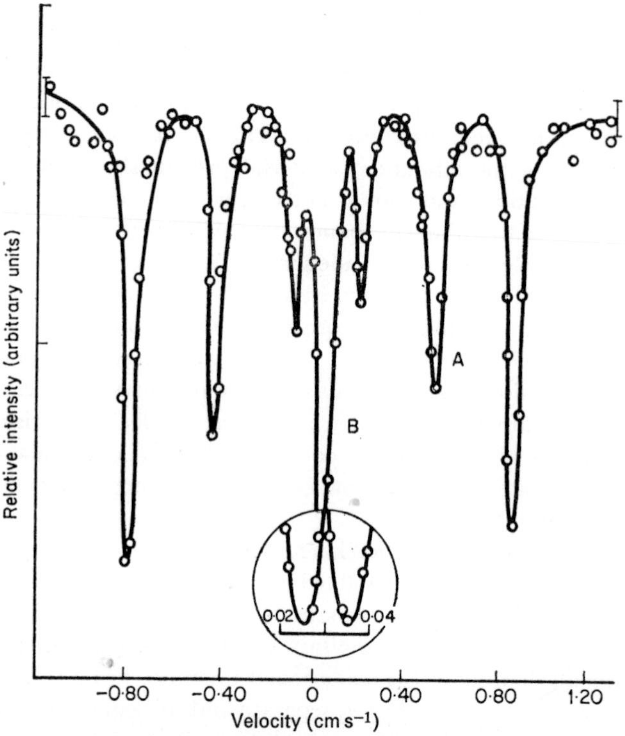

FIGURE 10.3. *A typical Mössbauer spectrum of a mixture of ZnO, Fe_2O_3(A) and $ZnFe_2O_4$(B). Inset, peak due to $ZnFe_2O_4$ enlarged to show quadrupole splitting. The isomer shift is quoted relative to natural iron and has been determined using a sodium nitroprusside standard. The error bars indicate $\pm\sigma$.* (*Reproduced with permission from Duncan, J. F., MacKenzie, K. J. D., and Stewart, D. J.*, Symposia of the Faraday Soc., *1967*, **1**, *105*.)

obtained is shown in Figure 10.3. The hyperfine spectrum is due to Fe_2O_3 and the central line to $ZnFe_2O_4$. It is therefore possible to compare the reaction rates by X-ray diffraction and Mössbauer spectra *independently*, as shown in Figure 10.4. When this is done, the measured rates are in satisfactory agreement only after long periods and/or at high temperature. The conclusion is therefore that even when the oxygen atoms are properly fixed as expected in the product, the cations are still mobile for very long periods before they achieve any stable situation which can be described as 'crystalline'. In this sense, therefore, the formation of $ZnFe_2O_4$ from the component oxides must be regarded as a two-stage reaction involving at least two transition steps, one for fixation of the oxygen lattice and the other

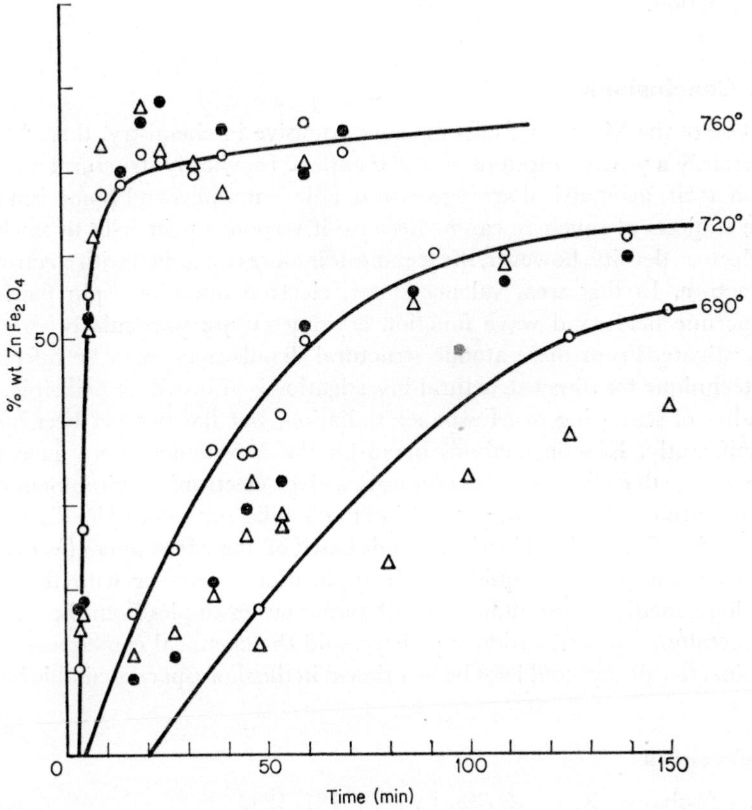

FIGURE 10.4. *Percentage $ZnFe_2O_4$ in reaction mixture estimated from integrated Mössbauer line intensity ●, Mössbauer peak intensity △, and X-ray scattering ○, plotted against time of reaction. (Reproduced with permission from Duncan, J. F., MacKenzie, K. J. D., and Stewart, D. J.,* Symposia of the Faraday Soc., *1967*, **1**, *113.)*

for fixation of the cation lattice. Closer study of reactions such as this, and especially a consideration of the entropy changes involved, suggests that oxygen movement is involved in cation rearrangement and *vice versa*, and that the transition step cannot be more clearly defined than by assuming a whole range of configurations through which the system can pass in forming the products, and which may be different for different individual atoms. Mössbauer line-broadening, which usually decreases as time and temperature are increased, confirm that there is normally a multitudinous array of cation sites of different energy in such a system. Clearly, then, the naive assumption of a *single* transition state is, in these cases, an unhelpful abstraction.

D. Conclusions

Use of the Mössbauer effect is now extensive in chemistry, there being currently a yearly output of over 200 papers. Inevitably structural aspects such as the geometrical arrangement of adjacent atoms and isomerism are the subjects of much research. Because it responds primarily to nuclear s-electron density, however, the technique is more concerned with electronic structure. In this area, valence states, electron donation, spin pairing, hyperfine fields and wave function asymmetry are particularly easy to investigate. From these atomic structural details may often be inferred. A technique for direct structural investigation is afforded, in principle, by studies of scattering of Mössbauer radiation, but has not yet been used significantly. Kinetic methods based on the Mössbauer effect have the advantage that the rate of transfer from a given electronic environment can be quantitatively studied. In addition to the solid state examples discussed in Section C, part (g), kinetic methods based on the Mössbauer effect have also been applied to liquids. Such applications, together with those in geology, medicine and industry, with numerous examples from the realms of metallurgy, coordination chemistry, and the chemical consequences of radioactive decay, could not be mentioned in the short space available here.

References

1. Mössbauer, R. L., *Z. Physik.*, 1958, **151**, 124.
2. Duncan, J. F., 'Nuclear Radiation Resonance (Mössbauer Effect) in Chemistry', Chapman and Hall, forthcoming publication.
3. Brady, P. R., Wigley, P. R. F., and Duncan, J. F., *Rev. Pure Appl. Chem. (Australia)*, 1962, **12**, 165.
4. Duncan, J. F., and Cook, G. B., 'Isotopes in Chemistry', Clarendon Press, Oxford, 1968.
5. Golding, R. M., Mok, K. F., and Duncan, J. F., *Inorg. Chem.*, 1966, **5**, 774; Duncan, J. F., Golding, R. M., and Mok, K. F., *J. Inorg. Nuclear Chem.*, 1966, **28**, 1114.
6. Bailey, R. E., and Duncan, J. F., *Inorg. Chem.*, 1967, **6**, 1444.
7. Duncan, J. F., and Trotter, K. W., forthcoming publication.

8. Bailey, R. E., Sylva, R. N., and Goodwin, R. A., personal communication.
9. Housley, R. M., and Nussbaum, R. H., *Phys. Rev.*, 1965, **138**, A753.
10. Black, P. J., *Nature*, 1965, **206**, 1223; Black, P. J., Longhorn, G., and O'Connor, D. A., *Proc. Phys. Soc.*, 1964, **83**, 925, 937; Moon, P. B., and Black, P. J., *Nature*, 1960, **188**, 481; Black, P. J., and Duerdoth, I. P., *Proc. Phys. Soc.*, 1964, **84**, 169.
11. Duncan, J. F., MacKenzie, K. J. D., and Stewart, D. J., *Symposia of the Faraday Soc.*, 1967, **1**, 103.

CHAPTER 11

Non-Stoichiometry in Fluorite Structures

L. E. J. ROBERTS

A. Introduction

The ideal fluorite structure of CaF_2 may be thought of as a simple cubic array of anions with a cation in the centre of each alternate cube, with a coordination number of eight. It has long been recognized that the resulting face-centred cubic structure of cations is essentially preserved over wide variations of the overall composition from MX_2. The evidence is of two types. Oxides of transition metals having this structure—the actinide and rare earth oxides—exist with compositions of $MO_{1\cdot6}$ to $MO_{2\cdot5}$, and single phase rare-earth fluorides from $MO_{2\cdot0}$ to $MO_{2\cdot25}$ at least; secondly, both oxides and fluorides readily form 'anomalous' solid solutions with compounds of cations of different valency, thus necessarily disturbing the cation:anion ratio of the crystal as a whole, while preserving statistically the fluorite structure, or a structure formed from fluorite with a small distortion.

The non-stoichiometric fluorite structures are of particular interest because the characteristic type of disorder is unique. The defects introduced into the lattice to account for the wide departure from the ideal composition are confined to the anion sublattice; the cation sites are all occupied, except possibly at temperatures approaching the melting points, and the cations do not move far from the ideal fluorite positions. Vacancies appear on the anion sublattice at low anion:cation ratios, and anions are accommodated in interstitial sites at high ratios. The concentrations of such 'defects' can be so large, however, that this terminology is misleading. It is apparent that the structure can tolerate large numbers of cations with a coordination number that differs from eight. Some of the evidence for local atomic configurations and for the longer-range ordering effects that occur at lower temperatures is reviewed in this chapter. The special case of the non-stoichiometric hydrides with a fluorite structure is not dealt with.

The high degree of disorder on the anion sublattice is accompanied by a high mobility of the anions, a feature which dominates the chemistry of the oxides at temperatures below 1500°C. This is illustrated by some representa-

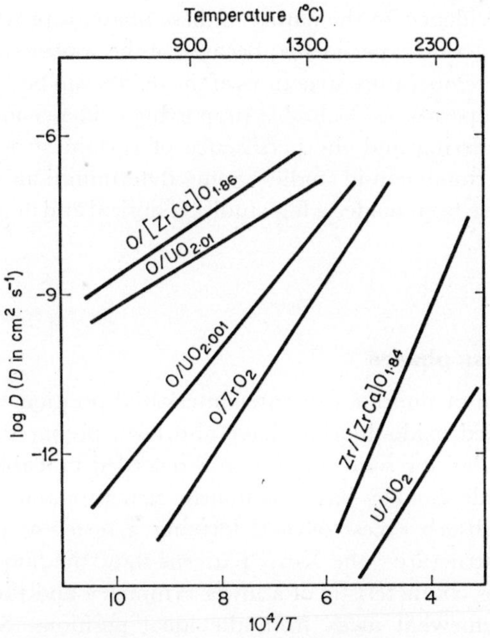

FIGURE 11.1. *Oxygen and cation diffusion coefficients for the fluorite-type oxides* UO_2, ZrO_2 *and* $(Zr, Ca)O_{2-x}$. (*Reprinted with permission from* Science of Ceramics, *1968*, **4**, *329*.)

tive values of the diffusion coefficients of oxygen and metal ions in fluorite-type oxides which are collected in Figure 11.1. Oxygen diffusion is very rapid, with an activation energy of 25–30 kcal mole^{-1} for interstitial or vacancy structures; this is probably a fortuitous circumstance, since the atomic movements concerned are unlikely to be the same. Close to stoichiometry, the energy of activation of oxygen diffusion is higher (~65 kcal mole^{-1}) owing to the need to form intrinsic defects, but oxygen diffusion is still many orders of magnitude faster than metal diffusion and remains so up to the melting point. These figures explain why it is so difficult to quench-in the high-temperature structure of the non-stoichiometric phases: oxygen can diffuse a mean distance of 1 micron in 1 second even at 900°C and so phase-disproportionation reactions depending on variations in oxygen composition are very rapid. Conversely, the cation sublattice is virtually frozen below 1200°C, at which temperature movement of 0·1 micron would take a given cation about 300 hours. Phase separations and ordering processes depending on cation movement will be very sluggish; different preparations may contain different concentrations of cation disorder, depending upon the thermal history.

Most of the evidence for the nature of these phases is provided by X-ray or neutron diffraction experiments; because of the considerations outlined above, the high temperature structures of the oxides can be studied reliably only at high temperatures. Valuable supporting evidence for phase limits, the onset of ordering and the persistence of certain structural features comes from thermodynamic studies. Some determinations of conduction mechanisms have been made: a few studies of optical and magnetic properties are available.

B. Anion-excess phases

Either oxygen or fluorine can enter interstitial positions in the fluorite lattice and mixed oxide-fluorides have also been prepared. The compositions of some limiting solid solutions are recorded in Table 11.1.

All the actinide dioxides have the fluorite structure and both UO_2 and PaO_2 readily absorb excess oxygen forming a series of compounds of closely related structure;[1] the X-ray patterns show the fluorite reflections split into groups characteristic of a lower symmetry and the metal atoms have moved somewhat away from the ideal positions. Solid solutions containing these oxides, such as $(U,Th)O_2$ or $(Pa,Th)O_2$, also accommodate excess oxygen readily in interstitial positions but the crystals often retain cubic symmetry, presumably because the metal cations are unable to rearrange to form an ordered array. It is interesting that NpO_2 does not show the same property; such 'non-stoichiometry' on the oxygen-rich side as has been reported has been shown to be due to surface processes of chemisorption,[2] though a higher oxide, Np_3O_8, is easily prepared. The lattice constant of NpO_2 is smaller than that of UO_2 and the higher oxidation states of neptunium are less stable; however, neptunium(IV) is not easily oxidized even in solid solutions where spatial considerations would be less important. The reason may be the comparative instability of neptunium(V) in such environments; the oxygen-rich protactinium oxides must contain protactinium(V) and there is some, but not conclusive, evidence that the fluorite structures contain uranium(V) rather than uranium(VI).

TABLE 11.1. *Anion-excess phases with fluorite or distorted fluorite structures*

Fluorite phase	Dissolved phase	Limiting composition	Structure	Ref.
CaF_2	ThF_4	$(Ca,Th)F_{2\cdot48}$	Cubic	a
LaOF	LaF_3	$LaO_{0\cdot55}F_{1\cdot88}$	Cubic	b
ThO_2	ThF_4	$\sim ThO_{1\cdot6}F_{0\cdot8}$	Cubic	c
UO_2	O_2	$UO_{2\cdot27}$	Cubic	d
	O_2	$UO_{2\cdot33}$	Tetragonal	d
$U_{0\cdot5}Th_{0\cdot5}O_2$	O_2	$U_{0\cdot5}Th_{0\cdot5}O_{2\cdot32}$	Cubic	e
PaO_2	O_2	$PaO_{2\cdot5}$	Tetragonal	f
SmF_2	F_2	$SmF_{2\cdot25}$	Cubic	g
		$\sim SmF_{2\cdot47}$	?	

(a) Zintl, E., and Udgard, A., *Z. anorg. Chem.*, 1939, **240**, 150.
(b) Klemm, W., and Klein, H. A., *Z. anorg. Chem.*, 1941, **248**, 147.
(c) D'Eye, R. W. M., *J. Chem. Soc.*, 1958, 196.
(d) Thermodynamic and Transport Properties of Uranium Dioxide and Related Phases, I.A.E.A., Vienna, 1965.
(e) Anderson, J. S., Edgington, D. N., Roberts, L. E. J., and Wait, E., *J. Chem. Soc.*, 1954, 3324.
(f) Roberts, L. E. J., and Walter, A. J., *Physico-Chemie du Protactinium*, C.N.R.S., Paris, 1966, 51.
(g) Catalano, E., Bedford, R. G., Silviera, V. G., and Wickman, H. H., *J. Phys. and Chem. Solids*, 1969, **30**, 1613.

(*a*) *Uranium oxides*

As a result of numerous investigations, which have been reviewed in detail elsewhere,[3] the upper boundary of the UO_{2+x} phase has been established with fair certainty up to 1700°. The next higher oxide, the cubic U_4O_9 phase, which is stable to about 1123°C, can be regarded as an ordered superstructure of the UO_{2+x} phase. The lower boundary of the fluorite phase is $UO_{2\cdot00}$ at temperatures below 1400°C but at higher temperatures, and under strongly reducing conditions, the boundary moves to the oxygen-deficient side.[4] The relevant portion of the phase diagram is shown as Figure 11.2. The region of the phase diagram around the triple point has recently been investigated again without introducing any major changes.[5]

Oxidation at low temperatures leads to the production of a number of additional phases, all tetragonal and related to the fluorite structure by the inclusion of interstitial oxygen. The thermodynamic stability of the tetragonal phases at low temperatures has not been determined; it is certain that they are all unstable above 650°C, disproportionating into U_4O_9 and $UO_{2\cdot6}$—a non-stoichiometric phase related to U_3O_8. The compositions and composition ranges of the tetragonal phases are also somewhat uncertain; the most probable values, based on several investigations,[1, 6] are collected in Table 11.2. The first product of the oxidation of UO_2 at low temperatures is normally α-U_3O_7, formed as a layer which

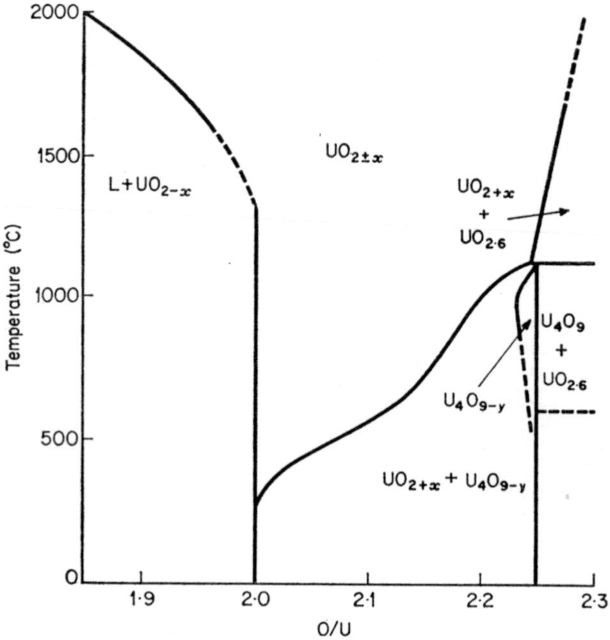

FIGURE 11.2. *Portion of the uranium–oxygen phase diagram.* (*Reprinted with permission from* Proc. Brit. Ceramic Soc., *1967*, **8**, *201.*)

TABLE 11.2. *Tetragonal oxides between U_4O_9 and U_3O_8*

	Cell edge Å	c/a	Composition O/U	Stability range °C
α-U_3O_7	5·46	0·99 or 1·01	2·26	<250
γ_1-U_3O_7	5·37–5·38	1·030–1·033	2·33	<460
γ_2-U_3O_7	5·39–5·40	1·01 –1·017	⩽2·30	<600

thickens as oxidation proceeds, and the final product is usually γ_1-U_3O_7, having $c/a = 1{\cdot}033$. The existence of an oxide $UO_{2\cdot37}$ with a structure of lower than tetragonal symmetry has been reported[8] but the characteristic features of the diffraction patterns have also been interpreted as due to a strained tetragonal lattice.[9]

(*b*) *Crystal structure of U_4O_9*

The major superlattice lines which appear on the X-ray[9] or neutron diffraction patterns[10] of any annealed sample of U_4O_9 can be accounted for by a cubic structure (space group $I\bar{4}3d$) with a lattice parameter $a = 4a_0$, a_0 being the side of the pseudo-fluorite cell with a cell edge 5·44 Å, compared to 5·470 Å for UO_2. The true unit cell therefore comprises 64 sub-cells and contains 832 atoms; the complete structure is unknown but the atomic positions of the interstitial oxygen in a statistical cell have been found and are described in a later section.

In fact, there seem to be more than one structure of U_4O_9. Belbeoch *et al.*, have reported a reversible cubic-rhombohedral transition at 65°C, the room temperature structure being rhombohedral and having $\alpha = 90{\cdot}078°$.[11] This transition was reversible for those samples that showed the phenomenon and the transition temperature is in the region where anomalies in the specific heat and thermal expansion have been reported. The intensity of the superstructure lines also increases above the transition temperature.[11, 12] This transformation was, however, observed only on samples of U_4O_9 which had been annealed at temperatures below 900°C; samples annealed at a higher temperature remained apparently cubic but showed the lowered intensity of superstructure lines at room temperature. The authors ascribe this difference to an effect of crystallite size. A rhombohedral distortion of a cubic cell can take place in four equivalent (111) directions; in large crystallites, such as are produced by annealing above 1000°C, the rhombohedral structure might nucleate in different areas of the crystal in the four possible ways and the statistical result would be that the total crystal symmetry appeared cubic. Very small crystallites (~0·1 μm) might contain one domain only and therefore the true symmetry would be apparent.

There have been other reports of more complex structures of U_4O_9. Neutron diffraction maxima corresponding to a cubic superlattice with $a = 8a_0$ have been reported for a U_4O_9 single crystal at room temperature and at 120°C;[13] the elucidation of the complete structure of such a cell (6656 atoms) would be formidable indeed. Electron microscopic examination of single crystal foils of U_4O_9 revealed the presence of non-periodic domain boundaries when the diffraction pattern corresponded to the presence of only one single crystal.[14] There were indications of four different types of domain. Since these patterns were presumably taken at room temperature, it is possible that they refer to the four rhombohedral orientations postulated by the French authors. The superlattice reflections determined by electron diffraction at 20°C were in agreement with a cell of $a = 4a_0$, space group $I\bar{4}3d$.[15] However, electron diffraction patterns of most specimens heated above 450°C showed marked changes, which occurred over a range from 500 to 800°C in times of up to 4 hours, and corresponded to a partial loss of order, since many of the superlattice lines disappeared. The range of transformation temperatures was related to decreasing values of y in the composition U_4O_{9-y}; it is well established that the U_4O_9 phase is itself non-stoichiometric (Figure 11.2) and it is proposed that the phase diagram should be modified to take account of an order-disorder transformation in U_4O_{9-y} at high temperatures. This conclusion must await confirmation in other experiments where the composition can be closely controlled, and an explanation of the reported persistence of the superstructure lines to 1000°C in one case; superstructure lines have also been reported at high temperatures in the X-ray diffraction experiments.[11]

Andresen[16] has pointed out the similarity in the neutron diffraction patterns of U_4O_9 and α-U_3O_7 and preliminary work has revealed the presence of superlattice lines in the diffraction patterns of the tetragonal γ_1- and γ_2-U_3O_7 phases.[3] The true unit cells of these phases are probably at least as large as that of U_4O_9, but they have not been determined.

(c) *Protactinium oxides*

Protactinium dioxide can be oxidized as readily as UO_2 and the end product of oxidation in air is the pentoxide, Pa_2O_5, which is white, diamagnetic and clearly a compound of protactinium(V). Pa_2O_5 has been reported in several crystalline forms. The structure stable up to 1000°C is tetragonal, having $a = 5{\cdot}429$ Å and $c/a = 1{\cdot}014$;[17] a doubled cell has also been reported.[16] Heating in air above 1050°C yields a hexagonal oxide and at least one other structure is formed above 1240°C, a rhombohedral distortion of a fluorite cell with $\alpha = 89{\cdot}65°$. The hexagonal high-temperature form, having $a = 3{\cdot}817$ Å and $c = 13{\cdot}22$ Å, may also be derived from a fluorite cell.[16]

Oxidation of PaO_2 (fluorite, $a_0 = 5{\cdot}509$ Å) or reduction of Pa_2O_5 leads to a series of intermediate oxides with structures closely related to fluorite, and excess oxygen in interstitial positions, as was proved by a comparison of density measurements with the volumes of the unit cells. The structures that were characterized at room temperature are listed in Table 11.3. The cell dimensions given are recorded for convenience in terms of the distortions of a fluorite cell and refer to the statistical sub-cell only: the true unit cells of the intermediate oxides have not been determined.

The entire series resembles the uranium oxide series quite closely, and it is reasonable that the larger size of the PaO_2 lattice allows the incorporation of interstitital oxygen up to a composition of $PaO_{2{\cdot}5}$. It is notable that one well-defined structure, the tetragonal $MO_{2{\cdot}33}$ structure, occurs in both series. A stable oxide of composition Pa_4O_9 ($PaO_{2{\cdot}25}$) does not seem to exist and the precise patterns of ordering are different in the two series; this is not surprising in view of the different ionic sizes involved.

There are indications of a non-stoichiometric PaO_{2+x} phase but the characteristics have not been established. The behaviour of the ternary system $PaThO_2$ on oxidation (see below) leads to the confident prediction that a PaO_{2+x} phase is stable at high temperatures.

TABLE 11.3. *The oxides of protactinium*

Composition O/Pa	Symmetry	Cell dimensions Å	Formation temperature (°C)
2·00	Cubic	$a = 5{\cdot}509$	
2·18–2·21	Cubic	$a = 5{\cdot}473$	
2·33	Tetragonal	$a = 5{\cdot}425$, $c/a = 1{\cdot}026$	
2·40–2·42	Tetragonal	$a = 5{\cdot}449$, $c/a = 0{\cdot}99$	
2·42–2·44	Rhombohedral	$a = 5{\cdot}449$, $\alpha = 89{\cdot}65°$	>800
2·50	Tetragonal	$a = 5{\cdot}429$, $c/a = 1{\cdot}014$	<1000
2·50	Hexagonal	$a = 3{\cdot}187$, $c = 13{\cdot}22$	1000–1200
2·50	Rhombohedral	$a = 5{\cdot}424$, $\alpha = 89{\cdot}76°$	>1240

(*d*) *Atomic structures of UO_{2+x} and U_4O_9*

Very significant information regarding the atomic arrangements in the non-stoichiometric UO_{2+x} phase and in U_4O_9 can be deduced from the neutron diffraction studies of Willis.[10] Single crystals of composition $UO_{2\cdot13}$ were studied at 800°C, a temperature at which this composition should be truly single phase (Figure 11.2); the compositions of the U_4O_9 crystals were also checked by making thermodynamic measurements. The neutron diffraction patterns of UO_{2+x} remained characteristic of fluorite; no additional reflections appeared. Thus the space group of UO_{2+x} is the same as that of UO_2 (Fm3m); the extra oxygen atoms are distributed at random so that the symmetry properties of the statistical cell, obtained by superimposing all the actual cells, are equivalent to that of UO_2. An analysis of the intensities of the reflections allowed the positional coordinates and occupation numbers of sites in the statistical cell to be determined.

A least-squares analysis showed that the uranium atoms remained on the fluorite positions, that the occupation number of oxygen on the fluorite-type sites fell appreciably from unity, and that two different types of interstitial site became occupied. The new sites may be described relative to the holes at $(\frac{1}{2},\frac{1}{2},\frac{1}{2})$ in a fluorite cell drawn with one uranium at (0,0,0)—these are the holes at the centre of those cubes of 8 oxygen atoms which are not occupied by a uranium atom. These $(\frac{1}{2},\frac{1}{2},\frac{1}{2})$ holes are not occupied; one of the actual interstitial positions for oxygen in UO_{2+x} is situated about halfway between the $(\frac{1}{2},\frac{1}{2},\frac{1}{2})$ hole and the centre of the line joining two normal oxygen sites (the O′ position) and the other about halfway from the $(\frac{1}{2},\frac{1}{2},\frac{1}{2})$ hole to the nearest oxygen site (the O″ position as in Figure 11.3). The site parameters and occupation numbers for $UO_{2\cdot12}$ are summarized in Table 11.4.[3]

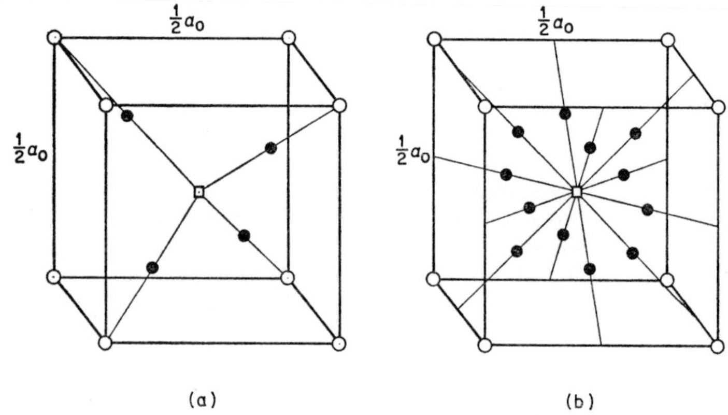

FIGURE 11.3. *Positions for interstitial oxygen atoms in U_4O_9 and UO_{2+x}. (a) The four equivalent O″ positions (●) relative to a cube of 8 fluorite-type oxygen atoms (○) surrounding the empty interstice (□) in the centre. (b) The 12 equivalent O′ positions (●) in a similar array of 8 fluorite-type oxygen atoms (○).* (*Reprinted with permission from Willis, B. T. M.,* J. Physique, *1964,* **25**, *437.*)

A similar analysis was applied to the fundamental (i.e. non-superlattice) reflections of U_4O_9. The restriction to fundamental reflections means that the results apply only to the 'composite cell' obtained by superimposing the 64 UO_2-type sub-cells in the true unit cell. Thus the details of the ordered structure will be lost but the 'temperature factors' recorded indicated that the displacement of the atoms from the statistical positions found is not large. Within these limitations, it was shown that the interstitial oxygen atoms are in similar positions to those in UO_{2+x} and the site parameters and occupation numbers are given in Table 11.5.[3]

These results establish beyond doubt the close relation between the UO_{2+x} and U_4O_9 structure, and geometrical considerations lead to some conclusions regarding the local atomic arrangements. The O′ and O″ sites

TABLE 11.4. *$UO_{2\cdot12}$: Least-squares results for statistical cell*

Atom	Coordinates in statistical cell			Contribution *m* to formula UO_m	Temperature factor $Å^2$
	x	*y*	*z*		
Uranium	0	0	0	—	1·18 ± 0·02
Oxygen O	*u*	*u*	*u*	1·87 ± 0·03	1·45 ± 0·04
Oxygen O′	0·5	*v*	*v*	0·08 ± 0·04	1·8 ± 1·4
Oxygen O″	*w*	*w*	*w*	0·16 ± 0·06	2·0 ± 1·6
Oxygen O‴	0·5	0·5	0·5	–0·02 ± 0·02	2·0 (fixed)

$u = 0{\cdot}267 \pm 0{\cdot}001$, $v = 0{\cdot}35 \pm 0{\cdot}01$, $w = 0{\cdot}41 \pm 0{\cdot}01$.

TABLE 11.5. *U_4O_9: Least-squares results for composite cell*

Atom	Coordinates in composite cell			Contribution *m* to formula UO_m	Temperature factor $Å^2$
	x	*y*	*z*		
Uranium	0	0	0	—	0·56 ± 0·04
Oxygen O	0·25	0·25	0·25	1·77 ± 0·02	1·57 ± 0·08
Oxygen O′	0·5	*v*	*v*	0·29 ± 0·05	1·25 ± 0·70
Oxygen O″	*w*	*w*	*w*	0·19 ± 0·04	1·60 ± 0·90

$v = 0{\cdot}372 \pm 0{\cdot}005$, $w = 0{\cdot}378 \pm 0{\cdot}005$.

cannot be occupied at random; in order to preserve credible interatomic distances, each O′ atom must displace two O atoms from normal fluorite ($\frac{1}{4},\frac{1}{4},\frac{1}{4}$) positions and each O″ atom must displace one such atom. This leads to a natural explanation for the appearance of vacancies on the ($\frac{1}{4},\frac{1}{4},\frac{1}{4}$) positions as O′ and O″ positions become occupied; an extra O atom goes to an O′ site, displacing two O atoms from their normal positions to O″ positions. This would lead to a ratio of numbers of atoms O′:O″:(O) of 1:2:2, where (O) is a vacancy on a normal lattice site. There is just enough room to insert a second O′ atom in a neighbouring equivalent site without displacing more lattice oxygen atoms, giving a ratio O′:O″:(O) of 2:2:2. These two local configurations or defects are the simplest possible which are compatible with the data, and are illustrated in Figure 11.4. The site occupation factors are not known with sufficient accuracy to allow a definite choice of model and, of course, more extended and complicated concentrations of defects than those of Figure 11.4 are possible. However, the U–O distances must be correct; each O′ atom is placed between 2 uranium nearest neighbours at 2·18 Å; this may be compared with the

U–O distance in UO_2, which is 2·37 Å, with 2·07 Å for the O–U–O–U strings in U_3O_8 or δ-UO_3, and with 1·90 Å as the closest U–O distances in uranates in which 'uranyl' groups can be identified. The interatomic distances are thus compatible with the presence of uranium(V) in the oxidized phase, a conclusion which is in line with the results of conductivity,

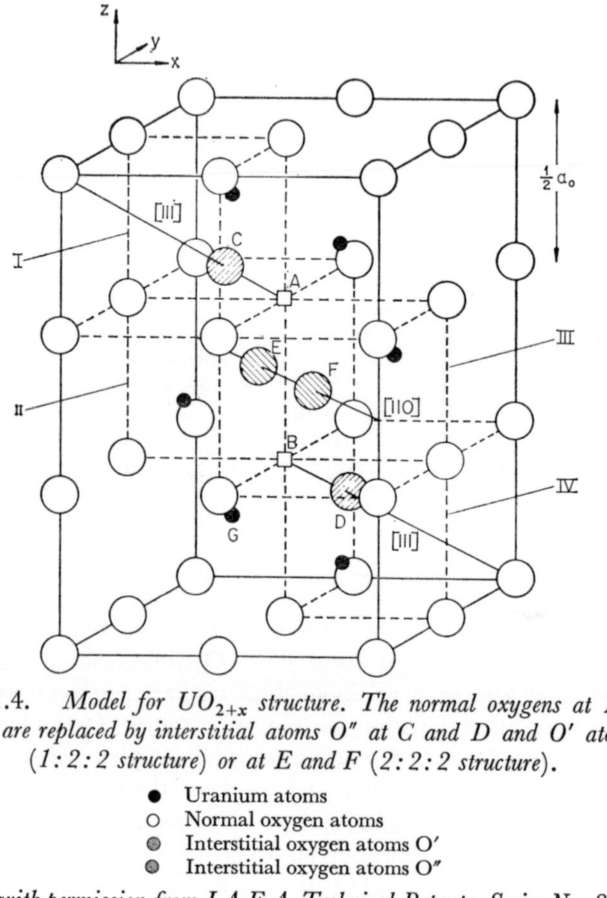

FIGURE 11.4. *Model for UO_{2+x} structure. The normal oxygens at* A *and* B *in $UO_{2\cdot00}$ are replaced by interstitial atoms O″ at C and D and O′ atoms at E (1:2:2 structure) or at E and F (2:2:2 structure).*

- ● Uranium atoms
- ○ Normal oxygen atoms
- Interstitial oxygen atoms O′
- Interstitial oxygen atoms O″

(*Reprinted with permission from I.A.E.A. Technical Reports, Series No. 39, 1965.*)

thermo-electric power and Hall coefficient measurements[19] and with the small amount of magnetic evidence which is available.[3, 20]

Several attempts have been made to correlate the structural model with the large amount of thermodynamic data for the UO_{2+x} and U_4O_{9-y} phases.[3] Entropy measurements indicate that U_4O_9 is an ordered phase compared to UO_{2+x}.[21] A calculation of the partial molal entropy of oxygen in UO_{2+x} based on the 2:2:2 model discussed above, with the assumption that two of the four uranium(V) ions introduced into the lattice to compensate for the two oxygen ions in O′ positions are on the two uranium sites which are nearest neighbours to the O′ sites, and the other two uranium(V) ions are distributed at random in the crystal, showed reasonable agreement with the variation of the experimental values from $UO_{2\cdot02}$ to $UO_{2\cdot13}$.[22] The divergence of the calculated from the experimental curves at compositions above $UO_{2\cdot13}$ was thought to be due to the difficulty of calculating the number of available sites at high concentrations of interstitial oxygen. Atlas[23] introduced a more sophisticated theory of defect interactions which allows for the varying energies of defect clusters with distance and, on the basis of a 2:1:2 model with only one uranium(V) trapped close to the O′ oxygen, succeeded in reproducing the entropy and enthalpy curves well from $UO_{2.04}$ to $UO_{2\cdot20}$. He pointed out, however, that the same good agreement can be obtained on the basis of a 3:2:3 structure and, indeed, on the basis of chains of defects of greater (and variable) length provided that a sufficient number of uranium(V) ions are assumed to be independent. The structural model is thus seen to be compatible with the thermodynamic evidence but cannot be further refined as a result of this type of calculation.

Finally, mention should be made of atomistic relaxation processes that have been postulated to explain internal friction measurements of small UO_{2+x} crystals. An activation energy of about 0·5 eV was calculated, which is too low to represent oxygen diffusion and too high to represent the migration of charge carriers.[24] A tentative explanation based on the reorientation of an interstitial complex such as that shown in Figure 11.4 was given, and there were indications of effects due to different types of complex in the same crystal.

(*e*) *Oxide solid solutions*

The behaviour towards oxidation of solid solutions of UO_2 or PaO_2 in oxides such as ThO_2, ZrO_2 or PuO_2, which cannot themselves be oxidized, is very closely parallel with that of the pure oxides. So long as the concentration of the 'inert' diluent is sufficiently high, the solid solutions absorb oxygen into interstitial positions but the symmetry remains cubic even at low temperatures: the ordering processes are clearly inhibited when a proportion of the cations cannot take part in them and, as we have seen, cation rearrangements are very slow.

The cell constant of these solid solutions decreases on oxidation, as does that of UO_2 itself. The change in cell constant per mole of uranium dioxide originally present—i.e. $\Delta a/y$ for $U_yTh_{1-y}O_2 \rightarrow U_yTh_{1-y}O_{2+x}$—depends only on the oxidation number (mean valency) reached by the uranium and is independent of y.[25] Very similar values of $\Delta a/y$ for a given oxidation number are found for $UO_2 \rightarrow UO_{2+x}$, indicative of the same type of atomic arrangement in both cases. The cell constants of $U_yTh_{1-y}O_{2+x}$ pass through a minimum and expand as the oxidation number of the uranium exceeds 5. The behaviour is contrasted with that of $Pa_yTh_{1-y}O_{2+x}$ in Figure 11.5.

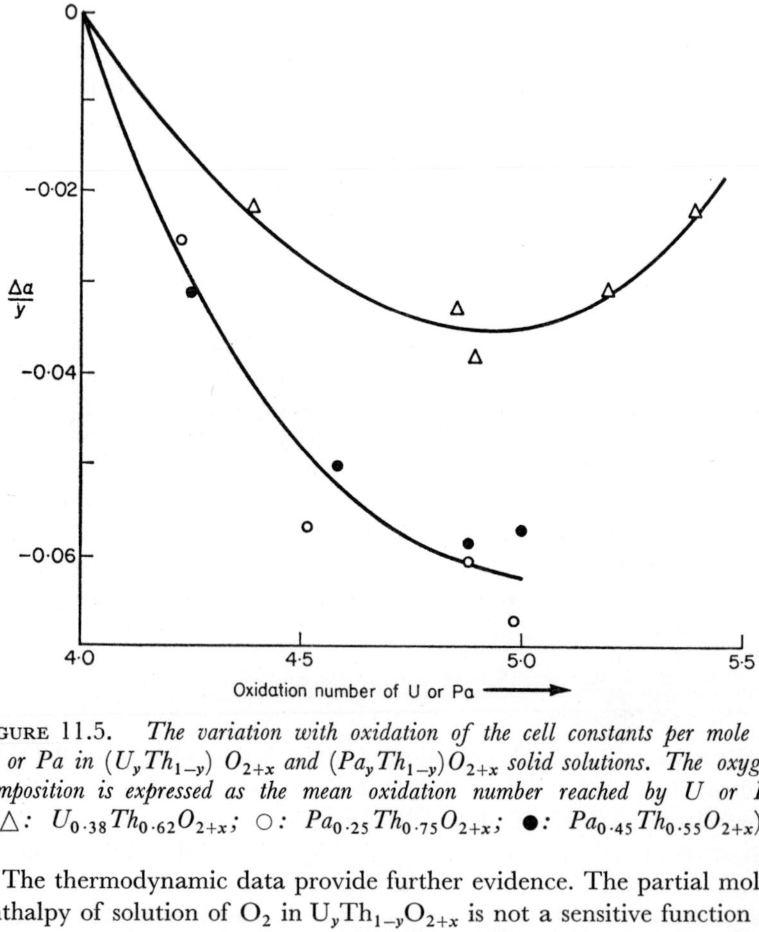

FIGURE 11.5. *The variation with oxidation of the cell constants per mole of U or Pa in* $(U_yTh_{1-y})\,O_{2+x}$ *and* $(Pa_yTh_{1-y})O_{2+x}$ *solid solutions. The oxygen composition is expressed as the mean oxidation number reached by U or Pa* ($\triangle$: $U_{0\cdot38}Th_{0\cdot62}O_{2+x}$; $\circ$: $Pa_{0\cdot25}Th_{0\cdot75}O_{2+x}$; $\bullet$: $Pa_{0\cdot45}Th_{0\cdot55}O_{2+x}$).

The thermodynamic data provide further evidence. The partial molal enthalpy of solution of O_2 in $U_yTh_{1-y}O_{2+x}$ is not a sensitive function of oxygen content, x, and, when compared at a given value of x, the values show only a small change with uranium concentration y, down to $y = 0{\cdot}2$.[22] Thus, the bonding energy of the interstitial oxygen is remaining approximately constant. A sharp reduction in $(-\Delta\bar{H})$ occurs, however, at lower values of y, and where the number of uranium atoms having one uranium nearest neighbour would also fall rapidly with decreasing uranium concentration. It seems, then, that the interstitital oxygen is going to positions such as O′ (Figure 11.4) which are between two uranium atoms, themselves nearest neighbours, and that much less energetically favourable sites are occupied only when sites between two uranium atoms are not available.

(*f*) *Fluoride systems*

Many cases of anomalous fluorite solid solutions are known in which fluorides of tri- or quadrivalent metals dissolve in the alkaline earth fluorides, CaF_2, SrF_2 or BaF_2.[26] In all those cases which have been investigated in detail, the excess fluorine has been shown to occupy interstitial positions. In these cases where the ionic radii of the divalent and trivalent cations are approximately the same, the f.c.c. lattice parameter increases with fluorine concentration. When the trivalent cation is much smaller than the divalent cation, the f.c.c. lattice parameter decreases with fluorine concentration as is the case for the uranium oxides[27] with excess oxygen.

The non-stoichiometric behaviour of some divalent rare-earth fluorides has been recognized for some time; a recent study has established the nature of the defects.[28] Both SmF_2 and EuF_2 form single fluorite phases from MF_2 to about $MF_{2{\cdot}25}$, the density increasing regularly as expected if the fluorine is entering interstitial positions and the lattice constants decreasing with fluorine concentration (Figure 11.6). The crystallographic nature of the phase or phases formed in the composition range $2{\cdot}25 < F/M < 2{\cdot}45$ has not been determined but the diffraction patterns are closely related to fluorite patterns, and the structure must be a distorted fluorite structure. It is noteworthy that all these compounds were prepared from the melt at about 1500°C and were very well crystallized after slow cooling. The X-ray and density determinations were all carried out at room temperature. The cubic fluorite phases can therefore be preserved to room temperature in these cases; either they are truly stable with respect to MF_2 and higher fluorides, or the disproportionation reactions are much more sluggish.

One neutron diffraction study of a solid solution containing excess fluorine has recently been reported.[29] Powder neutron diffraction patterns of a solid solution of YF_3 in CaF_2, $(Ca/Y)F_{2{\cdot}10}$, were measured at room temperature and analyzed by a least-squares comparison of observed and calculated structure factors to give the parameters of the average statistical cell. The results are shown in Table 11.6. Comparison with Table 11.4, the results for $UO_{2{\cdot}12}$, shows a striking similarity. The F′ and F″ positions are almost identical with the O′ and O″ positions and the apparent disagreement in the site occupation numbers for the two types of site may not be significant.

Considerations similar to those described above lead to atomic arrangements for the fluorides also as shown in Figure 11.4. Some difficulties remain: the F′–F′ distance in a 2:2:2 complex would be only 2·02 Å, which is very short compared to the normal ionic diameter of 2·5–2·7 Å and this

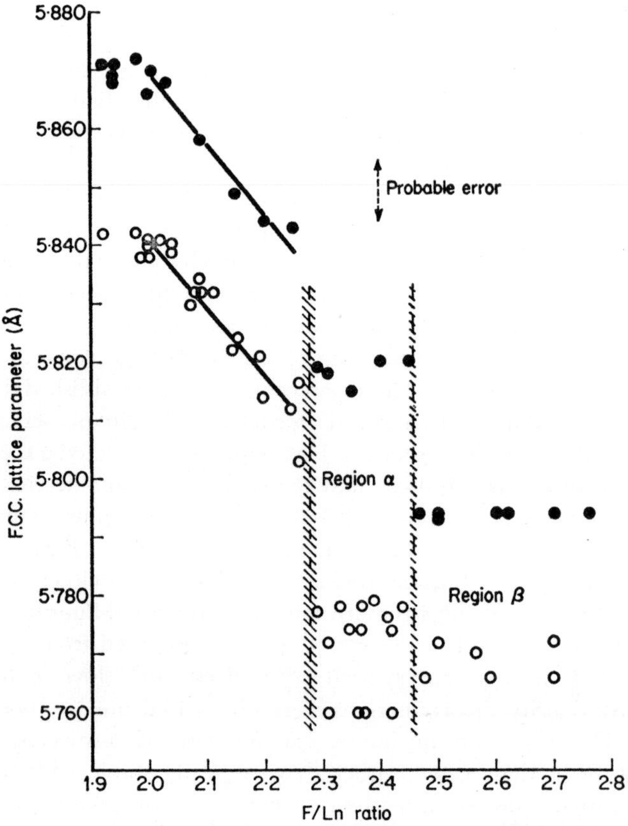

FIGURE 11.6. *The variations in lattice parameter of the samarium and europium fluorides with fluorine/metal ratio. The structures in the regions α and β are complex and distorted but the results are plotted as if the structure remained fluorite.* (*Reprinted with permission from Catalano, E., Bedford, R. G., Silviera, V. G., and Wickman, H. H.,* J. Phys. and Chem. Solids, *1969,* **30**, *1613.*)

arrangement does not seem possible. Even in the simple 2:1:2 complex F′–F′ distances as short as 2·24 Å occur. It is clear that more structural work is required before the actual atomic arrangement can be found.

These studies suggest that the structure of all the fluorite compounds containing excess anions is based on asymmetric interstitial arrays, which have been variously called 'complexes' or 'domains' and Figure 11.4 represents the two simplest variants. This interstitial structure may be a characteristic of the fluorite lattice and the persistence of this type of defect domain is the reason for the rather similar limiting compositions reached in very dissimilar systems; the concentration of interstitials which can be accommodated is simply a function of the volume available in the lattice. The size of the domains present in the high-temperature non-stoichiometric

TABLE 11.6. *$(Ca/Y)F_{2\cdot10}$: Structure parameters of average cell*

	Coordinates			Contribution to $2+x$ in $(Ca/Y)F_{2+x}$
	x	y	z	
Lattice F	0·25	0·25	0·25	1·88 ± 0·04
Interstitial F′	0·5	v	v	0·14 ± 0·03
Interstitial F″	w	w	w	0·08 ± 0·03

$v = 0{\cdot}36 \pm 0{\cdot}01$, $w = 0{\cdot}42 \pm 0{\cdot}01$.

phases cannot be smaller than the 2:1:2 structure (2 vacancies, 1 interstitial on O′ and 2 on O″) but the domains may be larger and may well be variable in size; the indications from the thermodynamic data are that they are not very large. It is clear that a number of complex structures is possible by long-range ordering of such domains at low temperatures, and the existence of a variety of closely related ordered structures is not surprising.

C. Anion deficient phases

A large literature exists on those rare-earth oxide systems where the cation can easily be oxidized to the quadrivalent state—Ce, Pr and Tb—much of which is due to Eyring and his school. The plutonium oxides have been intensively studied because of their importance in the nuclear energy industry. All these systems show wide ranges of stoichiometry at high temperatures, with ordered phases at low temperatures. Ternary systems again exist with fluorite structures and wide composition ranges. The 'stabilized' zirconias, in which the cubic form of ZrO_2 is preserved at low temperatures by the addition of divalent or trivalent oxides, have similar structures. A very short summary of the main structural features of the binary systems follows, together with some discussion of the structures that may persist in the non-stoichiometric fluorite phases, though the evidence available is not as direct as for the anion-excess phases.

(*a*) *Rare-earth oxides*

The dioxides are all normal fluorite compounds; the sesquioxides exist in one or more of the three structures called A, B or C. The body-centred cubic C-form is the usual low-temperature structure; Ce_2O_3 and Pr_2O_3 are hexagonal A-type above about 600°C and Tb_2O_3 is monoclinic B-type above 1400°C. The relationships between the fluorite, A, B and C structures have been discussed by Eyring and Holmberg.[30] For the present purpose, the most significant is the relation between the fluorite and the type C structure: the metal positions remain almost unchanged while the MO_8 cubes become MO_6 coordination groups by dropping oxygen atoms along the body diagonal of the fluorite cell. Chains of the MO_6 octahedra can be recognized lying along all four ⟨111⟩ directions throughout the lattice. One of these linear strings of octahedrally coordinated M^{3+} ions is illustrated in Figure 11.7. Only the oxygen atoms that coordinate metal atoms are shown; these

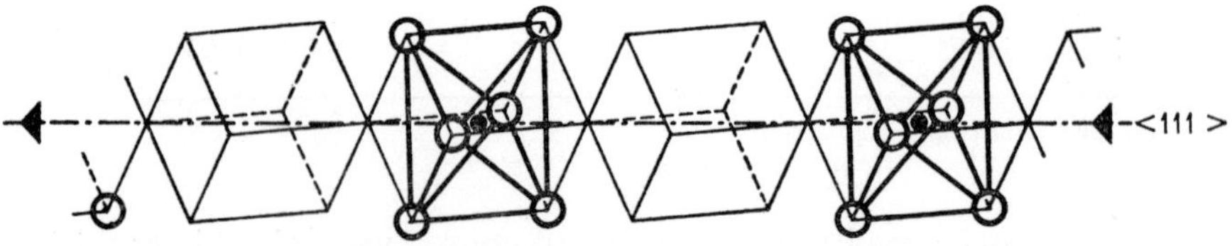

FIGURE 11.7. *A linear 'string' of octahedrally coordinated metal* (R^{3+}) *ions in* C *type* R_2O_3. (*Reprinted with permission from Hyde, B. G., and Eyring, L.,* Proc. 4th Conference Rare-Earth Research, *1964, 646.*)

○ = oxygen ion.
● = lanthanide ion.

form distorted octahedra which are joined by an empty cube containing two more oxygen vacancies.

A series of ordered, intermediate phases exist at low temperature in each of the systems Ce—O, Pr—O and Tb—O. The general sequence can be understood by reference to the very precise compositions of Hyde, Bevan and Eyring[31] for the praseodymium oxides, given in Table 11.7, and to the corresponding phase diagram, Figure 11.8. In addition to the five intermediate phases which dissociate at temperatures of 1050 to 450°C, there are two non-stoichiometric phases which are truly stable at high temperatures. The α-phase exists over a very wide composition range of PrO_x, $2{\cdot}00 \geqslant x \geqslant 1{\cdot}72$ and the σ-phase from $x = 1{\cdot}6$ to $1{\cdot}7$.

The homogeneity ranges of the five intermediate oxides are very narrow, suggesting that they are highly ordered, and the measured stoichiometries bracket closely the ideal formulae shown in the fourth column of Table 11.7. These formulae appear to constitute part of a homologous series of general formula Pr_nO_{2n-2} and the values of n are recorded in the last column of

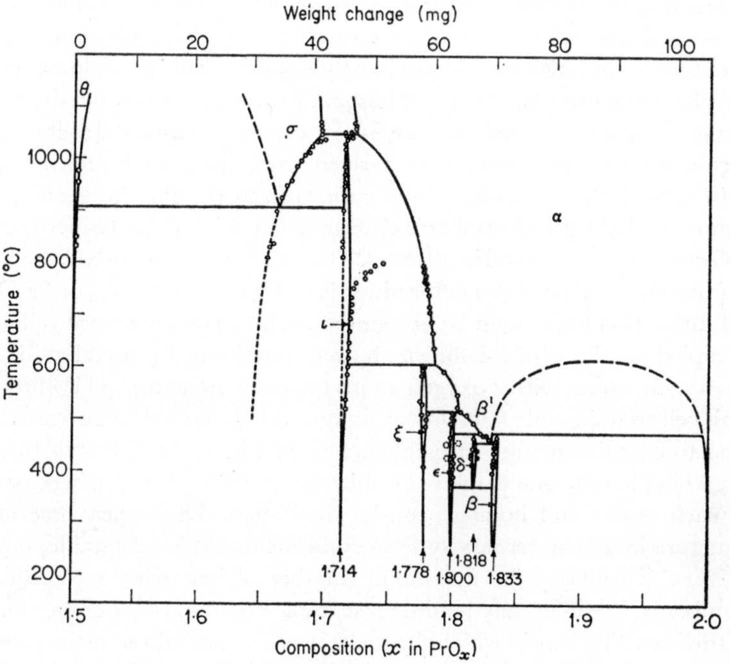

FIGURE 11.8. *Portion of the praseodymium-oxygen phase diagram.* ○, *Experimental points from the isobars; solid line, phase boundaries, stable; broken line, phase boundaries, stable, assumed; dashed line, phase boundaries, metastable.* (*Reprinted with permission from Hyde, B. G., Bevan, D. J. H., and Eyring, L.,* Phil. Trans., *1966,* **259A**, *584.*)

TABLE 11.7. *Stoichiometry of praseodymium oxides*

Phase symbol	x in PrO_x	Experimental range of x	Stoichiometric formula	Value of n
ϕ	1·500	1·500	Pr_2O_3	4
i	1·714	1·713–1·719	Pr_7O_{12}	7
ξ	1·778	1·776–1·778	Pr_9O_{16}	9
ϵ	1·800	1·799–1·801	Pr_5O_9	10
δ	1·818	1·817–1·820	$Pr_{11}O_{20}$	11
β	1·833	1·831–1·836	Pr_6O_{11}	12
α	2·000	—	PrO_2	∞

the table. The data available for the cerium and terbium oxides fit into the same general pattern; the compositions of all the intermediate phases that have been characterized in the three oxide systems can be fitted to the formula M_nO_{2n-2} with values of n from 5 to 12, and 16, but oxides having all values of n are not found in each system.[32]

Much of the earlier structural work carried out on these systems suffers because of the difficulty of attaining well-annealed samples of one single structure to investigate: this was hardly possible until the main features of the phase diagrams had been established. Recent X-ray powder diffraction studies[32] have confirmed that the β-, δ-, ϵ- and ζ-phases of Table 11.7 are highly ordered superstructures derived from the parent fluorite lattice with only slight distortion of the cations from the fluorite position; the symmetry of the pseudo-cells was determined as triclinic and superstructure reflections were observed in all cases. The only intermediate oxide for which the true unit cell has been determined is the i-phase, $PrO_{1.714}$ or Pr_7O_{12}.[33] All diffraction lines could be accounted for by a rhombohedral cell which occupied a volume of the fluorite lattice containing 7 praseodymium and 14 oxygen atoms, with 2 oxygen atoms removed along the $\langle 111 \rangle$ direction. This cell contains only 6- or 7-coordinated cations, and the structure can be seen to contain 'strings' of octahedra as in Figure 11.7, but in this case stretching in only one of the $\langle 111 \rangle$ directions, with all the 'strings' parallel to each other, and homogeneously distributed. The appearance of this structure in several ternary systems confirms that it is very stable, possibly because it contains the maximum number of ions with a coordination number of 7, which may be the most stable coordination number for ions of this size. The way in which the inclusion of a 'pair' of vacancies produces one 6- and six 7-coordinated ions is further illustrated in Figure 11.9; this arrangement is the characteristic repeating pattern of the M_7O_{12} structure.

A simple fluorite pattern was obtained by quenching some samples of composition $PrO_{1.826}$ after long annealing at 468°C,[32] at which temperature the structure is that of the disordered α-phase (Figure 11.8). No superstructure lines were detected, while the fluorite lines were broadened. The pattern was thus not a superposition of those of equal amounts of $PrO_{1.818}$ and $PrO_{1.833}$; some features suggested that the disordered phase contains very small units of the ordered structures in random orientation to each other. The σ-phase (Figure 11.8), which is certainly another non-stoichiometric phase above 910°C but which is not miscible with the α-phase, is thought to be body-centred cubic and therefore related to C-type Pr_2O_3; a similar phase exists in the same composition range ($1{\cdot}65 < O/M < 1{\cdot}70$) above 600°C in the cerium-oxygen system and a possible model for the structure is discussed below.

(*b*) *Ternary oxide phases*

Further evidence that bears on the problem of the structure of the non-stoichiometric phases has arisen from studies of solid solutions found in systems of the type MO_2 (fluorite)—M_2O_3. The existence of fluorite-type phases and of rare-earth type C structures over wide composition ranges has been recognized for some time. A recent paper by Bevan *et al.*[34] contains references to much of the earlier work and records a careful re-examination of several of these systems.

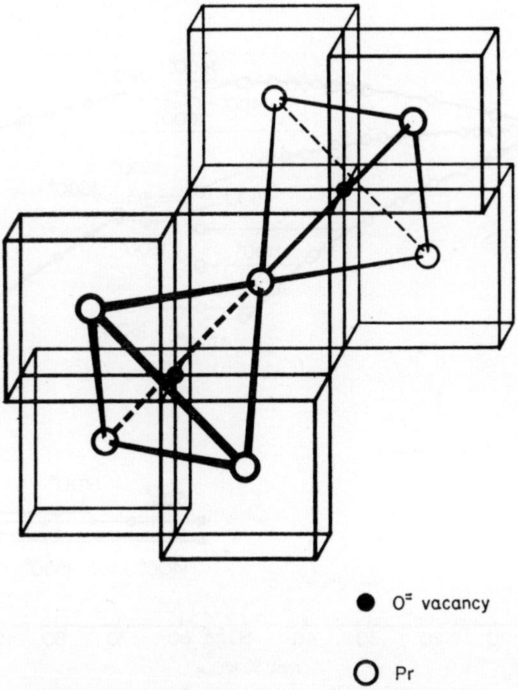

FIGURE 11.9. *Spatial relation between an octahedrally coordinated cation and the six nearest-neighbour 7-coordinated cations in the M_7O_{12} structure.* (*Reprinted with permission from Hyde, B. G., and Eyring, L.,* Proc. 4th Conference Rare-Earth Research, *1964, 648.*)

The most detailed results refer to the CeO_2-Y_2O_3 system but the characteristic features are repeated in the CeO_2-Gd_2O_3, CeO_2-Dy_2O_3 and CeO_2-Yb_2O_3 systems; a plot of cell edge versus composition for these systems is shown in Figure 11.10. The cell edge plotted is that of the pseudo-fluorite cell and is half the true cell edge for the b.c.c. type C solid solutions. There is always a miscibility gap between the fluorite and type C solid solutions, of a size which depends on temperature. The fluorite solid solution in the CeO_2-Y_2O_3 system extends to 42 mol% $YO_{1.5}$ at 1100°C and to 55 mol% $YO_{1.5}$ at 1700°C; the type C solid solution extends from Y_2O_3 down to 76 mol% $YO_{1.5}$ at 1700°C but as far as 62 mol% $YO_{1.5}$ at 1100°C. The cell constant of the type C solid solutions is a linear function of composition in all systems, but this is not true for the fluorite solid solutions. The fluorite cell-parameter data for the CeO_2-Y_2O_3 system was actually fitted to two equations, one from 0–10 mol% $YO_{1.5}$ and the other from 10–55 mol% $YO_{1.5}$ and in both ranges the cell constant varied linearly with the square of the mole fraction of $YO_{1.5}$.

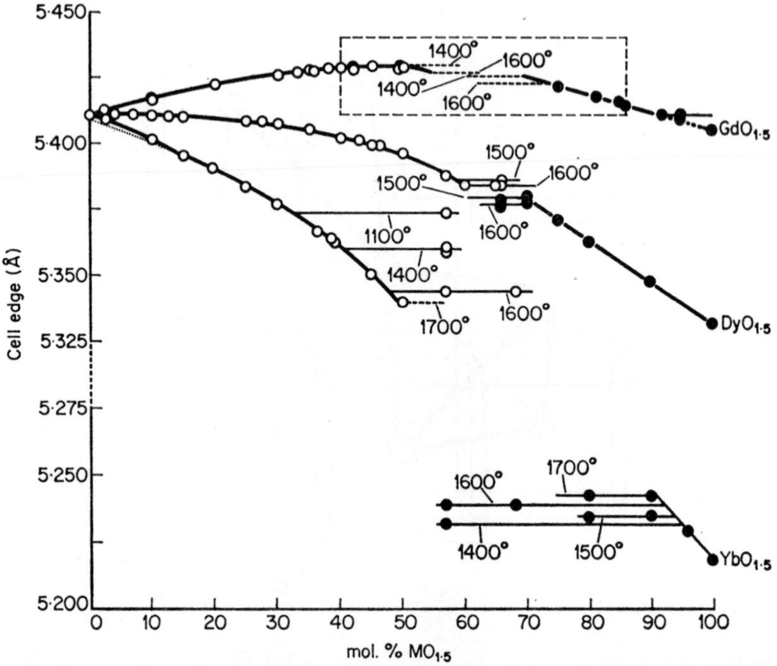

FIGURE 11.10. *Plot of cell edge against composition for several CeO_2-M_2O_3 systems.* (*Reprinted with permission from Bevan, D. J. M., Baker, W. W., Martin, R. L., and Parks, T. C.,* Proc. 4th Conference Rare-Earth Research, *1964, 462.*)

○ = fluorite-type phase.
● = type C phase.

Bevan *et al.*[34] and others[35] noted the appearance of extra weak, diffuse lines in the diffraction patterns of the fluorite phases; these appeared at 20 mol% $YO_{1.5}$ in the CeO_2-Y_2O_3 system, at higher $YbO_{1.5}$ contents in the CeO_2-Yb_2O_3 system and at 40 per cent $NdO_{1.5}$ in the CeO_2-Nd_2O_3 system. These diffuse lines appear at positions where strong type C reflections would be expected. It is this circumstance that has led previously to the idea that there is a continuous transition between fluorite and type C structures, but the existence of a miscibility gap in all cases that have been carefully explored seems well proven.

The appearance of the diffuse superstructure lines in the fluorite patterns, however, definitely suggests some form of long-range ordering in this solid solution range. The extra lines can all be indexed on the basis of a b.c.c. cell with a cell edge twice that of the fluorite cell. But the character of the extra lines is different from that of true type C reflections; they are diffuse and their intensities relative to those of neighbouring, sharp fluorite lines are much weaker than for the corresponding diffraction lines in the type C patterns. The change from 'fluorite plus superstructure' to a true type C pattern was sharp and easily recognized.

The accurately linear cell-dependence on composition of the type C solid solutions indicates both that the two cations are randomly distributed on the cation positions and that the oxygen in excess of the M_2O_3 composition is randomly distributed in interstitial positions in the type C structure; there is space to accommodate oxygen in tetrahedral holes in this open structure. It is not likely that the non-linear variation of cell parameter and diffuse extra reflections found in the fluorite phases are due to partial ordering of the two cations. The regular variation over a wide composition range argues against such a hypothesis. Further, when the overall compositions of a series of UO_2-Y_2O_3 solid solutions were brought to $(U_xY_{1-x})O_2$ by partial oxidation, the variation of cell constant with yttrium content was accurately linear, though a non-linear variation was found in the fully reduced state.[36] It is very probable, therefore, that the cations are randomly distributed over the cation sublattice of the fluorite structure, and that the non-linear variation of cell constant with composition—i.e. with the concentration of anionic 'vacancies'—and the appearance of diffuse reflections that indicate some degree of ordering are essentially due to ordering of the 'vacancies', presumably causing the cations to move away from the ideal fluorite sites.

Bevan *et al.*[34] demonstrated that the phase limits which they determined were the limiting compositions at equilibrium at the temperatures used during annealing. Cationic migration at the high temperatures could take place given enough time and clear evidence of this was obtained in the diphasic regions; metastable structures could be quenched if sufficient time had not been allowed for the system to reach equilibrium. But the X-ray results refer to samples quenched to room temperature and, although the cation distribution remains undisturbed on cooling, the oxygen distribution may not. A study of the structures at high temperatures would be interesting.

(*c*) *Metastable structures and hysteresis*

The account of the praseodymium and other rare-earth oxide systems given above refers to the equilibrium states of these systems. In fact, equilibrium is not always easily attained even in the binary systems, although both electron and oxygen migration processes are known to be rapid. A number of very stable phases can be produced which are not in true equilibrium[31, 37] and complex hysteresis loops have been traced during some sequences of oxidation and reduction, though others proceed normally.

There is a recurrent and regular pattern in this behaviour: it is the formation of an ordered from a disordered phase that is difficult; the reverse reaction takes place easily. A particularly well-documented case is the behaviour on cooling of the PrO_x system in the range $1{\cdot}67 < x < 1{\cdot}73$. Reference to the phase diagram (Figure 11.8) shows that the reaction concerned is the oxidation of the non-stoichiometric σ-phase to the well-ordered rhombohedral i-phase, Pr_7O_{12}. A single crystal of Pr_7O_{12} consists of the 'strings' of octahedra illustrated in Figure 11.7 oriented in one direction throughout the crystal. If the i structure nucleates in the σ-phase on multiple sites, the trigonal axes will be randomly distributed parallel to all four $\langle 111 \rangle$ axes of the cubic σ-crystal. As these nuclei grow, their boundaries will meet and adjacent microdomains of i may enclose microdomains of σ: complete conversion of the crystal to any one orientation of i will be inhibited. Such a crystal is truly diphasic and has been termed a 'pseudophase'.

The structural evidence, though incomplete, is also consistent with the idea that all the intermediate ordered phases contain these 'strings' of octahedrally coordinated trivalent ions, probably parallel to each other except for the b.c.c. σ and C-type structures. The general formula follows if the basic unit demands the removal of two oxygen ions from the cubic environment of $1/n$th of the cations, leading to the formula $MO_{2-2/n}$ or M_nO_{2n-2}. It is natural to assume that 'strings' of this type, perhaps variable in length and randomly oriented, occur at high temperatures in the non-stoichiometric α-phase. It has been suggested[37] that the appearance of the diffuse C-type superstructure lines in the fluorite solid solutions is due to the development as near neighbours of the randomly oriented 'strings', with the consequent production in overlap areas of a structure approximating to the type C structure. Complete conversion of a cubic α-phase with randomly oriented strings to the true C structure with a coherent arrangement of strings running throughout the crystal, is extremely difficult, however, and the miscibility gaps between both the α- and σ-phases in the binary systems and between the fluorite and type C solid solutions in the ternary systems can be readily understood.

(*d*) *The plutonium oxides*

The PuO_2-Pu_2O_3 system has been investigated by several techniques but some doubts remain concerning the phase diagram. It seems certain, however, that the only phases which exist at room temperature are PuO_2 (f.c.c.) and $PuO_{1\cdot52}$ (b.c.c. type C) and hexagonal Pu_2O_3. No other compounds of intermediate composition have been discovered. The type C phase may be the stable low-temperature form of Pu_2O_3. In addition, there is another cubic phase of composition $\sim PuO_{1\cdot61}$ which is stable above 300°C.[38]

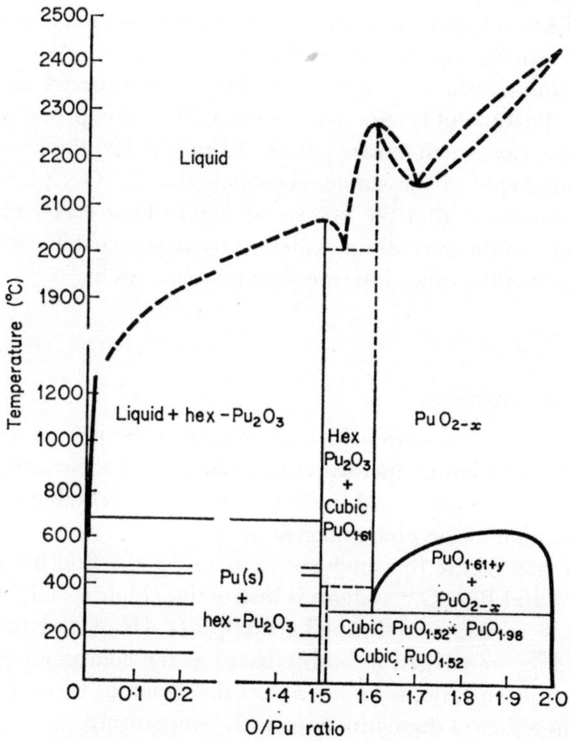

FIGURE 11.11. *Portion of a plutonium-oxygen phase diagram. (Reprinted with permission from I.A.E.A. Technical Reports, Series No. 79, 1967.)*

A critical assessment of all the known data[39] led to the phase diagram reproduced in Figure 11.11. A wide non-stoichiometric fluorite phase analagous to the α-phase in the rare-earth systems certainly exists at high temperatures. The phase diagram as drawn shows this phase continuous to $PuO_{1\cdot61}$; this implies that $PuO_{1\cdot61}$ is face-centred cubic. Another possibility, equally likely, is that $PuO_{1\cdot61}$ is itself non-stoichiometric and that a narrow miscibility gap exists between this phase and PuO_{2-x} above 650°C. This alternative would be required if $PuO_{1\cdot61}$ were body-centred, on the evidence given above of the immiscibility between fluorite and true type C phases. Such a miscibility gap, if narrow, is difficult to determine and might have been missed. The diffraction patterns obtained above 300°C showed only strong f.c.c. lines for $PuO_{1\cdot61}$, but weak superlattice reflections could have been missed. Some preparations of $PuO_{1\cdot61}$ are reported to show b.c.c. lines at room temperature after quenching but such experiments are open to some doubt regarding the efficiency of the quench and the extra reflections could be the 'fluorite superstructure' lines discussed above and not indicative of a true type C structure. More detailed work is required before this question can be regarded as resolved.

Notwithstanding these doubts, the absence of ordered intermediate phases in the Pu_2O_3-PuO_2 region does seem certain and provides a striking contrast to the rare-earth oxide systems. In particular, the absence of the plutonium analogue of the stable, rhombohedral M_7O_{12} phases is noteworthy; this indicates that the 'strings' shown in Figure 11.7 cannot form in parallel lines in the plutonium oxides, a circumstance that could account for the absence of the other intermediate phases as well.

(*e*) *Thermodynamic evidence*

Thermochemical measurements of oxygen pressures in equilibrium with the oxides at high temperatures have confirmed the existence of wide non-stoichiometric ranges and, indeed, have been valuable evidence for the determination of the phase diagrams.

A comparison of the thermodynamic data for the fluorite α-phases in the Ce—O[40] and Pu—O[41] systems is instructive. Figure 11.12 shows plots of the change in the partial molal enthalpy ($-\Delta\bar{H}_{O_2}$) and partial molal entropy ($-\Delta\bar{S}_{O_2}$) of oxygen in both systems as the composition, expressed as the anion/cation ratio, is varied across the α-phase; there was virtually no change in either of these functions with temperature.

There are clearly similar changes taking place in both oxide systems as the composition is reduced from $MO_{2\cdot0}$ to $MO_{1\cdot88}$: the energy required to create an 'oxygen vacancy' in the solid decreases very sharply and the phase becomes highly disordered (a change of over 55 entropy units in $\Delta\bar{S}(O_2)$ is extremely large). The minima in the ($-\Delta\bar{H}_{O_2}$) and ($-\Delta\bar{S}_{O_2}$) curves for CeO_{2-x} indicate some approach to a more ordered configuration in the composition range $CeO_{1\cdot82}$ to $CeO_{1\cdot77}$, though the measurements were made at temperatures where these compositions are well within the non-stoichiometric α-phase, which extends to $CeO_{1\cdot72}$. The oxides $CeO_{1\cdot81}$ and $CeO_{1\cdot78}$ are stable intermediate oxides at lower temperatures, and it may be inferred that the thermodynamic measurements indicate some approach to a partial ordering within the non-stoichiometric range at about the same compositions. In contrast, the PuO_{2-x} data show no similar minima at the same compositions, and there are no similar intermediate phases at room temperature.

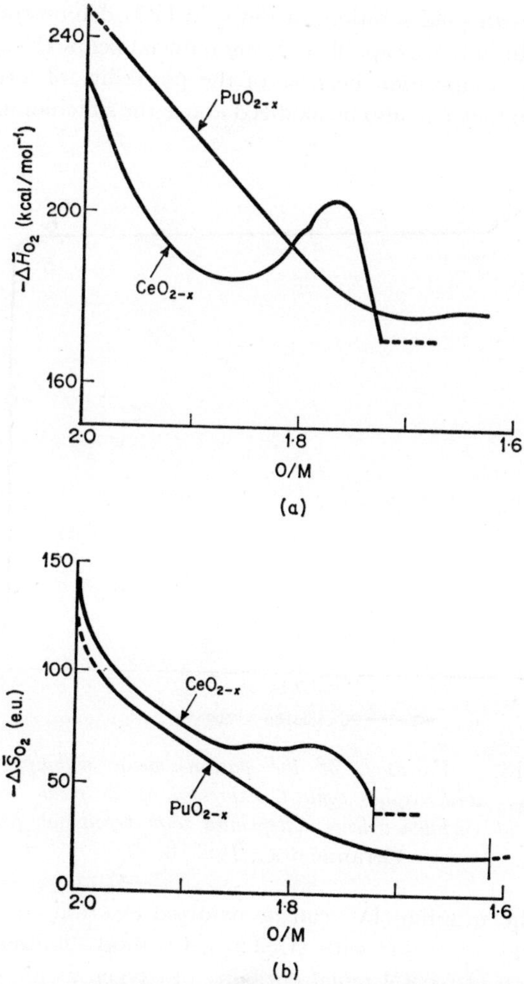

FIGURE 11.12. *(a) Partial molal enthalpy of oxygen in* CeO_{2-x} *and* PuO_{2-x} *as a function of composition. (b) Partial molal entropy of oxygen in* CeO_{2-x} *and* PuO_{2-x}. *(Reprinted with permission from* Proc. Brit. Ceramic Soc., *1967*, **8**, *214.)*

It is worth noting that the very large changes of partial molal entropy and enthalpy values for oxygen in the pure oxide systems almost certainly reflect the local ordering possible in these crystals because the mobility of electrons enables any configuration of trivalent and tetravalent ions to be readily assumed. Such large enthalpy and entropy variations do not occur when oxygen is removed from a similar crystal if free interchange of cations of different valencies is inhibited. Thermodynamic results for oxygen in equilibrium with solid solutions of PuO_2 in UO_2 demonstrate this point. Such solid solutions are capable of being reduced below the stoichiometric $(U_xPu_{1-x})O_2$ composition because of the possibility of forming plutonium(III), and they can also be oxidized above the stoichiometric composi-

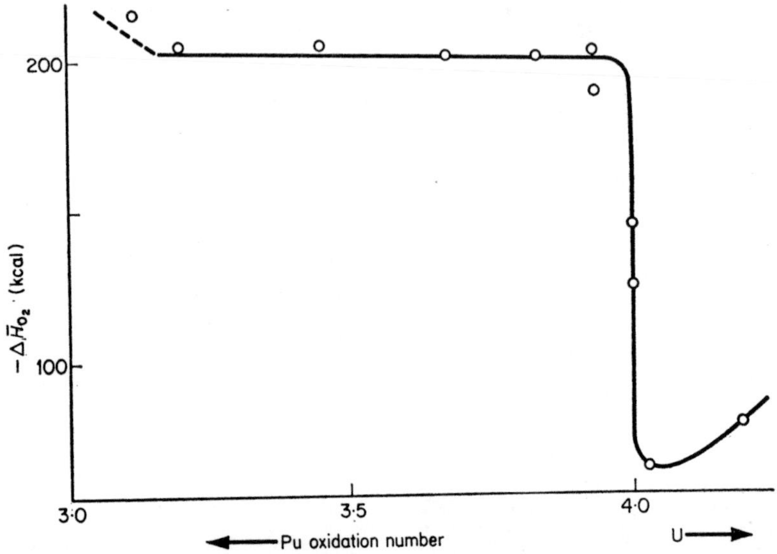

FIGURE 11.13. *Variation of the partial molal enthalpy of oxygen in* $(Pu_yU_{1-y})O_{2+x}$ *with oxygen content, expressed as the mean oxidation number of the cation of variable valence.* (*Reprinted with permission from* Proc. Brit. Ceramic Soc., *1967*, **8**, *215*.)

tion since the uranium(IV) can be oxidized to uranium(V) or (VI) in the same way as occurs with $(U_xTh_{1-x})O_2$ solid solutions (see above). The variation of partial molal enthalpy of oxygen as the composition is altered is shown in Figure 11.13: the values were independent of the actual composition of the crystal for PuO_2 contents of less than 30 mol% PuO_2 when plotted against the oxidation number (mean valency) of the cation of variable valence. There is a very large change in enthalpy as the stoichiometric composition is reached. Such a result is not surprising, since elementary energy considerations lead to the expectation that the energy required to form a 'vacancy' in the fluorite lattice is very much larger than the energy required to remove an oxygen ion from an interstitial position. However, it is significant that the value of $(-\Delta\bar{H}_{O_2})$ alters very little as further oxygen is removed from the crystal and the plutonium oxidation number is reduced to nearly three, in very sharp contrast to the results in Figure 11.12 for the pure binary oxide systems.[22]

The partial molal entropy changes as the mixed crystals are reduced are also small and easily explicable on a simple model of the removal of oxygen from similar environments taken at random throughout the crystal; perhaps the loss of oxygen is from a site between two plutonium ions with reduction of both from Pu^{4+} to Pu^{3+} to preserve local charge compensation. No cooperative rearrangement is possible since the U^{4+} ions cannot be reduced to U^{3+} ions under the conditions and temperatures used and so cannot take part in such processes. It is only when the distribution of trivalent and quadrivalent ions can readily alter that local ordering can occur and be reflected in the large enthalpy and entropy changes of Figure 11.12.

The thermodynamic data are thus in agreement with the hypothesis that the structures of the anion-deficient non-stoichiometric phases reflect local ordering and, in the case of the rare-earth oxides, may contain domains similar in structure to the ordered intermediate phases. But such data can only be regarded as supporting evidence and certainly not as proof.

(*f*) *The 'stabilized' zirconias*

Zirconia, ZrO_2, has a complex monoclinic structure at room temperature and transforms to a tetragonal structure allied to fluorite at 950°C. This phase-transformation reaction causes a change in density and consequently artefacts of pure zirconia crack on thermal cycling. Zirconia readily forms solid solutions with oxides of alkaline-earth or trivalent metals and such additions suppress the transformation to the monoclinic structure, thus 'stabilizing' the high temperature structure of the doped zirconia to room temperature; the stabilized zirconias are widely used as refractories.

Rare-earth oxides modify the tetragonal structure of ZrO_2 in a regular way depending on the ionic radius of the rare-earth ion; with increasing amounts of rare-earth oxide, the c/a ratio of the tetragonal phase decreases until the symmetry becomes cubic. The limiting cubic compositions contain about 8 mol% of Yb_2O_3 and about 13·5 mol% of Nd_2O_3,[42] and are stable on cooling to room temperature. With increasing concentration of rare-earth oxide additional lines characteristic of a doubled cell appear in the 'fluorite' range and the situation may be very similar to that of the CeO_2-M_2O_3 systems discussed above. There are also a number of other compounds in the complete ternary systems; for example, three intermediate phases were characterized in the ZrO_2-Sc_2O_3 system between two solid solution regions, one of Sc_2O_3 in ZrO_2 (f.c.c.) and the other of ZrO_2 in Sc_2O_3 (b.c.c.). Interestingly, one of these, a rhombohedral phase of composition (89 per cent ZrO_2, 11 per cent Sc_2O_3) is said to decompose at 600°C, at which temperature no cation movement is possible, and therefore would be an ordered superstructure of the fluorite lattice dependent upon long-range ordering of the oxygen only.

The most complete data for a cubic solid solution range exist for the ZrO_2-CaO system. Earlier work has been summarized by Carter and Roth,[43] who redetermined the phase limits of the cubic phase on samples which had been sintered at 2050°C and annealed for longer periods at lower temperatures. Carter and Roth reported the phase boundaries of the cubic phase above 900°C to be 9·8 to 19·6 mol% CaO. At temperatures up to 1600°C, there is no doubt that the structure is fluorite with oxygen missing from anion sites, though one report exists of a change of defect type at higher temperatures.[44]

Several investigators agree that the electrical conductivity of cubic lime-stabilized zirconia is due solely to oxygen transport; there is a maximum in the conductivity at any temperature at a composition close to the lower phase boundary of the solid solution. The major reason for the decline in conductivity at higher CaO contents, and thus higher vacancy concentrations, seems to be an increase in the activation energy necessary for migration. Of more immediate interest is the decrease in the conductivity of any one specimen on long annealing. The conductivity reached a stable minimum after 36 hours at 1100°C, or 2700 hours at 904°C, and the original value was restored on heating to 1400°C. The conductivity remained ionic and the activation energy of the conduction process was unchanged.

These changes in electrical properties can definitely be ascribed to an ordering process[43] which results in another structure distinct from that of ZrO_2 or $CaZrO_3$. The fully disordered structure can be preserved by quenching from 1400°C; it is fluorite with a random distribution of calcium and zirconium on the cation sites. However, the neutron diffraction patterns show diffuse scattering due to random displacements of the oxygen atoms from the ideal fluorite positions and the intensities of the neutron diffraction lines could be explained by moving the oxygen ions about 0·29 Å from the ideal sites parallel to $\langle 111 \rangle$ directions. The ordered structure which forms slowly at ~1000°C differs from the disordered in that the oxygen ions occupy the same positions but in an ordered way; the diffuse scattering has largely disappeared and been replaced by new diffraction peaks which are characteristic of a more complex structure. The oxygen coordination about the cations is highly distorted and similar to the coordination in monoclinic ZrO_2.

The slow process which produces the ordered structure suggests a cation migration as the rate-determining step and the evolution of the ordered structure may therefore be dependent on an ordering of cations and 'vacancies'. Complex diffraction effects observed in some single crystals of calcia-stabilized ZrO_2 do suggest the appearance in one crystal of zones of different composition. At compositions near the boundaries accepted for the cubic phase, there was some evidence for the actual precipitation of a phase of a new structure and the phase diagram may be more complex than has been thought.

These experiments provide valuable evidence for the actual structure of disordered anion-deficient fluorite phases, and further examination by neutron diffraction methods would be most rewarding. However, direct evidence of the structures of the non-stoichiometric binary phases can be obtained only from diffraction experiments carried out at temperatures where such phases are truly stable.

References

1. Roberts, L. E. J., *Quart. Rev.*, 1961, **15**, 442.
2. Rand, M. H., and Jackson, E. E., *U.K.A.E.A. Reports*, 1963, AERE-R-3636, R-3635.
3. 'Thermodynamic and Transport Properties of the Uranium Dioxide and Related Phases', I.A.E.A., Vienna, 1965.
4. Edwards, R. K., and Martin, A. E., Symposium on Thermodynamics related to Nuclear Materials, I.A.E.A., Vienna, 1965, p. 423.
5. Kotlar, A., Gerdanian, P., and Dodé, M., *J. Chim. phys.*, 1967, **64**, 862; Gerdanian, P., and Dodé, M., *J. Chim. phys.*, 1965, **62**, 171.
6. Perio, P., Thesis, University of Paris, 1955; Hoekstra, H. R., Santoro, A., and Siegel, S., *J. Inorg. Nuclear Chem.*, 1961, **18**, 154.
7. Westrum, E. F., Jr., and Grønvold, F., *J. Phys. and Chem. Solids*, 1962, **23**, 39.
8. Belbeoch, B., Piekarski, C., and Perio, P., *J. Nuclear Materials*, 1961, **3**, 60.
9. Belbeoch, B., Piekarski, C., and Perio, P., *Acta Cryst.*, 1961, **14**, 837.
10. Willis, B. T. M., *Nature*, 1963, **197**, 755; *J. Physique*, 1964, **25**, 431.
11. Belbeoch, B., Boiveneau, J. C., and Perio, P., *J. Phys. and Chem. Solids*, 1967; **28**, 1267; Belbeoch, B., and Boiveneau, J. C., *Bull. Soc. fr. Minéral, Crystallogr.*, 1967, **90**, 558.
12. Naito, K., Ishie, T., Hamaguchi, Y., and Oshima, K., *Solid State Comm.*, 1967, **5**, 349.
13. Masaki, N., and Doi, K., *Acta Cryst.*, 1968, **B24**, 1393.
14. Delavinguette, P., and Amelinckx, S., *J. Nuclear Materials*, 1966, **20**, 130.
15. Blank, H., and Ronchi, C., *Acta Cryst.*, 1968, **A24**, 657.

16. Andresen, A. J., Symposium on Reactor Materials, Stockholm, 1959.
17. Roberts, L. E. J., and Walter, A. J., *Physico-Chimie du Protactinium*, C.N.R.S., Paris, 1966, 51.
18. Stchouskoy, T., Pezerat, H., Bouissières, G., and Muxart, R., *Compt. rend.*, 1964, **259**, 3016.
19. Aronson, S., Rulli, J. E., and Schaner, B. E., *J. Chem. Phys.*, 1961, **35**, 1382; Nagels, P., Devreese, J., and Denayer, M., *J. Appl. Phys.*, 1964, **35**, 1175.
20. Leask, M. J. M., Roberts, L. E. J., Walter, A. J., and Wolf, W. P., *J. Chem. Soc.*, 1963, 4788.
21. Roberts, L. E. J., *Advances in Chemistry Series*, 1963, **39**, 66.
22. Roberts, L. E. J., and Markin, T. L., *Proc. Brit. Ceramic Soc.*, 1967, **8**, 201.
23. Atlas, L. M., *J. Phys. and Chem. Solids*, 1968, **29**, 1349.
24. Nagels, P., Van Lierde, W., De Batist, R., Denayer, M., De Jonghe, L., and Geves, R., *Thermodynamics*, I.A.E.A., Vienna, 1965, **2**, 311.
25. Anderson, J. S., Edgington, D. N., Roberts, L. E. J., and Wait, E., *J. Chem. Soc.*, 1954, 257.
26. Short, J., and Roy, R., *J. Phys. Chem.*, 1963, **67**, 1860; Zintl, E., and Udgard, A., *Z. anorg. Chem.*, 1939, **240**, 150; Ketelaar, J. A. A., and Willems, P. J. H., *Rec. Trav. chim.*, 1937, **56**, 29.
27. D'Eye, R. W. M., and Martin, F. S., *J. Chem. Soc.*, 1957, 1847.
28. Catalano, E., Bedford, R. G., Silviera, V. G., and Wickman, H. H., *J. Phys. and Chem. Solids*, 1969, **30**, 1613.
29. Cheetham, A. K., Fender, B. E. F., Steele, D., Taylor, R. I., and Willis, B. T. M., *Solid State Comm.*, 1970, **8**, 171.
30. Eyring, L., and Holmberg, B., *Advances in Chemistry Series*, 1963, **39**, 46.
31. Hyde, B. G., Bevan, D. J. M., and Eyring, L., *Phil. Trans.*, 1966, **259A**, 583.
32. Sawyer, J. O., Hyde, B. G., and Eyring, L., *Bull. Soc. chim. France*, 1965, 1190.
33. Eyring, L., and Baenziger, N. C., *J. Appl. Phys. Supplement*,. 1962, **33**, No. 1, 428.
34. Bevan, D. J. M., Baker, W. W., Martin, R. L., and Parks, T. C., *Proc. 4th Conference Rare-Earth Research*, 1964, 441.
35. McCullough, J. D., and Britton, J. D., *J. Amer. Ceram. Soc.*, 1952, **74**, 5225.
36. Anderson, J. S., Ferguson, I. F., and Roberts, L. E. J., *J. Inorg. Nuclear Chem.*, 1955, **1**, 340; Ferguson, I. F., and Fogg, P. G. T., *J. Chem. Soc.*, 1958, 196.
37. Hyde, B. G., and Eyring, L., *Proc. 4th Conference Rare-Earth Research*, 1964, 623.

38. Gardner, E. R., Markin, T. L., and Street, R. S., *J. Inorg. Nuclear Chem.*, 1965, **27**, 541.
39. 'The Plutonium–Oxygen and Uranium–Plutonium–Oxygen Systems', I.A.E.A., Vienna Tech. Rep. Ser. 79, 1967, p. 18.
40. Bevan, D. J. M., and Kordis, J., *J. Inorg. Nuclear Chem.*, 1964, **26**, 1509.
41. Markin, T. L., and Rand, M. H., *Thermodynamics*, I.A.E.A., Vienna, 1966, 145.
42. Lefèvre, J., *Ann. Chim. (France)*, 1963, [13] **8**, 117; Collongues, R., Lefèvre, R., Perez y Jorba, M., and Queyroux, F., *Bull. Soc. chim. France*, 1962, 149; Fehrenbacher, L. L., Jacobson, L. A., and Lynch, C. T., *Proc. 4th Conference Rare-Earth Research*, 1964, 687.
43. Carter, R. E., and Roth, W. L., General Electric Research Laboratory Report No. 63-RL-3479M, 1963; Report No. 67-C-308, 1967; 'E.M.F. Measurements in High Temperature Systems', *Proc. Symp. Inst. Mining and Met.*, 1967, 125.
44. Diness, A. H., and Roy, R., *Solid State Comm.*, 1965, **3**, 123.

CHAPTER 12

Some Aspects of Solvation

H. J. V. TYRRELL

A. Introduction

A simple fluid such as a liquid rare-gas can best be described in Bernal's term[1] as a heap of randomly arranged atoms. Such a heap has, paradoxically, a 'structure' characterized by a range of coordination numbers and controlled very largely by the repulsive forces between the particles. The radial distribution curve of such a fluid shows a marked peak corresponding to the nearest neighbours of the central atom; subsequent peaks for second-nearest neighbours and so on rapidly become indistinguishable as the distance from the central atom increases. For all atoms the correlation time τ_c, defined approximately as the time required for the atom to rotate through one radian or to move a distance equal to its diameter, will be the same. In more complex fluids, particularly those where strong specific interactions, such as hydrogen bonding, can occur, the orientation of one particle with respect to its neighbours will not be random, some configurations corresponding to slightly lower energies than others. The most extreme example of this is, of course, water. Its physical properties are unusual for a substance of low molecular weight, and attempts to interpret these properties in terms of association have a long history. In their simplest form they involve an equilibrium between monomeric water H_2O and 'bulky' or 'ice-like' water $(H_2O)_n$; a very full account appears in a recent book[2] on the structure of water. Even so, study of molecular motions in water by measuring the dielectric relaxation time[2–4] or proton magnetic resonance relaxation times[5] show that they can be described in terms of a single relaxation time. For example, the variation of the dielectric constant of liquid water with frequency can be so described though there may be a very slight improvement in the interpretation by assuming a very small spread of relaxation times. This dielectric relaxation time τ_d is about 2×10^{-11} s at 0°C compared with a value of about 10^{-5} s for ice. It falls steadily with temperature with an enthalpy of activation of about 4 kcal mole^{-1}; this is very close to the enthalpy of activation for viscosity and self diffusion.[6] The correlation time τ_c calculated from the spin-lattice relaxation time for protons (see section D) in water is of the same order.[7] The lifetime of any structural entities which may exist in water, or any other structured solvents, will therefore be less than this time, which is itself the time required for about 100 inter-particle collisions.

The concept of solvation has developed very largely from work on aqueous solutions, particularly of electrolytes. Nowadays the importance of solvation phenomena in determining the properties of non-aqueous solvents and solutions, and of mixed solvents is well recognized[8,9] but the fundamental principles are largely unchanged. Solvation is recognized experimentally when the properties of the solution differ from those expected from the properties of the solvent and the solute considered alone. Such a statement implies that we know what to expect when the solution is formed; this is itself a major problem. A well-known example where solvation is invoked to explain experimental observations is the series of limiting equivalent ionic conductances for the alkali metal cations,[10] shown in Table 12.1.

TABLE 12.1 *Limiting equivalent conductances and crystallographic radii of alkali metal cations in water at 25°C.*

Ion	Li^+	Na^+	K^+	Rb^+	Cs^+
λ^0_+	$38{\cdot}6_8$	50·10	73·50	$77{\cdot}8_1$	$77{\cdot}2_6$
r_+, Å	0·60	0·95	1·33	1·48	1·69

The fact that the smallest ion moves most slowly under the influence of an electric field is interpreted by assuming that the high charge density round the small ions causes solvent (water) molecules to be firmly attached to them, and to move with them in the electric field gradient. However, the limitation imposed on the orientational movements of molecules in such a solvation sphere is not large. No separate peaks corresponding to solvent molecules in the solvation sheath and in the bulk solvent can be observed in the proton magnetic resonance spectrum of lithium salts in water or even at −110°C in methanol.[11] The correlation time τ_c for methanol solutions of lithium salts (assuming a solvation number of four for the lithium ion) is 4×10^{-12} s compared with 10^{-12} s in pure solvent. Even though the lithium ion behaves as if it is surrounded by a number of closely bound solvent molecules, those solvent molecules are not, on the average, greatly restricted in their movements and must undergo rapid exchange with the solvent molecules in the bulk solvent.

The range of properties which can be studied to obtain information about solvation is very wide and there can be no doubt that weak specific interactions between solute and solvent molecules are of extreme importance even though the time for which individual unlike molecules interact may be very small. The existence of solid solvates of definite composition must have inspired the concept of a solvation number, the number of solvent molecules (S) bound round a given solute molecule (A), and of thermodynamic equilibrium constants for solvation equilibria of the form:

$$A + nS \rightleftharpoons AS_n$$

For solutions of ions in water many tables of ionic hydration numbers have been proposed, based on different types of experimental observation.[12] The terms 'primary solvation' referring to solvent molecules in the immediate neighbourhood of the solute molecule, and 'secondary solvation' referring to all other effects occurring at a rather greater distance are commonly used but are by no means universally accepted as being valid or useful. An influential version of this is that due mainly to Frank,[13,14] in which an ion is considered to be surrounded by three zones which, at least conceptually, are distinct. In the first, near the ions, solvent molecules are held firmly around the ion. At a distance from the ion the solvent is in its normal state, and in an intermediate zone, the solvent molecules are less ordered than in bulk water because of the geometrical disparity between the arrangement of the solvent molecules close to the ion and that characteristic of the very distant solvent molecules. In addition, non-polar solutes, and the non-polar parts of organic ions may provide some stabilization for the solvent structure on the average. Exchange between solvent molecules in different zones must be rapid to explain the single correlation times τ_d or τ_c observed. Such a model can provide an extremely flexible tool for rationalizing experimental data since different experiments measure different things. Bulk thermodynamic properties of the solution will reflect the overall balance between 'structure-building' and 'structure-breaking'. The self-diffusion coefficient of the solute or the ionic mobility of an ion will mirror more closely the effect of the solute on the neighbouring solvent molecules, as will the absorption spectrum of the solute. A good example of the use of the latter is afforded by the changes in the iodine absorption spectrum with solvent, some yielding violet solutions with an absorption spectrum similar to that of iodine vapour, while others give brownish solutions which can be interpreted in terms of charge transfer complexes.

The concept of a primary solvation sheath of ordered solvent molecules in contact with a central solute particle is by no means universally accepted.[15,16] Aqueous electrolyte solutions are often more fluid than the pure solvent, a fact which can be 'explained' by attributing 'structure-breaking' properties to the solvated ion. Samoilov[16] on the other hand has argued that the solvent molecules close to such an ion become more mobile than in the bulk solvent, i.e. have shorter correlation times. Such molecules are said to be in a state of 'negative solvation'. The concept of a primary solvation sheath would correspond to 'positive solvation', that is, to solvent molecules which in the vicinity of the ion have longer correlation times than in the solvent itself.

While the bulk properties of a solution may depend strongly on solvent–solute interactions, their quantitative interpretation in such terms requires knowledge of the properties of an analogous system in which such interactions do not occur. The problems associated with this requirement are illustrated here for some thermodynamic properties, and for certain transport properties. The most direct insight into solvation phenomena might be expected to come from spectroscopic studies. The effect of solvent–solute interaction on the spectrum of dissolved species may be altered, as in the case of the nitrate ion in concentrated calcium nitrate solutions where solvation seems to alter the symmetry properties of the anion,[17] or new vibrational frequencies may appear which can be assigned to vibrations within the solute–solvent complex;[18,19] with respect to the dependence of vibrational spectra on solvation phenomena a more detailed account appears in the next chapter. Nuclear magnetic resonance spectroscopy can in a few favourable cases show the presence of separate solvated species in a solution, and, even where this does not prove possible, the technique can be used in a variety of ways to study solvation phenomena. Some of these use concepts which are identical with those used in the interpretation of transport coefficients and the method is therefore discussed in some detail. Inevitably, this chapter must be somewhat selective in view of the protean character of the subject; a more detailed review of solvation in electrolytes has been given by Conway.[20]

B. Thermodynamic aspects of solvation

Raoult's Law defines an ideal solution. This predicts that the partial vapour pressure p_i of any component in a solution is related to its molar fraction N_i and its vapour pressure in the pure state p_i^0 by the equation:

$$p_i = p_i^0 N_i \tag{12.1}$$

There have always been two extreme views on the problem of explaining the departures from this relationship found for real liquids. One is to assume that they arise because of chemical interactions between components, either *association* between like species or *solvation* where interaction between molecules of different chemical species is important. In the case of electrolytes, the initial reluctance of the scientific world to accept the idea of ionization at all was tempered by the belief that electrolytes were only partly dissociated, and that this dissociation could be expressed in terms of a definite mass-action law equilibrium constant. This concept was applied to non-electrolyte solutions by Dolezalek[21] who sought to explain all deviations from ideality exclusively in terms of specific solvent–solvent, solute–solute and solvent–solute interactions. The other extreme view is to assume that non-ideality can be attributed to non-specific interactions arising from differences in the forces between the different types of species. This idea is at the heart of the theory of electrolytes where the importance of ion–ion interactions is regarded as paramount. The conflict between the protagonists of these two extreme views led to some notable polemics in the early years of the century, but it is now clear that the views can be reconciled. Non-specific interactions can explain much, but in many cases weak chemical interactions are also important. For example, in some solvents solutions of iodine are violet, in others, brown or red. The thermodynamic and other properties of the first group can be adequately explained in terms of non-specific interactions using regular solution theory,[22] while those of the second group imply the existence of specific chemical forces between the iodine molecule and the surrounding solvent.

In order to evaluate the role of such chemical forces in determining the thermodynamic properties of any system it is first essential to know how these properties are affected by the non-specific inter-molecular forces. Dolezalek assumed that, in the absence of specific interactions, all solutions would behave according to Raoult's Law. This could be so only if all molecular species had the same volume and identical inter-molecular forces. Such a model requires that the heat of mixing should be zero, a very restrictive condition which is not true even for mixtures of imperfect gases. The van der Waals' equation parameters, a and b, are different for different gases, and van Laar[23] was able to calculate from this model the heat of mixing of two unexpanded fluids in terms of the van der Waals' constants for each. This calculation, excluding all reference to specific interactions, predicted positive (endothermic) heats of mixing; its quantitative application was hindered by the inadequacies of the van der Waals' equation on which it was based. However, the treatment was later independently improved by Scatchard,[24] and by Hildebrand and Wood[25] using the following assumptions:

(i) the molecular distribution is random and independent of the temperature;

(ii) the energy of the system is the sum of the mutual interaction energies of pairs of molecules, essentially the hypothesis of 'pair-wise additivity' common to most theories of liquids;
(iii) the volume change on mixing is zero;
(iv) the energy of interaction per unit volume between molecules of species i and species j is taken to be c_{ij}, and it is assumed that:

$$c_{ij}^2 = c_{ii} c_{jj} \tag{12.2}$$

For pure liquids the quantity c_{ii} is identified with the energy of vaporization per unit volume, and on this basis the energy of mixing E^{M} is given by:

$$E^{\mathrm{M}} = (N_1 v_1 + N_2 v_2)(\delta_1 - \delta_2)^2 \phi_1 \phi_2 \tag{12.3}$$

where N_i, v_i, ϕ_i are respectively the molar fraction, molar volume and volume fraction of component i. The energy of vaporization per unit volume of component i is written δ_i^2. This quantity δ_i occupies a central position in the theory and is termed the 'solubility parameter'. If there is no volume change on mixing then E^{M} can be identified as H^{M}, and the partial excess molar enthalpy can then be calculated. If the excess entropy of mixing is assumed to be zero, the partial excess molar free energy is equal to the partial excess molar enthalpy; all the thermodynamic properties of a 'regular' solution can therefore be calculated from the composition and the solubilty parameters.

This assumption that the excess entropy of mixing is zero means that the molecular distribution remains random even though the interaction energies between the different species change, and is the basis of the theory of regular solutions. In a sense, a regular solution is an impossible one. If interaction energies are different, configurations of lowest energy will be favoured, and the excess entropy of mixing must differ from zero. In practice, if the preferred configurations are only of slightly lower energy than the remainder, the excess entropy will be so close to zero that the model can be expected to represent many real systems where c_{ij} lies between c_{ii} and c_{jj}. Much of the development of regular solution theory depended on measurements of the solubility of a solute such as iodine in a variety of solvents. According to the theory, if the activity of the solid solute phase is a_2^{s}, and the molar fraction of the solute at saturation is N_2^{s} (ϕ_1 being the volume fraction of component 1),

$$\ln a_2^{\mathrm{s}} = \ln N_2^{\mathrm{s}} + \frac{v_2 \phi_1^2 (\delta_2 - \delta_1)^2}{RT} \tag{12.4}$$

Differentiation of this equation with respect to temperature shows that:

$$R \frac{\partial \ln N_2^{\mathrm{s}}}{\partial \ln T} = R(\ln a_2^{\mathrm{s}} - \ln N_2^{\mathrm{s}}) \tag{12.5}$$

i.e. a plot of $R\partial \ln N_2^s/\partial \ln T$ against $R \ln N_2^s$ should be linear with unit slope if the solutions are regular. Figure 12.1 shows results for iodine in a number of solvents. Solvents giving violet solutions lie on or near the 'regular solution line', and spectroscopic examination shows that, where small deviations occur, e.g. in dichloroethylene and dichloroethane, there are small changes in the absorption spectrum. Those solvents giving reddish-brown solutions, where the absorption spectrum differs markedly from that found for regular solutions, also show large deviations from regular solution behaviour as can be seen in Figure 12.1. These and many other observations show that the regular solution model can provide a reasonable description of the thermodynamic properties of many solutions where no specific chemical interaction forces are present. Such solutions are non-ideal in that they do not obey Raoult's Law. When specific chemical forces are believed to be operative their magnitude can be assessed from the deviation of their thermodynamic properties not from the ideal solution model but from the regular solution model. For example, iodine dissolved in chloroform ($\delta = 9{\cdot}2$) gives a violet solution, and the solution is 'regular'. Benzene, which has an identical solubility parameter, gives a brown solution which is non-regular. If the difference between the thermodynamic properties of benzene–iodine solutions and of chloroform–iodine solutions is taken as a measure of the degree to which the following reaction occurs:

$$C_6H_6 + I_2 \rightarrow [C_6H_6, I_2]$$

it is possible to calculate an equilibrium constant which agrees closely with the deductions made from a spectroscopic study of the system. It is therefore possible to resolve the old quarrel between Dolezalek and van Laar. The latter was correct in considering that deviations from ideal behaviour can arise from non-specific interactions, while Dolezalek's view that solvation and association can occur and their extent be deduced from the external thermodynamic properties is also correct. Quantitatively, solvation occurs when $c_{ij}^2 \gg c_{ii}c_{jj}$, while association takes place if either c_{ii} or c_{jj} is large. On this view the transition between systems where non-specific forces are solely responsible for deviations from ideal behaviour, and those where solvation is primarily responsible must be gradual. This is borne out both by plots of the form shown in Figure 12.1, and by other experimental data.

The most commonly measured thermodynamic parameters for a binary solution are the *activities* of the two components, obtained from vapour pressure measurements and in other ways. The composition of the solution can be expressed in terms of the molar fraction, molality or molarity; to calculate the first two it is necessary to assume that the solute is not solvated. The activity coefficient on the appropriate scale is obtained by dividing the experimental activity by the appropriate concentration. Clearly, if the

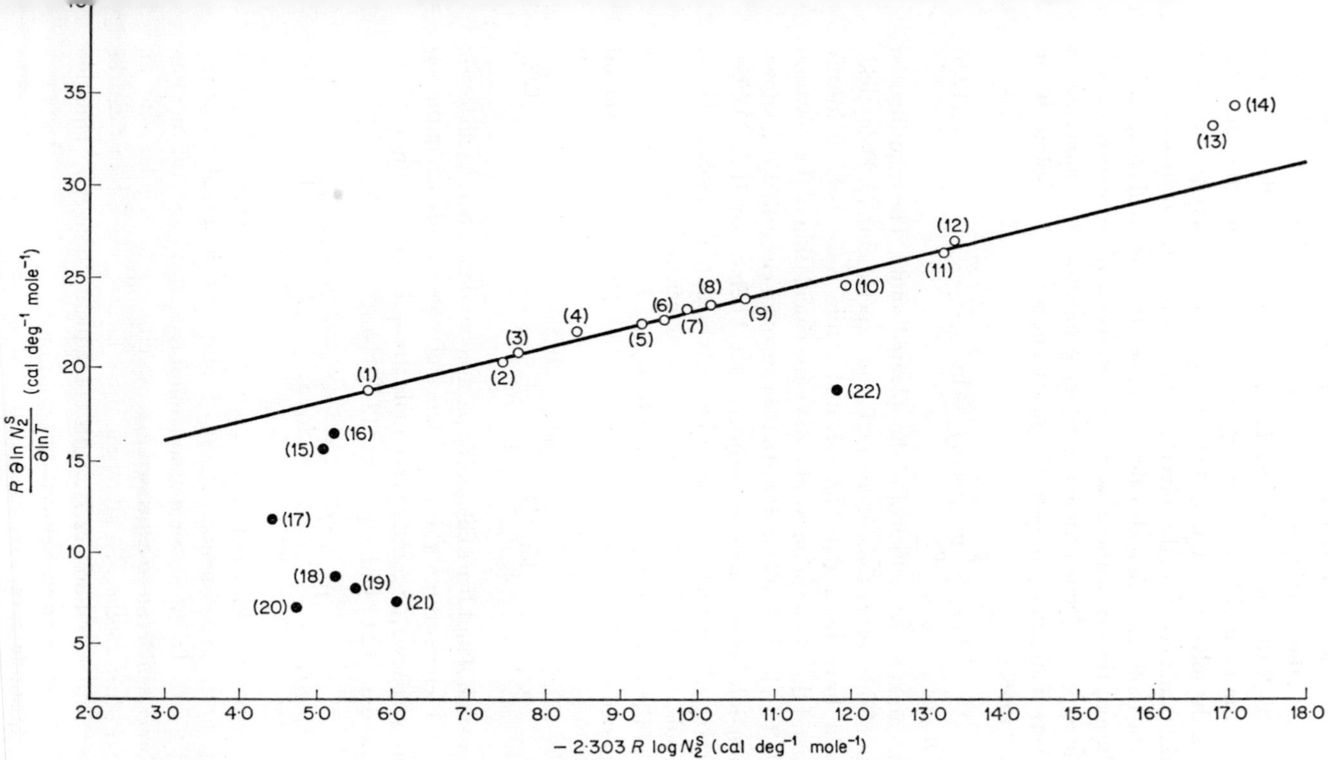

FIGURE 12.1. *Regular and non-regular solutions of iodine. Solutions showing regular behaviour lie on or near the full line which has unit slope. Violet solutions are marked ⊙, non-violet solutions ●. (From: Hildebrand, J. H., and Scott, R. L., 'Regular Solutions', Prentice-Hall, 1962, Appendix 4.)*

Solvents are: *(1)* CS_2, *(2)* $CHCl_3$, *(3)* $TiCl_4$, *(4)* CCl_4, *(5)* *c.* C_6H_{12}, *(6)* *c.* $(CH_3)_8Si_4O_4$, *(7)* *n*-C_7H_{16}, *(8)* *i*-C_8H_{18}, *(9)* $(CH_3)_3C.C_2H_5$, *(10)* $CCl_2F.CCl_2F$, *(11)* *c.*$C_6Cl_2F_4$, *(12)* $(C_3F_7.CO_2.CH_2)_4C$, *(13)* *c.* $C_6F_{11}.CF_3$, *(14)* C_7F_{16}, *(15)* 1,2-$C_2H_4Br_2$, *(16)* C_6H_6, *(17)* 1,3,5-$(CH_3)_3C_6H_3$, *(18)* *p*-$C_6H_4(CH_3)_2$, *(19)* $CHBr_3$, *(20)* $(C_2H_5)_2O$, *(21)* C_2H_5OH, *(22)* H_2O.

solute is *solvated* the activity coefficient of the solvated solute on either the mole fraction or the molality scale (but not that on the molarity scale) will differ from this 'experimental' value. The difference can readily be calculated, since the total Gibbs free energy of the solution and the chemical potential of the solvent are unaffected by solvation. For example, let a_A be the experimental activity of the solvent, and $f_{\pm}$ the activity coefficient of the solute on the mole fraction scale calculated from the chemical composition of the solution, that is, without assuming solvation. If the solute is a salt giving ν ions per molecule, these ν ions being solvated by h molecules of solvent, there being, in all, S molecules of solvent per mole of solute, it can be shown[26] that:

$$\ln f_{\pm} + \frac{h}{\nu} \ln a_A = \ln f'_{\pm} - \ln \frac{S+\nu-h}{S+\nu} \qquad (12.6)$$

$f'_{\pm}$ is the mean activity coefficient of the solvated solute. The quantities on the left-hand side, other than h, can be found experimentally. Dolezalek's view would have been that the solvated solute must behave ideally ($f'_{\pm} = 1$), and this assumption would allow the calculation of the solvation number h. This is too simple a view. In the case of an electrolyte it is known that the ion–ion forces contribute to the non-ideal behaviour. If the Debye–Hückel theory, developed to deal with such forces, applies to solvated ions, then $f'_{\pm}$ could be calculated as a function of concentration, and a better estimate of the solvation number could be made. Robinson and Stokes[26] have suggested that the familiar form of the Debye–Hückel equation should be used:

$$\log f'_{\pm} = -\frac{A|z_1 z_2| I^{1/2}}{1 + Ba^0 I^{1/2}} \qquad (12.7)$$

In this equation A and B are calculable parameters dependent on dielectric constant and temperature, a^0 is the distance of closest approach of the solvated ions, valency z_1, z_2, and I is the ionic strength of the medium. Combining equation (12.6) with equation (12.7) gives:

$$\log f_{\pm} = -\frac{A|z_1 z_2| I^{1/2}}{1 + Ba^0 I^{1/2}} - \frac{h}{\nu} \log a_A - \log\left(\frac{S+\nu-h}{S+\nu}\right) \qquad (12.8)$$

This is an equation with two unknown parameters, a^0 and h, which can be selected to fit the experimental values of $f_{\pm}$ and a_A. It is arguable whether this is adequate. In the absence of ion–ion forces, there are still the non-specific forces which can be expressed in terms either of the regular solution, or of some other model. To allow for these factors, extra terms can be introduced into equation (12.7), with consequent changes in equation (12.8). Each new term introduced into (12.7) leads to the introduction of another adjustable parameter into (12.8) with consequent loss of realism

Robinson and Stokes[26] used (12.8) to calculate hydration numbers for a large number of aqueous electrolytes from experimental activity coefficients. The results, while not too unreasonable, are by no means altogether convincing, and a really acceptable quantitative treatment of the effect of solvation upon thermodynamic properties of electrolytes is certainly not yet available.

C. Transport coefficients and solvation

For a two-component solution it is possible to define and measure three diffusion coefficients. The mutual diffusion coefficient D_{ij} measures the rate at which two components diffuse into one another. In addition there are two self-diffusion (or intra-diffusion[26a]) coefficients D_{ii}, D_{jj} which measure the rate at which suitably 'labelled' molecules of i or j distribute themselves through the solution. All three are a measure of the ease with which the molecules of the components move past one another and it might be expected that a suitable analysis of their magnitudes could provide evidence for the presence, or otherwise, of solvated species in the solution. To do this in a precise manner it is convenient to introduce the concept of 'frictional coefficients' which relate the gradient of chemical potential of a component j, grad $\boldsymbol{\mu}_j$, to its diffusion velocity $\mathbf{u}_j$ relative to that of another component. These can be defined in several ways,[27] a useful definition for the present purpose being that due to Klemm:[28]

$$\text{grad}\,\boldsymbol{\mu}_j = -\sum_k r_{jk} N_k(\mathbf{u}_j - \mathbf{u}_k) \tag{12.9}$$

The coefficients r_{jk} are, dimensionally, frictional coefficients per mole and are independent of the reference plane. Three such coefficients are required to describe diffusion and self-diffusion in a binary mixture, and it is easy to show[29, 30] that, if the components are labelled with the suffixes 0, 1, and if N_i represents the molar fraction of component i:

$$(N_0 D_{11} + N_1 D_{00})\left(\frac{\partial \ln a_1}{\partial \ln N_1}\right)_{T,P} = D_{01}\left[\frac{N_0 r_{01}}{N_0 r_{01} + N_1 r_{11}} + \frac{N_1 r_{01}}{N_1 r_{01} + N_0 r_{00}}\right] \tag{12.10}$$

Equation (12.10) takes the simple form:

$$D_{01} = (N_0 D_{11} + N_1 D_{00})\left(\frac{\partial \ln a_1}{\partial \ln N_1}\right)_{T,P} \tag{12.11}$$

when equation (12.12) is valid (cf. equation 12.2):

$$r_{01}^2 = r_{00} r_{11} \tag{12.12}$$

Loflin and McLaughlin[30] have shown that, for mixtures of fluids obeying a

Lennard–Jones type potential function, with the usual combining rules for obtaining the values of the energy (ϵ) and diameter (σ) parameters for mixtures, namely,

$$\epsilon_{01}^2 = \epsilon_{00}\,\epsilon_{11} \tag{12.13}$$

$$\sigma_{01} = \tfrac{1}{2}(\sigma_{00} + \sigma_{11}) \tag{12.14}$$

equation (12.12) implies that the two molecular species have similar diameters and masses. This condition is more restrictive than that (equation (12.2)) imposed on a regular solution, but it might be expected that major deviations from equation (12.11) could be interpreted in terms of solvation or association phenomena. This has been done, in slightly different terms, by Dullien[31] for mixtures of benzene with alcohols. This work has shown that the addition of benzene to an alcohol breaks down the associated structure of the alcohol because of the stronger interaction between the benzene and alcohol molecules. Similar qualitative conclusions can be reached from examination of the excess thermodynamic functions of these systems,[32] and of their Soret coefficients.[33]

A detailed attempt to interpret diffusion coefficient measurements for moderately concentrated aqueous electrolyte solutions in terms of hydration numbers has been made by Robinson and Stokes[34] on the basis of equation (12.11). They made the following assumptions:

(i) the thermodynamic factor ($\partial \ln a_1 / \partial \ln N_1$) should be calculated in terms of the hydrated solute;

(ii) D_{00} the self-diffusion coefficient of water in the solution always takes the value it possesses in pure water, namely, D_{00}^0;

(iii) the self-diffusion coefficient D_{11} of the solute can be replaced by ($D_{01}^0 + \Delta$), the limiting value of the mutual diffusion coefficient at infinite dilution (D_{01}^0) corrected for electrophoretic effects (Δ) by Onsager's method;[34]

(iv) that there is a need to correct for the change in the macroscopic viscosity of the system with concentration, by introducing the ratio of the viscosity (η^0) of the solvent to that of the solution (η).

For 1:1 electrolytes it was concluded that

$$D_{01} = (D_{01}^0 + \Delta)\left(1 + \frac{\partial \ln \gamma_\pm}{\partial \ln m}\left[1 + 0{\cdot}036m\left(\frac{D_{00}^0}{D_{01}^0} - h\right)\right]\eta^0/\eta\right) \tag{12.15}$$

The activity coefficient $\gamma_\pm$ is the normal mean activity coefficient measured on the molality scale. It is obvious that it is very hard to judge the exact status of this equation, and of the rather low solvation numbers (h) derived from it. Equation (12.11) on which it is based is severely restricted in its validity when applied to real systems, and the physical significance of the supplementary assumptions is obscure.

The mutual diffusion coefficient of a solute at infinite dilution D^0_{01} is related to the frictional coefficient in such a solution (r^0_{ij}) in a very simple way:

$$D^0_{01} = \frac{RT}{r^0_{01}} \tag{12.16}$$

In simple terms r^0_{01} might be expected to increase with the viscosity of the solvent η^0 and the radius a_1 of the diffusing solute. This can be shown to be true for the motion of spherical particles in a hydrodynamic continuum, a system for which Stokes[35] was able to show that:

$$r^0_{01} = 6\pi N\eta^0 a_1 \frac{1 + 2\eta^0/\beta a_1}{1 + 3\eta^0/\beta a_1} \tag{12.17}$$

where β is the coefficient of sliding friction between the spherical particle and the continuum. The two limiting cases are ($N =$ Avogadro Number)

$$\beta = 0: r^0_{01} = 4\pi N\eta^0 a_1 \tag{12.18}$$

$$\beta = \infty: r^0_{01} = 6\pi N\eta^0 a_1 \tag{12.19}$$

Combination of equations (12.16) and (12.19) leads to the usual form of 'Stokes-Einstein equation':

$$D^0_{01} = RT/6\pi N\eta^0 a_1 \tag{12.20}$$

This equation should be valid for *large* spherical molecules moving in a medium of small molecules, e.g. a polymer particle in a low molecular weight solvent. If the Stokes-law radius a_1 is greater than that expected from the molar volume of the solute, then the latter is assumed to be solvated. This would also be true if the solute were non-spherical, and equation (12.20) can be modified to deal with ellipsoidal species.[36] The contributions of solvation and of departure from spherical shape to the observed diffusion coefficient can then be estimated. This technique has been extensively used in the interpretation of diffusion measurements on protein solutions. For smaller solutes, Sutherland[37] suggested that the 'coefficient of sliding friction' β should approach zero and that a more appropriate equation than (12.20) would be:

$$D^0_{01} = \frac{RT}{4\pi N\eta^0 a_1} \tag{12.21}$$

Clearly, the uncertainty about the correct value of the numerical factor to use leads to some uncertainty about values of a_1 calculated in this way from diffusion coefficient measurements. Consequently comparison of Stokes-law radii with molecular radii derived from, say, molar volume measurements with a view to obtaining *quantitative* information about solvation for small molecules must also be of doubtful value. A method of overcoming this difficulty has been proposed by Stokes *et al.*[38, 39] By measuring the limiting mutual diffusion coefficient for a solute in solvents of different molar volumes it was found that $D^0_{01}\eta^0$ was a linear function of solvent molar volume (v_0) for solvents of a similar class, i.e.

$$D^0_{01}\eta^0 = \alpha + \beta v_0 \qquad (12.22)$$

where α and β are constants. β varied according to the class of solvent employed, but α was approximately the same for all types of solvent. It corresponded to the limiting case of solute diffusion in a solvent medium of zero molar volume, i.e. to a hydrodynamic continuum. Equation (12.20) would be expected to apply to this limiting case, and it would therefore be possible to obtain a reliable measure of the radius of the solute. For carbon tetrachloride as the solute, the value obtained is identical with that found from the radial distribution function derived from X-ray scattering. A less detailed study for iodine shows that, except in toluene, the ratio of the diffusion coefficient D^0_{ij} for carbon tetrachloride to that for iodine in a given solvent is approximately the inverse of the ratio of the solute radii. In toluene, but not in benzene or dioxan, the diffusion coefficient is lower than expected from the value for carbon tetrachloride in this solvent. Since charge transfer solute-solvent complexes are formed in all three solvents, most markedly in dioxan, the magnitude of the limiting mutual diffusion coefficient does not, on this analysis, seem to be a reliable indicator of solvation effects.

When a fluid is subjected to a shearing stress, flow occurs and, provided the Reynolds number is not exceeded, this flow has a streamline form. If a particle is suspended in the fluid it will, in general, be subject to forces which are different at different points on its surface, and will therefore rotate or tumble, thereby absorbing energy from the fluid. Such energy absorption will mean that a greater force must be applied to the fluid to achieve a given flow rate, i.e. the coefficient of viscosity of a suspension will be greater than that of the fluid itself. The effect on the viscosity will be proportional (to a first approximation) to the number of suspended particles per unit volume, and to their size. For small volume fractions ϕ of spherical particles, Einstein[40] showed that:

$$\eta = \eta^0(1 + 2{\cdot}5\phi) \qquad (12.23)$$

At larger volume fractions a modified equation,[41] which takes account of the fact that at higher concentrations the flow patterns round each particle are not independent, is:

$$\ln \eta/\eta^0 = 2{\cdot}5\phi/(1 - Q\phi) \qquad (12.24)$$

where Q is an interaction constant which can only be calculated approximately. For non-spherical particles the energy absorption due to tumbling in the streamline flow will be greater than for spherical particles and it is necessary to replace the factor 2·5, which appears in equations (12.23) and (12.24), by a quantity θ dependent on the axial ratio of the particle.[42] This has a minimum value of 2·5 for spherical particles and increases as the axial ratio departs from one:

$$\ln \eta/\eta^0 = \theta\phi/(1 - Q\phi) \tag{12.25}$$

Suppose that when the suspended particle rotates in the streamlines it carries with it molecules of solvent. These solvent molecules can be defined as being within the solvation sheath of the solute molecule. The volume fraction of the solute at a solute molarity c is:

$$\phi = cV_h \tag{12.26}$$

where V_h is the molar volume (l mole^{-1}) of the solute together with the solvent molecules which rotate with it. If the axial ratio of the solute is known then equation (12.25) can give V_h from measurements of the viscosity as a function of concentration.

Non-electrolytes obey equation (12.25) fairly well up to quite high concentrations. For example, at 25°C the viscosity of aqueous solutions of sucrose is given by:[42]

$$\ln \eta/\eta^0 = \frac{0{\cdot}875c}{1 - 0{\cdot}231c} \tag{12.27}$$

The sucrose molecule has an axial ratio of about 0·5, which corresponds to $\theta = 2{\cdot}85$. Hence $V_h = 307$ ml mole^{-1}. This compares with 212 ml mole^{-1} for the apparent molar volume of sucrose in water, assuming that the sucrose is not solvated. The difference suggests that sucrose has about 5 molecules of water firmly attached to it.

As in the case of diffusion there are difficulties associated with the application of equations based on a purely hydrodynamic theory to smaller molecules and ions. For very dilute salt solutions the relative viscosity is a linear function of $c^{1/2}$:

$$(\eta - \eta^0)/\eta^0 = Ac^{1/2} \tag{12.28}$$

This term in $c^{1/2}$ arises from ion–ion interaction and A can be calculated both for single salts and for mixtures from the theory of electrolytes. At higher concentrations an extended equation applied:[43]

$$(\eta - \eta^0)/\eta^0 = Ac^{1/2} + Bc + Dc^2 \tag{12.29}$$

the most important term being the linear one. B can be identified with

$2{\cdot}5V_h$ for large solutes from equations (12.23) and (12.26). As the molar volume of the solute approaches that of the solvent, the numerical factor should become less than the Einstein limit of 2·5 for small, nearly spherical molecules. A is, of course, zero for non-electrolyte solutions.

For aqueous solutions of many simple salts the B coefficient is negative[15] (e.g. potassium, rubidium, and caesium halides) at 25°C. This is not observed for solutions in non-associated solvents, or at higher temperatures, and is most simply explained in terms of a partial breakdown of the local structure in the solvent due to the presence of the solute. The experimental B coefficients are associated with the effect of the solute molecules in the flow pattern of the solvent and must contain a positive contribution proportional to V_h. Superimposed on this, structural breakdown gives rise to a lower solution viscosity than would be expected; this effect appears as a negative contribution to the viscosity B coefficient which in some cases evidently outweighs the normal positive contribution. For other solutes in associated solvents there is evidence of B coefficients greater than would be expected even from the volume V_h of the solvated solute. Such 'abnormally large' coefficients are taken to indicate a stabilization by the solute of the solvent structure. There is a good deal of other evidence drawn from many fields for the view that this can occur in structured solvents like water. Both effects can be considered in terms of changes in the 'effective structural temperature' of the solvent. Structural breakdown corresponds to a higher, and structural reinforcement to a lower, structural temperature. The viscosity will therefore change, just as the viscosity of the solvent changes with temperature. Since this decreases exponentially with increasing temperature, changes in the 'structural temperature' due to the introduction of a solute will have the greatest effect on the viscosity at low temperatures. The effect on the measured B coefficients will accordingly fall off as the temperature increases. Thus for structure-breaking solutes which can give negative B coefficients, the negative contribution should decrease with increasing temperature, the rate of decrease being greatest at low temperatures. Hence $\mathrm{d}B/\mathrm{d}T$ should be positive for such solutes. Conversely any positive contribution to B due to the presence of structure-forming solutes should decrease with temperature, $\mathrm{d}B/\mathrm{d}T$ being greatest at low temperatures. These changes are illustrated in Figure 12.2, which shows the variation of the experimental B coefficients for ammonium chloride and some tetraalkylammonium chlorides.[44]

Quite clearly, simple interpretations of the magnitude of experimental B coefficients in terms of solvation may be difficult. The experimental coefficient will, in general, be made up of several contributions:

(i) that due to the size of the unsolvated solute (positive and probably almost temperature-independent);

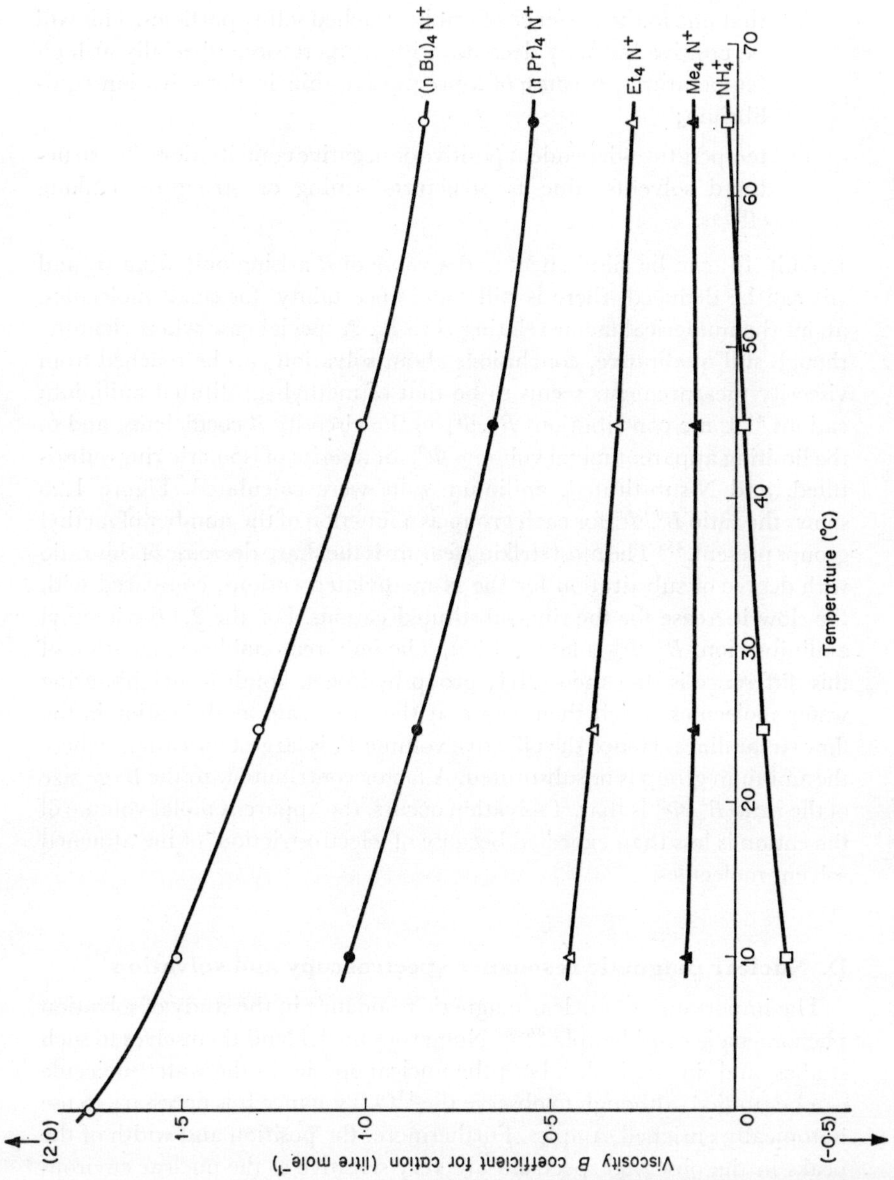

FIGURE 12.2. *Viscosity* B *coefficients for tetraalkylammonium cations, and* ($\times$*10*) *for the ammonium ion and their variation with temperature.* ○$(nBu)_4N^+$; ● $(nPr)_4N^+$; △ Et_4N^+; ▲ Me_4N^+. (*From Kay, R. L., Vituccio, T., Zawoyski, C., and Evans, D. F.*, J. Phys. Chem., *1966*, **70**, *2336.*)

□$(B_+ \times 10)NH_4^+$.

(*From Kaminsky, M.*, Discuss. Faraday Soc., *1957*, **24**, *171.*)

(ii) that due to the presence of firmly attached solute particles. This will be positive but may decrease with temperature, especially at high temperatures, because of a progressive shift in the solvation equilibrium;

(iii) temperature-dependent positive or negative contributions (for structured solvents) due to structure-forming or structure-breaking effects.

Even if (iii) can be eliminated, and a value of B arising only from (i) and (ii) can be deduced, there is still some uncertainty, for small molecules, about the numerical factor relating B to V_h. A special case where definite, though still qualitative, conclusions about solvation can be reached from viscosity measurements seems to be that of methyl-substituted anilinium cations.[45] Ionic contributions B_+, Φ_+^0 to the viscosity B coefficients, and to the limiting apparent molal volumes Φ^0, for a series of isomeric ring-substituted, and N-substituted, anilinium salts were calculated. Figure 12.3 shows the ratio B_+^0/Φ_+^0 for each group as a function of the number of methyl groups present.[45a] The most striking feature is the sharp decrease of this ratio with degree of substitution for the N-methylated cations, compared with the slow increase for the ring-substituted cations. For the 2,4,6-trimethyl anilinium ion, B_+/Φ_+^0 is large (4·05). The only reasonable explanation of this difference is that the —NH_3^+ group hydrogen-bonds to neighbouring water molecules which then rotate at the same rate as the cation in the flow streamlines. Hence the effective volume V_h is largest for cations where the aminium group is unsubstituted. A factor contributory to the large size of the ratio B_+/Φ_+^0 is that, if solvation occurs, the apparent molal volume of the cation is less than expected because of 'electrostriction' of the attached solvent molecules.

D. Nuclear magnetic resonance spectroscopy and solvation

The importance of nuclear magnetic resonance in the study of solvation phenomena is considerable.[46–48] Numerous nuclei lend themselves to such studies, and, in particular, both the nuclear species in the water molecule can be studied, although to observe the ^{17}O resonance it is necessary to use isotopically enriched samples. Furthermore, the position and width of the peaks in the observed spectrum are very sensitive to the nuclear environment, and even weak solvent–solute interactions should therefore be detectable. The addition of a solute to a solvent possessing a characteristic nuclear magnetic resonance spectrum commonly leads to small changes in the position of the solvent peaks. These can be partially accounted for by changes in the bulk magnetic susceptibility of the solution with solute concentration, but there can be an additional shift associated with solvent–solute interactions. The most detailed studies have been undertaken on electrolytes dissolved in water using the proton or the ^{17}O resonance peaks.

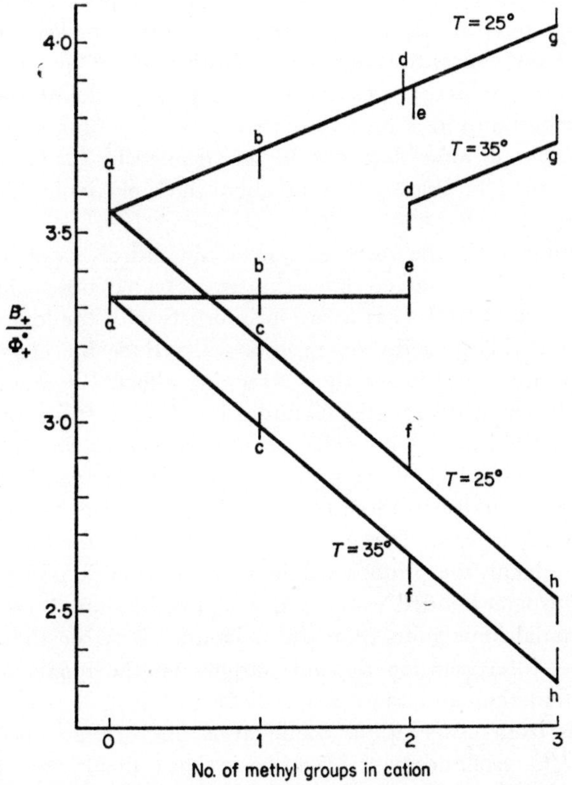

FIGURE 12.3. *The ratio B_+/Φ^0_+ for some* Me-*substituted anilinium cations as a function of the number of substituent groups.* (B_+: *viscosity B coefficient for cation.* Φ^0_+: *limiting apparent molal volume of cation. Both coefficients calculated on assumptions given in reference 45.*) (*a*) *Anilinium,* (*b*) *4-Methylanilinium,* (*c*) *N-Methylanilinium,* (*d*) *2,6-Dimethylanilinium,* (*e*) *3,4-Dimethylanilinium,* (*f*) *NN-Dimethylanilinium,* (*g*) *2,4,6-Trimethylanilinium,* (*h*) *NNN-Trimethylanilinium.*

The dissolved ions affect the electron density distribution in the solvent molecules either by electrostatic interactions (polarization), or by the formation of relatively strong chemical bonds with solvent molecules, or by altering the nature of the interactions within the bulk solvent itself. If the ion is paramagnetic, the unpaired electron can interact with the solvent nuclei to produce an additional shift (Knight shift). For aqueous solutions of diamagnetic ions at normal temperatures a single proton magnetic resonance peak from the solvent is observed. After correcting for bulk susceptibility changes, the shift $(\delta - \delta^0)$ from the peak position (δ^0) observed in the pure solvent is approximately linear in molar concentration, at least at low concentrations. The quantity $(\delta - \delta^0)/c$ extrapolated to zero concentration may be positive, corresponding to a shift towards higher fields, or negative, and is an approximately additive function of the ions present. Tables of ionic shifts can be constructed by the familiar method of making some plausible assumption about the molal ionic shift of a single ion.[49–52]

The significance of the observed shifts is not entirely clear. The fact that only a single peak is observed shows that any solvated ion species must have a short lifetime ($<10^{-8}$ s). If a cation interacts with the lone pair on the oxygen atom, this polarization process will reduce the electron density near the protons, and hence their screening effect. Resonance will then occur at a lower field strength. A similar low-field shift is produced by an anion if its effect is to repel the charge cloud round the proton towards the oxygen atom. However, if the proton becomes embedded in the electron cloud of the anion, it becomes more effectively shielded, and hence should show a high-field shift. If the ion tends to destroy the structure of the associated solvent, the protons will be more shielded in monomeric water than in hydrogen-bonded water and a high-field shift could result. The observed molal ionic shifts represent a balance between these and other effects. The interpretation depends largely on the model adopted[51, 53] for the interaction and there seems little prospect of reaching definite conclusions from observations made at a single temperature. Parallel studies on ^{17}O resonance peaks[54] emphasize the difficulties of apportioning the observed shifts among the various possible causes. Iodide ions, generally regarded as strong structure breakers, produce, as would be expected, a large upfield shift in the proton magnetic resonance spectrum, but, against expectation, the largest downfield shift given by any halide ion in the ^{17}O resonance. Apparently the breaking of hydrogen bonds has far less effect on the ^{17}O resonance than on the proton resonance. The molal shifts in both types of spectrum are dependent on temperature, and analysis of the effects of temperature has been used to obtain solvation numbers from proton magnetic resonance shifts.[55, 56] For example, Knapp, Waite and Malinowski[55] write the total chemical shift δ for a salt giving ν_+ positive and ν_- negative ions with h_+ and h_- water molecules attached, and characteristic molal ionic chemical shifts δ_+, δ_-, as

$$\delta = \frac{(\nu_+ h_+ + \nu_- h_-)m}{55{\cdot}51}(\delta_s - \delta^0) + \delta^0 \qquad (12.30)$$

with

$$\delta_s = (\nu_+ h_+ \delta_+ + \nu_- h_- \delta_-)/(\nu_+ h_+ + \nu_- h_-) \qquad (12.31)$$

and δ^0 the chemical shift for the pure solvent. If δ is plotted against temperature for a series of concentrations all the curves are found to cross at one point, i.e. at this temperature δ is the same for all solutions irrespective of concentration and hence,

$$\delta_s = \delta^0 \qquad (12.32)$$

If δ_s is then assumed to be independent of temperature then $(\nu_+ h_+ + \nu_- h_-)$, the 'solvation number' per mole of salt, can be calculated from equation (12.30). The validity of the results depends on the accuracy of the assumptions and the method has no more validity than any of the many others which have been proposed on the basis of other types of experiment.

The most unequivocal evidence for solvation is the appearance of extra peaks (ascribed to 'bound' solvent) in the spectrum of the solution as compared with the solvent. For this to occur the rate of exchange of the resonating nucleus between the two environments must be slow compared with the frequency difference between the chemical shifts. The area under the peaks represents the number of protons 'bound' and the number 'free'. Assuming that the anion is unsolvated, the solvation number of the cation can be found, provided that the salt concentration is known.[57–59] It is rare for multiple proton resonances to be observed in aqueous solutions at room temperature even when the water molecule exchange rate is low as in the case of the strongly hydrated Al(III), Be(II) and Ga(III) ions at room temperature.[60–63] However, it has been found that, for solutions of such ions, separate ^{17}O resonance signals can be observed for 'bound' and 'labile' water when a suitable paramagnetic ion species is added.[60] For this to occur, the residence time of the water molecules (not the protons) near the diamagnetic ion must be long, while that for water molecules near the paramagnetic ion must be short. The reason for the observed separation is that the water molecules not under the influence of the diamagnetic ion, are more strongly influenced by the unpaired electron of the paramagnetic species. For a given concentration of paramagnetic ion the observed shift of the resonance peak is proportional to the number of water molecules in the two states. Connick and Fiat[61] obtained hydration numbers of 5·9 and 4·2 for Al(III) and Be(II) chlorides from a study of the peak areas. Alei and Jackson[64] obtained similar values for these cations by comparing the shift $(\delta_p - \delta^0)$ in the resonance frequency of the 'free' water from that observed for pure water, due to the paramagnetic ion alone, and that $(\delta - \delta^0)$ due to a solution of both the paramagnetic ion and the cation under study. If h_+ water molecules are associated with this cation, concentration m, and the paramagnetic ion concentrations in the two cases are m_p^0 and m_p respectively, then

$$h_+ = \frac{55{\cdot}5}{m}\left\{1 - \frac{m_p}{m_p^0}\frac{(\delta_p - \delta^0)}{(\delta - \delta^0)}\right\} \tag{12.33}$$

The choice of paramagnetic ion is important since it is desirable to obtain a large shift with minimum broadening; Dy(III) has been found particularly suitable for this purpose. The method is not confined to diamagnetic ions but can be used with paramagnetic ions also, provided these have solvent molecules with sufficiently long residence times associated with them. Such an ion is the aquated Cr(III) ion which exchanges water very slowly from the solvation shell;[65] no ^{17}O resonance spectrum from solutions of this ion alone is observed because of the paramagnetic broadening. However, the shift of the labile water in Dy(III) solutions is increased when Cr(III) salts are added because some of the water becomes locked into the solvation sphere of the Cr(III) ion. The value obtained[64] in this way for the hydration number was rather high (6·8) because the transverse electron relaxation time of Cr(III) is less than that of Dy(III). Under these conditions equation (12.33) tends to break down.

The non-appearance of separate proton resonance peaks in these and other cases where the residence time of water near the cation is known to be long, is due to the lability of the protons themselves. This exchange rate decreases with temperature but for separate proton resonance peaks to be observed before the solution freezes it may be necessary to work with mixed solvents. In acetone–water mixtures at –40°C, salts of Be(II), Mg(II). Al(III) and Ga(III), show separate peaks corresponding to 'bound' and 'free' water, from the areas of which solvation numbers can be calculated.[59, 66] Some additives like acetone and tetramethyl urea do not compete with water for solvation sites, others like dimethyl sulphoxide[67] and N-methylacetamide[68, 69] do. This opens the way to quantitative studies of selective solvation in mixed solvents. One of the more remarkable observations has been that of separate proton resonance peaks for solvated Al(III) species in water-acetonitrile mixtures at 27°C with between one and six water molecules attached to the cation.[70] For water-acetonitrile ratios (R) of less than 4, separate peaks for bound and free MeCN are found. When R exceeds 5·3 a 'free' water peak appears which rapidly broadens and merges with the 'bound' water sign as R is increased further. Some representative solvation numbers found in this very direct way are shown in

Table 12.2; they represent that number of molecules associated with the cation (assuming no anion solvation) which remain in its vicinity for more than $\sim 10^{-4}$ s.

When the residence time of a proton, or of an ^{17}O nucleus, in the vicinity of the solute is less than 10^{-4} second, information on solvation has to be sought either through a study of molal shifts (see above) or through an examination of relaxation times. In the steady magnetic field H_0 an isolated nucleus with spin quantum number I, magnetic moment μ, has available to it $(2I+1)$ energy levels each separated by energy $\mu H_0/I$. For a set of such nuclei the distribution among these levels will be of the normal Boltzmann form. Absorption of an intense pulse of radiation of frequency corresponding to $\mu H_0/I$ results in promotion of some nuclei to higher states, the new distribution among the possible spin levels corresponding to a 'spin temperature' higher than that of the surrounding 'lattice'. If **M** is the vector sum of the nuclear magnetic moments per unit volume, the energy absorption corresponds to a change in $\mathbf{M}_z$, the component of **M** lying along the direction of H_0. When the existing radiation is removed the excited nuclei will lose energy to the lattice at a rate proportional to T_1^{-1}, where T_1 is termed the spin-lattice, thermal, or longitudinal relaxation time. The process continues until the energy distribution among the spin levels corresponds to the normal lattice temperature, i.e. it is a process of thermal equilibration between the spins and the lattice, which is accompanied by a change in the longitudinal component of **M**. For energy loss from the excited nucleus to occur it is necessary for the fluctuations in the local magnetic field, due to the molecular motions of the neighbouring nuclei, to oscillate at the nuclear resonance frequency. Molecular motions occur in times of the order of 10^{-11} s while the resonance frequency is usually between 10 and 100 mHz (time per cycle $10^{-7}-10^{-8}$ s). Hence the probability of a sufficiently strong fluctuation of the correct frequency is small, and T_1 is comparatively long $(10-10^{-3}$ s) except in the presence of paramagnetic species. The presence of the unpaired electron leads to a greater intensity of the local field fluctuation, and so to values of 10^{-4} s or less for T_1. In general T_1 depends on the resonant frequency[71] and hence on H^0. If the only important relaxation mechanism is the magnetic dipole–dipole interaction between pairs of protons, as for example in an aqueous solution of a diamagnetic salt, T_1 has two components. One is due to the mutual effect of the proton pair within a single molecule $(T_1)_{intra}$, and the other to the interaction of a nucleus in one molecule with nuclei in other molecules $(T_1)_{inter}$. In the special case of highly fluid liquids like water both components become independent of frequency, and are inversely proportional to the correlation time τ_c (page 296). For intramolecular relaxation τ_c itself is inversely proportional to a rotational diffusion constant, while for

TABLE 12.2. *Some cation hydration numbers derived from n.m.r. data*

Cation	Solvent	Anion	Temperature range	Method	Hydration number
Al(III)	Water	ClO_4^-	–50°C	p.m.r; peak area[a]	6·0
	Water	Br^-	–58° to –62°C	p.m.r; peak area[a]	5·8
	Water	Cl^-	–47°C	p.m.r; peak area[a]	6·0
	Water + Co^{2+}	$Cl^- + ClO_4^-$	~20°C	^{17}O; peak area[d]	5·9
	Water + Dy^{3+}	ClO_4^-	~20°C	^{17}O; peak shift[e]	6·0
	Water/Acetone	NO_3^-	–40°C	p.m.r; peak area[a]	6·0
	Water/Acetone	Br^-	–54°C	p.m.r; peak area[a]	5·8
Ga(III)	Water	ClO_4^-	–62°C	p.m.r; peak area[a]	6·0
	Water + Co^{2+}	ClO_4^-	35°C	^{17}O; peak shift[g]	$6·28 \pm 0·26$
		ClO_4^-	35°C	^{17}O; peak area[g]	$5·89 \pm 0·20$
	Water + Mn^{2+}	NO_3^-	48·5°C	p.m.r; linewidth[b]	6·1
		NO_3^-	2°C	p.m.r; linewidth[b]	6·0
	Water/Acetone	ClO_4^-	–35° to –63°C	p.m.r; peak area[a]	5·9
	Water/Acetone	Cl^-		p.m.r; peak area[c]	6 to 1 acc. to acetone conc.
In(III)	Water	Cl^-	–59°C	p.m.r; peak area[a]	4·7
	Water/Acetone	ClO_4^-	–89° to –99°C	p.m.r; peak area[a]	5·9
Be(II)	Water	Cl^-	–50°C	p.m.r; peak area[a]	4·5
	Water + Co^{2+}	Cl^-	20°C	^{17}O; peak area[d]	4·2
	Water + Dy^{3+}	ClO_4^-	20°C	^{17}O; peak shift[e]	3·8
	Water/Acetone	Cl^-	–62°C	p.m.r; peak area[a]	4·2
Mg(II)	Water/Acetone	ClO_4^-	–78° to –82°C	p.m.r; peak area[a]	5·8
	Water + Mn^{2+}	NO_3^-	43°C	p.m.r; linewidth[f]	3·8

Ca(II)	Water + Mn^{2+}	NO_3^-	31°C	p.m.r; linewidth[f]	4·3
Sr(II)					5·0
Ba(II)					5·7
Zn(II)					3·9
Cd(II)					4·6
Hg(II)					4·9
Pb(II)					5·7

[a] Fratiello, A., Lee, R. E., Nishida, V. M., and Schuster, R. E., *J. Chem. Phys.*, 1968, **48**, 3705.
[b] Swift, T. J., Fritz, O. G., and Stephenson, T. A., *J. Chem. Phys.*, 1967, **46**, 406.
[c] Fratiello, A., Lee, R. E., and Schuster, R. E., *Chem. Comm.*, 1969, 37.
[d] Connick, R. E., and Fiat, D. N., *J. Chem. Phys.*, 1963, **39**, 1349.
[e] Alei, M., and Jackson, J. A., *J. Chem. Phys.*, 1964, **41**, 3402.
[f] Swift, T. G., and Sayre, W. G., *J. Chem. Phys.*, 1965, **44**, 3569.
[g] Fiat, D. N., and Connick, R. E., *J. Amer. Chem. Soc.*, 1966, **88**, 4754.

intermolecular relaxation, it is inversely proportional to a translational diffusion coefficient. If the liquid is regarded as a hydrodynamic continuum both these diffusion coefficients can be shown to be proportional to T/η, where T is the absolute temperature and η the macroscopic viscosity of the system. Since

$$1/T_1 = (1/T_1)_{\text{inter}} + (1/T_1)_{\text{intra}} \qquad (12.34)$$

the reciprocal, $1/T_1$, of the measured longitudinal relaxation time should, on these assumptions, be a function of η/T. Other relaxation mechanisms, e.g. quadrupole coupling important in deuterated compounds, may be operative, but the equations of $1/T_1$ maintain the same general form. These relaxation times can be measured in several ways including the spin-echo method.[5]

The solvent molecules in the vicinity of a solute may or may not have the same freedom of movement as in the bulk solvent. Hence the experimental value of $(1/T_1)$ is contolled by a correlation time which is an average value of the correlation times of nuclei in solvent molecules inside and outside the sphere of influence of the solute. For an m-molal solution of a salt giving ν_+ cations and ν_- anions per mole, Engel and Hertz[5] write:

$$\tau_c = \left[1 - \frac{m}{55{\cdot}5}(\nu_+ h_+ + \nu_- h_-)\right]\tau_c^0 + m\nu_+ h_+ \tau_c^+/55{\cdot}5 + m\nu_- h_- \tau_c^-/55{\cdot}5 \qquad (12.34)$$

τ_c^0 is the correlation time for the bulk solvent, and τ_c^+, τ_c^-, correlation times for the anions and cations respectively. The numbers of solvent molecules influenced respectively by the cation and the anion are h_+, h_-. At the present, τ_c, τ_c^0, etc. cannot be calculated from $(1/T_1)$, and Engel and Hertz proceeded by analogy. Since $\tau_c = f(\eta/T)$ it would be expected that, by analogy with equation (12.29),[5, 72]

$$1/T_1 = (1/T_1^0)\,[1 + B'm + D'm^2] \qquad (12.35)$$

The B coefficient in (12.29) is an approximately additive property of the ions present, and it might be expected that B' should therefore also be divisible into ionic contributions. For a series of diamagnetic salts, Engel and Hertz have shown[5] that the B' values run, on the whole, parallel to the viscosity B coefficients. Structure-breaking salts have negative B' values, i.e. $1/T_1$ and hence τ_c are less than in water. The effect of the solute is, on the average, to give greater rotational and translational freedom to the solvent molecules especially, presumably, to those close to the solute. This is Samoilov's 'negative hydration',[16] though no decision can be made on the basis of these studies about one vital question, namely, whether the mobile solvent molecules are in the immediate neighbourhood of the ion, or outside a strongly oriented zone of solvent centred on the solute molecule (cf. page 298).

A resonance absorption peak is characterized both by its position and by its width. In condensed phases, a nucleus is subjected both to the steady field H^0 and to a local field which varies from nucleus to nucleus within a range H_{local}. Hence the assembly of nuclei will resonate over a narrow range of frequencies centred on that corresponding to H^0. The local variability of the magnetic field also means that the Larmor precessional frequency ($\mu I/h$) varies from nucleus to nucleus over a range of about $10^4\,s^{-1}$. Nuclear spins in phase at any instant therefore go out of phase in a time of the order of 10^{-4} s. Energy exchange between identical nuclei takes place only when they are in the correct phase relation, and this spin-exchange also contributes to the line broadening. Other factors are the non-homogeneity of H^0, nuclear electric quadrupole interaction, a very small value of T_1, and the presence of paramagnetic ions in the system. It is necessary to introduce a spin–spin or transverse relaxation time T_2 whose reciprocal is the rate constant governing the approach to equilibrium of the transverse components $\mathbf{M}_x$, $\mathbf{M}_y$ of the nuclear magnetization. When the intensity of the applied radio-frequency field is small, it can be shown that the resonance peak should theoretically be Lorenzian in shape (damped oscillator) with

$$1/T_2 = 2\pi(\nu_0 - \nu_{1/2}) \qquad (12.36)$$

where ν_0 is the frequency of the peak maximum and $\nu_{1/2}$ is the frequency at which the signal is half the maximum value. Although the experimental peaks are not always strictly Lorenzian, the width at half height is assumed to be proportional to $1/T_2$. This quantity is an average relaxation time dependent on more than one factor. However the relative importance of any given factor varies with temperature and a careful analysis of the variation in line widths with temperature can lead to information on the rate of chemical exchange for solvent molecules in a solvation sphere, provided that a value for the solvation number is available.[61, 73]

Proton relaxation times, as measured from line-width changes, vary with solute concentration and have been used in an attempt to determine hydration numbers of several cations and anions in aqueous alkali halide solutions. One procedure used is very simple.[74] The difference in the line width for a solution of concentration c and for the pure solvent is plotted against c. The plot is linear up to a concentration c' when there is a distinct departure from linearity. This is taken to be the salt concentration at which *all* water molecules are associated with either the ion or the cation. Not surprisingly the hydration numbers found in this way are very high and the method is not very satisfactory. A method which has a more convincing theoretical basis is that developed by Swift and Sayre.[75] If a probe species

Y is added to a salt solution for which all solvated species have lifetimes of less than 10^{-4} s, then the rate of disappearance of Y is given by:

$$-\mathrm{d}\ln[Y]/\mathrm{d}t = k_a[H_2O]_a + k_b[H_2O]_b \qquad (12.37)$$

where suffix 'a' refers to 'free' water and 'b' to 'bound' water. If Y reacts much more rapidly with 'free' than with 'bound' water the rate constant k_a is much greater than k_b and equation (12.34) becomes:

$$-\mathrm{d}\ln[Y]/\mathrm{d}t = k_a[H_2O]_a \qquad (12.38)$$

A suitable probe is the hexaquomanganous ion, $Mn(H_2O)_6^{2+}$. The rate of exchange of water between the solvent shell of the manganous ion is assumed to obey equation (12.38) and to be equal to the exchange rate for protons. This rate is the reciprocal of the lifetime of a proton in the solvation sphere of the Mn(II) ion, τ_{Mn}, which can theoretically be obtained from proton line-width data. In practice the matter is more complex. Hence, an empirical comparative method has been worked out, and later shown to be theoretically justified provided certain conditions are fulfilled. A solution AB contains a standard ion (Al(III), Be(II), NH_4^+, or H^+, as the nitrate) with known solvation number N_A, and a small concentration of manganous salt. A solution A′B contains another cation, also as the nitrate, with solvation number $N_{A'}$, and the same concentration of manganous salt. Similar solutions A, A′ without added manganous ions are also examined. The line widths W_{AB}, $W_{A'B}$, W_A, $W_{A'}$, are measured for these four solutions, and the following empirical equation is used:*

$$\frac{W_{AB} - W_A}{W_{A'B} - W_{A'}} = \frac{[H_2O]_{A'B} - N_{A'}[A']}{[H_2O]_{AB} - N_A[A]} \qquad (12.39)$$

Cross comparison of the standard ions shows that internal consistency can be obtained over a very wide range of experimental conditions using the following standard solvation numbers:

$$Al^{3+}(6),\ Be^{2+}(4),\ NH_4^+(0),\ H^+(0)$$

The method has been challenged since equation (12.39) is purely empirical[76, 77] and the results obtained are not always consistent with more direct n.m.r. methods which are undoubtedly preferable when they are applicable. A somewhat different form of this technique has been employed to examine the solvation of paramagnetic ions in mixed solvents[78] where the line broadening caused by the paramagnetic ion on the p.m.r. spectrum of water alone, and in an admixture with a second solvent, are compared. The method has been used to show that Co^{2+} and Ni^{2+} ions are always preferentially solvated by water in water-dioxan mixtures but not in water-dimethylformamide or water-dimethyl sulphoxide mixtures.

* $[H_2O]_{A'B}$, $[H_2O]_{AB}$ are water concentrations in the solutions A′B, AB respectively, and [A′], [A] the respective cation concentrations.

E. Conclusions

The quantitative study of solvation phenomena was originally dominated by two concepts, namely the stoichiometry of the solvation complex and the thermodynamics of the solvation reaction. Both concepts are essentially derived from experience with chemical reactions of the conventional type, and are therefore most successful where relatively strong solute–solvent interactions are concerned. One class of these which has attracted much attention is that of charge-transfer interactions, and in this field the classical concepts are undoubtedly valid. As the solvent–solute interaction becomes weaker it becomes increasingly difficult to separate the weak chemical forces causing the specific interactions we call solvation, from the non-specific interactions responsible for many ion-solvent interactions or for the behaviour of regular solutions. The stoichiometry of the solvation complex then becomes essentially a function of the solution property used to measure it. Furthermore, a solvation equilibrium, like any other, can be regarded as a ratio of two rates, namely,

(i) the rate at which solvent molecules leave the bulk solvent to come under the influence of the solute (k_1),

(ii) the rate at which they leave the sphere of influence of the solute to rejoin the bulk solvent (k_2).

If k_2/k_1 is large, the interaction is thermodynamically weak. Since k_1 is essentially independent of the solute, k_2 governs the thermodynamic stability of the solvated species. A large k_2 corresponds to a short 'residence time' in the solvation sphere for any given solvent molecule. Thus, even the least equivocal quantitative evidence for solvation, the appearance of separate peaks for 'bound' and 'unbound' nuclei in the nuclear magnetic resonance spectrum of a solution, applies only to those solvated species which have a lifetime in excess of about 10^{-4} s. In such a situation, it is clearly politic to examine solvation phenomena by as many techniques as possible and to preserve a healthy scepticism about the validity of the precise conclusions to be drawn from any one of them.

References

1. Bernal, J. D., *Proc. Roy. Soc.*, 1964, **280A**, 299.
2. Eisenberg, D., and Kaufmann, D., 'The Structure of Water', Oxford University Press, 1969.
3. Collie, C. H., Hasted, J. B., and Riston, D. M., *Proc. Phys. Soc.*, 1948, **60**, 145.
4. Grant, E. H., Buchanan, T. J., and Cook, H. F., *J. Chem. Phys.*, 1957, **26**, 156.
5. For example, Engel, G., and Hertz, H. G., *Ber. Bunsengesellschaft Phys. Chem.*, 1968, **72**, 808.

6. Tyrrell, H. J. V., 'Diffusion and Heat Flow in Liquids', Butterworth, London, 1961.
7. Hertz, H. G., 'Progress in NMR Spectroscopy', Pergamon, Oxford, 1967, p. 159.
8. Parker, A. J., *Quart. Rev.*, 1962, **16**, 163.
9. Franks, F. (ed.), 'Physico-Chemical Processes in Mixed Aqueous Solvents', Heinemann, London, 1967.
10. Robinson, R., and Stokes, R. H., 'Electrolyte Solutions', Butterworth, London, 1959, Appendices 3.1; 6.2.
11. Nakamura, S., and Meiboom, S., *J. Amer. Chem. Soc.*, 1967, **89**, 1765.
12. For example, Bockris, J. O'M., *Quart. Rev.*, 1949, **3**, 173.
13. Frank, H. S., and Evans, M. W., *J. Chem. Phys.*, 1945, **13**, 507.
14. Frank, H. S., and Yang-Wen, Wen, *Discuss. Faraday Soc.*, 1957, **24**, 133.
15. Gurney, R. W., 'Ionic Processes in Solution', McGraw-Hill, New York, 1953.
16. Samoilov, O. Ya., *Discuss. Faraday Soc.*, 1957, **24**, 141; 'Structure of Aqueous Electrolyte Solutions and the Hydration of Ions', Consultants Bureau Enterprises Inc., New York, 1965.
17. Irish, D. E., and Walrafen, G. E., *J. Chem. Phys.*, 1967, **46**, 378.
18. Hester, R. E., and Plane, R. A., *Inorg. Chem.*, 1964, **3**, 768.
19. de Silveira, A., Marques, M. A., and Marques, N. M., *Mol. Phys.*, 1965, **9**, 271.
20. Conway, B. E., *Ann. Rev. Phys. Chem.*, 1966, **17**, 481.
21. Dolezalek, F., *Z. phys. Chem.* (*Leipzig*), 1908, **64**, 727.
22. Hildebrand, J. H., and Scott, R. L., 'The Solubility of Non-Electrolytes', 3rd edn., Dover Publ. Inc., 1964.
23. van Laar, J. J., *Z. phys. Chem.* (*Leipzig*), 1910 **72**, 823; 1913, **83**, 599.
24. Scatchard, G., *Chem. Rev.*, 1931, **8**, 321.
25. Hildebrand, J. H., and Wood, S. E., *J. Chem. Phys.*, 1933, **1**, 817.
26. Ref. 10, p. 39.
27. Tyrrell, H. J. V., Thomas Graham Memorial Symposium, University of Strathclyde, September 1969, in press.
28. Klemm, A., *Z. Naturforsch.*, 1953, **8a**, 397.
29. Tyrrell, H. J. V., *J. Chem. Soc.*, 1963, 1599.
30. Loflin, T., and McLaughlin, E., *J. Phys. Chem.*, 1969, **73**, 186.
31. Dullien, F. A. L., *Trans. Faraday Soc.*, 1963, **59**, 856.
32. Rowlinson, J. S., 'Liquids and Liquid Mixtures', Butterworth, London, 1959, p. 179 *et seq.*
33. Farsang, G., and Tyrrell, H. J. V., *J. Chem. Soc.* (*A*), 1969, 1839.
34. Ref. 10, p. 309 *et seq.*
35. Stokes, G. G., Mathematical and Physical Papers, Volume 3 (especially Section IV, p. 55), Cambridge University Press, London, 1903.
36. Perrin, F., *J. Phys. Radium*, 1936, **7**, 1.
37. Sutherland, W., *Phil. Mag.*, 1905, **9**, 781.

38. Stokes, R. H., Dunlop, P. J., and Hall, J. R., *Trans. Faraday Soc.*, 1953, **49**, 886.
39. Hammond, B. R., and Stokes, R. H., *Trans. Faraday Soc.*, 1955, **51**, 1641.
40. Einstein, A., *Ann. Physik*, 1906, **19**, 289; *ibid.*, 1911, **34**, 591.
41. Vand, V., *J. Phys. Chem.*, 1948, **52**, 277, 300, 314.
42. Stokes, R. H., and Mills, R., 'Viscosity of Electrolytes and Related Properties', Pergamon, Oxford, 1965.
43. Jones, G., and Dole, M., *J. Amer. Chem. Soc.*, 1929, **51**, 2950.
44. Kay, R. L., Vituccio, T., Zawoyski, C., and Evans, D. F., *J. Phys. Chem.*, 1966, **70**, 2336; Evans, D. F., Cunningham, G. P., and Kay, R. L., *ibid.*, p. 2974; Kaminsky, M., *Discuss. Faraday Soc.*, 1957, **24**, 171.
45. Robertson, C. T., and Tyrrell, H. J. V., *J. Chem. Soc.* (*A*), 1969, 1938.
45a. Robertson, C. T., M.Phil. Thesis, University of London, 1969.
46. Burgess, J., and Symons, M. C. R., *Quart. Rev.*, 1968, **22**, 276.
47. Silver, B. L., and Luz, Z., *Quart. Rev.*, 1967, **21**, 458.
48. Hinton, J. F., and Amis, E. S., *Chem. Rev.*, 1967, **67**, 367.
49. Shoolery, J. W., and Alder, B. J., *J. Chem. Phys.*, 1955, **23**, 805.
50. Hertz, H. G., and Spaltoff, W., *Z. Elektrochem.*, 1959, **63**, 1096.
51. Hindman, J. C., *J. Chem. Phys.*, 1962, **36**, 1000.
52. Butler, R. N., *Chem. and Ind.*, 1969, 456.
53. Bergquist, M. S., and Forslind, E., *Acta Chem., Scand.*, 1962, **16**, 2069.
54. Luz, Z., and Yagil, G., *J. Phys. Chem.*, 1966, **70**, 554.
55. Knapp, P. S., Waite, R. O., and Malinowski, E. R., *J. Chem. Phys.*, 1968, **49**, 5459.
56. Chmelnick, A. M., and Fiat, D. N., *J. Chem. Phys.*, 1968, **49**, 2101.
57. Swinehart, J. H., and Taube, H., *J. Chem. Phys.*, 1962, **37**, 1759.
58. Matwiyoff, N. A., *Inorg. Chem.*, 1966, **5**, 788.
59. Fratiello, A., Lee, R. E., Nishida, V. M., and Schuster, R. E., *J. Chem. Phys.*, 1968, **48**, 3705.
60. Jackson, J. A., Lemons, J. F., and Taube, H., *J. Chem. Phys.*, 1960, **32**, 553.
61. Connick, R. E., and Fiat, D. N., *J. Chem. Phys.*, 1963, **39**, 1349.
62. Fiat, D. N., and Connick, R. E., *J. Amer. Chem. Soc.*, 1966, **88**, 4754.
63. Baldwin, H. W., and Taube, H., *J. Chem. Phys.*, 1960, **33**, 206.
64. Alei, M., and Jackson, J. A., *J. Chem. Phys.*, 1964, **41**, 3402.
65. Hart, J. P., and Taube, H., *J. Chem. Phys.*, 1950, **18**, 757.
66. Fratiello, A., Lee, R. E., and Schuster, R. E., *Chem. Comm.*, 1969, 37.
67. Fratiello, A., Lee, R. E., Nishida, V. M., and Schuster, R. E., *Inorg. Chem.*, 1969, **8**, 69.
68. Hinton, J. F., and Amis, E. S., *Chem. Comm.*, 1967, 100.
69. Hinton, J. F., Amis, E. S., and Mettetal, W., *Spectrochim Acta*, 1969, **25A**, 119, 709.
70. Supran, L. D., and Sheppard, N., *Chem. Comm.*, 1967, 832.

71. For example, Pople, J. A., Schneider, W. G., and Bernstein, H. J., 'High Resolution Nuclear Magnetic Resonance', McGraw-Hill, New York, 1959; Abragam, A., 'The Principles of Nuclear Magnetism', Oxford University Press, 1961.
72. Eisenstadt, M., and Friedman, H. L., *J. Chem. Phys.*, 1966, **44**, 1407; 1967, **46**, 2182.
73. Swift, T. J., and Connick, R. E., *J. Chem. Phys.*, 1962, **37**, 307.
74. Fabricand, B. P., Goldberg, S., Leifer, R., and Unger, S. G., *Mol. Phys.*, 1964, **7**, 425.
75. Swift, T. J., and Sayre, W. G., *J. Chem. Phys.*, 1965, **44**, 3567.
76. Meiboom, S., *J. Chem. Phys.*, 1967, **46**, 410.
77. Swift, T. J., and Sayre, W. G., *J. Chem. Phys.*, 1967, **46**, 410.
78. Frankel, L. S., and Stengel, T. R., *Canad. J. Chem.*, 1968, **46**, 3183.

CHAPTER 13

Vibrational Spectroscopic Studies of Complexes in Solution

D. N. WATERS

A. Introduction

The area of chemistry which is concerned with the nature and properties of complex ions in solution is one which is remarkable for the great diversity of the experimental methods which have contributed to its development. Many types of investigation have been concerned primarily with the determination of the equilibrium concentrations of species and the deduction of related thermodynamic quantities, such as the enthalpies and entropies of complex formation.[1] Other types of study, mainly spectroscopic, have been directed towards the elucidation of structural problems, such as the stereochemistry and bonding properties of particular complexes. Various other techniques have been brought to bear on the questions of ion solvation and interionic interactions in solution.[2] In this chapter the applications of the methods of vibrational spectroscopy to the study of solutions containing complex molecules and ions will be discussed. Although the techniques of infrared and Raman spectroscopy have been mainly used, with great success, in their role as tools for the investigation of molecular structure, it will be a principal aim of this chapter to illustrate their increasing importance both for the quantitative study of ionic equilibria, and for the understanding of solvation and ionic interaction phenomena.

Two principal types of complex are generally recognized. These are the common (Brønsted) acids and the complexes of metals with donor ligands. The behaviour of both in solution is characteristically described in terms of association or dissociation equilibria, often with essential solvent participation. Infrared and Raman methods have been applied to the study of these complexes in both aqueous and non-aqueous solvent media, although for the study of aqueous solutions the relatively feeble Raman scattering, as compared to the intense infrared absorption (especially at long wavelengths), of liquid water has resulted in the Raman method's finding by far the wider application.

We shall first consider the use of vibrational spectroscopy for the estimation of the concentrations of ionic species in solution. The method is based on the measurement of band intensities, and to a large extent is free from any considerations of molecular structure which arise in connection with the observed spectra. For the quantitative study of electrolyte solutions at high concentrations, where many of the 'classical' methods lose their validity because of non-ideal solution behaviour, the method of vibrational spectroscopy is one of the most important available. In the section that follows we shall examine some of the ways in which infrared and Raman methods have yielded information on solvent–solute and solute–solute interactions in ionic solutions. Finally we shall give a brief resumé of typical structural results on individual complex species which have been obtained from solution investigations. Work in all these areas is proceeding apace, and it is not possible to give more than a selection of results which pertain to each of the topics discussed. Space does not permit the consideration of the general theoretical principles upon which structural conclusions are based,[3] nor any detailed account of modern developments in experimental techniques,[4] without which, in many cases, the results quoted here could not have been obtained. It is hoped, however, that the examples cited will provide a reasonably balanced perspective of this extensive field, and at the same time highlight some of the points of growth and current interest enjoyed by infrared and Raman spectroscopy in their application to solution chemistry.

B. Quantitative equilibrium studies

The use of vibrational intensity measurements for the determination of molecular or ionic concentrations is in principle straightforward if it can be assumed that the spectrum of each species is unaffected by the other species present, and simply contributes additively to the spectrum of the solution. There is evidence that, in certain circumstances, intermolecular interactions can produce a significant effect on intensities, as is seen, for example, in comparisons of Raman scattering coefficients for organic molecules in gas and liquid phases[5] and in different solvents.[6] These 'dielectric' effects are probably more important for Raman than for infrared measurements,[6] but, even here, they appear not to be large unless considerable changes of environment are encountered during a series of observations. Thus, for measurements carried out in a given solvent medium, and in the absence of any specific interactions involving the species under consideration, the above assumption usually represents an adequate approximation. Even so, in experimental determinations of both infrared and Raman intensities, proper account must be taken of various instrumental factors in order to achieve meaningful results. In infrared work, the main consideration is that the instrumental slit width should be small compared with the true absorption band width,[7] a requirement which can usually be met without difficulty for solution measurements. In Raman investigations, the observed intensity is affected by several factors, among which variations of the intensity of the exciting radiation, differences of scattering geometry, and changes of refractive index and absorption coefficient of the sample with changes of solution composition are primarily important.[8] Fortunately, the need to make explicit, individual corrections for these effects can usually be circumvented by the use of an internal or external standard in the measurement procedure (although, for an external standard, the refractive index problem requires separate consideration[9]). For work on ionic solutions, the $\nu_1(a_1)$ line of the perchlorate ion is commonly adopted as an internal standard, it having been demonstrated[10] (for several aqueous systems) that line intensities in the spectrum of this species are not measurably affected by changes of the solution composition. For more detailed information on the practical aspects of Raman intensity measurement, the discussion by Irish[11] may be consulted.

(*a*) *Ionization of protonic acids*

The hydronium ion, H_3O^+ has been detected[12] by means of its infrared spectrum in aqueous solutions of several acids, for example, HCl, HBr, HNO_3, $HClO_4$, H_2SO_4 and H_3PO_4. The observed bands are strongly broadened by hydrogen bonding, however, and although intensity measurements on the spectrum of this ion have been suggested as a method for investigating acid ionization processes, it has been found advantageous to turn to the spectra of the undissociated acid molecules and their anionic dissociation products for quantitative interpretation of the equilibria.

The principles of the method are conveniently seen by reference to a few typical studies. In the case of nitric acid the fact of incomplete dissociation in concentrated aqueous solution was first shown qualitatively in early Raman investigations.[13] It was found that at low acid concentrations the spectrum corresponds to that of the known nitrate ion, but that, with increasing concentration, a new spectrum appears which is evidently due to the undissociated HNO_3 molecule. A quantitative examination of this system has been made by Krawetz,[14] who used Raman intensity measurements of the $\nu_1(a_1)$ line of the nitrate ion at 1049 cm^{-1} to determine the concentration of this species in nitric acid solutions of varying composition. Here, as in all similar applications of intensity measurements to the determination of concentrations, it was necessary to know the factor of proportionality relating intensity and concentration for the ion being determined. (Referred to a solution containing 1 mol litre^{-1}, this factor is usually termed the molar intensity for the species.) For the nitrate ion this was readily obtained from observations on sodium nitrate solutions. The degree of dissociation α of the acid thus followed, enabling an apparent ionization constant $k_m = \alpha^2 m/(1-\alpha)$ to be calculated. By combining these values with activity coefficient data derived from osmotic measurements, the thermodynamic ionization constant $K_m = k_m \gamma_{\pm}^2/\gamma_u$ was found by extrapolating a plot of $\log K_m \gamma_u$ against m to zero m. The symbols here have their usual significance, γ_u denoting the activity coefficient of the undissociated acid. The result, at 25°C, was $K_m = 25$ mol litre^{-1}.

Aqueous solutions of sulphuric acid have also been the subject of quantitative Raman measurements.[15] In this slightly more complicated system two stages of dissociation are possible, and it follows that, in order to determine the concentration of the three species H_2SO_4, HSO_4^- and SO_4^{2-}, molar intensity values for at least two of them are required. The value for the sulphate ion is easily obtained from intensity measurements on ammonium sulphate solutions, but those for the other two species are less directly accessible. The Raman spectra of the system show, however, that over a considerable range of concentration of the acid (0 to about 12 mol litre^{-1}) the only two species present in solution (apart from H_3O^+ and water) are SO_4^{2-} and HSO_4^-. Hence, the determination of sulphate in this concentration range also gives the concentration of bisulphate ion *and* the molar intensity of this latter species. Knowing then the values for both SO_4^{2-} and HSO_4^-, it is possible to extend the observations to solutions of higher concentration, and to determine the complete composition diagram for the system. An estimate of the thermodynamic ionization constant for the stage $HSO_4^- + H_2O \rightleftharpoons H_3O^+ + SO_4^{2-}$ has also been made from the results of this investigation.[16]

Similar methods have been applied to the study of several other acids in aqueous solution. Those which have been the subject of photoelectric Raman measurements include perchloric,[17] iodic,[18] selenic,[19] trifluoroacetic,[20] methanesulphonic[21] and some arylsulphonic acids.[22] A discrepancy between the results of n.m.r. and Raman measurements in the case of perchloric acid has caused difficulty, but is now explained[23] by the failure of the earlier n.m.r. measurements to allow for variation of the specific proton chemical shift of H_3O^+ with change of acid concentration. Occasional infrared studies of acid dissociation equilibria have also been made, including, for example, a determination[24] of the equilibrium constant for the boric acid ionization,

$$B(OH)_3 + 2H_2O \rightleftharpoons H_3O^+ + B(OH)_4^-$$

(*b*) *Metal complex formation*

The widespread occurrence of step-wise complex formation between a metal ion M and a ligand X, in accordance with the generalized equilibrium (13.1),

$$MX_{i-1} + X \rightleftharpoons MX_i \tag{13.1}$$

where $i = 1, 2, 3 \ldots$, has led to a number of vibrational intensity measurements directed towards the determination of both the compositions of complex ions and their concentrations in solution.

As a simple example of the determination of complex ion stoichiometries from intensity measurements we may consider the Raman work of Chantry and Plane[25] on solutions of complex cyanides. The single vibrational mode of the free cyanide ion is observed in solution at 2079 cm^{-1}. In solutions which also contain zinc or cadmium ions this line appears, along with a second line in the 'C—N stretching' region at 2152 cm^{-1} (Zn) or 2141 cm^{-1} (Cd), due to a cyano-complex. Knowledge of the molar intensity of the free cyanide line (obtained from measurements on KCN solutions) permits the concentration of free cyanide in solutions containing the complex to be determined, and thus, by difference, the concentration of bound cyanide. In this way the formulae of the complexes were established as $Zn(CN)_4^{2-}$ and $Cd(CN)_4^{2-}$ respectively. These were the only complexes detected in these systems. However, in similarly prepared solutions containing copper(I), silver(I) or mercury(II) ions, the appearance of additional lines in the C—N region suggests the presence of other complexes containing different numbers of cyanide ligands per metal atom. Nevertheless, the average number of ligands at each concentration is still calculable from the stoichiometric relations, and these data, together with a knowledge of the intensity trends accompanying variations of the cyanide concentration, permit unambiguous deductions of the various formulae to be made.

In other cases the stoichiometry of a particular complex has been established from the variations of vibrational band intensities (due to the complex) accompanying changes in the composition of the solution. The method of continuous variations[26] has been most widely used. In this method, solutions are prepared in which the *sum* of the concentrations C_M and C_X of metal and ligand respectively, is maintained constant, but their *ratio* is varied. If a single complex is formed from the reaction

$$M + nX \rightleftharpoons MX_n$$

it then follows (provided variations of activity coefficients can be neglected) that the concentration of the complex will be at a maximum in that solution for which $C_X/C_M = n$. If, therefore, a spectral line due only to the complex can be identified and its intensity in each solution determined, a plot of this intensity against the ratio C_X/C_M will exhibit a maximum at the value of C_X/C_M which corresponds to the stoichiometry of the complex. In this simple form the method has been applied to show, for example, that the interaction between calcium ions and nitrate ions in aqueous solution leads to a 1 : 1 complex (or ion pair, see below),[27] while cerium(IV) ions and nitrate ions interact under similar conditions apparently to yield a 1 : 3 complex.[28] There are, however, many cases where the application of Job's method is less straightforward. This arises particularly if more than one complex is formed between the metal and the ligand ion in the range of solution compositions encountered. If each complex predominates over some part of the composition range, the method as described may still yield reliable estimates of the successive values of n. More generally, a modified expression[29] can in principle be used to obtain n from the observed intensity behaviour, but difficulties arising from overlapping spectra of the various species can introduce ambiguities. Thus, although the method has been used in a number of other cases, the results obtained must frequently be interpreted with some caution.

Many studies have been concerned with the use of vibrational intensities for the determination of equilibrium constants in metal complex systems. The validity of results obtained in this way is often necessarily restricted to the conditions of relatively high ionic strength required for the solution measurements, but, in several cases, these have shown remarkable constancy even over extended concentration ranges.[30, 31] As in the essentially similar measurements on acid ionization equilibria, the method depends on the ability to determine a molar intensity value for a band characterizing each species whose concentration is desired. When only a single complex is involved, as in a reaction of the type $M + X \rightleftharpoons MX$, only one such molar intensity is required, since a knowledge of the concentration of a single participant at equilibrium suffices to determine the equilibrium constant. In this way, for example, Clarke and Woodward[32] have determined the degrees of dissociation of some methylmercuric salts in aqueous solution at various concentrations. The results are interestingly compared with ionization data for the corresponding acids. Similarly, Davis and Plane[33] have determined an association constant for $Cd(NO_3)^+$ in aqueous solution, in satisfactory agreement with the results of other measurements.

For equilibria involving several complexes, additional considerations arise. The molar intensity of the *ultimate* complex formed in a series of step reactions can often be readily determined, since, by progressively increasing the ligand concentration, the whole of the metal present may be converted to this form. An illustration is provided by the work of Nixon and Plane[30] on solutions of gallium bromide containing added bromide ion. By measuring the Raman intensity of the $\nu_1(a_1)$ line of the $GaBr_4^-$ complex for solutions of fixed gallium but increasing bromide ion concentration, it was shown that the intensity approached asymptotically a constant limiting value, evidently corresponding to the complete formation of this complex. From this limiting intensity the molar intensity of the species could be calculated. Studies of this kind have been made for a variety of other halide systems, including Zn^{2+}/Cl^-,[34] Zn^{2+}/Br^-,[35] Cd^{2+}/Br^- [35] and Ga^{3+}/Cl^-.[15] By infrared measurements Jones and Penneman[36,37] have similarly characterized the ultimate complexes formed between the cyanide ion and metals of groups Ib and IIb. In some cases, however, complete formation of the complex of highest ligand co-ordination number cannot be attained under solution conditions, and it is then not possible to determine a molar intensity by this procedure. Examples of such behaviour are furnished by the chloro-complexes of indium(III)[38,39] and thallium(III),[31,40] for which the ultimate species (MCl_6^{3-}) are known in the solid state, but are only partially formed in concentrated aqueous hydrochloric acid solutions.

The observations of Nixon and Plane (ibid.) on the gallium bromide system are of further interest in that they showed that the lower complexes $GaBr^{2+}$, $GaBr_2^+$ and $GaBr_3$ are not formed in appreciable concentrations in the aqueous solutions investigated. Thus (neglecting any contribution from the possible occurrence[41] of outer-sphere complexes such as $Ga(H_2O)_n^{3+}Br^-$), the equilibrium reduces essentially to

$$Ga^{3+} + 4Br^- \rightleftharpoons GaBr_4^-$$

and it is accordingly possible, using the determined molar intensity of the tetrabromo-complex, to evaluate the concentrations of the three species represented in this equilibrium in solutions in which the complex is incompletely formed. These measurements yield an equilibrium quotient for the reaction, and, by extending the observations to several temperatures, enthalpy and entropy changes have been calculated.

More generally, the lower complexes cannot be neglected. Occasionally, conditions may be found where each species is present in solution alone, or at most in the presence of the next higher complex only. By proceeding in a step-wise fashion from the intensity of the highest complex (determined as above) it may then be possible to determine the molar intensities for the other species and the various step equilibrium constants. Essentially this method was used by Jones and Penneman[36] in their evaluation of formation constants in the silver cyanide system. Most commonly, however, several complexes coexist in appreciable concentrations over a wide range of solution compositions, and since in addition these may have overlapping spectra, it becomes increasingly difficult to obtain reliable values of the molar intensities. This has been found to be the case for the majority of the halide systems which have been studied to date. In one attempt to overcome these difficulties, at least for Raman measurements, the assumption has been made[35] that the bonds in a series of step complexes are all strictly equivalent from the point of view of the contribution which they make towards the Raman intensities of the species. The Wolkenstein intensity theory[42] then leads to the result that the molar intensities of (the totally symmetric modes of) the successive complexes should bear a simple relation to each other, being directly proportional to the number of bonds they contain. It thus becomes necessary only to determine the molar intensity for one complex (e.g. the ultimate). Unfortunately there are at present insufficient data available to make possible any general appraisal of this interesting approach.

C. Structural aspects; solvent effects and solution interactions

We turn now to a consideration of some results of mainly structural significance which have accrued from infrared and Raman investigations of ionic solution systems. Many of these are of interest primarily for the information they give about individual complex ions which are present in solution—for example, their stereochemistry, bonding character, and structural relationships with other species. The interpretation of vibrational data in these terms is valid to the extent that a species under consideration can be regarded as a kinetically distinct entity, free from material interactions with the solvent or other ions in the medium. The discussion of observations of this kind will be taken up in the next section. Here, however, we wish to examine more closely the question of solution interactions and aspects of the behaviour of the solvent in relation to the stability and structure of species in solution. Vibrational methods have been increasingly used to study these effects in recent years, and although many valuable results have been secured, the potential of the methods for the investigation of certain types of behaviour is still largely unexplored. This therefore seems an opportune occasion to survey some of the initial achievements and results which hold promise in this extensive field.

(a) *Ion solvation*

Many common ionizing solvents display donor properties with respect to metal ions dissolved in them. As a result, an ion or molecule which may be formally represented as MX_i, is in general more accurately formulated in solution as MX_iS_{n-i}, where S denotes a molecule of solvent and n indicates the total coordination number of the metal atom. Step equilibria of the type given by equation (13.1) are likewise more adequately described by the following,

$$MX_{i-1}S_{n-i+1} + X \rightleftharpoons MX_iS_{n-i} + S \qquad (13.2)$$

$i = 1, 2, 3 \ldots n$, a situation which may be interpreted in terms of a competition between solvent molecules and ligand ions for coordination sites on the metal atom.[43] Changes of coordination number also commonly occur at one or more stages of the complex formation sequence, with accompanying changes in the steric disposition of the ligands.

Vibrational spectroscopy has provided a good deal of information concerning these aspects of ion solvation. Evidence for the coordination of solutes by solvent molecules is provided both by the observation of features attributable to metal–solvent vibrations in the spectra of solutions, and by the appearance of altered 'internal' vibrational features of the solvent or solute species. These effects are similar to those which have been studied extensively (particularly by infrared methods) in *solid* addition compounds prepared from a wide range of acceptor solutes in donor solvents. *Solution* systems, on the other hand, have received comparatively little attention in these respects, notably on account of the characteristically low intensities of metal–solvent bands in the Raman spectra of solutions.

The now well-established observations of metal–solvent vibrations in the Raman spectra of aqueous metal salt solutions have accordingly attracted especial interest. Mathieu[44] was the first to attribute weak bands near 400 cm^{-1} in the spectra of magnesium, zinc, and copper nitrates and sulphates in aqueous solution to the totally symmetric modes of the $M(H_2O)_6^{2+}$ cations. Similar bands were observed for beryllium, magnesium, zinc and aluminium salts, in both H_2O and D_2O, by da Silveira, Marques and Marques.[45] Other ions, including mercury(II), indium(III), gallium(III), and thallium(I) and (III) have been studied by Hester and Plane[46] and Spiro,[31] who found in all cases that the bands are polarized, thereby confirming the assignments. Goggin and Woodward[47] observed a low intensity band at 463 cm^{-1} in aqueous solutions of methylmercuric perchlorate, which they were able to assign to the water-mercury stretching vibration of the ion $CH_3 \cdot Hg \cdot OH_2^+$. The band observed for the hexa-aquozinc ion (390 cm^{-1}) has further been shown[34] to diminish in intensity as chloride ions are added to the solution, in accordance with the progressive replacement of bound water molecules by chloride ions.

Changes in the internal frequencies of the water molecule accompanying metal ion hydration have also been recorded. As might be expected from a structure involving oxygen coordination to the metal, the 'bending' frequency ν_2, of the water molecule is most affected. The displacements were measured[48] for a variety of metal ions in the infrared spectra of solutions, and, although of small magnitude (–3 to –14 cm^{-1} for the ions investigated), they show a roughly linear correlation with the ionic charge of the metal. For the triply charged chromium(III) ion, an analogous shift has been observed[49] in the position of the 980 nm overtone band of water in the near infrared. Changes of the intensity of this band were used to calculate the number of water molecules bound to Cr^{3+} ions in solution, that is, to estimate the hydration number of this ion. Although the investigation was of an exploratory nature, the method appears capable of extension to other ions which form strongly bound hydrates.

Turning to other solvent systems, the work of Loehr and Plane[50] on solutions of arsenic trichloride in a range of oxygen donor solvents has given results of interest. The simple four-line Raman spectrum of pure liquid $AsCl_3$ (point group C_{3v}) was found by these authors to undergo significant line broadening and displacement when the compound was dissolved in water, deuterium oxide, or any of several alcohols. A weak new line in the region of 600–700 cm^{-1} was also observed for each solution, and was identified as the arsenic-solvent vibrational mode. These spectral changes were not, however, observed in solutions in diethyl ether, which consequently appears as a much more weakly coordinating solvent towards arsenic trichloride.

Frequency displacements of the characteristic vibrations of groups such as $\rangle$C=O, $\geq$P=O or —C≡N caused by coordination of the oxygen or nitrogen donor function to dissolved species have been fairly extensively studied. A simple example is provided by the observations of Susz and Chalandon[51] on solutions of metal halides in various ketones, where the appearance of substantially reduced carbonyl frequencies demonstrates oxygen-coordination to the metal. Raman studies of solutions of $AlCl_3$ in thionyl chloride,[52] and of $AlCl_3$, $GaCl_3$, $SnCl_4$ and $SbCl_5$ in phosphorus oxychloride[53] have similarly shown the presence of M—O coordinate links. An investigation[54] of $SbCl_5$ in the mixed solvent system $POCl_3 + PO(CH_3)_3$ elegantly illustrates the competition between donors for a given acceptor solute: in this system the frequencies of the trimethylphosphine oxide are found to be strongly perturbed, while those of the phosphorus oxychloride remain undisplaced. Solutions of acceptor halides in nitrile solvents have been the subject of several investigations; the spectra frequently differ from the usual pattern in that the frequency of the C≡N stretching vibration is actually higher in the complexes than in the free solvent, an effect which has been examined theoretically from several points of view.[55]

The spectra of solutions of aluminium chloride in organic acid chlorides provide evidence for more than one type of solvent interaction. The structures of such solutions have been much studied because of their Friedel-Crafts activity.[56] It has been found, for example, that a solution in benzoyl chloride exhibits a lowering of the carbonyl frequency of the solvent, from 1773 to 1548 cm^{-1}, which represents typical behaviour for the formation of a donor–acceptor complex as in I (R = Ph). Solutions in acetyl chloride, however, show no band near this frequency, but instead a strong new band near 2200 cm^{-1}, which is attributed to the vibration of the (essentially triply bonded) carbon–oxygen function in the ionic formulation II ($R = CH_3$).

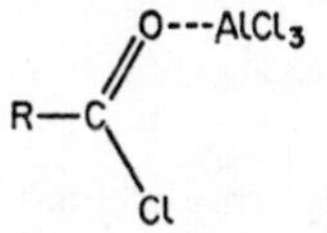

I

$$[R{-}C{\equiv}O]^+\,[AlCl_4]^-$$

II

Intermediate cases are also known, where it appears that an equilibrium between the two types of structure exists, the position of equilibrum depending on such factors as the nature of the solvent and the temperature. For example, a 1:1 mixture of acetyl chloride and aluminium chloride in chloroform exists predominantly[56] as the less polar donor–acceptor complex I ($R = CH_3$), while aluminium bromide dissolved in benzoyl chloride shows a temperature-dependent equilibrium between the two types of structure over the range 0–50°C, the ionic form prevailing at the higher temperature.[57]

(*b*) *Solvent effects on complex stability*

The observations just cited illustrate the importance of the solvent medium in determining the nature of the species present at equilibrium in reactions involving ionic participants. In step equilibria such as those suggested by equation (13.2) the position of equilibrium is likewise strongly dependent on the solvent, the significant properties of the latter being, in most cases, its donor strength towards the various species present, and its dielectric constant.[58] The effects are seen, for a simple system, in a comparison of the Raman spectra of solutions of zinc chloride in alcohols and water[34, 59] In 1-butanol the spectrum consists of a single line (305 cm^{-1}) attributable to the $ZnCl_2$ molecule, but in aqueous solution at moderate concentration other lines appear at lower frequencies as a result of auto-complex-formation, the predominant chloro-species now being $ZnCl_4^{2-}$. Methanol occupies an intermediate position, the principal solute species being the neutral molecule, but lines of the tetrachlorozincate ion also appearing with appreciable intensity.

In a Raman investigation of the complexes formed between indium(III) bromide and hydrobromic acid in aqueous solution, Woodward and Bill[60] observed broad, unresolved bands evidently due to the simultaneous presence of several complex species in the solution. Variation of the HBr concentration produced changes in the spectral band contours, but at no concentration was it found possible to obtain any single complex in isolation. However, when an aqueous solution containing $InBr_3$ and HBr was shaken with diethyl ether or methyl isobutyl ketone, and the spectrum of the organic extract examined, a well-defined four-line spectrum was observed, which could be readily assigned to the single tetrahedral complex $InBr_4^-$ present in the organic phase.

This valuable investigation has been the model for several subsequent studies. Woodward and Taylor[38,61] investigated the $InCl_3$—HCl, $FeCl_3$—HCl, $SnCl_2$—HCl and $SnBr_2$—HBr systems, and in each case found evidence for mixtures of complexes at all concentrations of the components in water, but were able preferentially to extract the singly charged halo-complexes ($M^{III}X_4^-$ or $M^{II}X_3^-$) into diethyl ether. Chloro- and bromo-complexes of copper(I) were studied by Creighton and Lippincott,[62] who similarly were able to isolate and characterize the linear ions $CuCl_2^-$ and $CuBr_2^-$ by ether extraction. Halo-complexes of mercury(II) were the subject of an investigation by Short, Waters and Morris,[63] who were able to isolate all three species HgX_2, HgX_3^- and HgX_4^{2-} ($X = Cl$, Br, I) in solution by an appropriate choice of conditions. The neutral molecules were obtained by dissolution of the appropriate salt, e.g. $HgCl_2$, in anhydrous tri-n-butyl phosphate (TBP); the trihalo-complexes were prepared by dissolving equimolar proportions of the anhydrous mercuric halide and anhydrous lithium halide in TBP; and the tetrahalo-complexes were obtained by extraction into TBP from aqueous solutions containing a high concentration of the corresponding halogen acid. Davies and Long[64] have further studied the chloro- and bromo-complexes of mercury(II) in TBP, and have found that the trihalo-complexes can also be obtained in this solvent by extraction from aqueous solutions containing LiX and HgX_2 in 1:1 molar ratio. These authors have also made similar studies of several other systems, including CdX_2—LiX, ($X = Cl$, Br or I),[65] TlX_3—LiX ($X = Cl$ or Br),[40] and AsX_3—LiX ($X = Cl$ or Br).[66] Although differences of detail in the behaviour of these systems was noted, the clear picture emerges that neutral species and complexes of low charge are stabilized in organic solvents, while species of higher charge are relatively much more favoured in aqueous solution.

An incidental feature of these investigations is that the same molecule or complex has sometimes been observed in both aqueous and organic media (e.g. $GaCl_4^-$ in water and TBP,[67,68] $TlBr_4^-$ in water, Et_2O, and TBP,[40] $HgCl_4^{2-}$ in water and TBP,[64] $AsCl_4^-$ in water and Bu_2O[66]). It has been commonly observed that vibrational bands are significantly narrower in the organic solutions, probably as a result of diminished hydrogen bonding (and, possibly, slower ligand exchange[20]) in these media as compared with an aqueous environment.

(c) *Ion-pair formation*

To a first approximation the criterion for a pair of chemically bound atoms to exhibit a Raman effect is that there should be some degree of electron sharing between them, i.e. that the bond should possess some covalent character. This was empirically recognized early in the historical development of Raman spectroscopy, first by Krishnamurti,[69] and has since received formal expression in various theoretical expositions of the effect, e.g. by Placzek[70] and, more recently, by Long and Plane.[71] Quantitative experimental evidence exists to show that the criterion has widespread validity.[72] On the other hand, it has also been realized that a pair of oppositely charged ions, held together by purely electrostatic forces, will polarize each other with the result that the polarizability of the assemblage will not simply be given by the sum of the individual ion polarizabilities, but will include a term which depends on the distance between them. Hence, even for this system, interionic vibration should lead to a non-vanishing Raman intensity. George, Rolfe and Woodward,[73] in an analysis of this problem, concluded that the intensity should be very small, of the order of 10^{-2} to 10^{-5} of that associated with the vibration of a normal covalent linkage. The experimental observations of Woodward and co-workers[73, 74] lend support to this conclusion. The study of aqueous solutions of thallous hydroxide and dimethylthallic hydroxide, in both of which systems ion-pairing is known to occur to an appreciable extent, has failed to reveal Raman lines attributable to the ion-pair vibration. We refer here to 'contact' or 'inner-sphere' ion pairs, that is, ions which are in direct contact with no intervening solvent molecules. It is to such ion pairs that the conclusions of George *et al.* (ibid.) directly relate. 'Outer-sphere' or 'solvent-separated' ion pairs, in which one or both ions are solvated, are expected to give a still lower intensity (on account of the increased distance between the ion centres) and, for these, there can be even less possibility of detecting a Raman feature.

Nevertheless, there is Raman evidence of a less direct kind which, in certain circumstances, can provide a positive indication of ion pairing, at least of the inner-sphere type. When one of the ions is polyatomic, the ion pairing interaction can lower the free-ion symmetry sufficiently to produce spectroscopically observable effects. Perchlorates, sulphates and nitrates have been principally studied. For a large number of metal perchlorates in concentrated aqueous solution Hester and Plane[10] could detect no significant variations of intensity or frequency in the anion spectrum, a result which is consistent with conclusions drawn from other physical measurements[75] that little association occurs with this anion under the conditions investigated. Many sulphates and nitrates, on the other hand, are known (from thermodynamic, conductimetric and spectrophotometric data[2, 75]) to be highly associated in aqueous solution, often at relatively low concentration. The absence, therefore, of any significant changes, either of frequency or intensity, in the anion spectrum of many metal sulphates in concentrated solution[10] must be taken to signify that the association, in these systems, is of the outer-sphere type. Exceptions occur, however, with indium[76] and scandium[77] sulphates, where the appearance of strongly perturbed anion frequencies argues for the presence of some inner-sphere penetration. In these cases also, weak, polarized Raman lines (at 255 cm^{-1} (In) and 280 cm^{-1} (Sc)), not present in the other sulphate solutions examined, were attributed to the metal–oxygen stretching vibrations of the $M—OSO_3^+$ 'complexes'. There seems little doubt, therefore, that these cation–anion bonds must possess a small degree of covalent character, and consequently that the term 'contact ion-pair' should be interpreted in a manner which admits this possibility. Results of thermodynamic and other measurements have similarly been interpreted[78] in terms of a continuous transition of structural types between purely electrostatic (Bjerrum) ion-pairs and complexes with finite covalent character.

The vibrational spectra of metal nitrates in solution are also consistent with two types of ion-pair. The interpretation of the spectra of these systems has recently been clarified by the observation[79] that the free nitrate ion itself in aqueous solution is perturbed from strict D_{3h} symmetry by hydrogen bonding with water molecules. The main manifestation of this distortion is a splitting of the ν_3(e) stretching mode, centred at approximately 1376 cm^{-1}, into two lines separated by about 56 cm^{-1}. This spectrum is seen for dilute solutions of alkali metal nitrates, and only small modifications are found for dilute solutions of several other nitrates, including, for example, zinc. However, the spectra of zinc nitrate solutions in which the concentration is high enough to force Zn^{2+} and NO_3^- ions into direct contact (this must occur when there are fewer than six molecules of water per molecule of the salt) show new lines which can only be due to an inner-sphere complex.[80, 81] The principal features are a further splitting of ν_3(e), into two lines some 170 cm^{-1} apart, and a splitting of ν_4(e), at 720 cm^{-1} in the free ion, into two components at about 720 and 750 cm^{-1}. As judged by the criterion of these spectral changes, many other metal nitrates appear to form inner-sphere complexes in solution with considerable readiness. Even in dilute solution, the nitrates of Ca(II), Cd(II), Hg(II), Cu(II),

In(III), Ce(IV), Sc(III), La(III), Bi(III) and Th(IV) all appear to be present, at least in part, as contact ion-pairs.[82] In some cases, though not in all, low intensity metal–oxygen bands have been observed for these complexes, indicating a weak covalence in the interaction.

The stereochemical aspects of the coordination of the nitrate ion in these complexes merit brief attention. For both unidentate and bidentate coordination, assumed to take the forms shown in (III) and (IV) respectively, the symmetry of the nitrate group is reduced from D_{3h} to C_{2v}, and the degeneracy of the e vibrations is resolved into components of symmetry

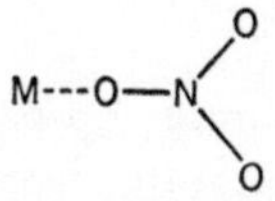

III

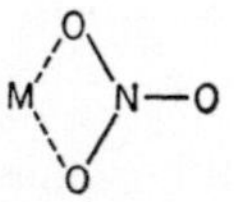

IV

types a_1 and b_1. Thus the *number* of lines appearing in the spectra of the complexes is not alone a sufficient criterion for distinguishing between the two structural configurations. However, force-constant calculations[83,84] for model systems of each type have shown that differences are to be expected in the *sense* of the observed splittings. In particular, for unidentate bonding the a_1 component derived from the e stretching mode (at about 1376 cm^{-1}) should lie at a lower frequency than the b_1 component. For bidentate bonding this order of frequencies should be reversed. Since Raman polarization measurements can differentiate between a_1 and b_1 vibrations (the former being polarized), a means for establishing the coordination geometry of the nitrate group is apparent. On this basis Raman measurements have shown[47,82] that unidentate coordination occurs to Ca^{2+}, Zn^{2+}, Cd^{2+}, Hg^{2+}, CH_3Hg^+ and In^{3+} in solution, while bidentate coordination is preferred by Sc^{3+}, La^{3+}, Ce^{4+}, Th^{4+} and Bi^{3+}.

It is well-known that ion-pairing interactions are promoted in solvents of low dielectric constant. Vibrational evidence (from the observation of anion perturbations) in fact exists in support of the attribution of ion-pair structures to ionic solutes in a variety of solvents of this type. Strong interaction between nitrate ions and many metals has been demonstrated in acetone[85] and acetonitrile.[86] Gallium dichloride and gallium dibromide, $Ga^+GaCl_4^-$ and $Ga^+GaBr_4^-$ respectively,[87] exhibit Raman spectra in benzene consistent with a close association between anions and cations.[88] Solutions of disodium dichromium decacarbonyl in dimethyl sulphoxide ($\epsilon = 46{\cdot}3$) give spectra characteristic of the free anion V, but in tetrahydrofuran ($\epsilon = 7{\cdot}4$) show perturbations suggestive of an interaction between cations and terminal carbonyl groups of the anion.

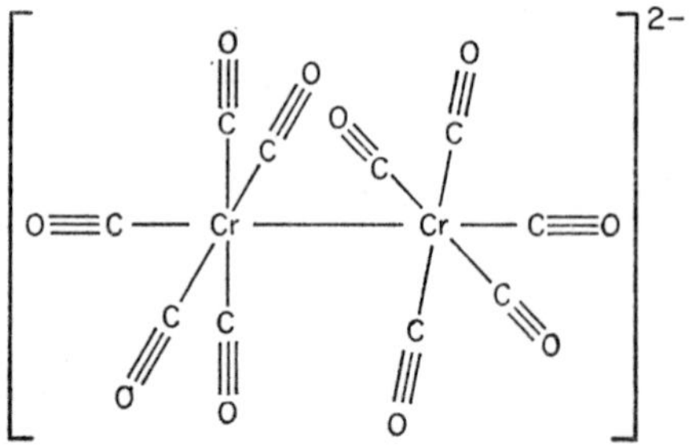

V

The vibration of an electrostatically bound ion-pair involves a first order variation of the dipole moment with the internuclear distance. Hence, in contrast to its weakness in the Raman effect, such a vibration should appear with high intensity in the infrared. Bands arising from this type of vibration have recently been demonstrated[90] in the far infrared spectra of some tetraphenylborates in organic solvents. The interionic vibration frequency of $Na^+BPh_4^-$ was located at 175 cm^{-1} in solutions of 1,4-dioxan, pyridine, piperidine and tetrahydrofuran, and that of $K^+BPh_4^-$ was observed at 133 cm^{-1}. These values, and the observation of an isotopic shift of the 198 cm^{-1} band of $NH_4^+PBh_4^-$ to 183 cm^{-1} for $ND_4^+BPh_4^-$, suggest that unsolvated ions are responsible for the absorption. This method of studying ion-pairs clearly holds promise for the future.

(*d*) *Long range interactions in solution*

Certain observations suggest that vibrational spectroscopy might become an important method for the investigation of interactions in solution between ions and molecules separated by distances greater than those associated with the primary coordination spheres. Thus small but significant frequency displacements in the spectra of polyatomic anions[91] and solvent molecules[92] which are known not to reside in the inner spheres of dissolved cations have been interpreted in terms of long-range cooperative interactions. As expected, these effects appear to be greatest with cations which are small and therefore highly polarizing, or which carry high charges. Perhaps the most striking manifestation of long range effects in the spectra of solutions, however, has been found in Raman intensity observations. For aqueous solutions containing the $In(H_2O)_6^{3+}$ cation, Hester and Grossman[93] noted a remarkable enhancement of the intensity associated with the metal–oxygen stretching vibration at reduced temperatures. The probable explanation, which is also consistent with the remarkably high viscosities shown by these solutions, is that ordered hydration spheres are formed around the tripositive cations extending several water molecules deep, and that, in accordance with conventional polarizability theory, the high intensity is derived from a symmetrical breathing mode of these large hydration clusters.

D. Structural aspects; individual complexes

The completeness and the rigour with which the vibrational spectrum of a given complex can be analyzed are in practice very variable. For species of moderately simple structure, for which complete spectral data are obtainable, it is often possible, from a consideration of the number and activity of the normal modes of vibration, unambiguously to establish the stereochemistry. Assignments of the frequencies may then provide the basis for force field calculations, leading to information about the electronic distribution and bonding in the species. For molecules of greater complexity this type of detailed analysis may not be possible. Even if only partial spectral data can be obtained, however, reliable deductions concerning the structure can still often be made, either from the observation of special relationships between the infrared and Raman spectra (e.g. the rule of mutual exclusion is useful in support of the attribution of centrosymmetric structures) or from comparisons with the spectra of species whose structures are already known.

In this section we shall survey a few results which illustrate the above procedures. The discussion will be confined to complexes representing the simpler structural types, and only a small selection of data, taken largely from recent solution investigations, will be given. Several more detailed reviews[94] are available to supplement the necessarily limited account given here.

(a) *Halo-complexes*

Some of the methods which have been used to obtain particular halo-complexes in solution for Raman investigation were outlined above. Complementary infrared data are in many cases also available, either from solution or solid-state studies, and structural conclusions and assignments are therefore mostly well established. Some representative results, which are restricted for convenience to chloro-complexes, are summarized in Table 13.1.

One of the major results of the vibrational spectroscopy of metal complexes has been the development of a greatly increased understanding of the factors which affect the strengths of metal–ligand bonds. In this the data for halo-complexes have made a principal contribution, since the

TABLE 13.1. *Vibrational frequencies of some chloro complexes, cm^{-1}*

	ν_1	ν_2	ν_3	ν_4	ν_5	Ref.
Triatomic Species ($D_{\infty h}$)						
$CuCl_2^-$	296					a
$ZnCl_2$	305					b
$CdCl_2$	290					c
$HgCl_2$	320					d
Tetra-atomic Species (C_{3v})						
$SnCl_3^-$	297	128	256	103		e
$SeCl_3^+$	430	206	415	172		f.
$TeCl_3^+$	399	170	385	150		f
Tetra-atomic Species (D_{3h})						
$CdCl_3^-$	265		287	90		c
$HgCl_3^-$	294					g
Penta-atomic Species (T_d)						
$MgCl_4^{2-}$	252		344			c
$ZnCl_4^{2-}$	288	116	298	130		h
$CdCl_4^{2-}$	260		281	92		c
$HgCl_4^{2-}$	266		255	111		g
BCl_4^-	405	190	670, 707	274		i
$AlCl^-$	349	146	493	183		j
$GaCl_4^-$	346	114	386	149		k
$InCl_4^-$	321	89	337	112		l
$TlCl_4^-$	307	72–102	295	72–102		m
$MnCl_4^{2-}$	256		278, 301	120		n
$FeCl_4^{2-}$	266	82	286	119		o
$FeCl_4^-$	330	114	378	136		l, o
$CuCl_4^{2-}$ (D_{2d})	274	118, 136	237, 278	118, 136		o, p
Hepta-atomic Species (O_h)						
$TiCl_6^{2-}$	320	271	316	183	173	q
$ZrCl_6^{2-}$	326	225	293	150	150	r
$HfCl_6^{2-}$	331	204	272	145	152	r
$ThCl_6^{2-}$	294	255	259		114	s
$NbCl_6^-$	335	240	333		170	r
$TaCl_6^-$	380	295	333		180	r
UCl_6^{2-}	299	237	262		121	s

a. Creighton, J. A., and Lippincott, E. R., *J. Chem. Soc.*, 1963, 5134.
b. Kecki, Z., *Spectrochim. Acta*, 1962, **18**, 1165; Irish, D. E., McCarroll, B., and Young, T. F., *J. Chem. Phys.*, 1963, **39**, 3436.
c. Davies, J. E. D., and Long, D. A., *J. Chem. Soc.* (*A*),, 1968, 2054.
d. Allen, G., and Warhurst, E., *Trans. Faraday Soc.*, 1958, **54**, 1786.
e. Woodward, L. A., and Taylor, M. J., *J. Chem. Soc.*, 1962, 407.

f. Sawodny, W., and Dehnicke, K., *Z. anorg. Chem.*, 1967, **349**, 169; Robinson, E. A., and Ciruna, J. A., *Canad. J. Chem.*, 1968, **46**, 3197.
g. Short, E. L., Waters, D. N., and Morris, D. F. C., *J. Inorg. Nuclear Chem.*, 1964, **26**, 902; Davies, J. E. D., and Long, D. A., *J. Chem. Soc.* (*A*), 1968, 2564.
h. Quicksall, C. O., and Spiro, T. G., *Inorg. Chem.*, 1966, **5**, 2232.
i. Creighton, J. A., *J. Chem. Soc.*, 1965, 6589.
j. Gerding, H., and Houtgraaf, H., *Rec. Trav. chim.*, 1953, **72**, 21; Jones, D. E. H., and Wood, J. L., *J. Chem. Soc.* (*A*), 1967, 1140; *Spectrochim. Acta*, 1967, **23A**, 2695.
k. Woodward, L. A., and Nord, A. A., *J. Chem. Soc.*, 1956, 3721.
l. Woodward, L. A., and Taylor, M. J., *J. Chem. Soc.*, 1960, 4473.
m Davies, J. E. D., and Long, D. A., *J. Chem. Soc.* (*A*), 1968, 2050.
n. Edwards, H. G. M., Ware, M. J., and Woodward, L. A., *Chem. Comm.*, 1968, 540; Edwards, H. G. M., Woodward, L. A., Gall, M. J., and Ware, M. J., *Spectrochim. Acta*, 1970, **26A**, 287.
o. Avery, J. S., Burbridge, C. D., and Goodgame, D. M. L., *Spectrochim. Acta*, 1968, **24A**, 1721.
p. Forster, D., *Chem. Comm.*, 1967, 113.
q. Clark, R. J. H., Maresca, L., and Puddephatt, R. J., *Inorg. Chem.*, 1968, **7**, 1603.
r. Davies, J. E. D., and Long, D. A., *J. Chem. Soc.* (*A*), 1968, 2560.
s. Woodward, L. A., and Ware, M. J., *Spectrochim. Acta*, 1968, **24A**, 921.

---

large number of results available has furnished many informative correlations among species of related structure. It is established, for example, that a dominant factor in this respect is the oxidation number of the metal. This is seen in comparisons of the stretching frequencies (or M—X force constants) of pairs of ions such as $Fe^{II}Cl_4^{2-}$ (ν_1, 266 cm^{-1}) and $Fe^{III}Cl_4^-$ (330 cm^{-1}) or $Ir^{III}Cl_6^{3-}$ (ν_3, 296 cm^{-1}) and $Ir^{IV}Cl_6^{2-}$ (333 cm^{-1}); or of complexes forming an isoelectronic series such as $Zn^{II}Cl_4^{2-}$ (ν_1, 288 cm^{-1}), $Ga^{III}Cl_4^-$ (346 cm^{-1}), $Ge^{IV}Cl_4$ (397 cm^{-1}) and $As^{V}Cl_4^+$ (422 cm^{-1}). For such series, where the bonding is predominantly of σ type, an increase of oxidation number is invariably associated with an increase of metal–ligand frequency. This behaviour, as first pointed out by Woodward,[95] is a manifestation of the contraction of the metal orbital which accompanies the progressive increase in nuclear charge. For transition element complexes MX_n of constant oxidation state only small changes in frequency are seen as the metal atom is varied within a given transition series. This is as expected, but it is of particular interest that such changes as are observed correlate well with the known order of ligand field stabilization energies for the series. The effect is seen[96] in the ions MnX_4^{2-} (d^5), FeX_4^{2-} (d^6), CoX_4^{2-} (d^7), NiX_4^{2-} (d^8) and ZnX_4^{2-} (d^{10}), (X = Cl, Br or I), for which the antisymmetric stretching frequency (ν_3, f_2) reaches a maximum at the d^7 configuration, corresponding to maximum LFSE for tetrahedral fields. Unfortunately, the corresponding behaviour in the symmetric stretching mode (ν_1, a_1) has not been demonstrated as yet, owing to the incompleteness of the Raman data for these (coloured) ions.

When π-bonding between metal and ligand is significant, variations attributable to differences of σ-bonding among related species may be overshadowed. This is seen, for example, in some frequency relations among hexafluorides and hexachloro-anions in the third transition series. For fluorine the possibility of π-bonding may be discounted. In the molecules WF_6 (d^0), ReF_6 (d^1), OsF_6 (d^2), IrF_6 (d^3) and PtF_6 (d^4), the metal–fluorine stretching frequencies (and force constants) decrease progressively, in parallel with the known increase in bond length. For the ions $ReCl_6^{2-}$ (d^3), $OsCl_6^{2-}$ (d^4), $IrCl_6^{2-}$ (d^5) and $PtCl_6^{2-}$ (d^6), however, the metal–chlorine stretching frequencies increase (and the bond lengths decrease), an observation which is consistent with an increase in π-bonding by back-donation of metal d electrons to vacant d orbitals of chlorine.

The data for halo-complexes illustrate one further important correlation, namely that an increase in the coordination number of a given metal leads generally to a decrease in the metal–ligand stretching frequencies. Series such as $HgCl_2$ (ν_1, 360 cm^{-1}), $HgCl_3^-$ (293 cm^{-1}), $HgCl_4^{2-}$ (270 cm^{-1}) and $SnCl_4$ (ν_1, 368 cm^{-1}), $SnCl_6^{2-}$ (311 cm^{-1}) show this effect, which can be rationalized in terms of decreased electron donation from ligand to metal in the higher complexes, in order to preserve approximate electroneutrality of the metal. (In occasional cases where the bonds formed are not all equivalent, exceptions to the behaviour have been noted.[97])

(b) *Aquo-, hydroxo- and oxo-complexes*

Raman observations on aquo-complexes were briefly discussed in Section C.

Of the hydroxo-complexes which have been investigated in aqueous solution, only a few, e.g. $B(OH)_4^-$, $Zn(OH)_4^{2-}$, $Sn(OH)_6^{2-}$, have been found to possess simple monomeric structures (Table 13.2). Solutions of arsenites have been shown[98] by means of a Job analysis of Raman intensities to contain an equilibrium between the species $As(OH)_3$, $AsO(OH)_2^-$, $AsO_2(OH)^{2-}$ and AsO_3^{3-} in the pH range 3·5–15. The aluminate ion appears[99] to be monomeric AlO_2^- at high pH, but polymeric, involving a network of octahedrally coordinated aluminium ions, at lower pH. Many heavy metal hydroxo-complexes are certainly polynuclear, and for some of these highly symmetrical 'cluster' structures have been indicated by the combined weight of Raman and other evidence. Thus Maroni and Spiro,[100] in Raman studies of Bi(III) and Pb(II) hydroxo-complexes in aqueous solution, have obtained spectra which support the presence of $Bi_6(OH)_{12}^{6+}$ and $Pb_4(OH)_4^{4+}$ as octahedral and tetrahedral cage structures respectively, in agreement with conclusions reached from solution X-ray scattering investigations. Other ions, inculding $Ta_6O_{19}^{8-}$, $V_{10}O_{28}^{6-}$, $Mo_7O_{24}^{4-}$ and $W_{12}O_{42}^{12-}$, have also been identified[101] in aqueous solutions of defined pH, on the basis of comparisons with the spectra of crystals of known structure.

TABLE 13.2. *Vibrational frequencies of some hydroxo-complexes and oxyanions, cm⁻¹*

		ν_1	ν_2	ν_3	ν_4	ν_5	Ref.
$B(OH)_4^-$	T_d	754	379	947	533		a
$Zn(OH)_4^{2-}$	T_d	470	300	420	300		b
$Sn(OH)_6^{2-}$	O_h	556					c
AlO_2^-	$D_{\infty h}$	628					d
AsO_3^{3-}	C_{3v}	752	340	680	340		e
VO_4^{3-}	T_d	827	340	780	340		f
SO_4^{2-}	T_d	981	451	1104	613		g
CrO_4^{2-}	T_d	847	349	890	370		h
MoO_4^{2-}	T_d	897	318	841	318		i
WO_4^{2-}	T_d	931	324	833	324		i
MnO_4^-	T_d	838	355	921	429		h
ClO_4^-	T_d	935	460	1110	630		j
BrO_4^-	T_d	801	331	878	410		k
IO_4^-	T_d	791	256	853	325		l

a. Edwards, J. O., Morrison, G. C., Ross, V. F., and Schultz, J. W., *J. Amer. Chem. Soc.*, 1955, **77**, 266.
b. Lippincott, E. R., Psellas, J. A., and Tobin, M. C., *J. Chem. Phys.*, 1952, **20**, 536.
c. Tobias, R. S., and Freidline, C. E., *Inorg. Chem.*, 1965, **4**, 215.
d. Carreira, L. A., Maroni, V. A., Swain, J. W., and Plumb, R. C., *J. Chem. Phys.*, 1966, **45**, 2216.
e. Loehr, T. M., and Plane, R. A., *Inorg. Chem.*, 1968, **7**, 1708.
f. Griffith, W. P., and Wickens, T. D., *J. Chem. Soc.* (*A*), 1966, 1087; Krebs, B., and Müller, A., *J. Mol. Spectroscopy*, 1967, **22**, 290.
g. Walrafen, G. E., *J. Chem. Phys.*, 1963, **39**, 1479.
h. Hendra, P. J., *Spectrochim. Acta*, 1968, **24A**, 125.
i. Busey, R. H., and Keller, O. L., *J. Chem. Phys.*, 1964, **41**, 215; Müller, A., Krebs, B., Kebabcioglu, R., Stockburger, M., and Glemser, O., *Spectrochim. Acta*, 1968, **24A**, 1831.
j. Redlich, O., Holt, E. K., and Bigeleisen, J., *J. Amer. Chem. Soc.*, 1944, **66**, 13; Hester, R. E., and Plane, R. A., *Inorg. Chem.*, 1964, **3**, 769.
k. Brown, L. C., Begun, G. M., and Boyd, G. E., *J. Amer. Chem. Soc.*, 1969, **91**, 2250.
l. Siebert, H., *Z. anorg. Chem.*, 1953, **273**, 21.

Such studies well illustrate the usefulness of Raman spectroscopy in a diagnostic role, even when the species concerned are much too complicated for complete vibrational analysis.

Most of the common oxo-anions (NO_3^-, ClO_4^-, SO_4^{2-}, etc.) have well-defined structures and their spectra present few problems of interpretation. Table 13.2 gives data for a few of these, including the newly prepared perbromate ion and the intensely coloured permanganate ion. The successful observation of the Raman spectrum of this latter species is one of the undoubted triumphs of the revolution which the introduction of the laser source has brought to Raman spectroscopy.

(c) *Oxo-anion complexes*

The weakly covalent complexes involving nitrate and sulphate as ligands were discussed above. Calculations of the frequency displacements and splittings to be expected in the anion spectrum accompanying complex formation have been made[83, 102] for other simple oxo-anions (perchlorate, phosphate, carbonate). There is no conclusive vibrational evidence for perchlorate complexes under solution conditions, but an example of (bidentate) phosphate-coordination has been found in a Raman investigation[103] of bis-diphosphatozinc in aqueous solution.

(d) *Ammine complexes*

Until recently, Raman spectral studies of complex ammines were comparatively few, although the use of infrared spectroscopy, mainly for characterizing the ligand vibrations, has been long developed.[104] The infrared-active skeletal stretching frequencies (ν_3, f_{1u}) of most simple hexammines (e.g. $Co(NH_3)_6^{3+}$) are, indeed, unusually weak and for this reason the assignments of the skeletal modes in these compounds were for a long time in doubt. Raman results, including polarization data, from aqueous solutions of several hexammines have proved valuable for clarifying the assignments (Table 13.3), which have also been confirmed, for the cobalt complex, by a recent analysis[105] of the low-temperature vibronic absorption spectrum of the salt $[Co(NH_3)_6][Co(CN)_6]$.

Of related interest are the chelate complexes formed between metal ions and ethylenediamine in solution. Krishnan and Plane[106] have studied the

TABLE 13.3. *Skeletal frequencies of some hexammines*

	ν_1 a_{1g}	ν_2 e_g	ν_3 f_{1u}	ν_4 f_{1u}	ν_5 f_{2g}	Ref.
$Co(NH_3)_6^{3+}$	495	440	476	332		a
$Ru(NH_3)_6^{3+}$	500	475	463	283, 263	248	b
$Rh(NH_3)_6^{3+}$	514	483	472	302, 287	240	b
$Ir(NH_3)_6^{3+}$	527	500	475	279, 264	262	b
$Pt(NH_3)_6^{4+}$	569	545				c

a. Shimanouchi, T., and Nakagawa, I., *Inorg. Chem.*, 1964, **3**, 1805; Haas, T. E., and Hall, J. R., *Spectrochim. Acta*, 1966, **22**, 988.
b. Griffith, W. P., *J. Chem. Soc.* (*A*), 1966, 899.
c. Clegg, D. E., and Hall, J. R., *Spectrochim. Acta*, 1967, **23A**, 263.

bis-ethylenediamine complexes of zinc, cadmium and mercury and compared the Raman-active metal–nitrogen stretching frequencies with those of the simple tetrammines $Zn(NH_3)_4^{2+}$, $Cd(NH_3)_4^{2+}$ and $Hg(NH_3)_4^{2+}$. In each case the frequencies for the chelate complexes were found to be appreciably higher than those for the tetrammines, a result which is consonant with stronger metal–nitrogen bonding in the chelates.

(*e*) *Cyano-complexes*

Values of the C≡N and M—C symmetric vibrational frequencies of some cyano-complexes of Group IB and IIB metals are given in Table 13.4.

TABLE 13.4. *Symmetric stretching frequencies of complex cyanides of Groups IB and IIB, cm^{-1}*

	ν_{CN}	ν_{MC}	Ref.
$Cu(CN)_2^-$	2106		a, b
$Cu(CN)_3^{2-}$	2094		a, b
$Cu(CN)_4^{3-}$	2094	288	a, b
$Ag(CN)_2^-$	2139	360	a, c
$Ag(CN)_3^{2-}$	2108		a, c
$Ag(CN)_4^{3-}$	2097		a, c
$Au(CN)_2^-$	2164	448	c
$Zn(CN)_4^{2-}$	2152	347	a, d
$Cd(CN)_4^{2-}$	2141	327	a, d, e
$Hg(CN)_2$	2190	412	a, f
$Hg(CN)_3^-$	2160	358	a, f
$Hg(CN)_4^{2-}$	2140	335	a, f

a. Chantry, G. W., and Plane, R. A., *J. Chem. Phys.*, 1960, **33**, 736; 1961, **35**, 1027.
b. Reisfield, M. J., and Jones, L. H., *J. Mol. Spectroscopy*, 1965, **18**, 222; Jones, L. H., *J. Chem. Phys.*, 1958, **29**, 463.
c. Jones, L. H., *Spectrochim. Acta*, 1963, **19**, 1675.
d. Poulet, H., and Mathieu, J. P., *Compt. rend.*, 1959, **248**, 2079.
e. Couture, L., and Mathieu, J. P., *Ann. Physique.*, 1948, [12] **3**, 521.
f. Woodward, L. A., and Owen, H. F., *J. Chem. Soc.*, 1959, 1055; Griffith, W. P., *J. Chem. Soc.*, 1964, 4070.

Data are particularly complete for the triad $Hg(CN)_2$, $Hg(CN)_3^-$ and $Hg(CN)_4^{2-}$, from which it is seen that ν_{MC} undergoes a progressive decrease with increasing coordination number of the metal, the expected effect if there is a simultaneous metal–ligand bond weakening; cf. the earlier discussion of halo-complexes. At the same time ν_{CN} is observed to approach the value found for the free cyanide ion (2079 cm^{-1}), and this behaviour also suggests a diminished bonding interaction between metal and ligand in the higher complexes.

Force-constant calculations on complex cyanides have been carried out by several authors[107] with the object of obtaining information on the nature of the M—C and C≡N bonds in these species. There are many interesting facets to the results of these calculations, but it must suffice here to observe that the evidence points generally to nearly pure σ-bonding between metal and ligand, a conclusion which finds independent support in the remarkably low Raman intensities of the M—C stretching modes found for most of the complexes which have been studied.

The somewhat exceptional facility with which cyano-complexes can be studied in aqueous solution by infrared methods has been referred to above. The possibility arises because of the relative transparency of water to infrared radiation in the 'C≡N stretching' frequency region. The technique has recently been used[108] to characterize the ion $Ni(CN)_5^{3-}$ from measurements on the system $Ni(CN)_4^{2-}/CN^-$ in water; the observation of an isobestic point in the C≡N absorption contour confirmed the formation of the pentacyano-complex, but no evidence for $Ni(CN)_6^{4-}$ was found.

We mention finally an unusual case of ambimorphism which has been discovered from Raman studies[109] of the eight-coordinate complexes $Mo(CN)_8^{4-}$ and $W(CN)_8^{4-}$. The spectra of the crystalline compounds $K_4M(CN)_8, 2H_2O$ ($M = Mo$ or W) were obtained in agreement with previous infrared spectra, and supported a dodecahedral structure (point group D_{2d}) for the anions. However, the spectra of aqueous solutions were found to be different from those of the solids and suggested a structure of higher symmetry, probably square-antiprismatic (D_{4d}). The results point clearly to the small differences of energy which can sometimes separate alternative configurations.

References

1. Rossotti, F. J. C., and Rossotti, H., 'The Determination of Stability Constants', McGraw-Hill, New York, 1961.
2. Nancollas, G. H., 'Interactions in Electrolyte Solutions', Elsevier, Amsterdam, 1966.
3. Wilson, E. B., Decius, J. C., and Cross, P. C., 'Molecular Vibrations', McGraw-Hill, New York, 1955. An introductory treatment is given by Woodward, L. A., 'Raman Spectroscopy', ed. H. A. Szymanski, Plenum Press, New York, 1967, p. 1.
4. Ferraro, J. R., 'Raman Spectroscopy', ed. H. A. Szymanski, Plenum Press, New York, 1967, p. 44; Hendra, P. J., and Stratton, P. M., *Chem. Rev.*, 1969, **69**, 325.

5. Schrötter, H. W., and Bernstein, H. J., *J. Mol. Spectroscopy*, 1964, **12**, 1.
6. Mirone, P., *Spectrochim. Acta.*, 1966, **22**, 1897; Fini, G., Mirone, P., and Patella, P., *J. Mol. Spectroscopy*, 1968, **28**, 144.
7. Russell, R. A., and Thompson, H. W., *Spectrochim. Acta*, 1957, **9**, 133.
8. Bernstein, H. J., and Allen, G., *J. Opt. Soc. Amer.*, 1955, **45**, 237; Rea, D. G., *ibid.*, 1959, **49**, 90; Wall, T. T., and Hornig, D. F., *J. Chem. Phys.*, 1966, **45**, 3424.
9. Vollmar, P. M., *J. Chem. Phys.*, 1963, **39**, 2236; cf. refs. 21 and 32.
10. Hester, R. E., and Plane, R. A., *Inorg. Chem.*, 1964, **3**, 769.
11. Irish, D. E., 'Raman Spectroscopy', ed. H. A. Szymanski, Plenum Press, New York, 1967, p. 224.
12. Falk, M., and Giguere, P. A., *Canad. J. Chem.*, 1957, **35**, 1195.
13. Rao, I. R., *Proc. Roy. Soc.*, 1930, **A127**, 279; Woodward, L. A., *Nature*, 1930, **126**, 58.
14. Krawetz, A. A., Thesis, University of Chicago, 1955. See ref. 15.
15. Young, T. F., Maranville, L. F., and Smith, H. M., 'The Structure of Electrolytic Solutions', ed. W. J. Hamer, Wiley, New York, 1959, p. 35.
16. Monk, C. B., 'Electrolytic Dissociation', Academic Press, London, 1961, Chapter 12.
17. Covington, A. K., Tait, M. J., and Wynne-Jones, W. F. K., *Proc. Roy. Soc.*, 1965, **A286**, 235; Heinzinger, K., and Weston, R. E., *J. Chem. Phys.*, 1965, **42**, 272.
18. Durig, J. R., Bonner, O. D., and Breazeale, W. H., *J. Phys. Chem.*, 1965, **69**, 3886.
19. Walrafen, G. E., *J. Chem. Phys.*, 1963, **39**, 1479.
20. Kreevoy, M. M., and Mead, C. A., *Discuss. Faraday Soc.*, 1965, **39**, 166.
21. Clarke, J. H. R., and Woodward, L. A., *Trans. Faraday Soc.*, 1966, **62**, 2226.
22. Bonner, O. D., and Torres, A. L., *J. Phys. Chem.*, 1965, **69**, 4109.
23. Duerst, R. W., *J. Chem. Phys.*, 1968, **48**, 2275.
24. Goulden, J. D. S., *Spectrochim. Acta*, 1959, **9**, 657.
25. Chantry, G. W., and Plane, R. A., *J. Chem. Phys.*, 1960, **33**, 736.
26. Job, P., *Ann. Chim. (France)*, 1928, [10] **9**, 113.
27. Irish, D. E., and Walrafen, G. E., *J. Chem. Phys.*, 1967, **46**, 378.
28. Miller, J. T., and Irish, D. E., *Canad. J. Chem.*, 1967, **45**, 147.
29. Katzin, L. I., and Gebert, E., *J. Amer. Chem. Soc.*, 1950, **72**, 5455; cf. ref. 11.
30. Nixon, J., and Plane, R. A., *J. Amer. Chem. Soc.*, 1962, **84**, 4445.
31. Spiro, T. G., *Inorg. Chem.*, 1965, **4**, 731.
32. Clarke, J. H. R., and Woodward, L. A., *Trans. Faraday Soc.*, 1966, **62**, 3022; *ibid.*, 1968, **64**, 1041.
33. Davis, A. R., and Plane, R. A., *Inorg. Chem.*, 1968, **7**, 2565.
34. Irish, D. E., McCarroll, B., and Young, T. F., *J. Chem. Phys.*, 1963, **39**, 3436.

35. Yellin, W., and Plane, R. A., *J. Amer. Chem. Soc.*, 1961, **83**, 2448.
36. Jones, L. H., and Penneman, R. A., *J. Chem. Phys.*, 1954, **22**, 965.
37. Penneman, R. A., and Jones, L. H., *J. Chem. Phys.*, 1956, **24**, 293; *J. Inorg. Nuclear Chem.*, 1961, **20**, 19.
38. Woodward, L. A., and Taylor, M. J., *J. Chem. Soc.*, 1960, 4473.
39. Hanson, M. P., and Plane, R. A., *Inorg. Chem.*, 1969, **8**, 746.
40. Davies, J. E. D., and Long, D. A., *J. Chem. Soc.* (*A*), 1968, 2050.
41. Morris, D. F. C., and Andrews, B. D., *Electrochim. Acta*, 1967, **12**, 41.
42. Wolkenstein, M., *Compt. rend. Acad. Sci. U.R.S.S.*, 1941, **32**, 185; Chantry, G. W., and Plane, R. A., *J. Chem. Phys.*, 1960, **32**, 319.
43. Katzin, L. I., *Transition Metal Chem.*, 1966, **3**, 56.
44. Mathieu, J. P., *Compt. rend.*, 1950, **231**, 896.
45. da Silveira, A., Marques, M. A., and Marques, N. M., *Compt. rend.*, 1961, **252**, 3983.
46. Hester, R. E., and Plane, R. A., *Inorg. Chem.*, 1964, **3**, 768.
47. Goggin, P. L., and Woodward, L. A., *Trans. Faraday Soc.*, 1962, **58**, 1495.
48. Nightingale, E. R., jun., 'Chemical Physics of Ionic Solutions', ed. B. E. Conway and R. G. Barradas, Wiley, New York, 1966, p. 87.
49. Duerst, R. A., and Taylor, J. K., *J. Res. Nat. Bur. Stand.*, 1964, **68A**, 625.
50. Loehr, T. M., and Plane, R. A., *Inorg. Chem.*, 1969, **8**, 73.
51. Susz, B. P., and Chalandon, P., *Helv. Chim. Acta*, 1958, **41**, 1332.
52. Long, D. A., and Bailey, R. T., *Trans. Faraday Soc.*, 1963, **59**, 594.
53. Kinell, P. O., Lindqvist, I., and Zackrisson, M., *Acta Chem. Scand.*, 1959, **13**, 1159; Gerding, H., Koningstein, J. A., and van der Worm, E. R., *Spectrochim. Acta*, 1960, **16**, 881.
54. Kinell, P. O., Lindqvist, I., and Zackrisson, M., *Acta Chem. Scand.*, 1959, **13**, 190.
55. Overend, J., and Scherer, J. R., *J. Chem. Phys.*, 1960, **32**, 1296; Bent, H. A., *Chem. Rev.*, 1961, **61**, 275; cf. ref. 56.
56. Cook, D., 'Friedel-Crafts and Related Reactions', Vol. 1, ed. G. A. Olah, Interscience, New York, 1963, p. 767.
57. Perkampus, H.-H., and Weiss, W., *Angew. Chem.*, *Internat. Edn.*, 1968, **7**, 70.
58. Drago, R. S., and Purcell, K. F., *Progr. Inorg. Chem.*, 1964, **6**, 271.
59. Kecki, Z., *Spectrochim. Acta*, 1962, **18**, 1165.
60. Woodward, L. A., and Bill, P. T., *J. Chem. Soc.*, 1955, 1699.
61. Woodward, L. A., and Taylor, M. J., *J. Chem. Soc.*, 1962, 407.
62. Creighton, J. A., and Lippincott, E. R., *J. Chem. Soc.*, 1963, 5134.
63. Short, E. L., Waters, D. N., and Morris, D. F. C., *J. Inorg. Nuclear Chem.*, 1964, **26**, 902.
64. Davies, J. E. D., and Long, D. A., *J. Chem. Soc.* (*A*), 1968, 2564.
65. Davies, J. E. D., and Long, D. A., *J. Chem. Soc.* (*A*), 1968, 2054.
66. Davies, J. E. D., and Long, D. A., *J. Chem. Soc.* (*A*), 1968, 1761.

67. Woodward, L. A., and Nord, A. A., *J. Chem. Soc.*, 1956, 3721.
68. Morris, D. F. C., Andrews, B. D., and Short, E. L., *J. Inorg. Nuclear Chem.*, 1966, **28**, 2436.
69. Krishnamurti, P., *Indian J. Phys.*, 1930, **5**, 113.
70. Placzek, G., *Handbuch der Radiologie*, 1934, **6**(**2**), 366.
71. Long, T. V., and Plane, R. A., *J. Chem. Phys.*, 1965, **43**, 457.
72. Hester, R. E., 'Raman Spectroscopy', ed. H. A. Szymanski, Plenum Press, New York, 1967, p. 101, and refs. therein.
73. George, J. H. B., Rolfe, J. A., and Woodward, L. A., *Trans. Faraday Soc.*, 1953, **49**, 375.
74. Goggin, P. L., and Woodward, L. A., *Trans. Faraday Soc.*, 1960, **56**, 1591.
75. 'Stability Constants of Metal-ion Complexes', comp. L. G. Sillen and A. E. Martell, Chemical Society, London, 2nd edn., 1964.
76. Hester, R. E., Plane, R. A., and Walrafen, G. E., *J. Chem. Phys.*, 1963, **38**, 249.
77. Stauch, B., and Komissarova, L. N., *Coll. Czech. Chem. Comm.*, 1967, **32**, 1484.
78. Davies, C. W., 'Ion Association', Butterworths, London, 1962, Chapter 7.
79. Irish, D. E., and Davis, A. R., *Canad. J. Chem.*, 1968, **46**, 943.
80. Hester, R. E., and Plane, R. A., *J. Chem. Phys.*, 1966, **45**, 4588.
81. Irish, D. E., Davis, A. R., and Plane, R. A., *J. Chem. Phys.*, 1969, **50**, 2262.
82. Mathieu, J. P., and Lounsbury, M., *Discuss. Faraday Soc.*, 1950, **9**, 196; Stauch, B., and Komissarova, L. N., *Z. Chem.*, 1966, **6**, 474; Cooney, R. P. J., and Hall, J. R., *Austral. J. Chem.*, 1969, **22**, 337; Knoeck, J., *Analyt. Chem.*, 1969, **41**, 2069; ref. 81 and refs. therein.
83. Brintzinger, H., and Hester, R. E., *Inorg. Chem.*, 1966, **5**, 980.
84. Hester, R. E., and Grossman, W. E. L., *Inorg. Chem.*, 1966, **5**, 1308.
85. Norwitz, G., and Chasan, D. E., *J. Inorg. Nuclear Chem.*, 1969, **31**, 2267.
86. Addison, C. C., Amos, D. W., and Sutton, D., *J. Chem. Soc. (A)*, 1968, 2285.
87. Woodward, L. A., Garton, G., and Roberts, H. L., *J. Chem. Soc.*, 1956, 3723; Woodward, L. A., Greenwood, N. N., Hall, J. R., and Worrall, I. J., *ibid.*, 1958, 1505; Garton, G., and Powell, H. M., *J. Inorg. Nuclear Chem.*, 1957, **4**, 84.
88. Kinsella, E., Chadwick, J., and Coward, J., *J. Chem. Soc. (A)*, 1968, 969.
89. Edgell, W. F., Pauuwe, N., *Chem. Comm.*, 1969, 285.
90. French, M. J., and Wood, J. L., *J. Chem. Phys.*, 1968, **49**, 2358.
91. Matwiyoff, N. A., Darley, P. E., and Movius, W. G., *Inorg. Chem.*, 1968, **7**, 2173; Irish, D. E., Chang, G., and Nelson, D. L., *Inorg. Chem.*, 1970, **9**, 425.
92. Hester, R. E., and Plane, R. A., *Spectrochim. Acta*, 1967, **23A**, 2289.

93. Hester, R. E., and Grossman, W. E. L., *Spectrochim. Acta*, 1967, **23A**, 1945.
94. Adams, D. M., 'Metal-Ligand and Related Vibrations', Arnold, London, 1967; Clark, R. J. H., 'Halogen Chemistry', Vol. 3, ed. V. Gutmann, Academic Press, London, 1967, p. 85; Hester, R. E., *Co-ordination Chem. Rev.*, 1967, **2**, 319; James, D. W., and Nolan, M. J., *Progr. Inorg. Chem.*, 1968, **9**, 195.
95. Woodward, L. A., *Trans. Faraday Soc.*, 1958, **54**, 1271; cf. Godnev, I. N., and Alexandrovskaya, A. M., *Optics and Spectroscopy*, 1961, **10**, 14.
96. Clark, R. J. H., and Dunn, T. M., *J. Chem. Soc.*, 1963, 1198; Sabatini, A., and Sacconi, L., *J. Amer. Chem. Soc.*, 1964, **86**, 17.
97. Clark, R. J. H., *Spectrochim. Acta*, 1965, **21**, 955.
98. Loehr, T. M., and Plane, R. A., *Inorg. Chem.*, 1968, **7**, 1708.
99. Carreira, L. A., Maroni, V. A., Swain, J. W., and Plumb, R. C., *J. Chem. Phys.*, 1966, **45**, 2216.
100. Maroni, V. A., and Spiro, T. G., *J. Amer. Chem. Soc.*, 1966, **88**, 1410; 1967, **89**, 45.
101. Aveston, J., and Johnson, J. S., *Inorg. Chem.*, 1964, **3**, 1051; Tobias, R. S., *Canad. J. Chem.*, 1965, **43**, 1222; Griffith, W. P., and Lesniak, P. J. B., *J. Chem. Soc.* (*A*), 1969, 1066.
102. Hezel, A., and Ross, S. D., *Spectrochim. Acta*, 1968, **24A**, 985; Goldsmith, J. A., and Ross, S. D., *ibid.*, p. 993.
103. Brintzinger, H., and Plane, R. A., *Inorg. Chem.*, 1967, **6**, 623.
104. Cotton, F. A., 'Modern Coordination Chemistry', ed. J. Lewis and R. G. Wilkins, Interscience, New York, 1960, p. 301.
105. Yeranos, W. A., *Inorg. Chem.*, 1968, **7**, 1259.
106. Krishnan, K., and Plane, R. A., *Inorg. Chem.*, 1966, **5**, 852.
107. Woodward, L. A., and Owen, H. F., *J. Chem. Soc.*, 1959, 1055; Chantry, G. W., and Plane, R. A., *J. Chem. Phys.*, 1961, **35**, 1027; Jones, L. H., *Inorg. Chem.*, 1963, **2**, 777; *J. Chem. Phys.*, 1965, **43**, 594.
108. Coleman, J. S., Peterson, H., and Penneman, R. A., *Inorg. Chem.*, 1965, **4**, 135.
109. Parish, R. V., Simms, P. G., Wells, M. A., and Woodward, L. A., *J. Chem. Soc.* (*A*), 1968, 2882; Hartman, K. O., and Miller, F. A., *Spectrochim. Acta*, 1968, **24A**, 669.

CHAPTER 14

Vibrational Studies of the Jahn–Teller Effect

J. A. CREIGHTON

A. Introduction

Jahn and Teller[1] showed in 1937 that all orbitally degenerate non-linear molecules are unstable with respect to certain types of distortion, and may be expected to distort spontaneously with consequent removal of the degeneracy. The normal coordinates q in which the instability occurs (hereafter described as Jahn–Teller active) depend on the symmetry Γ_e of the degenerate electronic state, and may be identified by the rule that the symmetry species of q must be one of the non-totally symmetric species contained in the symmetric square of Γ_e:

$$\Gamma_q \text{ must be contained in } [\Gamma_e^2](\Gamma_q \neq a_1) \tag{14.1}$$

Jahn and Teller noted that the instability is small in molecules in which the degenerate orbitals are not greatly involved in the bonding. Subsequent developments in the theory[2] have recognized that the instability then results not in a static distortion at normal temperatures, but in a dynamic effect in which the vibrations in the Jahn–Teller active coordinate are executed with large amplitudes of motion, and on average the structure remains undistorted.

These conclusions have in general been borne out by structural measurements. Degenerate states are common in electronically excited molecules and several detailed studies have been reported on excited states[3] and radical ions, particularly of aromatic molecules.[4] Most of the work has been on compounds of the transition elements in the solid state, however, as has been described in a recent extensive review.[5] Strong Jahn–Teller effects are expected at metal atoms which are degenerate in the σ-antibonding d_γ orbitals, i.e. in the following configurations: d^9, high-spin d^4, low-spin d^7 in octahedral environments, and (neglecting spin-orbit coupling which may remove the degeneracy in the $d(t_2)$ orbitals) d^8, d^9, high-spin d^3 and d^4, low-spin d^5 and d^6 in tetrahedral environments. Static distortions have not yet been observed in all these configurations through the lack of suitable compounds. Almost all known examples in pure substances are in octahedral and tetrahedral compounds of Cu(II) (d^9) and the high-spin octahedral compounds of Cr(II) and Mn(III) (d^4), where the effects normally give rise to stabilizations of several thousand wavenumbers. The ground state in octahedral compounds of Cu(II), Cr(II) and Mn(III) is E_g, and from (14.1) a tetragonal (e_g) distortion resulting in point symmetry D_{4h} is expected,[2] as is commonly found in these compounds.[6] In tetrahedral Cu(II) compounds where the orbital ground state is T_2, the compounds may be expected from (14.1) to distort along normal coordinates of either e or t_2 symmetry. Distortions of the former kind leading to point symmetry D_{2d} seem to be preferred and are observed in the $CuCl_4^{2-}$ and $CuBr_4^{2-}$ ions.[7] (The absence of a distortion of $NiCl_4^{2-}$ (d^8) in the triphenylmethylphosphonium salt[8] has been attributed to the dominant effect of spin-orbit coupling in removing the ground-state orbital degeneracy,[10] although in $(Et_4N)_2NiCl_4$ the anion symmetry is also D_{2d}.[9]) Static distortions in $NaNiO_2$ (octahedral Ni(III) (d^7))[11] and $(Nb_6I_8)I_3$ (degeneracy in a multicentre orbital of the cation)[12] have also been reported.

A wider range of degenerate metal ions and coordination environments are available if the metal ions are trapped as impurities in a host crystal, and numerous studies of degenerate ions in host lattices have been made using electron spin resonance spectroscopy.[5] Static distortions, which result in anisotropic *g*-tensors, have been detected in octahedral Cu(II) and Mn(III) in a wide range of host crystals, and also in octahedral Ni(III), Pt(III), Ni(I) and Ag(II) and tetrahedral Cr(II) trapped in lattices. It was first noted[13] in a study of Cu(II) in $[Zn(H_2O)_6]SiF_6$, however, that, although the *g*-tensor is anisotropic at low temperatures, at high temperatures it becomes isotropic, as a result of rapid interconversions between the three possible directions of the tetragonal distortion of the Cu(II) coordination sphere within the host crystal. This phenomenon of *pseudo-rotation* has been found in several dilute crystals containing degenerate metal atoms,[5] and it also occurs in pure compounds in which the ligands are not shared by adjacent degenerate metal atoms, and where the Jahn–Teller distortions are therefore not cooperative. The interconversion involves entirely *linear* motion of the ligands within the Jahn–Teller active coordinate and the structure does not pass through the undistorted configuration (in octahedral compounds the intermediate structures have D_{2h} symmetry).[2] The barrier to pseudorotation is therefore less than the full Jahn–Teller stabilization energy. Nevertheless, in substances with fairly strong Jahn–Teller effects the barrier is normally considerably greater than the zero-point energy of vibrational motion in the Jahn–Teller active coordinate, so that static distortions can be frozen in at low temperatures.

Weak Jahn–Teller effects, with stabilization energies of the order of a few hundred wavenumbers or less, occur at metal atoms which are degenerate in the weakly π-bonding or -antibonding d_ϵ orbitals. The configurations involved are octahedral d^1 and d^2, and tetrahedral d^1, low-spin d^3 and high-spin d^6 (in octahedral low-spin d^4 and d^5, and high-spin d^6 and d^7 spin-orbit coupling removes the ground state degeneracy,[14] and this is likely to dominate the weak Jahn–Teller effect). There is at present no definite evidence of static distortions at room temperature due to d_ϵ degeneracy, but a few cases are known of static distortions at very low temperatures, for example in Y(II) (d^1) in a CaF_2 matrix.[5] In the free molecules VCl_4, TcF_6 and ReF_6 where the Jahn–Teller stabilization energies have been determined, however,[15,16] the stabilization energies are of similar magnitude to the zero-point energy of the Jahn–Teller active vibrations, so that static distortions cannot be frozen into these molecules at low temperatures.

Much of the evidence for the Jahn–Teller effect has been obtained from solids, where the interpretation of measurements has not always been free from doubt, since crystal-packing effects may themselves cause distortions which may augment or modify the Jahn–Teller effect. Many of the results of vibrational studies to be described here are for molecules or ions in the gaseous state or in solution. Thus, although these studies have been limited by the lack of a wide range of degenerate compounds, and there are uncertainties in the interpretation of the anomalies observed in the spectra of such compounds, ambiguities arising from crystal-packing effects have normally been avoided.

B. Vibrational properties of electronically degenerate molecules

The vibrational spectroscopic properties of molecules with static Jahn–Teller distortions are predicted[17] to be those defined by the normal selection rules appropriate to the symmetry of the distorted structure. Molecules with dynamic Jahn–Teller coupling are predicted, however, to exhibit more complicated effects, which arise both from the splitting of the energy levels of the Jahn–Teller active vibrations by the vibronic coupling and also from the special selection rules which govern vibrational transitions in degenerate molecules.

The effect of first-order vibronic coupling on the energy levels of the doubly degenerate Jahn–Teller active vibrations of a molecule in an E state is shown in Figure 14.1. Each level is split into $v+1$ doubly degenerate components characterized by a quantum number $|j| = v + \frac{1}{2}, v - \frac{1}{2}, \ldots \frac{1}{2}$, where v is the vibrational quantum number.[18, 19, 20] The vibrational selection rule for transitions from the ground state is $\Delta|j| = 0, 1, 2$ and from excited states is $\Delta|j| = \pm 1, \pm 2$. Those with $\Delta|j| = 2$ are expected to be

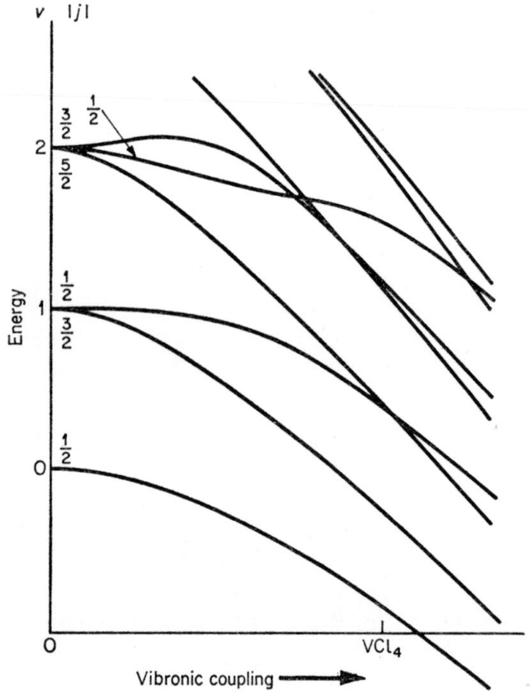

FIGURE 14.1. *The effect of vibronic coupling on the energy levels of a Jahn–Teller active e vibration of an electronically doubly degenerate molecule.* (*Reproduced with permission from ref. 19.*)

weak enough to be neglected, however, and the fundamental and overtone bands of the Jahn–Teller active vibrations are therefore expected to be doublets. Further splitting of the doubly degenerate vibronic levels, not shown in Figure 14.1, results from second-order vibronic coupling.[16] A diagram analogous to Figure 14.1 has been published for molecules in T states,[21] for which e and t_2 vibrations are Jahn–Teller active but in which the excited states of only the t_2 vibrations are split by first-order vibronic coupling.[22]

The special selection rules governing vibrational transitions in degenerate molecules arise from the possibility of simultaneous transitions between the components of the degenerate electronic state. In molecules in which only the effects of orbital degeneracy need be considered, it has been predicted by Child and Longuet-Higgins[20] that non-centrosymmetric molecules may possess an electric dipole moment (and hence a rotational absorption spectrum) which would be forbidden by symmetry if the molecule were non-degenerate, and that vibrations which are infrared- or Raman-forbidden for non-degenerate molecules may become active in degenerate molecules. In addition, some Raman bands of degenerate molecules may have depolarization ratios (ρ) greater than 6/7 (with respect to natural incident light), while others which are not due to a_1 vibrations may have $\rho < 6/7$, and still others, which for non-degenerate molecules are completely polarized, may have $\rho > 0$. Also for molecules with an anomalous dipole moment there may be an envelope of intense, irregularly spaced overtones of the Jahn–Teller active vibrations centred about the electronic Jahn–Teller splitting energy in both the Raman and infrared spectra of molecules with Jahn–Teller effects of intermediate strength. The symmetry rules which govern these phenomena and the results for all symmetric top point groups have been published,[20] and are summarized for common point groups in Table 14.1.

These conclusions are unaffected by the presence of electron spin if spin-orbit coupling is weak, as it is in molecules which have only spin degeneracy or in which the spin-orbit coupling constant is small, or in tetrahedral or octahedral molecules in orbital E states. If spin-orbit coupling is strong, the resulting Γ_8 states of octahedral and tetrahedral molecules possess anomalous properties similar to those described above and summarized in Table 14.1, while the Γ_6 and Γ_7 states have normal spectroscopic properties except that their Raman bands may have depolarization ratios greater than 6/7.

C. Experimental results: strong Jahn–Teller effects

(a) 4-*fold coordination: the* $CuCl_4^{2-}$ *ion*

In some salts of the composition M_2CuCl_4, such as the ammonium salt, the coordination around each copper atom is distorted octahedral, with a square-planar arrangement of four short Cu—Cl bonds together with two longer bonds to adjacent $CuCl_4^{2-}$ units.[24] Caesium and tetramethylammonium tetrachlorocuprate(II) have been found[7] to contain discrete $CuCl_4^{2-}$ ions, however, with D_{2d} symmetry and bond angles of ~101 and 127°. A single-crystal analysis of the polarized visible absorption spectrum of the $CuCl_4^{2-}$ ion in Cs_2CuCl_4, and in the host lattice Cs_2ZnCl_4, shows[25]

TABLE 14.1. *Infrared, Raman, and Jahn–Teller vibrations of electronically degenerate molecules in common point groups*

Point group	D_3	C_{3v}	D_{2d}	D_{3h}	D_{6h}	T_d		O_h		O_h^*	
Electronic state Γ_e	E	E	E	E′, E″	E_1, E_2	E	T	E	T	Γ_6, Γ_7	Γ_8
Infrared active											
(∥ band)	a_2, e	a_1, e	a_1, a_2, b_2	a_2'', e''	a_{2u}, e_{2u}						
(⊥ band)	a_1, a_2, e	a_1, a_2, e	e	a_1', a_2', e'	b_{1u}, b_{2u}, e_{1u}	t_1, t_2	a_1, e, t_1, t_2	t_{1u}, t_{2u}	$a_{2u}, e_u, t_{1u}, t_{2u}$	t_{1u}	$a_{1u}, e_u, t_{1u}, t_{2u}$
Raman active											
$0 \leqslant \rho \leqslant \frac{6}{7}$	—	—	b_1, b_2	e′	e_{2g}	a_1, e	—	a_{1g}, e_g	—	—	—
$\rho = \frac{6}{7}$	a_2	a_2	a_2	a_1'', a_2', a_2''	a_{2g}, b_{1g}, b_{2g}	a_2, t_1	a_2	a_{2g}, t_{1g}	a_{2g}	—	—
$\frac{6}{7} \leqslant \rho \leqslant 2$	—	—	e	e″	e_{1g}	t_2	t_1	t_{2g}	t_{1g}	e_g, t_{1g}, t_{2g}	a_{2g}, t_{1g}
$0 \leqslant \rho \leqslant 2$	a_1, e	a_1, e	a_1	a_1'	a_{1g}	—	a_1, e, t_2	—	a_{1g}, e_g, t_{2g}	a_{1g}	a_{1g}, e_g, t_{2g}
Jahn–Teller active	e	e	b_1, b_2	e′	e_{2g}	e	e, t_2	e_g	e_g, t_{2g}	—	e_g, t_{2g}

that the splitting of the copper $d(t_2)$ orbitals by the distortion is ~5000 cm^{-1}. This considerable distortion appears to be an intrinsic property of $CuCl_4^{2-}$ and not a consequence of crystal packing, since the visible absorption spectra of this ion in crystalline Cs_2CuCl_4, Cs_2ZnCl_4 and in solution are very similar. It is almost certain therefore that the distortion is a consequence of the Jahn–Teller effect.

In view of the magnitude of the distortion and since the Jahn–Teller splitting is much greater than kT at room temperature, normal vibrational properties characteristic of D_{2d} symmetry are expected. The splitting of degenerate vibrations as a result of a descent in symmetry from T_d to D_{2d} is indicated in Table 14.2. Only those vibrations derived from f_2 are infrared active while all vibrations are Raman active. The splitting of the f_2 stretching vibration is clearly seen in the infrared data for $(Et_4N)_2CuCl_4$ in solution in nitromethane[26] (Table 14.2). The corresponding Raman bands were too weak to be detected in the Raman spectrum of this solution, but the splitting is confirmed by the infrared spectra of the crystalline caesium, tetramethylammonium and tetraethylammonium salts, which also show the splitting of the f_2 bending vibration.[27] Further evidence of D_{2d} vibrational behaviour has come from a single crystal Raman study of Cs_2CuCl_4.[28] The relative magnitudes of the scattering tensor components for each of the observed bands were compared with those calculated for the various vibrational symmetry species of D_{2d}. The agreement was satisfactory for most of the bands, and enabled the assignment of the b_2 and e vibrations shown in Table 14.2 to be made.

Though the solid state measurements in Table 14.2 support the evidence for a static Jahn–Teller distortion provided by the solution infrared spectrum, the importance of the solution measurement and the inconclusive nature of splittings observed in the spectra of solids is illustrated by a comparison with the data for the homologous non-degenerate ion $ZnCl_4^{2-}$. The Raman spectrum of $ZnCl_4^{2-}$ in solution is characteristic of a regular tetrahedral structure with no evidence for splitting of the f_2 stretching or bending vibrations.[29] However, a Raman study of crystalline Cs_2ZnCl_4 shows these vibrations to be split by 21 and 12 cm^{-1} respectively,[28] presumably as a result of crystal-packing effects (though the crystal structure of this compound is not known). The corresponding splittings in Cs_2CuCl_4 are 27 and 19 cm^{-1}, and while it may be argued that the larger splittings in the copper compound are evidence of a Jahn–Teller effect, the difference is too small for this conclusion to be convincing.

(*b*) 6-*fold coordination*

The infrared spectrum of manganese(III) *tris*-acetylacetonate (high spin d^4) provides fairly clear qualitative evidence of a static Jahn–Teller

TABLE 14.2. *Vibrational frequencies of the $CuCl_4^{2-}$ ion, cm^{-1}*

Cs_2CuCl_4		$(Me_4N)_2CuCl_4$		$(Et_4N)_2CuCl_4$		$(Et_4N)_2CuCl_4$ (soln)		Assignment	
i.r.[a]	Ra	i.r.	Ra	i.r.	Ra	i.r.[a]	Ra	D_{2d}	T_d
	299		276		277		274 pol	a_1	a_1, ν(Cu—Cl)
288	280	281		267		278		b_2	f_2, ν(Cu—Cl)
255	253	237	232	248		237		e	
	140	145		136				e	f_2, δ(Cl—Cu—Cl)
	121	128		118				b_2	
	106							a_1, b_1	e, δ(Cl—Cu—Cl)

[a] Limit of observation 200 cm^{-1}.

distortion when compared with the spectra of the analogous chromic (d^3) and ferric (d^5) compounds (Figure 14.2).[30] Below 650 cm^{-1} metal *tris*-acetylacetonate complexes show bands whose frequencies vary with the nature of the metal. The assignment of all the bands in this region is not known with certainty, though bands near 450 cm^{-1} are probably due

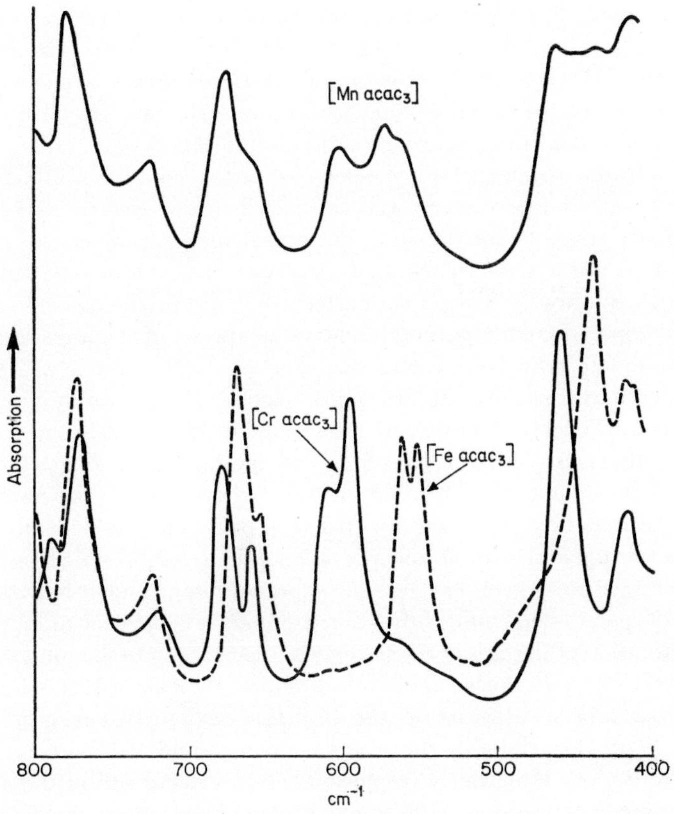

FIGURE 14.2. *The infrared spectra of manganese, chromium and iron* tris-*acetylacetonates.*[30]

to vibrations involving large contributions from M—O stretching, while those at 550–600 cm^{-1} are possibly due to bending motions of the chelate rings.[31] It is certainly clear, however, from their metal sensitivity that these bands arise from vibrations involving some motion of the oxygen atoms relative to the metal. Figure 14.2 shows that the frequencies of the metal-sensitive vibrations of iron(III) *tris*-acetylacetonate are lower than those of the chromium(III) compound, and this is correlated with the difference in the electron configurations $(t_{2g})^3$ $(e_g)^2$ and $(t_{2g})^3$ of these compounds, whereby the iron compound has two additional electrons in the antibonding e_g orbitals. In manganese(III) *tris*-acetylacetonate the electron configuration is $(t_{2g})^3$ $(e_g)^1$. If the e_g orbitals are split by a static Jahn–Teller distortion with the d_{z^2} orbital lower in energy, four of the Mn—O bonds will be similar in strength to the bonds in chromium(III) *tris*-acetylacetonate while two will be weaker as in the iron(III) compound, or *vice versa* if the $d_{x^2-y^2}$ orbital is lower in energy. In either case the spectrum of manganese(III) *tris*-acetylacetonate might be expected to resemble a superposition of the spectra of the chromium(III) and iron(III) compounds and this is borne out remarkably well in Figure 14.2.

The infrared spectra of these acetylacetonate complexes in solution, though broad and poorly resolved compared to the spectra of the solid compounds, suggest that the static distortion in manganese(III) acetylacetonate is not a crystal-packing or a cooperative solid state effect, but persists in solution.[30] This is supported by a recent discussion[32] of the various proposed assignments of the electronic spectrum of manganese(III) acetylacetonate both in the solid state and in solution, from which it is concluded that the e_g orbitals are split by 9000 cm^{-1} as a result of a tetragonal Jahn–Teller distortion. An approximate crystal field calculation indicates that this splitting is consistent with a difference in length between long and short bonds of 0·34 Å. However, this distortion was not confirmed by a determination of the crystal structure of the acetylacetonate, since within experimental error all the Mn—O bonds are of the same length and the molecular point group is D_3. This apparent disagreement between the results of spectroscopic and diffraction methods is believed[32] to be due to pseudorotation of the axis of the tetragonal distortion. On the time scale of less than 10^{-11} s the spectroscopic studies indicate a static distortion, but on a longer time scale the axis of the distortion oscillates between the three M—O bond directions in the acetylacetonate, and the time-averaged structure is then D_3. The isotropic temperature factors from the crystal structure determination are consistent with this picture. From the difference in length of the long and short bonds indicated by the spectroscopic data, the mean amplitude of motion of oxygen atoms accompanying pseudorotation is 0·23 Å, and this is comparable with the root-mean-square amplitude of 0·24 Å determined from the isotropic temperature factors.

Similar situations appear to arise in a number of *tris*-chelate compounds of copper(II). X-ray diffraction measurements show that the cations in $[Cu(en)_3]SO_4$ and $[Cu(bipy)_3]Br_2,6H_2O$ are undistorted with D_3 symmetry (en = ethylenediamine; bipy = 2,2′-bipyridyl),[33] and the *g*-values of the ions $[Cu(bipy)_3]^{2+}$ and $[Cu(phen)_3]^{2+}$ (phen = 1,10-

phenanthroline) in matrices of some crystalline salts of the analogous zinc *tris*-chelates have been found to be isotropic at 350 K.[34,35] At 90 K, however, the g-values of $[Cu(bipy)_3]^{2+}$ and $[Cu(phen)_3]^{2+}$ are anisotropic, and are consistent with tetragonal distortions of the cations. These observations are most reasonably attributed to pseudorotation, which is frozen in at low temperatures.[34,35] In agreement with this the electronic spectrum of $[Cu(phen)_3]^{2+}$ in solid solution in $[Zn(phen)_3](NO_3)_2$, $2H_2O$ is also consistent with a tetragonal distortion and indicates a Jahn–Teller splitting of 7000 cm^{-1}, though some details of the single crystal spectrum at 77 K are not yet fully understood.[34] The infrared spectrum of $[Cu(bipy)_3]^{2+}$ provides some support for an instantaneously distorted structure, since, compared with the spectra of its nickel(II) and zinc(II) analogues, there is additional splitting of ligand internal vibrations at 650–600 and 450–400 cm^{-1}, and the strong metal–ligand stretching band at 290 cm^{-1} in $[Ni(phen)_3]^{2+}$ and $[Zn(phen)_3]^{2+}$ is very weak and broad in the copper compound.[36] The spectrum of $[Cu(phen)_3]^{2+}$ does not resemble a superposition of the spectra of the nickel and zinc compounds, however, and the differences in the spectra of these compounds are not understood in detail. Nevertheless, the breadth of the metal–ligand stretching band in the copper *tris*-chelate is almost certainly due to overlapping components arising from the instantaneous presence of both strong and weak metal–ligand bonds.

The position is at present somewhat less clear in other six-coordinated compounds of copper(II) and of low-spin cobalt(II). The crystal structures of $K_2Pb[Cu(NO_2)_6]$, $K_2Ba[Co(NO_2)_6]$, and of a number of salts of the $[Cu(NH_3)_6]^{2+}$ ion are cubic at room temperature (in contrast to $K_2Ba[Cu(NO_2)_6]$ which is tetragonal), and the electron spin resonance spectra of these compounds (except $[Cu(NH_3)_6]Br_2$) are isotropic at this temperature.[37] The electronic spectra of all these compounds indicate tetragonal structures, however, with the e_g orbitals split by ~7500 cm^{-1} in the nitro-complexes and by ~10,900 cm^{-1} in $[Cu(NH_3)_6]^{2+}$. The observations are consistent with pseudorotation in these compounds. However, the infrared spectra of $K_2Pb[Cu(NO_2)_6]$, $K_2Pb[Co(NO_2)_6]$ and $K_2Ba[Co(NO_2)_6]$ are closely similar to the spectra of the analogous non-degenerate compounds $K_2Pb[Fe(NO_2)_6]$ and $K_2Ba[Fe(NO_2)_6]$ and provide no evidence for Jahn–Teller distortions in the copper(II) and cobalt(II) compounds. The effect of cooling on the behaviour of $K_2Pb[Cu(NO_2)_6]$ has also been studied but is not yet clear. Between room temperature and 235 K changes are observed in the X-ray powder data and electron spin resonance spectrum, but the patterns do not resemble those of the tetragonal compound $K_2Ba[Cu(NO_2)_6]$, while in the neutron-scattering powder pattern no change was observed down to 90 K.

D. Experimental results: weak Jahn–Teller effects

(*a*) 4-*fold coordination*

(i) *Vanadium tetrachloride.* From electron diffraction measurements on the vapour[38] VCl_4 is known to be a regular tetrahedral molecule with dimensions closely similar to those of the neighbouring d^0 molecule $TiCl_4$ (see Table 14.3). The ground state due to the single electron in the 3d(e)

TABLE 14.3. *Vibration frequencies, cm^{-1}, and dimensions, Å, of VCl_4 and $TiCl_4$*

	VCl_4		$TiCl_4$	
	Raman (soln[a])	Infrared (vapour)	Raman (soln[a])	Infrared (vapour)
ν_2	—		121 s	
ν_4	128 m	129[a]	137 m	138[a]
ν_1	383 vs, pol		389 vs, pol	
ν_3	475 w	488 vs	501 m	498·5 vs
$\nu_1 + \nu_4$		515 w, sh	528 w	526 m
$\nu_2 + \nu_3$		~578 w, br		617 s
$\nu_3 + \nu_4$		614 w		637 w
$\nu_1 + \nu_3$		874 w		886 w
$2\nu_3$		972 w		997 w
Bond lengths: M—Cl		$2{\cdot}138 \pm 0{\cdot}002$		$2{\cdot}170 \pm 0{\cdot}002$
Cl—Cl		$3{\cdot}485 \pm 0{\cdot}004$		$3{\cdot}539 \pm 0{\cdot}004$
Root-mean-square amplitudes: M—Cl		$0{\cdot}051 \pm 0{\cdot}002$		$0{\cdot}049 \pm 0{\cdot}003$
Cl—Cl		$0{\cdot}135 \pm 0{\cdot}004$		$0{\cdot}116 \pm 0{\cdot}004$

[a] Solution in CCl_4.

orbitals is therefore ^2E, and from (14.1) the molecule is unstable with respect to distortions in the bending coordinate of species e. Ligand field and molecular orbital calculations show,[39] however, that the magnitude of the distortion is likely to be comparable with the amplitude of the zero point vibration in this coordinate, so that in agreement with the observed regular tetrahedral structure, VCl_4 is expected to show a dynamic Jahn–Teller effect rather than a static distortion.

Vibrational evidence for a dynamic Jahn–Teller effect in VCl_4 is shown most clearly from a comparison of the spectral data with those of the adjacent non-degenerate molecule $TiCl_4$. As shown in Table 14.3, the frequencies and intensities of three of the four fundamentals, $\nu_1(a_1)$, $\nu_3(f_2)$ and $\nu_4(f_2)$, and of their combinations, are very similar for the two molecules, and their assignments may be regarded as well established.[38, 40, 41] However, the behaviour of the remaining fundamental $\nu_2(e)$ differs markedly for the two compounds. In the Raman spectrum of $TiCl_4$ (and of most other tetrahedral XY_4 molecules) ν_2 appears as a strong band slightly lower in frequency than ν_4, but for VCl_4 only one band, which must be assigned, at least in part, to ν_4 on the basis of its coincidence with an infrared band,[41b] is observed in this region. Moreover, in the infrared spectrum where the combination $\nu_2 + \nu_3$ is readily observed for $TiCl_4$, only a weak and very broad feature (~578 cm^{-1}) is observed in the corresponding region of the spectrum of VCl_4.

In order to account for the single Raman band in the bending region of VCl_4, it was first suggested[40] that ν_2 is coincident with ν_4. This assignment has since been shown[38] to be inconsistent, however, with the frequencies ν_1, ν_3 and ν_4 when these are considered in conjunction with the amplitudes of vibrational motion measured by electron diffraction. These frequencies and amplitudes may be used to make an estimate of ν_2, whence it has been concluded that either $\nu_2 = 97$ cm^{-1} or the large observed Cl—Cl amplitude is due to the participation of this e vibration in vibronic coupling, the effect of which is to split ν_2. The latter possibility seems the most likely since there is no obvious reason why ν_2 should be much lower than in $TiCl_4$, but in either event ν_2 is not coincident with ν_4, and it appears that ν_2 has not yet been observed in the Raman spectrum of VCl_4.

By assuming that the large Cl—Cl amplitude is due to vibronic coupling of the e vibration, the amplitudes and the frequencies ν_1, ν_3 and ν_4 have been used to calculate the vibronic coupling constant, which determines the energy levels of the e vibration.[38] The energy levels are shown in Figure 14.1, and the stabilization of the ground state is found to be 73 ± 43 cm^{-1}. The fundamental transition ν_2 is split into two components at 71 and 149 cm^{-1}, but several hot-band transitions from well populated, low-lying, excited levels of the vibration also occur in this region, as may be seen from Figure 14.1. Bands due to the e vibration are therefore likely to be broad, and this possibly accounts for the failure to observe ν_2 in the Raman spectrum and for the broad contour of the infrared band at ~578 cm^{-1} ascribed to $\nu_2 + \nu_3$. Second-order splitting of the vibronic levels possibly also contributes to a broadening of ν_2, as has been suggested for ReF_6.[16]

Although the failure to observe ν_2 in the Raman spectrum thus appears most likely to be a consequence of the Jahn–Teller effect, there remains the anomalous Raman depolarization ratios and the intense overtones of ν_2 predicted on theoretical grounds. The depolarization ratios of the ν_1, ν_3 and ν_4 bands are expected to be different from the normal values in non-degenerate molecules (see Table 14.1), and intense progressions of overtones of ν_2 are predicted to occur in the Raman spectrum with a maximum of intensity near the Jahn–Teller splitting energy (taken to be four times the stabilization energy) of ~300 cm^{-1}. Only qualitative estimates of the depolarization ratios of the ν_1, ν_3 and ν_4 bands have yet been made.[40] By eye these were not noticeably different from the corresponding ratios for $TiCl_4$, but this method is clearly unable to detect small effects. Strong overtones of ν_2 in the region of 300 cm^{-1} are certainly absent from the Raman spectrum however. The reason for this is not clear. As is possibly true for the ν_2 fundamental itself, the overtone transitions may be broadened by superimposed hot-band transitions from low-lying vibronic levels, and have thus escaped detection. A study of the Raman spectrum of VCl_4 with laser Raman instruments would possibly now detect these vibronic effects.

(ii) *The tetrachloroferrate(II) ion.* The $FeCl_4^{2-}$ ion, like its homologues $CoCl_4^{2-}$ and $NiCl_4^{2-}$, is regular tetrahedral in its triphenylmethylarsonium salt.[8] The magnetic moment of this ion of ~5·20 BM,[42] though somewhat larger than the expected spin-only value of 4·90 BM, shows that it has the high-spin d^6 configuration $(e)^3(t_2)^3$ and, like VCl_4, it is therefore in an orbital E state. The Mössbauer spectrum of this ion in $(Et_4N)_2FeCl_4$ has been studied,[43] and it was concluded from the temperature-dependence of the quadrupole splitting of the iron resonance below 190 K, that the d(e) orbitals are not exactly degenerate in this salt but are split by 185 cm^{-1}. Unfortunately the cause of this splitting is not yet clear since the crystal structure of the salt is not known. It is possibly due to a weak Jahn–Teller effect and indeed is of a similar magnitude to the Jahn–Teller splitting in VCl_4. However, in the related salts $(Et_4N)_2CoCl_4$ and $(Et_4N)_2NiCl_4$, whose crystal structures are known,[9] the non-degenerate anions are slightly distorted to D_{2d} symmetry, presumably as a result of crystal-packing, and it is probable therefore that in these salts a slight splitting of the d(e) orbitals also occurs. Raman data on the $FeCl_4^{2-}$ ion, which might resolve this question, are still incomplete. Only the strongest band ν_1 has been observed in the Raman spectrum of $(Et_4N)_2FeCl_4$ presumably because of experimental difficulties in recording spectra of the powdered solid,[44] and in the absence of the weaker ν_3 and ν_4 bands no conclusions can be reached on the intensity of the Jahn–Teller active e vibration ν_2.[44] A weak band in the infrared spectra of $(Me_4N)_2FeCl_4$ and $(Et_4N)_2FeCl_4$ has been assigned to either ν_2 or a lattice mode.[45] Its frequency is close to that of bands in the corresponding Mn(II) and Co(II) compounds and in no way anomalous for the Fe(II) compound, but since ν_2 is infrared-inactive in T_d and D_{2d} symmetries the possible assignment of this band to ν_2 must be regarded as tentative.

(b) 5-fold coordination: molybdenum and tungsten pentachlorides

Molybdenum pentachloride is monomeric and trigonal-bipyramidal in the vapour state,[46] and this is probably true also of tungsten pentachloride.

The point symmetry of these molecules is therefore D_{3h}, and the d orbitals are split into three levels of species $a' + e' + e''$ of which e'' is lowest in energy. The ground state of these d^1 configurations is therefore $^2E''$. The vibrational species of the molecules are $2a_1' + 2a_2'' + 3e' + e''$, of which $2a_1' + 3e' + e''$ are normally Raman active and $2a_2'' + 3e'$ are infrared active, but as a result of the degeneracy all the vibrations become allowed in both the Raman and infrared spectra (see Table 14.1). The Jahn–Teller active vibrations are the three vibrations of species e', and since these are both Raman and infrared active even in the absence of degeneracy, they should be readily observable in both types of spectra, in contrast to the situation in tetrahedral and octahedral molecules where the Jahn–Teller active fundamentals are allowed only in Raman scattering.

The Raman spectrum of molybdenum pentachloride vapour at 350°C shows no unusual features when compared to the vapour spectrum of the adjacent d^0 molecule $NbCl_5$.[47] The pattern of bands of both molecules resembles that of the molecule $SbCl_5$, whose assignment is well established, and the six observed bands were satisfactorily assigned to the six fundamentals which are normally Raman-active in the absence of electronic degeneracy. No broadening was found in the Jahn–Teller active e' fundamental bands and the predicted reversed polarization of the e'' fundamental was not observed. Additional bands present in the spectrum at higher temperatures (450°C) were not of frequencies consistent with the a_2'' fundamentals (rendered Raman-active as a consequence of the degeneracy), and they were satisfactorily attributed to combination and difference bands. The Raman spectrum of $MoCl_5$ thus provides no evidence for a Jahn–Teller effect in this molecule; possibly spin-orbit coupling is sufficiently strong here to dominate the Jahn–Teller effect.[23, 48]

In contrast, the infrared spectra of molybdenum and tungsten pentachloride in solution in CCl_4 differ markedly from the spectra of niobium and tantalum pentachlorides in the same solvent, and whereas the d^0 species show at least seven bands below 450 cm^{-1}, the d^1 molecules show only two definite bands in this region.[48] The explanation of this result is not entirely clear since the state of the pentachlorides in solution in CCl_4 is not known, except for niobium pentachloride which is dimeric[49] as in the solid. It has been pointed out, however, that association is likely to increase rather than decrease the number of observed infrared bands,[48] and on the assumption that the molybdenum and tungsten compounds are monomeric in solution, the observation of only two infrared bands has been attributed to Jahn–Teller effects, which broaden the e' vibrational bands of these compounds. The frequencies of the two observed bands are indeed in the correct region for their assignment to the Jahn–Teller inactive a_2'' vibrations. However, the conclusion that Jahn–Teller coupling is significant in $MoCl_5$ disagrees with the conclusion from the Raman spectrum, and in view of the uncertainty concerning the structures of such compounds in solution, confirmatory infrared studies of the monomeric molecules in the vapour state or in dilute inert matrices are clearly required.

(c) 6-*fold coordination*

(i) *Rare metal hexafluorides.* The effects of Jahn–Teller coupling on the vibrational spectra of degenerate molecules were first observed in the rare metal hexafluorides by Weinstock and Claassen[50] and it is in these molecules that the effects have since been studied in greatest detail. An extensive review of the studies by Weinstock and Goodman has been published;[16] only the main points are summarized here.

The rare metals form two series of hexafluorides, ReF_6, OsF_6, IrF_6, PtF_6 and TcF_6, RuF_6, RhF_6 in which the electron configurations increase from d^1 in steps of one d electron. These compounds give rise to a number of electronic bands in the visible and near infrared regions of the spectrum due to transitions between the components into which the $d(t_{2g})$ orbitals are split by the effects of spin–orbit coupling and electron–electron repulsions. It has proved possible to account quantitatively for all these bands by ligand field theory using only two parameters for each transition series, and thus to establish the octahedral symmetry of the compounds and determine their ground states[14] (except for RhF_6 whose absorption spectrum has not yet been investigated). The ground state of PtF_6 is non-degenerate (A_{2g}), while the four-fold Γ_8 degeneracy of $IrF_6(d^3)$ (and most probably also of RhF_6) is almost entirely due to the spin of the three unpaired $d(t_{2g})$ electrons (ground state essentially ^4S). These three compounds, along with the d^0 analogues WF_6 and MoF_6, are therefore not expected to show Jahn–Teller effects.[51] On the other hand, the Γ_8 ground states of ReF_6 and TcF_6 (d^1) and the $E_g + T_{2g}$ ground states of OsF_6 and RuF_6 (d^2) are derived from orbitally degenerate ^2P and ^3P states respectively, and Jahn–Teller effects are expected for these molecules.

The vibrational species of octahedral XY_6 molecules are $a_{1g} + e_g + 2f_{1u} + f_{2g} + f_{2u}$ of which the g species are Raman active and f_{1u} is infrared allowed. For molecules in Γ_8 or T_{2g} states $\nu_2(e_g)$ and $\nu_5(f_{2g})$ are the Jahn–Teller active vibrations, while for E_g states only ν_2 is Jahn–Teller active.

The infrared spectra of all these hexafluorides in the vapour states have been recorded.[16] Raman data are less complete largely as a result of experimental difficulties caused by the colours and photosensitivity of some of the compounds, but Raman spectra have been reported for WF_6, ReF_6, OsF_6, MoF_6 and TcF_6, including vapour spectra of ReF_6 and MoF_6, and also a partial spectrum of IrF_6 in solution. The vibrational spectra of the

d^3 and d^4 molecules IrF_6, RhF_6 and PtF_6 show, as expected, no unusual features when compared to the spectra of the d^0 molecules WF_6 and MoF_6, and the assignment of all the bands in terms of the six fundamentals of each molecule follows the pattern for WF_6 and MoF_6. However, the spectra of the d^1 and d^2 molecules ReF_6, TcF_6, OsF_6 and RuF_6, though showing normal bands due to ν_1, ν_3 and ν_4 and their combinations, are unusual

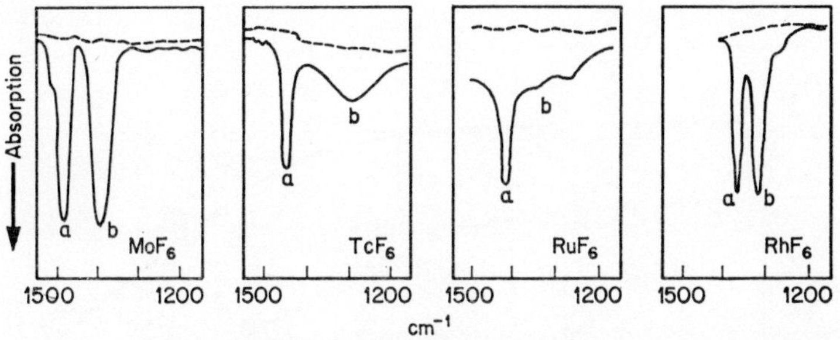

FIGURE 14.3. *Infrared band contours of* $\nu_1 + \nu_3$ *(a) and* $\nu_2 + \nu_3$ *(b) for the* 4d *transition series hexafluorides.*

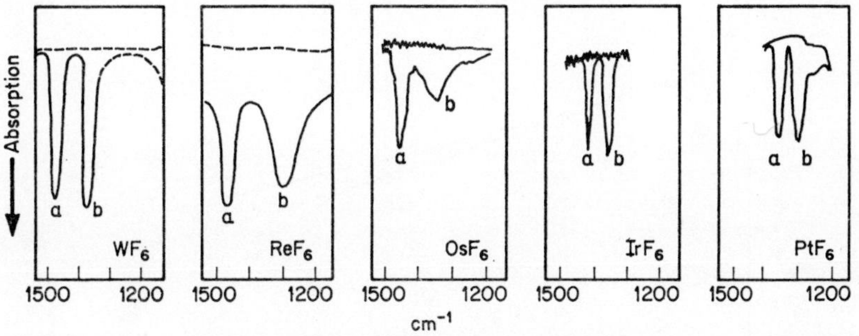

FIGURE 14.4. *Infrared band contours of* $\nu_1 + \nu_3$ *(a) and* $\nu_2 + \nu_3$ *(b) for the* 5d *transition series hexafluorides.*

(*Reproduced with permission from ref.* 52.)

in that the fundamental and combination bands of the Jahn–Teller active vibrations ν_2 and ν_5 are broad and of low peak intensity. This is shown in Figures 14.3 and 14.4, which portray the combination bands $\nu_1 + \nu_3$ and $\nu_2 + \nu_3$ for hexafluorides of the second and third transition series.[52] It is seen that for the d^0, d^3 and d^4 hexafluorides of both series, these combinations are prominent bands of roughly equal intensity, but for the d^1 and d^2 molecules $\nu_2 + \nu_3$ is a broad band of low peak absorbance compared to $\nu_1 + \nu_3$. Broadening of the ν_2 and ν_5 fundamental bands also occurs in the Raman spectra of the d^1 and d^2 molecules, and is such that ν_2 of OsF_6 and both ν_2 and ν_5 of TcF_6 could not be detected as fundamentals.[16]

In addition to these anomalous contours, the frequencies of the centres of bands involving ν_2 and ν_5 are also unusually low for the d^1 and d^2 hexafluorides compared to the corresponding bands of the non-vibronic molecules (see Figure 14.5), unlike ν_1, ν_3, ν_4 and ν_6 which undergo a smooth

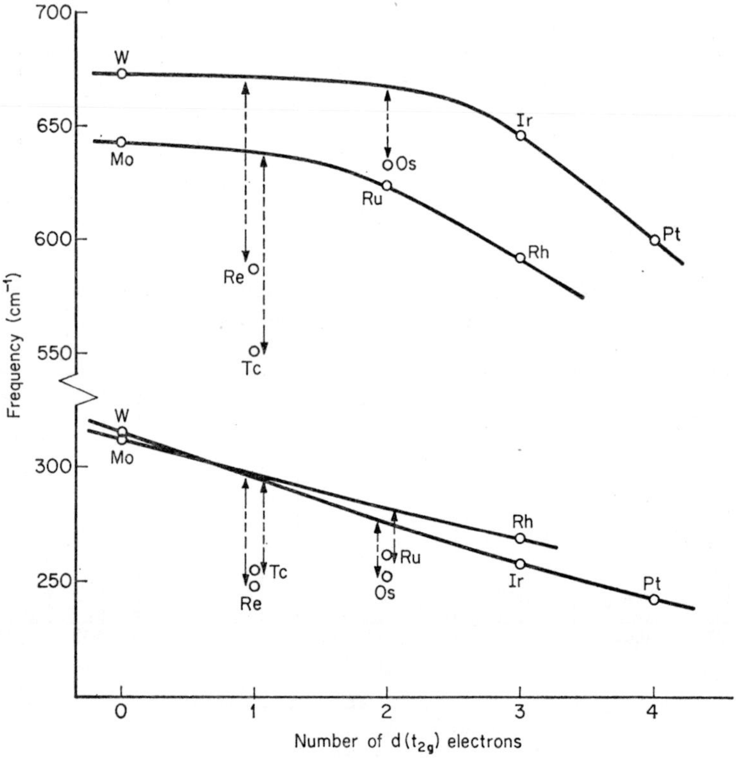

FIGURE 14.5. *The frequencies ν_2 and ν_5 of the* 4d *and* 5d *transition series hexafluorides.*

variation with the number of d electrons throughout the entire series of hexafluorides. The occurrence of unexpectedly low frequencies and broad band contours in only the Jahn–Teller active vibrations of the nine hexafluorides leaves little doubt that these effects are due to Jahn–Teller coupling, and some success has been achieved, particularly for ReF_6, in a detailed assignment of the spectra in terms of the predicted vibronic transitions.[16]

The energy levels of the e_g vibration (ν_2) of ReF_6 are split by vibronic coupling in the same manner as is shown in Figure 14.1. It is seen that the $0 \rightarrow 1$ transition is split into two components approximately symmetrically placed on either side of the undisplaced frequency ν_2^0. It is reasonable to assume (see Figure 14.5) that the observed ν_2 fundamental and combination bands involve the *lower* component of ν_2 (587 cm^{-1}). The upper component, which Figure 14.5 shows to be ~168 cm^{-1} higher in frequency, is not easily recognized in the spectra because of its accidental coincidence with ν_1 (755 cm^{-1}), as a result of which the Raman fundamental and infrared combination bands of this upper component are concealed by the stronger fundamental and combination bands of ν_1. Even $\nu_1 \pm \nu_6$ (F_{2u}) becomes infrared allowed in degenerate octahedral molecules (see Table 14.1), and the assignment of bands whose frequencies are appropriate to the upper components of $\nu_2 \pm \nu_6$ ($F_{1u} + F_{2u}$) in the infrared spectrum of the vapour is therefore not unambiguous (though their presence in the spectra is in accord with the proposed coincidence since the intensity acquired by $\nu_1 \pm \nu_6$ is not known, and may be small). However, confirmatory evidence for this coincidence of ν_1 with the upper component of ν_2 comes from the unusually broad contour of $\nu_1 + \nu_3$ (see Figure 14.3) when compared with this band of the non-vibronic hexafluorides, as a result of overlapping by the upper component of $\nu_2 + \nu_3$. The first-order splitting of ν_2 of 168 cm^{-1} is in agreement with that (130–200 cm^{-1}) estimated from considerations of the electrostatic interactions between the $d(t_{2g})$ electron and the fluorine atoms.[53] The breadth of the observed ν_2 fundamental and combination bands are attributed to second-order vibronic effects which further split the lower component of ν_2 (see Figure 14.6).

The splitting of bands involving ν_5 is expected to be more complex than for ν_2 since, in addition to two components expected for the $0 \rightarrow 1$ transition, the first excited state of this vibration is sufficiently low to be well populated at room temperature, and five components (one of which coincides with a $0 \rightarrow 1$ component) are expected for the $1 \rightarrow 2$ transition.[16] There is little evidence for combinations of ν_5 in the infrared spectrum, but if it is assumed that the Raman band most obviously associated with ν_5 (246 cm^{-1})—which Figure 14.5 shows to be anomalously low in frequency—is due to the *lower* component of the $0 \rightarrow 1$ transition, the other components may be calculated. As shown in Figure 14.6, where the calculated frequencies are indicated beneath the Raman spectrum of the vapour,[54] these other components have possibly just been detected.

Fewer bands involving ν_2 and ν_5 have been detected for TcF_6, OsF_6 and RuF_6 and the assignment of vibronic components for these compounds is much less complete. For TcF_6, whose vibronic levels are similar to those of ReF_6, the upper component of $\nu_2 + \nu_3$, estimated from Figure 14.5 to be

1475 cm^{-1}, is just discernible as a shoulder at this frequency on $\nu_1 + \nu_3$ (1453 cm^{-1}).[16] For OsF_6 the pattern of vibronic levels is more complex. With the vibronic coupling constants determined from the assignments for ReF_6, the excited states of ν_2 and ν_5 have been shown to be split into a number of closely spaced levels. Bands involving ν_2 and ν_5 are therefore expected to be broad envelopes of unresolved components, as indeed is found. An unexpected broadening has also been observed, however, in the Jahn–Teller inactive fundamental $\nu_4(f_{1u})$ of OsF_6, whereby the PQR structure of the fundamental could not be resolved in the infrared spectrum.

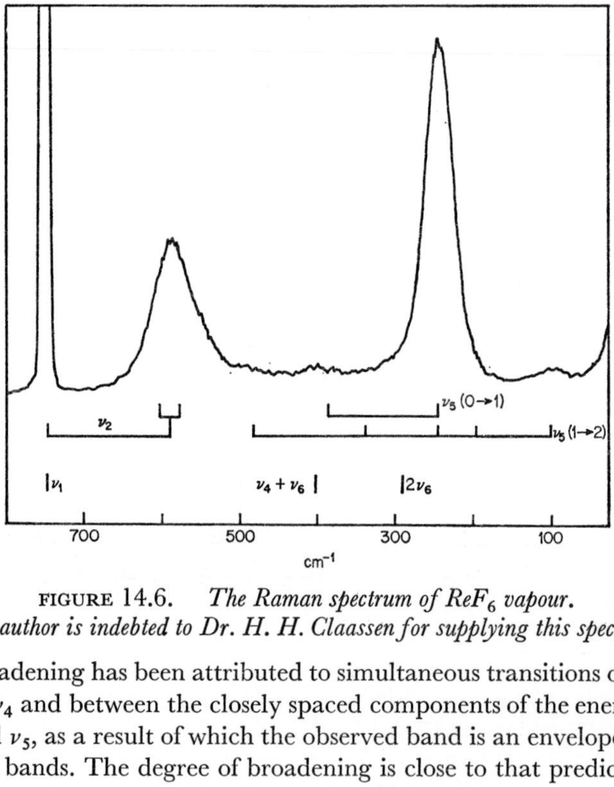

FIGURE 14.6. *The Raman spectrum of ReF_6 vapour.*
(*The author is indebted to Dr. H. H. Claassen for supplying this spectrum.*)

This broadening has been attributed to simultaneous transitions occurring both in ν_4 and between the closely spaced components of the energy levels of ν_2 and ν_5, as a result of which the observed band is an envelope of combination bands. The degree of broadening is close to that predicted from the estimated energy levels of ν_2 and ν_5.[16] For RuF_6 combinations of ν_2 and ν_5 are scarcely discernible in the infrared spectrum, and the position of ν_2 is uncertain because of overlapping impurity bands in the region of ν_2 and ν_3. However, curve resolution of the spectrum is believed[16] to show that ν_2, unlike ν_5, is unshifted from its interpolated value (Figure 14.5). This behaviour is consistent only with a substantial splitting of the $E_g + T_{2g}$ electronic state, with T_{2g} as the ground state.

No data have yet been published on the depolarization ratios of Raman bands of the degenerate hexafluorides, nor on the infrared absorption in the region of the f_{2u} fundamental ν_6. Definite conclusions on the detection of the predicted infrared activity of $\nu_1 \pm \nu_6(f_{2u})$ cannot be reached for ReF_6 because of the overlapping upper component of $\nu_2 \pm \nu_6$, as already noted, but for the other degenerate hexafluorides there appears to be no evidence for this activity.

(ii) *Hexachloro- and hexabromometallate anions.* Of the large number of known transition element hexachloro- and hexabromo- anions, relatively few have the configurations d^1 and d^2 which, by analogy with the rare metal hexafluorides, are expected to be orbitally degenerate, and the Jahn–Teller active vibrations of these few have not yet been investigated. Some interesting results have been reported, however, by Woodward and Ware[55] on the Raman intensities of the ν_2 vibrations of the non-degenerate ions $ReCl_6^{2-}$ and $ReBr_6^{2-}$ ($5d^3$), $OsCl_6^{2-}$ ($5d^4$) and UCl_6^{2-} ($5f^2$), and Table 14.4 lists the qualitative Raman intensities for these and related ions in solution. (Different relative intensities, particularly of ν_5, occur when these ions are examined in crystalline salts.) For most hexachloro-complexes so far examined ν_2 appears as a Raman band of medium intensity and for $PtCl_6^{2-}$ ($5d^6$), whose electronic structure is nearest to those of $ReCl_6^{2-}$ and $OsCl_6^{2-}$, ν_2 is in fact very strong.[56] For $ReCl_6^{2-}$ and $OsCl_6^{2-}$, however, ν_2 is too weak to have yet been detected as a fundamental in the Raman spectrum, and for $ReBr_6^{2-}$ and UCl_6^{2-} this band is also weaker than in the related ions $PtBr_6^{2-}$ and $ThCl_6^{2-}$.[55] In addition, although the frequencies ν_1 and ν_5 of $ReCl_6^{2-}$, $OsCl_6^{2-}$, $ReBr_6^{2-}$ and UCl_6^{2-} are close to those of the related ions $PtCl_6^{2-}$, $PtBr_6^{2-}$ and $ThCl_6^{2-}$, ν_2 is appreciably lower for the former ions (see Table 14.4). In this respect, therefore, the d^3, d^4 and f^2 ions resemble the orbitally degenerate hexafluorides.

Like the d^3 and d^4 hexafluorides IrF_6 and PtF_6, $ReCl_6^{2-}$, $ReBr_6^{2-}$ and $OsCl_6^{2-}$ are not expected to show normal Jahn–Teller effects, and the same is true of UCl_6^{2-}, which is known from magnetic susceptibility measurements to have an A_1 ground state.[57] However, Child[51] has shown that a weaker *pseudo*-Jahn–Teller effect can occur in molecules with low-lying excited states, which have the same spin multiplicity as the ground state, and which are mixed with the ground state by the vibrational motion. The vibrations participating in this effect are those whose species are contained in the direct product of the symmetry species Γ_o and Γ_e of the ground and excited states:

$$\text{i.e. } \Gamma_q \text{ must be contained in } \Gamma_o \times \Gamma_e \qquad (14.2)$$

For $OsCl_6^{2-}$ (A_2 ground state) the upper E and T_1 components of the ^3P ground term are estimated[51] to reduce ν_2 by ~5 cm^{-1} and ν_5 by ~10 cm^{-1}, while the E component of the $^3H_4$ ground term of UCl_6^{2-}, 1100 cm^{-1} above

TABLE 14.4. *Raman frequencies, cm^{-1}, and intensities of hexahalogeno–anions*

	$TiCl_6^{2-}$	WCl_6	$ReCl_6^{2-}$	$OsCl_6^{2-}$	$PtCl_6^{2-}$	$SnCl_6^{2-}$
ν_1	316 vs	409 vs	346(10)	346(10)	344(7)	311(10)
ν_2	—[a]	317 m	[275][b]	[274][b]	320(10)	299(2)
ν_5	171 m	209 m	159(5)	165(5)	162(6)	158(4)
ν_2/ν_1	—	0·78	0·79	0·79	0·93	0·74

	$ThCl_6^{2-}$	UCl_6^{2-}	$ReBr_6^{2-}$	$PtBr_6^{2-}$	$SnBr_6^{2-}$	PtF_6^{2-}
ν_1	294 s	299 s	213(10)	207(7)	185(10)	600(10)
ν_2	255 m	237 w	174(2)	190(10)	138(3)	576(3)
ν_5	114 m	121 m	104(8)	97(6)	95(3)	210(2)
ν_2/ν_1	0·87	0·79	0·82	0·92	0·75	0·96

[a] Obscured by cation band, intensity at most weak.
[b] Frequency calculated from $\nu_2 + \nu_3$.

the ground state,[57] possibly affects ν_2 of this ion. For $ReCl_6^{2-}$ and $ReBr_6^{2-}$, however, the ground state is essentially ^4S, and vibronically induced coupling with the nearby doublet states is spin-forbidden. Jahn–Teller effects in $ReCl_6^{2-}$ and $ReBr_6^{2-}$ can therefore arise only as a result of a small ^2P contribution to the ground state due to spin-orbit coupling, and, as for IrF_6 (d^3),[51] the effect of this is likely to be negligible.

In view of this conclusion for $ReCl_6^{2-}$ and $ReBr_6^{2-}$, it is worth noting that, as shown by studies on d^6 complexes, other factors may also influence the intensities and frequencies of ν_2. Woodward and Ware[58] have suggested that the striking differences in the relative intensities of ν_2 and ν_1 of $PdCl_6^{2-}$, $PtCl_6^{2-}$ and $PtBr_6^{2-}$ from the normal pattern shown by $SnCl_6^{2-}$ and PtF_6^{2-} may be a consequence in the former ions of $d_\pi - d_\pi$ bonding involving metal $d(t_{2g})$ donor orbitals, and which is an obvious feature distinguishing these groups of ions. The qualitative intensities recently determined[59] for the d^0 species WCl_6 and $TiCl_6^{2-}$ are similar to those of $SnCl_6^{2-}$ and are therefore in accord with this suggestion. It has alternatively been suggested by Albrecht and Taylor,[60] however, that the high intensity of ν_2 and ν_5 of $PdCl_6^{2-}$, $PtCl_6^{2-}$ and $PtBr_6^{2-}$ relative to ν_1 may be a consequence of the resonance Raman effect, which was shown by them to intensify selectively the ν_2 and ν_5 bands of d^6 ions. The normal intensity of ν_2 and ν_5 relative to ν_1 observed for PtF_6^{2-} was attributed to the much greater separation of the Raman exciting line from electronic absorption bands of this ion.

The frequencies of ν_2, when taken relative to ν_1 to compensate for differences in mass and in valency states, also vary considerably among the d^0 and d^6 species (see Table 14.4), and are much higher for $PdCl_6^{2-}$, $PtCl_6^{2-}$ and $PtBr_6^{2-}$ than for all the other transition element hexahalogeno-species. The reason for this is again not clear, but with the newly available data for WCl_6 it now appears that the anomalies in the ν_2 frequencies are in $PdCl_6^{2-}$, $PtCl_6^{2-}$ and $PtBr_6^{2-}$ rather than in $ReCl_6^{2-}$, $ReBr_6^{2-}$ and $OsCl_6^{2-}$. A full understanding of these frequency and intensity measurements must probably await data on other hexahalogeno-complexes of this series.

(iii) *Vanadium hexacarbonyl.* Vanadium hexacarbonyl (d^5) is isomorphous with its octahedral d^6 homologue $Cr(CO)_6$.[61] Its low-spin configuration is confirmed by magnetic susceptibility measurements in solution, and its orbital ground state is therefore T_{2g}.

Like $Cr(CO)_6$, $V(CO)_6$ shows a single CO stretching band (1980 cm^{-1}, f_{1u}) in the infrared spectrum of the vapour. However, unlike this band of $Cr(CO)_6$ in which the PQR structure is readily seen, the CO stretching band of $V(CO)_6$ is broad, with no visible structure.[62] Moreover, in solution, for which this band of $V(CO)_6$ is also broad, the corresponding bands of all the Group VI hexacarbonyls are very sharp. An obvious difference of $V(CO)_6$ from the Group VI hexacarbonyls is the degeneracy of the ground state, and the broadening of its CO stretching vibration has been attributed[62] to the Jahn–Teller effect although it occurs in a Jahn–Teller inactive u vibration. The broadening is possibly due to a small *static* distortion at normal temperatures which slightly splits the degeneracy of the f_{1u} ν(C—O) vibration. This implies a fairly strong Jahn–Teller effect, and indeed degeneracy in the metal $d(t_{2g})$ orbitals in metal carbonyls is likely to have a greater effect than in, for example, the rare metal hexafluorides because of the important part played by the t_{2g} electrons in metal–ligand π-bonding. Alternatively, however, the Jahn–Teller effect may be dynamic at room temperature, and, as in OsF_6, where the PQR structure of $\nu_4(f_{1u})$ could not be resolved,[16] the broad ν(C—O) band may then be an envelope of combinations of the CO stretching vibration with transitions between closely spaced vibronic levels of the e_g and t_{2g} Jahn–Teller active vibrations. Consistent with either possibility the *g*-values measured by electron spin resonance indicate a small static tetragonal distortion at 1·3 K.[63]

(iv) *Xenon hexafluoride and related species.* Although xenon hexafluoride is electronically non-degenerate, as indicated by its diamagnetism in the vapour state,[64] its vibrational properties are reminiscent of those of a degenerate molecule. Electron diffraction measurements[65] of Gavin and Bartell indicate that the molecule possibly has O_h symmetry with very large amplitudes of the f_{1u} bending vibration ν_4, but that more probably it has a slight C_{3v} distortion, and oscillates between the eight such configurations in which the lone pair orbital projects through one of the faces of the co-ordination octahedron (pseudorotation among different linear combinations of the components of the f_{1u} bending coordinate). For the former structure ν_4 is estimated to be ~60 cm^{-1} from the diffraction data and for the latter ~10 cm^{-1}, but in either case it is very low in frequency. The pseudorotating structure is supported by recent infrared measurements on XeF_6 in an argon matrix at 4 K, in which four bands in the 'Xe—F stretching' region (614, 592, 573, 527 cm^{-1}) are consistent with a molecule frozen in a minimum energy C_{3v} structure; combination bands appearing at slightly higher temperatures indicate $\nu_4(0 \rightarrow 1) = 7 \pm 5$ cm^{-1}.[66] At normal temperatures, however, the infrared spectrum of the vapour shows only two broad 'stretching' bands (612, 520 cm^{-1}).[67]

It has recently been suggested[68] that the stereochemical influence of lone pairs of electrons may be regarded as a consequence of the *pseudo*-Jahn–Teller effect. This approach identifies the normal coordinates in which instability would occur if molecules were not distorted by lone pairs, and the lower symmetry normally found is attributed to a *pseudo*-Jahn–Teller interaction between the highest occupied orbital, which contains the lone pair, and low-lying empty orbitals.

In XeF_6 the lone pair occupies an a_{1g} antibonding orbital, and the lowest empty orbital is a nearby antibonding t_{1u} orbital. From (14.2), vibrations in the t_{1u} coordinates are therefore *pseudo*-Jahn–Teller active. If the coupling with the electronic motion is strong, as is likely for the t_{1u} bending motion since this displacement effectively makes room for the lone pair, a static distortion in this coordinate will take place leading to C_{3v} (or C_{2v}) symmetry, and vibrational motion in this t_{1u} bending coordinate will occur with very low frequency, as observed. For TeF_6, of course, in which the a_{1g}^* orbital is empty, no *pseudo*-Jahn–Teller effect occurs, and the symmetry is O_h.

This theory, in addition to rationalizing the structure and vibrational properties of XeF_6, is also in accord with observations on the hexahalogeno-anions of selenium and tellurium. X-ray structural studies have so far indicated that these ions have O_h symmetry. However, accurate measurements on K_2TeBr_6 have shown that the amplitude of motion of the bromine atoms perpendicular to the Te—Br bonds is greater than that along the bonds, possibly because of abnormally large amplitudes of the bending motions of the $TeBr_6^{2-}$ anions.[69] The Raman spectra of $SeCl_6^{2-}$, $TeCl_6^{2-}$ and $TeBr_6^{2-}$ in crystalline salts show the normal pattern of three sharp bands $\nu_1(a_{1g})$, $\nu_2(e_g)$ and $\nu_5(f_{2g})$, which are characteristic of regular octahedral structures.[70] In the infrared spectra of these salts, however, the f_{1u} stretching vibrations ν_3 give rise to unusually broad bands compared to those bands of other hexahalogeno-anions, while there is doubt whether the f_{1u} bending vibrations ν_4 have yet even been observed. In the region expected for ν_4 by reference to Raman-active bending frequencies ν_5, only weak bands were observed whose frequencies were strongly dependent on the nature of the cations, and, by analogy with XeF_6, ν_4 is probably of abnormally low frequency.

E. Conclusion

Detailed vibrational studies have been carried out on very few molecules or ions with strong Jahn–Teller effects, but measurements on the $CuCl_4^{2-}$ ion and on manganese(III) *tris*-acetylacetonate in solution and in the solid state show that, as predicted, the vibrational spectra are characteristic of the expected distorted structures. However, the vibrational spectra give information only on the instantaneous structures (with lifetimes as short as 10^{-11} s), and although X-ray diffraction shows that the $CuCl_4^{2-}$ ion is statically distorted in crystalline Cs_2CuCl_4, the time-averaged structure of manganese(III) *tris*-acetylacetonate is undistorted as a result of pseudorotation of the tetragonal axis.

Studies on compounds with weak Jahn–Teller effects have been somewhat more numerous, since there are a number of very simple molecules and ions of this type which may be studied in the vapour state or in solution. For these compounds the Jahn–Teller active vibrations give rise to unusually broad bands with low peak intensities, and in some cases, for example TcF_6 or VCl_4, the fundamentals have not yet been observed. Because of the broadening the predicted splitting of the Jahn–Teller active vibrations has not been easy to observe, but there is good evidence that it has now been detected in ReF_6 and TcF_6 in the infrared combination bands of the Jahn–Teller active vibrations. There is at present no accurate published data with which to test the modified rules for Raman depolarization ratios, but the predicted progressions of overtones of the Jahn–Teller active vibrations and the appearance of normally forbidden bands in the infrared and Raman spectra of degenerate molecules have not yet been observed. However, the Jahn–Teller active vibrations of octahedral and tetrahedral molecules, and their overtones, are allowed only in Raman scattering, and because of the colour of most of these degenerate compounds, their Raman spectra have in the past been difficult to record. With the recent substantial improvements in Raman spectrometers such effects may now be detected.

References

1. Jahn, H. A., and Teller, E., *Proc. Roy. Soc.*, 1937, **A161**, 220.
2. Liehr, A. D., *Progr. Inorg. Chem.*, 1962, **3**, 281; 1962, **4**, 455; Liehr, A. D., *J. Phys. Chem.*, 1963, **67**, 389; Longuet-Higgins, H. C., *Adv. Spectroscopy*, 1961, **2**, 429.

3. Herzberg, G., *Discuss. Faraday Soc.*, 1963, **35**, 7.
4. de Groot, M. S., Hesselmann, I. A. M., and van der Waals, J. H., *Mol. Phys.*, 1966, **10**, 241; Nieman, G. C., *J. Chem. Phys.*, 1969, **50**, 1660, 1674, and references therein.
5. Sturge, M. D., *Solid State Phys*,. 1967, **20**, 91.
6. Orgel, L. E., and Dunitz, J. D., *Nature*, 1957, **179**, 462.
7. Morosin, B., and Lingafelter, E. C., *J. Phys. Chem.*, 1961, **65**, 50; Morosin, B., and Lingafelter, E. C., *Acta Cryst.*, 1960, **13**, 807.
8. Pauling, P., *Inorg. Chem.*, 1966, **5**, 1498.
9. Stucky, G. D., Folkers, J. B., and Kistenmacher, T. J., *Acta Cryst.*, 1967, **23**, 1064.
10. Liehr, A. D., and Ballhausen, C. J., *Ann. Phys.* (*New York*), 1959, **6**, 134.
11. Dyer, L. D., Borie, B. S., and Smith, G. P., *J. Amer. Chem. Soc.*, 1954, **76**, 1499.
12. Bateman, L. R., Blount, J. F., and Dahl, L. F., *J. Amer. Chem. Soc.*, 1966, **88**, 1082.
13. Bleaney, B., and Bowers, K. D., *Proc. Phys. Soc.*, 1952, **A65**, 667.
14. Moffitt, W., Goodman, G. L., Fred, M., and Weinstock, B., *Mol. Phys.*, 1959, **2**, 109.
15. Morino, Y., and Uehara, H., *J. Chem. Phys.*, 1966, **45**, 4543.
16. Weinstock, B., and Goodman, G. L., *Adv. Chem. Phys.*, 1965, **9**, 169.
17. Thorson, W. R., *J. Chem. Phys.*, 1958, **29**, 938.
18. Longuet-Higgins, H. C., Öpik, U., Pryce, M. H. L., and Sack, R. A., *Proc. Roy. Soc.*, 1958, **A244**, 1; Struck, C. W., and Herzfeld, F., *J. Chem. Phys.*, 1966, **44**, 464.
19. Uehara, H., *J. Chem. Phys.*, 1966, **45**, 4536.
20. Child, M. S., and Longuet-Higgins, H. C., *Phil. Trans.*, 1961, **254A**, 259.
21. Caner, M., and Englman, R., *J. Chem. Phys.*, 1966, **44**, 4054.
22. Moffitt, W., and Thorson, W. R., *Phys. Rev.*, 1957, **108**, 1251.
23. Child, M. S., *Phil. Trans.*, 1962, **255A**, 31.
24. Willett, R. D., *J. Chem. Phys.*, 1964, **41**, 2243.
25. Ferguson, J., *J. Chem. Phys.*, 1964, **40**, 3406; Sharnoff, M., and Reimann, C. W., *ibid.*, 1965, **43**, 2993.
26. Forster, D., *Chem. Comm.*, 1967, 113.
27. Adams, D. M., and Lock, P. J., *J. Chem. Soc.* (*A*), 1967, 620; also references 44 and 45.
28. Beattie, I. R., Gilson, T. R., and Ozin, G. A., *J. Chem. Soc.* (*A*), 1969, 534.
29. Morris, D. F. C., Short, E. L., and Waters, D. N., *J. Inorg. Nuclear Chem.*, 1963, **25**, 975.
30. Forman, A., and Orgel, L. E., *Mol. Phys.*, 1959, **2**, 362.
31. Nakamoto, K., McCarthy, P. J., Ruby, A., and Martell, A. E., *J. Amer. Chem. Soc.*, 1961, **83**, 1066; Gillard, R. D., Silver, H. G., and Wood, J. L., *Spectrochim. Acta*, 1964, **20**, 63.

32. Davis, T. S., Fackler, J. P., and Weeks, M. J., *Inorg. Chem.*, 1968, **7**, 1994.
33. Palmer, R. A., and Piper, T. S., *Inorg. Chem.*, 1966, **5**, 864, and references therein.
34. Kokoska, G. F., Reimann, C. W., Allen, H. C., and Gordon, G., *Inorg. Chem.*, 1967, **6**, 1657.
35. Allen, H. C., Kokoska, G. F., and Inskeep, R. G., *J. Amer. Chem. Soc.*, 1964, **86**, 1023.
36. Inskeep, R. G., *J. Inorg. Nuclear Chem.*, 1962, **24**, 763.
37. Isaacs, N. W., and Kennard, C. H. L., *J. Chem. Soc.* (*A*), 1969, 386; Bertrand, J. A., and Carpenter, D. A., *Inorg. Chem.*, 1966, **5**, 514; Elliott, H., and Hathaway, B. J., *ibid.*, 1966, **5**, 885; Elliott, H., Hathaway, B. J., and Slade, R. C., *ibid.*, p. 669; Hathaway, B. J., Dudley, R. J., and Nicholls, P., *J. Chem. Soc.* (*A*), 1969, 1845.
38. Morino, Y., and Uehara, H., *J. Chem. Phys.*, 1966, **45**, 4543.
39. Ballhausen, C. J., and de Heer, J., *J. Chem. Phys.*, 1965, **43**, 4304; Lohr, L. L., Jr., and Lipscomb, W. N., *Inorg. Chem.*, 1963, **2**, 911.
40. Dove, M. F. A., Creighton, J. A., and Woodward, L. A., *Spectrochim. Acta*, 1962, **18**, 267.
41. (a) Grubb, E. L., and Belford, R. L., *J. Chem. Phys.*, 1963, **39**, 244; (b) Creighton, J. A., Green, J. H. S., and Kynaston, W., *J. Chem. Soc.* (*A*), 1966, 208.
42. Gill, N. S., *J. Chem. Soc.*, 1961, 3512.
43. Gibb, T. C., and Greenwood, N. N., *J. Chem. Soc.*, 1965, 6989.
44. Avery, J. S., Burbridge, C. D., and Goodgame, D. M. L., *Spectrochim. Acta*, 1968, **24A**, 1721.
45. Sabatini, A., and Sacconi, L., *J. Amer. Chem. Soc.*, 1964, **86**, 17.
46. Ewens, R. V. G., and Lister, M. W., *Trans. Faraday Soc.*, 1938, **34**, 1358.
47. Beattie, I. R., and Ozin, G. A., *J. Chem. Soc.* (*A*), 1969, 1691.
48. Bader, R. F. W., and Huang, K. P., *J. Chem. Phys.*, 1965, **43**, 3760.
49. Kepert, D. L., and Nyholm, R. S., *J. Chem. Soc.*, 1965, 2871; Werder, R. D., Frey, R. A., and Günthard, Hs. H., *J. Chem. Phys.*, 1967, **47**, 4159.
50. Weinstock, B., and Claassen, H. H., *J. Chem. Phys.*, 1959, **31**, 262.
51. Child, M. S., *Mol. Phys.*, 1960, **3**, 605.
52. Weinstock, B., Claassen, H. H., and Chernick, C. L., *J. Chem. Phys.*, 1963, **38**, 1470.
53. Child, M. S., and Roach, A. C., *Mol. Phys.*, 1965, **9**, 281.
54. Claassen, H. H., Malm, J. G., and Selig, H., *J. Chem. Phys.*, 1962, **36**, 2890.
55. Woodward, L. A., and Ware, M. J., *Spectrochim. Acta*, 1964, **20**, 711; 1968, **24A**, 921.
56. Woodward, L. A., and Creighton, J. A., *Spectrochim. Acta*, 1961, **17**, 594.
57. Hutchison, C., and Candela, G., *J. Chem. Phys.*, 1957, **27**, 707; Satten, R. A., Schreiber, C. L., and Wong, E. Y., *ibid.*, 1965, **42**, 162.

58. Woodward, L. A., and Ware, M. J., *Spectrochim. Acta*, 1963, **19**, 775.
59. Creighton, J. A., *Chem. Comm.*, 1969, 163.
60. Albrecht, A. C., and Taylor, K. A., see reference 83 of Hester, R. E., 'Raman Spectroscopy', ed. H. A. Szymanski, Plenum Press, New York, 1967, p. 101.
61. Calderazzo, F., Cini, R., and Ercoli, R., *Chem. and Ind.*, 1960, 934.
62. Haas, H., and Sheline, R. K., *J. Amer. Chem. Soc.*, 1966, **88**, 3219.
63. Pratt, D. W., and Myers, R. J., *J. Amer. Chem. Soc.*, 1967, **89**, 6470.
64. Code, R. F., Falconer, W. E., Klemperer, W., and Ozier, I., *J. Chem. Phys.*, 1967, **47**, 4955.
65. Gavin, R. M., and Bartell, L. S., *J. Chem. Phys.*, 1968, **48**, 2460; Bartell, L. S., and Gavin, R. M., *ibid.*, p. 2466.
66. Kim, H., *Bull. Amer. Phys. Soc.*, 1968, **13**, 425.
67. Jortner, J., and Rice, S. A., *Chem. Rev.*, 1965, **65**, 199.
68. Bartell, L. S., *J. Chem. Educ.*, 1968, **45**, 754; Pearson, R. G., *J. Amer. Chem. Soc.*, 1969, **91**, 4947.
69. Brown, I. D., *Canad. J. Chem.*, 1964, **42**, 2758.
70. Beattie, I. R., and Chudzynska, H., *J. Chem. Soc.* (*A*), 1967, 984; Barrowcliffe, T., Beattie, I. R., Day, P., and Livingston, K., *ibid.*, p. 1810; Creighton, J. A., and Green, J. H. S., *ibid.*, 1968, 808.

CHAPTER 15

Spectroscopic Properties of the Diatomic Oxides of the Transition Elements

R. F. BARROW

A. Introduction

The gaseous monoxides of the transition metals present interesting problems in a number of related fields. The group contains some of the most stable diatomic molecules known, so that they are often of importance at high temperatures, and it is indeed through mass-spectrometric studies of high temperature equilibria that their energies of dissociation have most commonly been measured. Their stability and open-shell electron configurations lead to the appearance of allowed electronic transitions in the visible region of the spectrum and some of these are observed in the spectra of not too hot stars. The electronic states involved are often of high multiplicity, and analyses of the transitions offer challenges to conventional high resolution spectroscopy, while the states themselves pose theoretical problems in the formulation of their energy levels. The lighter members ScO, TiO, VO are yet sufficiently simple for *ab initio* calculations of their electronic structures to be made. Important results have recently followed studies in matrix-isolation spectroscopy, and this technique has supplemented information from gas-phase spectroscopy about nuclear magnetic hyperfine structure which proves to be important in the ground states of some of these molecules.

A more general account[1] of the spectroscopy of diatomic transition element molecules contains a list of references up to the end of 1965 and includes some later work. Theoretical work on these molecules was discussed very completely by Carlson and Claydon[2] in the same volume. For the most part, therefore, only later references supplementing those given in references 1 and 2 will be given here. Hyperfine structures in these molecules have not been reviewed before, and for this reason some prominence is given to the subject in this chapter.

B. Methods of study

What are in principle the simplest experiments to give information about the fundamental property of these molecules, their stability, are the studies of the high-temperature equilibria

$$MO_g = M_g + O_g$$

or, using a reference molecule,

$$M_1O_g + M_2 = M_2O_g + M_1$$

In either case, the partial pressures are estimated from measurements of ion current in a mass spectrometer:[3] in the second type of experiment, M_1O, say, is taken as a reference molecule, and errors tend to cancel. The most precise (Third Law) treatment of the results demands knowledge of the thermodynamically important states or sub-states of the gaseous oxides, so that the full power of the method is developed only when reliable calculations of the high-temperature entropies can be made.

Conventional gas-phase spectroscopic studies, in emission or absorption, have limitations imposed by the fact that the observed transitions are often between states of high multiplicity, and that ordinary methods suitable for these molecules lead naturally to the development of high rotational temperatures. What in fact is required is the excitation at low rotational temperatures of molecules which appear *in equilibrium* only at high temperatures. Flash photolysis seems promising here, and a number of oxide spectra have been observed in this way,[4] but hitherto only at low resolution.

The electronic spectra may be drastically simplified by observation of the matrix isolated molecule at low temperature. Absorption spectra, especially with the comparison of $M^{16}O$ and $M^{18}O$, may indicate the positions of 0–0 transitions and give information about upper-state vibration frequencies; ground-state vibration frequencies may be observed in the infrared.[5] Fluorescence studies may give further information about ground-state frequencies and, for example, absorption in a singlet system may be followed either by triplet–triplet or by triplet–singlet fluorescence.[6] Electron spin resonance observations on matrix-isolated molecules have given important information about the characters of the ground states of odd-electron molecules, in spite of the complications arising from the random orientation of molecules in the lattice.[7]

Matrix Hartree–Fock calculations have so far been reported[2] for ScO, TiO and VO: these have proved both stimulating and reliable. The possibility of some extension to heavier molecules seems to be dictated only by time and inclination: it is to be hoped that more of these calculations will be made.

C. Properties of the ground states

The values known at present for the internuclear distances, force constants and dissociation energies for the transition metal oxides are collected together in Table 15.1. The most stable molecules are the oxides of the metals in Groups IV and V. Force constants and internuclear distances run more or less parallel, and in the first series VO has the biggest force constant and shortest internuclear distance: in respect of dissociation energy the maximum may lie at TiO. Stability often increases down the groups, and TaO and ThO have particularly high energies of dissociation. The oxides of groups VII and VIII are markedly less stable, and in Group I, although gaseous CuO and AgO are known, AuO is not. Perhaps it is significant here that although gold is known to form stable diatomic molecules with something like a third of the elements in the Periodic Table, the most stable seem to be those with rather electropositive elements like Al; thus AuO, AuS* and AuF are not known, although AuSe, AuTe and AuCl are, so that the electronegative character of gold may be important in its stable diatomic compounds.

The two ionic species, NbO^+ and TaO^+ seem both to be somewhat more stable than the respective parent neutral molecules (however, information about both species is scanty, and it is not impossible that the spectrum assigned to NbO^+ arises in fact from NbN).

Although many spectra have been observed, very little certain analysis has been done for the oxides of the lanthanides, and no rotational analyses have been performed: for PrO and GdO what seem to be reliable vibrational analyses lead to ground-state force constants which are, as might be expected, equal for the two molecules.

In the case of the lanthanides, an interesting parallel has been drawn[8] between the behaviour of the dissociation energies and the heats of sublimation of the metals, which follow the same general pattern from praseodymium to ytterbium: see Figure 15.1. For example, the heat of sublimation of samarium is low in comparison with that of gadolinium because, whereas in gadolinium $4f^7\,5d\,6s^2$ solid goes to $4f^7\,5d\,6s^2$ gas, in samarium $4f^5\,5d\,6s^2$ solid vaporizes to gaseous atoms of a different configuration, namely $4f^6\,6s^2$. Ames, Walsh and White 'correct' the observed dissociation energies

* Recent mass-spectrometric work has shown that AuS is quite stable, with $D_0^0 = 98{\cdot}7 \pm 6$ kcal mol^{-1} (Gingerich, K. A., *J. Chem. Soc., D*, 1970, 580).

of these oxides by adding on the energy separations $4f^{n-1}\ 5d - 4f^{n}$ in M^{2+}, to obtain a regular series of slowly decreasing values of D_0^0 from lanthanum to lutetium. Similar behaviour is shown by the monosulphides of the rare-earths.[9]

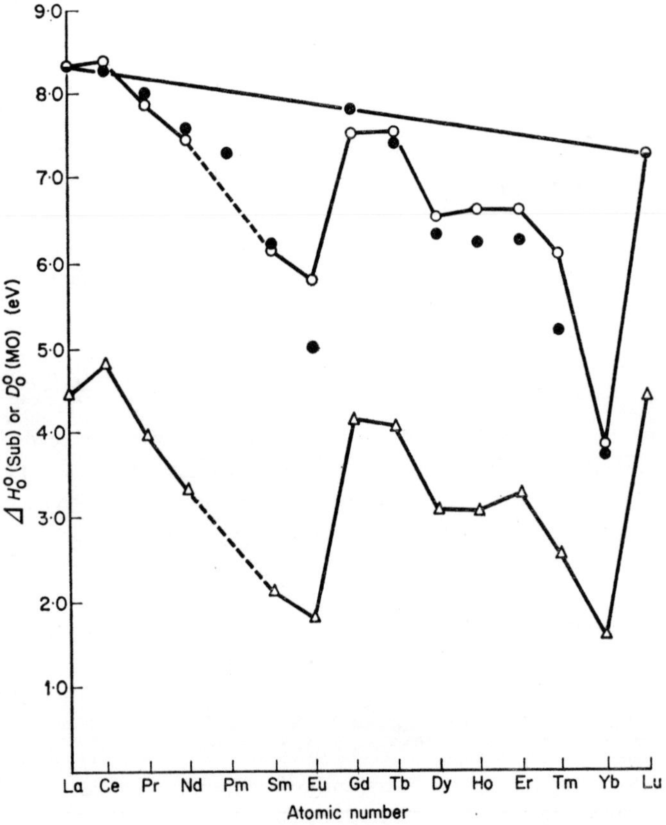

FIGURE 15.1. *Variation with atomic number of the dissociation energy of the gaseous monoxides of the lanthanides and of the heat of sublimation of the metals. The calculated values of D_0^0 are obtained by subtracting the energy separations* $(4f^{n-1}5d - 4f^{n})$ *in* M^{2+}.

(*Reproduced with permission from Ames, L. L., Walsh, P. N., and White, D.,* J. Phys. Chem., *1967*, **71**, *2708.*)

○ = observed D_0^0
● = calculated D_0^0
△ = observed ΔH_0^0

Considerations of probable electron configurations are often helpful, both in technical spectroscopy and also in providing a basis for predicting the characters of ground states for use in the statistical calculation of thermodynamic functions. In all these oxides, it seems to be a reasonable

TABLE 15.1. *Properties of the ground states of the oxides*

	Sc	Ti	V	Cr	Mn	Fe	Co	Ni	Cu
Z	21	22	23	24	25	26	27	28	29
r, Å	1·668	1·620	1·589	1·627	—	1·626[d]	—	—	[1·728][e]
k_e, mdyn Å^{-1}	6·560	7·183	7·351	5·819	5·142	5·680	5·35	5·04	2·96
D_0^0, kcal mol^{-1}	160·4 ± 2·0[a]	167·4 ± 2·3[b] 157·8 ± 1·0	148·8 ± 2·5[a]	103·3 ± 5[c]	85·4 ± 4	90·3 ± 3	86·4 ± 3	86·5 ± 5	62·7 ± 3
	Y	Zr	Nb	Mo	Tc	Ru	Rh	Pd	Ag
Z	39	40	41	42	43	44	45	46	47
r, Å	1·790	1·711	1·691	—	—	1·7	—	—	2·000
k_e, mdyn Å^{-1}	5·803	7·650	7·86	—	—	6·32	—	—	1·971
D_0^0, kcal mol^{-1}	170·4 ± 2·5[a]	181 ± 5	180 ± 2·5	116 ± 15	—	123 ± 7[f]	97 ± 7[f]	67 ± 7[f]	—
	La	Ce							
Z	57	58							
r, Å	1·825	1·816[h]							
k_e, mdyn Å^{-1}	5·565	—							
D_0^0, kcal mol^{-1}	190·3 ± 2·5[a]	193·0 ± 3							
	Lu	Hf	Ta	W	Re	Os	Ir	Pt	Au
Z	71	72	73	74	75	76	77	78	79
r, Å		[1·724]	1·687	—	—	—	—	1·727[n]	—
k_e, mdyn Å^{-1}	6·115	8·23	9·179	9·79[j]	—	—	—	6·307	—
D_0^0, kcal mol^{-1}	166·7 ± 3[g]	182·6 ± 6	197 ± 12	160 ± 10	—	—	84 ± 7[k]	88 ± 6[f]	—

TABLE 15.1.—*continued*

Z	Th 90	Pa 91	U 92						
r, Å	1·840	—	—						
k_e, mdyn Å^{-1}	9·179	—	—						
D_0^0, kcal mol^{-1}	198 ± 5[l]	—	181·4 ± 2[l]						
Z	Pr 59	Nd 60	Sm 62	Eu 63	Gd 64	Tb 65	Dy 66	Ho 67	Er 68
r, Å	—	—	—	—	—	—	—	—	—
k_e, mdyn Å^{-1}	5·912	—	—	—	5·913	—	—	—	—
D_0^0, kcal mol^{-1}	181·7[g]	172·5[g]	142·1[g]	133·8[g]	173·0[g]	173·4[g]	150·4[g]	152·7[g]	151·7[g]
Z	Tm 69	Yb 70							
r, Å	—	—							
k_e, mdyn Å^{-1}	—	—							
D_0^0, kcal mol^{-1}	139·3[g]	⩽88·3[g]							
	NbO^+ 41	TaO^+ 73							
r, Å	[1·570][m]	1·667							
k_e, mdyn Å^{-1}	—	9·639[i]							
D_0^0, kcal mol^{-1}	—	—							

Values of r in parentheses are of r_0, rather than r_e.

[a] Coppens, P., Smoes, S., and Drowart, J., *Trans. Faraday Soc.*, 1967, **63**, 2140.
[b] Drowart, J., Coppens, P., and Smoes, S., *J. Chem. Phys.*, 1969, **50**, 1046.
[c] As reference (1), recalculated by P. Goldfinger.
[d] Barrow, R. F., and Senior, M., *Nature*, 1969, **223**, 1359.
[e] Lagerqvist, A., and Uhler, U., *Z. Naturforsch.*, 1967, **22b**, 551.
[f] Norman, J. H., Staley, H. G., and Bell, W. E., *Advances in Chemistry Series*, 1968, **72**, 101, with estimated corrections to 0 K by P. Goldfinger.
[g] Ames, L. L., Walsh, P. N., and White, D., *J. Phys. Chem.*, 1967, **71**, 2707.
[h] Clements, R. M., D.Phil., Thesis, University of Oxford, 1969.
[i] Cheetham, C. J., and Barrow, R. F., *Trans. Faraday Soc.*, 1967, **63**, 1835.
[j] Weltner, W., and McLeod, D., *J. Mol. Spectroscopy*, 1965, **17**, 276.
[k] Drowart, J., and Goldfinger, P., *Angew. Chem., Internat. Edn.*, 1967, **6**, 581, based on Norman, J. H., Staley, H. G., and Bell, W. E., *J. Chem. Phys.*, 1965, **42**, 1123.
[l] Goldfinger, P., 1969, personal communication.
[m] Dunn, T. M., and Rao, K. M., *Nature*, 1969, **222**, 266. The bands described may arise from NbN rather than NbO^+.
[n] Nilsson, C., Scullman, R., and Mehendalé, N., *J. Mol. Spectroscopy*, 1970, **35**, 177.

assumption to begin by considering them as $M^{2+}O^{2-}$, although the charge separation is probably by no means complete. Thus the lowest configuration in ScO may be written simply as (σ_{4s}) which gives the one state $^2\Sigma^+$ which is the ground state. YO and LaO are similar.[5] The ground state of LuO may also be expected to be $^2\Sigma^+$, though this is not yet known.

In TiO, the ground state is $(\delta_{3d})(\sigma_{4s})\ ^3\Delta$, but neither the $^1\Delta$ state of this configuration, nor the $^1\Sigma^+$ state from $(\sigma_{4s})^2$ is very highly excited (see Fig. 15.2), and in ZrO and HfO the ground states are $\sigma^2\ ^1\Sigma^+$. In the next group, the ground state of VO[7,10] is certainly $\delta^2_{3d}\ \sigma_{4s}\ ^4\Sigma^-$, and that of NbO[11] seems also to be $^4\Sigma^-$: however, in TaO the ground state is $\delta\sigma^2\ ^2\Delta$. So, in Groups IV and V, as we go from top to bottom, σ_{ns} increases in stability in comparison with $\delta_{(n-1)d}$. Very little is known about the oxides of later groups. In CrO a transition in which the lower state is probably the ground state has been analyzed as $^5\Pi$—$^5\Pi$: this would fit with a ground state of configuration $\delta^2\sigma\pi$ and the transition either $\sigma 4p \leftrightarrow \sigma 4s$ or $\pi 4p \leftrightarrow \pi 3d$. (In the absence of detailed calculations, these labels are used here very loosely: for example what is indeed largely σ_{4s} in ScO becomes largely σ_{3d} in ZnO.)

Bands which involve the ground state of MnO are known, but their rotational structure has not yet been analyzed. Their appearance at low resolution suggests a high multiplicity Σ—Σ transition. If so, the ground state of MnO may be $\delta^2\sigma\pi^2\ ^6\Sigma^+$. Some of the structure of a band system of FeO in the orange region has been analyzed,[12] but at longer wavelengths each band shows a great deal of additional structure: again this looks like a high-multiplicity Σ—Σ transition, and the ground state of FeO may be $^5\Sigma$ or $^7\Sigma$. (In the isoelectronic molecule MnF, there is quite good evidence that the ground state is $^7\Sigma$.) Analysis of the CuO bands is incomplete, but it seems likely that the ground state is $^2\Pi$ (as in AgO): however, if this is so, the configuration cannot correspond to a simple hole in the $\pi 3d$ electron shell, for the spin–orbit coupling constant increases markedly in the series CuO—CuS—CuSe. Bands of CoO and of NiO are known, but no analyses have been made: probably the ground state of CoO is a quartet and that of NiO a triplet.

These considerations are summarized in Table 15.2. One point which emerges clearly from this discussion is that for Groups IV–VIII one can at present do no better than to give *lower limits* for the statistical weights to be used in calculations of the thermodynamic properties of these molecules.

D. Electronic transitions

Definite conclusions[13] can be drawn about the nature of the two lowest known excited states in the molecules ScO, YO and LaO, respectively A $^2\Pi$ and B $^2\Sigma^+$. There is a mutual perturbation between these two states

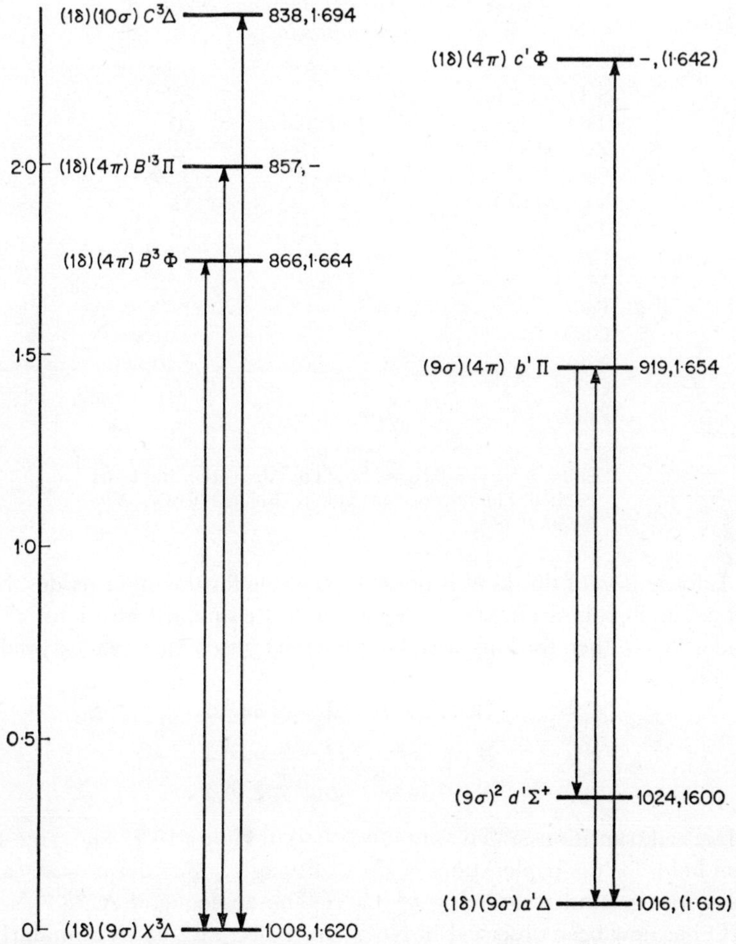

FIGURE 15.2. *States and observed transitions in TiO. Vibration frequencies,* cm^{-1}*, and internuclear distances, Å, are also given: values in parentheses are of* r_0*; other values are of* r_e*. The designations of the orbitals follow Carlson and Claydon*[2]*:* 9σ *is largely* 4s, *Ti;* 10σ *is largely* 4p, *Ti;* 4π *is largely* 4p, *Ti;* 1δ *is largely* 3d, *Ti.*

and it is this which is largely responsible for the observed Λ-doubling in A $^2\Pi_{1/2}$ and the spin splitting in B $^2\Sigma^+$. Moreover the splitting constants are very close to what is calculated on the assumption of the case of 'pure precession' for a p electron. Thus the evidence is that these states are largely

$$\sigma^* \text{np} \quad {}^2\Sigma^+$$
$$\pi^* \text{np} \quad {}^2\Pi$$

TABLE 15.2. *Ground states and electron configurations*

ScO, YO, LaO	σns	$^2\Sigma^+$
TiO	$(\delta 3d)\ (\sigma 4s)$	$^3\Delta$
ZrO, HfO, ThO	σ^2	$^1\Sigma^+$
CeO	? $\sigma\phi$	? $^3\Phi$
VO, NbO	$\delta^2\,\sigma$	$^4\Sigma^-$
TaO	$\delta\sigma^2$	$^2\Delta$
CrO	? $\delta^2\,\sigma\pi$	? $^5\Pi$
MnO	? $\delta^2\,\sigma\pi^2$	? $^6\Sigma^+$
FeO	—	? $^5\Sigma$, ? $^7\Sigma$
CoO	—	? quartet
NiO	—	? triplet
CuO[a]	$d^{10}\,\pi^{-1}$	$^2\Pi_i$
ZnO	d^{10}	$^1\Sigma^+$

[a] Shirk, J. S., and Bass, A. M., 'The Absorption and Laser Excited Fluorescence of Matrix Isolated CuO', *NBS Report 10 053.*

Information of this kind is not yet available for the other oxides, but if, for example, it is assumed that in TiO there are similar transitions, $\pi^* \leftrightarrow \sigma$ and $\sigma^* \leftrightarrow \sigma$, then the long wavelength triplet systems are exactly predicted as follows:

$$\mathrm{C}\ldots\delta\sigma^*,\quad {}^3\Delta - \mathrm{X}\,\delta\sigma\,{}^3\Delta$$

$$\mathrm{B}\ldots\delta\pi^*,\quad {}^3\Pi - \mathrm{X}\,\delta\sigma\,{}^3\Delta$$

$$\mathrm{A}\ldots\delta\pi^*,\quad {}^3\Phi - \mathrm{X}\,\delta\sigma\,{}^3\Delta$$

Observed transitions in TiO are illustrated in Figure 15.2. The same situation holds for the triplet states of ZrO, although $\ldots\delta\sigma\,{}^3\Delta$ is now an excited state, the ground state being $\sigma^2\ {}^1\Sigma^+$.† The analogue of A $^3\Phi - \mathrm{X}^3\Delta$ in ZrO has now been observed in NbO^+ (or NbN) at 2·05 eV. Similarly the two analysed systems of VO may be written

$$\mathrm{C}\ldots\delta^2\,\sigma^*\quad {}^4\Sigma^- - \mathrm{X}^4\Sigma^-$$

$$\mathrm{B}\ldots\delta^2\,\pi^*\quad {}^4\Pi\ - \mathrm{X}^4\Sigma^-$$

Evidence for other molecules is incomplete and it is quite uncertain whether this simple situation holds for oxides of higher groups. The few values available of excitation energies are summarized in Table 15.3. It seems that the $\sigma^* - \pi^*$ separation is rather constant at ~0·5 – 0·6 eV, and it is interesting that the $\delta\pi^*\,{}^3\Pi - {}^3\Phi$ separations in TiO and ZrO are nearly the same and equal to about 0·25 eV.

† For a recent discussion of the low-lying states of TiO and ZrO, see Brewer, L., and Green, D. W., *High Temperature Science*, 1969, **1**, 26.

TABLE 15.3. *Excitation energies, eV*

Molecule	$\sigma^* \leftarrow \sigma$	$\pi^* \leftarrow \sigma$	$\sigma^* - \pi^*$
ScO	2·550	2·045	0·505
YO	2·572	2·047	0·525
LaO	2·212	1·620	0·592
TiO	2·399	2·002 $^3\Pi$; 1·749 $^3\Phi$	0·523
ZrO	2·672	2·200 $^3\Pi$; 1·952 $^3\Phi$	0·596
VO	2·160	1·562	0·598
NbO	2·643	?1·86	?0·78
CrO	(2·048)	?1·38	?0·67
MnO	(2·219)	—	—
FeO	(2·442)	—	—

Singlet states of TiO provide another estimate of the separation $\pi^* - \sigma$, for $\delta\pi^*\,{}^1\Phi$ is found to lie at 2·212 eV above the lowest singlet state $\delta\sigma\,{}^1\Delta$. The singlet–triplet separation for $\delta\sigma\,{}^1\Delta - {}^3\Delta$ is only about 0·07 eV, but for $\delta\pi^*\,{}^1\Phi - {}^3\Phi$ it is about 0·5 eV.

However, even in TiO these simple considerations seem to fail in respect of the state b $^1\Pi$. This is assigned the configuration $\sigma\pi^*$ or (9σ) (4π) rather than (1δ) (4π) for b $^1\Pi$ is observed in transition to $(9\sigma)^2\,\mathrm{d}\,{}^1\Sigma^+$, and (1δ) (4π) would require a presumably weak two-electron jump. There are then two states (of the configurations considered here) which have not yet been observed: (9σ) (4π) $^3\Pi$, at about 1 eV, and (1δ) (4π) $^1\Pi$ at 2·5 – 3 eV.

If simple generalizations do not seem to suffice for a preliminary discussion of the states of a comparatively simple molecule like TiO, it may be expected that the states of more complex molecules like HfO, CeO, ThO will defy analysis, at least at present. In the heavier molecules, states will often tend towards case-c, where, for example, states like $^3\Pi_1$, the $\Omega = 1$ component of $^3\Sigma^-$ and $^1\Pi$, all look alike. Quite a lot is now known about the states of the molecules CeO, ThO and TaO, and the ground state of TaO is established as $^2\Delta_r$, but the observed states of the other molecules are not yet understood. This is perhaps not surprising in a situation in which multiplicity has little meaning and where multiplet separations may be of the same order as singlet–triplet or doublet–quartet intervals or as the separations between different orbital energies. The absorption spectrum of CeO, for example, is dominated by 0–0 bands, which even at temperatures as low as 2000 K arise from at least eight lower states. Five of these may be the states $\ldots\phi\sigma\,{}^3\Phi$, $^1\Phi$, $\ldots\sigma^2\,{}^1\Sigma^+$, but it seems that at least three more states or sub-states lie low enough to give rise to absorption bands.

E. Hyperfine structure

In a few cases, notably ScO, LaO, VO, NbO and NbO^+ (or NbN), hyperfine structure effects arise which are large enough to appear in optical spectra at the resolutions commonly employed for the analysis of rotational

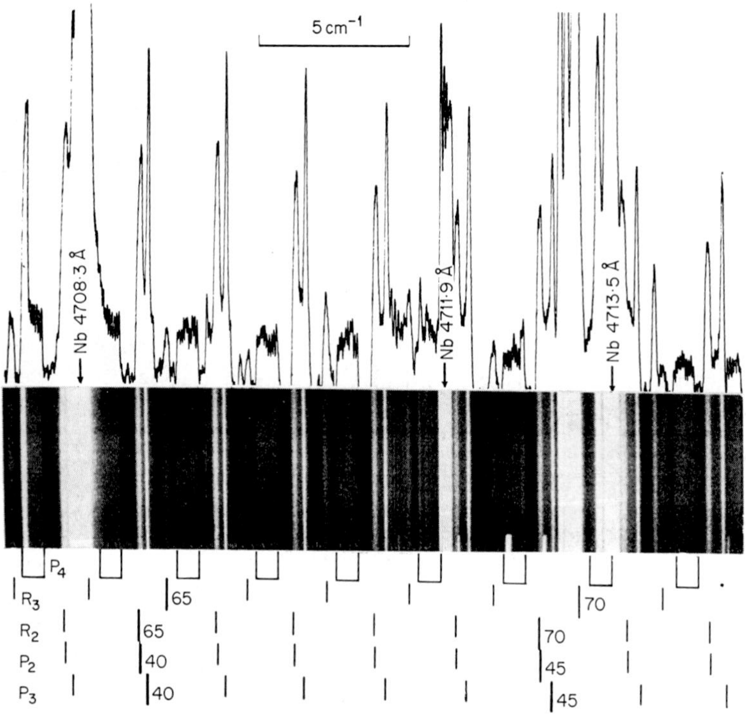

FIGURE 15.3. *A small part of the 0–0 band (4688Å) of the blue system of NbO. The spectrum is contaminated with some atomic Nb lines in addition to those marked, but it shows the completely resolved hyperfine structure clearly. Only one of the four possible widely spaced branches is shown in this section of the spectrum, although one member of the* P_1 *branch can be seen at the extreme right-hand edge of the trace and photograph. The absolute numbering, as given (in terms of* J-$\frac{1}{2}$*), is not yet conclusive but is probably correct for the narrowly spaced branches. The first member of the* P_4 *branch shown in the figure is probably* P_4 (39). *(The auther is greatly indebted to Dr. T. M. Dunn and Dr. K. M. Rao for this spectrogram.)*

structure in electronic band systems. Hyperfine structure has also been observed at far higher resolution in the electron spin resonance spectra of trapped molecules. In all the cases so far studied the dominant interaction is the magnetic one, and the large effects arise in molecular ground states where there is an unpaired σ electron for which the contact term depending upon $\Psi^2(0)$ can be large. Such hyperfine structure effects are important because they throw direct light upon the nature of some of the orbitals involved in these molecules.

Three coupling cases[14] may be expected to be of practical importance:

(i) a_β, in which the electronic angular momenta, orbital and spin, are tightly coupled to the molecular axis, but the nuclear spin is coupled to $\mathbf{J}$;

(ii) $b_{\beta J}$, in which the electron spin couples to the total angular momentum, apart from electron or nuclear spin,

$$\mathbf{J} = \mathbf{N} + \mathbf{S}$$

and $\mathbf{I}$ couples weakly to $\mathbf{J}$;

(iii) $b_{\beta S}$, in which the nuclear spin couples to the electron spin,

$$\mathbf{G} = \mathbf{I} + \mathbf{S}$$

and $\mathbf{G}$ couples weakly with $\mathbf{N}$,

$$\mathbf{F} = \mathbf{N} + \mathbf{G}$$

The resulting energies, W_{hfs}, are:

$$\text{case } a_\beta \qquad [a\Lambda + (b+c)\Sigma]\frac{\Omega}{J(J+1)}\mathbf{I}.\mathbf{J}$$

$$b_{\beta J} \qquad b\left[\frac{J(J+1) - N(N+1) + S(S+1)}{2J(J+1)}\right]\mathbf{I}.\mathbf{J}$$

$$b_{\beta S} \qquad \frac{b}{2}[G(G+1) - I(I+1) - S(S+1)]$$

where

$$\mathbf{I}.\mathbf{J} = \tfrac{1}{2}[F(F+1) - J(J+1) - I(I+1)]$$

(The case of a $^2\Sigma$ state in mixed coupling, $b_{\beta S} - b_{\beta J}$, has been treated by Radford[15]: Atkins[16] has discussed the situation in $\Omega = \frac{1}{2}(^2\Pi_{1/2})$ case-c states.) The coupling coefficients a, b and c are given by

$$a = 2\frac{\beta\mu_I}{I}\left[\frac{1}{r^3}\right]_{\text{av}}$$

$$b = 2\frac{\beta\mu_I}{I}\left[\frac{8\pi}{3}\Psi^2(0) - \frac{3\cos^2\theta - 1}{2r^3}\right]_{\text{av}}$$

$$c = 3\frac{\beta\mu_I}{I}\left[\frac{3\cos^2\theta - 1}{r^3}\right]_{\text{av}}$$

where β is the Bohr magneton.

The large effects with which we are here concerned are associated with the presence of electrons in penetrating orbitals, i.e. s and, in heavy atoms, $p_{1/2}$, electrons, for which $\Psi^2(0)$ may be large: in this situation the dominant terms are in

$$b' = \frac{16\pi\beta\mu_I}{3I}\Psi^2(0).$$

Particularly simple e.s.r. spectra have been observed[5] for the molecules ScO, YO and LaO in Ne or Ar matrices at 4 K, for non-rotating molecules in $^2\Sigma$ states behave quite analogously to $^2S_{1/2}$ atoms. The diagonal matrix element of the hyperfine interaction is

$$\tfrac{1}{6}(3b+c)\,[F(F+1)-I(I+1)-S(S+1)]$$

which becomes identical with the expression for atoms when A is written for $b+c/3$. In ^{45}Sc, $I = 7/2$, and all the transitions $m_s = +1$, $m_I = 0$ have been observed for the rotationless molecule. Experimentally accessible fields correspond to an intermediate case in which the Breit-Rabi treatment[17] applies, and this leads to transitions given by

$$h = -\frac{am_I}{1-a^2/4} \pm \left[\frac{a^2 m^2{}_I}{(1-a^2/4)^2} + \frac{1-a^2(I+\frac{1}{2})}{1-a^2/4}\right]^{1/2}$$

where

$$a = A/\nu \text{ and } A \simeq b' \text{ (ignoring the distinction between } b \text{ and } b')$$

and

$$h = g\beta H/\mathbf{h}\nu$$

H being the magnetic field at resonance and ν the frequency (fixed in these experiments at 9·432 GHz). The transitions are illustrated in Figure 15.4. The spectrum of ScO isolated in Ar at 4 K is as follows[5]:

Transition, m_F	H, gauss obs.	calc.
−3–−4	587	557
−2–−3	733	721
−1–−2	1000	991
0–−1	1433	1446
+1– 0	2168	2172
+2–+1	3150	3169
+3–+2	4323	4353
+4–+3	5633	5640

The calculated values of H are with $g = 2{\cdot}00$ and $A = 2{\cdot}01$ GHz.

The importance of the results for ScO, YO and LaO is that they afford proof that the ground states of these molecules are indeed $\sigma\ ^2\Sigma^+$. Had they been, for example, $\delta\ ^2\Delta$, the g-tensor would have been expected to be highly anisotropic and to have differed considerably from the free spin value. At the same time, the hyperfine interaction with the nucleus would also have been expected to be anisotropic, with the isotropic contact term very small.

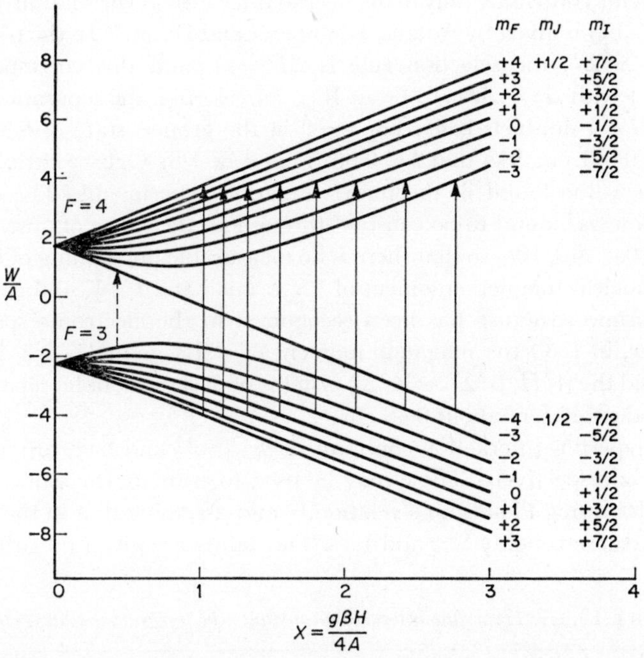

FIGURE 15.4. *Splitting of energy levels of a system with $S = \frac{1}{2}$ and $I = 7/2$ in external magnetic field.*
(*Reproduced with permission from Weltner, W., McLeod, D., and Kasai, P. H.*, J. Chem. Phys., *1967*, **46**, 3172.)

The characteristic feature of the electronic spectra of ScO and LaO in the gas phase is the appearance of doublets, essentially independent of N or J, in both the analyzed transitions, $B\,^2\Sigma^+$, $A\,^2\Pi - X\,^2\Sigma^+$. The B—X system of ScO has been studied in most detail.[18] The ground state, in which the spin splitting is undetectably small, seems to be a good example of $b_{\beta S}$ coupling: with $I = 7/2$ and $G = 3$ or 4, the pattern of levels is:

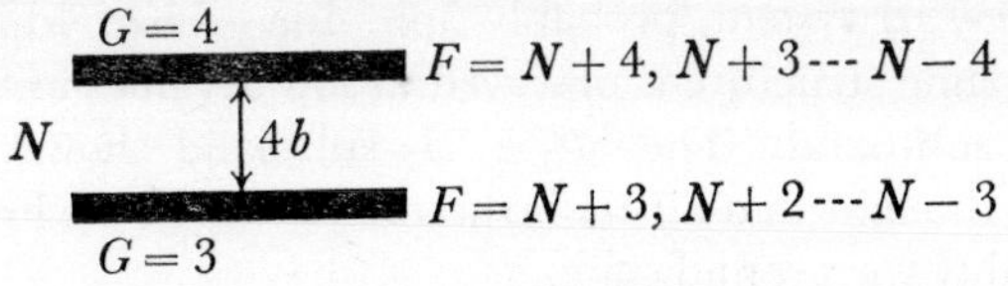

The situation in the excited states $B^2\Sigma^+$ is quite different. These show comparatively large spin splittings, i.e. large separation of the F_1 levels, with $J = N + \frac{1}{2}$, from the F_2 levels, with $J = N - \frac{1}{2}$. Thus **S** couples to **N**, to give **J**, and the **J.I** and **N.G** interactions are very small. In fact, no hyperfine structure is observed in the upper state and its hyperfine components contribute only to the overall linewidth. A correlation diagram for $b_{\beta S} - b_{\beta J}$ is given by Adams, Klemperer and Dunn.[18] In the transition $B^2\Sigma^+ - X^2\Sigma^+$, the selection rule is $\Delta F = \pm 1$, and this corresponds to $\Delta N = \pm 1$ with $\Delta S = \Delta I = 0$. From W_{hfs} for case $b_{\beta S}$, the separation of the $G = 3$, $G = 4$ doublets is $4b$. For $v = 0$ in the ground state of ScO, $4b = 0{\cdot}250 \pm 0{\cdot}001\ cm^{-1}$ so that $b = 0{\cdot}0625\ cm^{-1}$ or 1·87 GHz, a little smaller than the value found in the matrix-isolation experiments. The doublet separation was found to be constant throughout the range of observations of N, $\sim 10 \leqslant N \leqslant 100$, so that there is no measurable uncoupling of **I** and **S**.

The nuclear magnetic moment of ^{89}Y is small and $I = \frac{1}{2}$, so that for YO no hyperfine structure has been recognized in the electronic spectrum. However, in LaO the magnetic moment of ^{139}La, with $I = 7/2$, is again large, and the $A^2\Pi$, $B^2\Sigma^+ - X^2\Sigma^+$ systems both show doublet separations, independent of J, of about $0{\cdot}48\ cm^{-1}$.

The hyperfine interaction constants in $^2S_{1/2}\,Sc^{2+}$ and La^{2+} are 5·04 and 8·10 GHz respectively. These may be used to estimate the values of b in the oxides, using Goudsmit's relation,[17] and assuming that in the oxides, the metals are roughly Sc^+ and La^+. The values are given in Table 15.4.

TABLE 15.4. *Hyperfine interaction constants, b, for ground states, cm*$^{-1}$

Molecule	(i) E.s.r.	(ii) Electronic spectra	(iii) Estimated from M^{2+}
ScO	0·0670	0·0625	0·075
YO	0·0068	—	—
LaO	0·130	0·12	0·12
VO	0·0797	0·080	—
NbO	—	?0·19	—

The NbO^+ molecule (or perhaps NbN: positive identification has not yet been made) provides a spectacular example of hyperfine structure.[19] A single 0–0 sequence of emission bands at 16000–17000 cm^{-1} has been assigned to a $^3\Phi - {}^3\Delta$ system, probably $5p\pi$, $4d\delta \rightarrow 5s\sigma$, $4d\delta$. Very large, $\sim 1\ cm^{-1}$, hyperfine structure is observed at low J values in the $^3\Phi_4 - {}^3\Delta_3$ and $^3\Phi_2 - {}^3\Delta$ sub-bands. The $^3\Phi_3 - {}^3\Delta_2$ sub-band shows no hyperfine structure. It is thought that all the structure resides in the lower state, $^3\Delta$, which is probably the ground state.

It is believed that these observations can be interpreted in terms of case-a_β coupling in the ground state. The absence of splitting in $^3\Delta_2$ when $\Sigma = 0$ shows that $a \ll (b+c)$, and, as before, probably $b \gg c$. Then

$$W_{\text{hfs}} = \tfrac{1}{2}\frac{b\Sigma\Omega}{J(J+1)}\mathbf{I}.\mathbf{J}$$

where the factor $1/2(=1/2S)$ arises from the fact that although the total spin is 1, only one electron contributes significantly to the interaction. In this case the hyperfine splittings are expected to decrease with increasing J as $\mathbf{I}$ becomes more and more nearly perpendicular to the molecular axis, and this is observed. However, the interval rule is not obeyed, and the brief report at present available draws attention to other anomalies which have yet to be explained.

Two systems of gaseous VO have been the subject of rotational analysis:[10] $B\,^4\Pi - X\,^4\Sigma^-$ and $C\,^4\Sigma^- - X\,^4\Sigma^-$. Both systems show lines broadened by unresolved nuclear hyperfine structure, and, with some reservations in respect of $B\,^4\Pi$, the hyperfine structure arises largely in the ground state, which is a case-b $^4\Sigma^-$ state from the configuration $\delta^2\sigma$. The observed effects may be explained if the coupling is taken to be $b_{\beta J}$. Assuming as before that the dominant interaction is the contact interaction between the $\sigma 4s$ electron and the ^{51}V nucleus (with $I = 7/2$, $\mu = 51{\cdot}4$ Å), we deduce that

$$W_{\text{hfs}} = \frac{b}{3}\left[\frac{J(J+1)+S(S+1)-N(N+1)}{2J(J+1)}\right]\mathbf{I}.\mathbf{J}$$

where the factor 1/3 is the spin factor, $1/2S$.
Then for

$$\begin{aligned} &F_1, \quad J = N+\tfrac{3}{2}\\ &F_2, \quad J = N+\tfrac{1}{2}\\ &F_3, \quad J = N-\tfrac{1}{2}\\ &F_4, \quad J = N-\tfrac{3}{2}\end{aligned}$$

we have

$$W(F_1) = \frac{b}{2N+3}\mathbf{I}.\mathbf{J}$$

$$W(F_2) = \frac{b(N+9/2)}{3(2N+1)(N+3/2)}\mathbf{I}.\mathbf{J}$$

$$W(F_3) = -\frac{b(N-7/2)}{3(2N+1)(N-\frac{1}{2})}\mathbf{I}.\mathbf{J}$$

$$W(F_4) = -\frac{b}{2N-1}\mathbf{I}.\mathbf{J}$$

For $N \geqslant 20$, the overall structure width in F_1 and F_4 becomes equal to about $3{\cdot}5b$, and in F_2 and F_3 equal to about $1{\cdot}2b$. This explains the most striking feature of the VO bands, namely the occurrence of both broad and comparatively sharp branches. In $C\,^4\Sigma^- - X\,^4\Sigma^-$, of the eight branches observed, R_1, P_1, R_4 and P_4 lines are broad, with apparent widths of about $0{\cdot}4\ cm^{-1}$, while the $F_2 - F_2$ and $F_3 - F_3$ branches have much smaller apparent widths, about $0{\cdot}12\ cm^{-1}$. From the apparent widths, $b \approx 0{\cdot}1$ cm^{-1}, but this is an overestimate, as Doppler broadening also contributes.

However, in the ground state of VO (and observed in levels $v = 0$, 1, 2 and 3) there is a highly unusual perturbation between the F_2 and F_3 components. Ordinarily the selection rule $\Delta J = 0$ holds rigidly, but here J is not a good quantum number, and it is rather $\Delta F = 0$ which applies. The perturbing matrix element contains as the only unknown the hyperfine interaction constant b, and by matching observed and calculated $F_2 - F_3$ separations it has been possible to derive a fairly precise estimate of b: $0{\cdot}080 \pm 0{\cdot}005\ cm^{-1}$. There may be other examples of this type of perturbation, but the conditions for its existence will not often be satisfied.

The blue bands of NbO (^{93}Nb has $I = \frac{9}{2}$, $\mu = 61{\cdot}4$ Å) also seem to arise from a $^4\Sigma^- - {}^4\Sigma^-$ transition and absorption pictures show the same pattern of broad and sharp line branches. In some beautiful emission spectrograms (Figure 15.3), the ten hyperfine components of the broad 'lines' are resolved. Rough measurements give a limiting overall width at high N of about $0{\cdot}86\ cm^{-1}$, and from this, b may be estimated to be about $0{\cdot}19$ cm^{-1}, using the case-$b_{\beta J}$ formulae.

Interesting results have recently been obtained[7] from a study of the electron spin resonance spectrum of randomly oriented VO in an argon matrix at 4 K. The zero-field fine structure splitting is large (but not determined), and the spectrum is interpreted in terms of the transition $M_s = +\frac{1}{2} \leftrightarrow -\frac{1}{2}$ with an effective spin Hamiltonian

$$\mathscr{H} = g_{11}\beta H_z S_z + 2g_\perp \beta(H_x S_x + H_y S_y)$$
$$+ A_{11} I_z S_z + 2A_\perp (I_x S_x + I_y S_y)$$

in which S is taken to be $\frac{1}{2}$. The analysis gives $g_{11} = 2{\cdot}0023 \pm 0{\cdot}0010$, $g_\perp = 1{\cdot}9804 \pm 0{\cdot}0010$, $A_{11} = 714{\cdot}7 \pm 0{\cdot}5$ MHz and $A_\perp = 837{\cdot}1 \pm 0{\cdot}5$ MHz. Resolution of A_{11} and $A_\perp$ into isotropic and dipolar parts yields

$$A_{iso} = 796{\cdot}3\ \text{MHz};\ A_{dip} = -40{\cdot}8\ \text{MHz}$$

Now $A_{iso} \simeq \frac{1}{3}b$ where b is the hyperfine interaction constant introduced in the discussion of the spectrum of the gaseous molecule. We have $3A_{iso}/\mathbf{c} \simeq b = 0{\cdot}0797\ cm^{-1}$, in excellent agreement with the value obtained from the ground-state perturbation in the gas-phase spectrum. For the dipolar constant, Kasai gives

$$(A_{\text{dip}})_{\text{VO}} \simeq (2/3)\,(A_{\text{dip}})_{3d\delta} = -(4/21) g_e \beta_e g_n \beta_n \langle 1/r^3 \rangle_{3d}$$

$\langle 1/r^3 \rangle_{3d}$ is estimated at about 1·7 a.u. from the levels of V^{2+} in $3d^2 4s^2\ {}^4F$ to give $(A_{\text{dip}})_{\text{VO}} = -46$ MHz in reasonable agreement with the observed value.

The various estimates of b discussed here are collected together in Table 15.4. They are *roughly* proportional to the product of nuclear charge and nuclear magnetogyric ratio as is shown in Figure 15.5, as would be expected

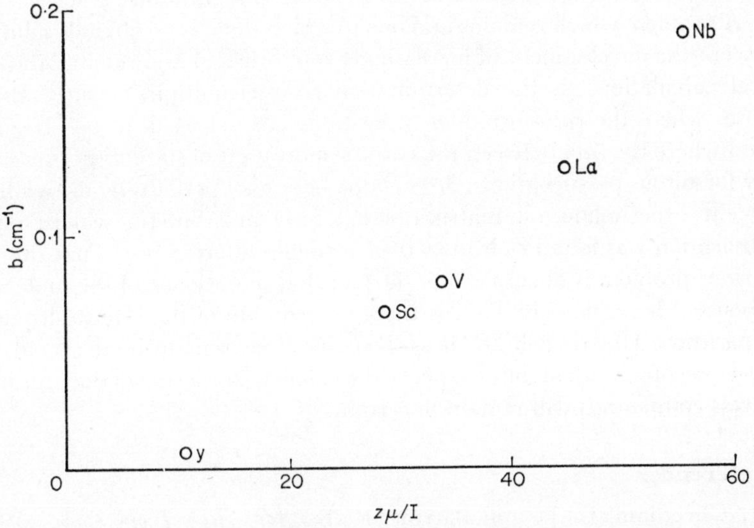

FIGURE 15.5. *Variation of nuclear hyperfine interaction constant with the product of nuclear magnetogyric ratio and nuclear charge.*

for not too heavy atoms with single s electrons outside closed shells. This is remarkable experimental confirmation of the view already put forward on theoretical grounds that the outer σ orbitals in the ground states of these molecules are essentially ns atomic orbitals.

Other oxides which may show large nuclear hyperfine structure include those of ^{55}Mn, ^{59}Co, ^{141}Pr, ^{175}Lu[11] and ^{181}Ta.

F. Conclusions

One aim of a brief review of this kind is to indicate the profitable directions of future work. The present results of matrix Hartree–Fock calculations have proved useful and stimulating, and it may be hoped that they can be extended at least to other oxides, CrO and MnO, of the first transition series. The rotational structure of high-multiplicity transitions may best be studied in high resolution by flash photolysis or, perhaps, flash discharge techniques: however, some theoretical models to help the analyst are lacking, and, for example, the energy levels in $^5\Sigma$, $^6\Sigma$ and $^7\Sigma$ states have been discussed only in cases where the spin–orbit coupling is negligibly small. Sometimes ground states may be identified in conventional emission studies combined with the results of low-temperature matrix-isolation experiments. As has been shown, in favourable cases, electron spin resonance studies of matrix-isolated molecules may provide very direct experimental evidence about electron configurations.

A problem which remains and one to which there is no obvious solution, except the development of increasingly sophisticated and reliable theoretical calculations, is the determination of intermultiplicity intervals, in cases where the spin-forbidden transitions are too weak to be observed, and where the link between the two (or more) sets of states is not revealed by fortuitous perturbations. Some hope here may perhaps be drawn from recent experiments on matrix-isolated SnO and SnS in which singlet absorption was found to lead to triplet–singlet fluorescence. That this is a serious problem is illustrated by the fact that in only one of the molecules discussed here, namely TiO, is there an estimate of the intermultiplicity separation. However, in so far as these questions remain unanswered, the spectroscopic, and, in some respects, the valency, properties of the transition metal compounds will remain uncertain.

References

1. Cheetham, C. J., and Barrow, R. F., *Adv. High Temp. Chem.*, 1967, **1**, 7.
2. Carlson, K. D., and Claydon, C. R., *Adv. High Temp. Chem.*, 1967, **1**, 43.
3. Grimley, R. T., 'The Characterization of High Temperature Vapors', ed. J. L. Margrave, Wiley, New York, 1967, p. 195.
4. Callear, A. B., and Norrish, R. G. W., *Proc. Roy. Soc.*, 1960, **A259**, 304.
5. Weltner, W., McLeod, D., and Kasai, P. H., *J. Chem. Phys.*, 1967, **46**, 3172.
6. Smith, J. J., and Meyer, B., *J. Mol. Spectroscopy*, 1968, **27**, 304.
7. Kasai, P. H., *J. Chem. Phys.*, 1968, **49**, 4979.
8. Ames, L. L., Walsh, P. N., and White, D., *J. Phys. Chem.*, 1967, **71**, 2708.
9. Smoes, S., Coppens, P., Bergman, C., and Drowart, J., *Trans. Faraday Soc.*, 1969, **65**, 682.
10. Richards, D., and Barrow, R. F., *Nature*, 1968, **217**, 842; 1968, **219**, 1244.

11. Dunn, T. M., 1969, personal communication.
12. Barrow, R. F., and Senior, M., *Nature*, 1969, **223**, 1359.
13. Berg, R. A., Wharton, L., Klemperer, W., Büchler, A., and Stauffer, J. L., *J. Chem. Phys.*, 1965, **43**, 2416.
14. Frosch, R. A., and Foley, H. M., *Phys. Rev.*, 1952, **88**, 1337.
15. Radford, H. E., *Phys. Rev.*, 1964, **136**, 1571A.
16. Atkins, P. W., *Proc. Roy. Soc.*, 1967, **A300**, 487.
17. See for example, Kuhn, H. G., 'Atomic Spectra', 2nd edn., Longmans, London, 1969.
18. Adams, A., Klemperer, W., and Dunn, T. M., *Canad. J. Phys.*, 1968, **46**, 2213.
19. Dunn, T. M., and Rao, K. M., *Nature*, 1969, **222**, 266.

CHAPTER 16

Vibrational Studies of Metal–Metal Bonding

M. J. WARE

A. Introduction

The first observation of a metal–metal stretching fundamental was made by Dr. L. A. Woodward[1] and it is characteristic that one of his earliest contributions to Raman spectroscopy should now stand as a starting point for surveying a large and rapidly growing area of the subject. In 1934 he demonstrated the diatomic nature of the mercurous ion by detecting an intense line at 169 cm^{-1} in the Raman spectrum of aqueous mercurous nitrate. However, only in recent years has it become possible to apply the methods of vibrational spectroscopy to other molecules containing metal–metal bonds. By 1966 a handful of compounds had been studied, with the results as summarized in Table 16.1; but in the three years subsequently, over a hundred compounds were examined. The reasons for this sudden growth are readily traced to advances of instrumental technique and the increasingly widespread synthesis of new compounds. It is only since the development of Fourier transform spectroscopy in the far infrared and the excitation of Raman spectra at relatively long wavelengths by lasers that the majority of metal–metal (M—M) bonded compounds have become tractable to study by vibrational spectroscopy. The former technique ensures that infrared spectra may now be obtained routinely down to 50 cm^{-1} or so, and the latter advance has largely surmounted the problems of obtaining Raman spectra from small quantities of coloured or photosensitive materials in the solid state. Owing to the diminishing difficulty of obtaining data, the growth in the synthetic chemistry of metal–metal bonds, and indeed in the *recognition* of metal–metal bonding in

TABLE 16.1. *Metal–metal stretching frequencies (known up to 1966)*

Bond	ν(M—M), cm^{-1}	Compound	Reference
Hg—Hg	169	$Hg_2(NO_3)_2$ aqueous	a
Hg—Hg	170	Hg_2Cl_2 solid	b
Sn—Sn	190	$Sn_2(CH_3)_6$ liquid	c
Hg—Co	152(a_1)	$Hg[Co(CO)_4]_2$ solution	d
Cd—Co	161(a_1)	$Cd[Co(CO)_4]_2$ solution	d
Cd—Cd	183	$Cd_2(AlCl_4)_2$ melt	e
Re—Re	128	$Re_2(CO)_{10}$ solid, solution	f, g

a. Ref. 1.
b. Poulet, H., and Mathieu, J. P., *J. Chim. phys.*, 1963, **60**, 442.
c. Brown, M. P., Cartmell, E., and Fowles, G. W. A., *J. Chem. Soc.*, 1960, 506.
d. Ref. 27.
e. Ref. 11.
f. Cotton, F. A., and Wing, R. M., *Inorg. Chem.*, 1965, **4**, 1328.
g. Lewis, J., Manning, A. R., Miller, J. R., Ware, M. J., and Nyman, F., *Nature*, 1965, **207**, 142.

previously known compounds, is now beginning to be matched by an increasing knowledge of their vibrational properties—to the mutual benefit of the preparative chemist and the inorganic spectroscopist.

Recent reviews[2] list over a thousand references to the preparation and characterization of compounds containing metal–metal bonds, but the majority of these fall into five broad categories:

(i) Simple polyatomic metal species

Many diatomic molecules have been characterized in the gas phase at elevated temperatures by ultraviolet emission spectroscopy; vibrational data so obtained are particularly accurate and bond-stretching force constants [k(M—M)] are readily obtained from the harmonic vibration frequencies. Much of this work antedates the direct vibrational studies and provides a valuable point of reference, for example, in evaluating the significance of force constants obtained by approximate means. Table 16.2 summarizes these data, many of which may be found in Herzberg's[3] excellent compilation. The development of gas-phase Raman spectroscopy at high temperatures (see Chapter 5) may enable the vibrational fundamentals of many more such species to be observed directly.

(ii) Binary metal carbonyls and their derivatives

These compounds, of formula $M_x(CO)_y$, are currently receiving much attention because they afford an opportunity to study the chemistry of

TABLE 16.2. *Bond parameters of diatomic metal species in the gas phase*

Molecule	ν(M—M)† (cm^{-1})	k(M—M) (mdyn $Å^{-1}$)	D_0(M—M)‡ (kcal $mole^{-1}$)	Reference
$^{7}Li_2$	351·43	0·255	26	a
$^{23}Na_2$	159·23	0·172	17·8	a
$^{39}K_2$	92·64	0·098	11·8	a
$^{85}Rb_2$	57·28	0·082	11·1	a
$^{133}Cs_2$	41·99	0·069	10·4	a
NaK	123·29	0·131	14·3	a
NaRb	106·64	0·121	13·1	a
RbCs	49·41	0·075	—	a
Cu_2	266·1	1·325	45·5	b
Ag_2	192·4	1·176	37·6	b
Au_2	190·9	2·113	51·5	b
CuAg	(250)		40·7	b
CuAu	(250)		54·5	b
AgAu	(210)		47·6	b
Sn_2	(300)		45·8	c
SnCu	(215)		41·4	c
SnAg	(190)		31·6	c
SnAu	(200)		57·5	c

† Harmonic frequencies, but figures in parentheses are approximate, being estimated from Gordy's rule.

‡ Bond dissociation energies from mass spectroscopic data.

a. Ref. 3.

b. Ackerman, M., Stafford, F. E., and Drowart, J., *J. Chem. Phys.*, 1960, **33**, 1784.

c. Ackerman, M., Drowat, J., Stafford, F. E., and Verhaegen, G., *J. Chem. Phys.*, 1962, **36**, 1557.

metals in very low oxidation states, and have a relevance to catalytic and metallurgical problems. Their structures, reaction kinetics and thermodynamic properties are widely studied, but the methods of synthesis have not yet been systematized and preparative methods remain rather arbitrary.

(iii) Non-transition metal–metal carbonyl derivatives

The metal carbonyl fragment, $M(CO)_y$, has proved to be an excellent building unit for the synthesis of molecules containing bonds between the transition metal and metals of the main groups. Synthesis is usually systematic and may often be achieved by simple metathesis, e.g.:

$$(C_6H_5)_3SnCl+NaMn(CO)_5 \rightarrow NaCl+(C_6H_5)_3SnMn(CO)_5$$

and the resulting metal–metal bonds often show remarkable stability, e.g.:

$$(C_6H_5)_3SnMn(CO)_5+Cl_2 \rightarrow Cl_3SnMn(CO)_5+C_6H_5Cl$$

Certain post-transition metal groups, especially XHg— and X_3Sn—, where X is a halogen, show such versatility towards transition metals in low oxidation states that they have come to be regarded as ligands in the same light as the halogens or pseudohalogens. Conversely, post-transition metal derivatives may be fully substituted by metal carbonyl groups giving interestingly symmetrical molecules such as $Zn[Mn(CO)_5]_2$, $Tl[Co(CO)_4]_3$ and $Sn[Co(CO)_4]_4$, which commend themselves strongly to the vibrational spectroscopist.

(iv) Metal halides, alkyls, alkoxides and hydroxides

A good many apparently simple metal derivatives have proved on careful examination not to be monomeric but to contain several formula units in the molecule; the metal atoms usually lie at the apices of a regular polyhedron, with respect to which the ligands may occupy bridging (polyhedron edges or faces) or terminal positions. When, as is usually the case, bridging ligands are present the extent of metal–metal 'bonding' is often an open question and different physical methods demand varying criteria for determining the presence of such an interaction.

The most notable examples of metal cluster formation are to be found among the early transition metals as halides, e.g. Mo_6Cl_{12}, $Re_3Cl_{12}^{3-}$, and in the Main Groups III and IV as hydroxides and alkoxides, e.g. $Tl_4(OEt)_4$, $Pb_4(OH)_4^{4+}$.

(v) Bridged species: carboxylates

Although the presence of monodentate ligands in bridging positions has been noted in other categories, we refer here to systems in which the metal–metal interaction is strongly augmented, or even wholly sustained by bidentate ligands linking the two metal atoms. Carboxylato-complexes of formula $M_2(RCO_2)_4$ are important in magnetochemical studies, but compounds of this type are the least tractable to vibrational analysis and little more will be said of them in this account.

To see the role of vibrational spectroscopy in perspective, it is instructive to survey briefly the physical techniques that can be used to study the metal–metal bond. Methods that are applicable to the solid state have proved most useful; almost all our accurate knowledge of structural parameters, in particular the lengths of metal–metal bonds, has come from X-ray diffraction; however the number of structure determinations is still relatively small; sensitivity or instability of the materials and the difficulty in obtaining good single crystals account for some of the difficulties. In the case of certain elements, notably iron and tin, Mössbauer spectroscopy (see Chapter 10) can contribute useful information on the nature of the metal environment and involvement in bonding. Methods based on the presence of incomplete d-electron shells, e.g. magnetic and e.s.r. spectroscopy, have only a limited application: the majority of metal–metal compounds, and particularly those in low formal oxidation states, obey the 'effective atomic number' rule in having no unpaired electrons.

Owing to the relatively low volatility of M—M compounds, methods requiring a sample in the gas phase, such as microwave spectroscopy or electron diffraction, have found little application as yet, but the operation of these techniques at higher sample temperatures holds promise.[4] Mass spectrometry, on the other hand, has proved a very useful technique[5] for characterizing such compounds in circumstances where the other methods would fail, and the measurement of appearance potentials has provided approximate values for the dissociation energies of a number of M—M bonds. It is clear that vibrational spectroscopy has a useful contribution to make to the study of M—M bonded species: its strength lies in the relative ease in acquiring data, particularly from the solid state, and the straightforward method of interpretation; its chief weakness lies in the essentially qualitative nature of the structural information obtained and the well-known dangers of an incautious interpretation of the vibrational spectra of complex molecules, in which there is a temptation to make the facts support a burden of interpretative inference that they cannot justly bear.

The aims of a vibrational study are, first, to provide a rapid diagnostic test for the presence of metal–metal bonding in newly synthesized compounds—evidence that can be valuable to the direction of current preparative work; second, to furnish information about the probable geometry of the metal–metal framework, which provides a useful pointer to the prosecution of rigorous structure determination—for instance the decision to carry out an X-ray crystallographic study, and third, to attempt the computation of molecular force fields in the hope that metal–metal bond-stretching force constants may prove useful parameters in describing the nature of the metal–metal bond.

The purpose of this chapter is to note the characteristics of M—M stretching frequencies as they appear at present, to describe and interpret the important features in the vibrational spectra of selected M—M bonded compounds, showing their relevance to structural inorganic chemistry, and to discuss the present problems inherent in the attempt to convert metal–metal vibrational frequencies into meaningful force constants.

B. Characteristics of metal–metal vibrations

At the outset, it is necessary to define the circumstances in which it may be valid to refer to certain normal modes of a molecule as metal–metal vibrations, and thus imply that they may be considered as approximately localized 'group vibrations', largely independent of the motions of the rest of the molecule. It appears at present that M—M stretching fundamentals fall within the frequency range 50 to 300 cm^{-1}, and the great majority of these within the narrower range 100–250 cm^{-1}. Table 16.3 lists typical ν(M—M) values for a range of metals. Other molecular motions of the same symmetry species and comparable frequency will cause the normal modes of vibration to take on a 'mixed' character, in which none can be simply identified as pure ν(M—M) vibrations. Of the ligands commonly encountered, the stretching and bending motions of the heavier halogen atoms, which fall below 300 cm^{-1}, are particularly effective in this respect, and the mixing increases with the mass of the halogen. Thus for M—M species containing coordinated iodine atoms, rough normal co-ordinate analyses show complete mixing of the M—M and M—I motions in the low-frequency normal modes. Although, for reasons of convenience, I use the term 'ν(M—M)', it must be understood that such simple descriptions are, at best, an approximation, which in the event of substantial vibrational mixing cannot be sustained. Nevertheless, when only light atoms are coordinated to the metal it is useful, and not too inaccurate, to refer to M—M vibrations for the purposes of qualitative interpretation. However, even light atoms may give rise to very low frequency bending modes—coordinated carbon monoxide is notable in this respect—so normal coordinate analysis is desirable in all cases.

The most notable feature of ν(M—M) is their appearance[6] as fundamentals of very high intensity in the Raman spectrum—subject, of course, to the symmetry selection rules. This intensity is often an order of magnitude greater than that of any other feature in the spectrum; consequently the ease of observation and identification makes Raman spectroscopy the dominant technique. While infrared spectra often supply valuable complementary information, there are greater difficulties in unambiguously identifying the ν(M—M) fundamentals. The reasons for the remarkably high Raman intensity are readily understood in a qualitative way. Several empirical equations have been proposed to relate Raman intensity of stretching fundamentals to the parameters of the bonded atoms, in particular, the correlations[7] discovered by Woodward and Long, and Long and Plane. While differing in their explicit form, they agree in attributing high Raman intensity to a high degree of covalent character in the relevant bonds and to large polarizabilities of the bonded atoms: both characteristics are present to a marked degree in M—M bonds, where the atomic numbers are high and electronegativity differences are small. By the same token, even for heteronuclear M—M bonds changes in the electric dipole moment during vibration may be rather small and the infrared-active

TABLE 16.3. *Typical metal–metal stretching frequencies*

Bond	ν(M—M), cm^{-1}	Compound	Bond	ν(M—M), cm^{-1}	Compound
		Homonuclear			
Zn—Zn	175	Zn/$ZnCl_2$	Mn—Mn	157	$Mn_2(CO)_{10}$
Cd—Cd	183	$Cd_2(AlCl_4)_2$	Re—Re	128	$Re_2(CO)_{10}$
Hg—Hg	169	$Hg_2(OH_2)_2^{2+}$	Ru—Ru	186, 152	$Ru_3(CO)_{12}$
Ga—Ga	233	$Ga_2Cl_6^{2-}$	Os—Os	161, 120	$Os_3(CO)_{12}$
Ge—Ge	270	Ge_2H_6	Co—Co	177	$Co_2(CN)_{10}^{6-}$
Sn—Sn	190	$Sn_2(CH_3)_6$	Rh—Rh	225, 200 176, 128	$Rh_4(CO)_{12}$
Pb—Pb	116	$Pb_2(CH_3)_6$	Ir—Ir	208, 164, 131	$Ir_4(CO)_{12}$
Te—Te	122	$Te_2(CH_3)_2$	Ni—Ni	190	$Ni_2(CN)_6^{4-}$
		Heteronuclear			
Zn—Mn	259, 170	$Zn[Mn(CO)_5]_2$	Ge—Co	240	$Cl_3GeCo(CO)_4$
Zn—Co	265, 170	$Zn[Co(CO)_4]_2$	Sn—Mo	172	$Me_3SnMo(CO)_3Cp$
Cd—Mn	209, 160	$Cd[Mn(CO)_5]_2$	Sn—W	169	$Me_3SnW(CO)_3Cp$
Cd—Fe	226, 216	$[CdFe(CO)_4]_n$	Sn—Mn	182	$Me_3SnMn(CO)_5$
Cd—Co	216, 159	$Cd[Co(CO)_4]_2$	Sn—Fe	185	$Me_3SnFe(CO)_2Cp$
Hg—Cr	186	$Hg[Cr(CO)_3Cp]_2$	Sn—Co	176	$Me_3SnCo(CO)_4$
Hg—Mo	178	$Hg[Mo(CO)_3Cp]_2$	Sn—Rh	210	$Rh_2Cl_2(SnCl_3)_4^{4-}$
Hg—W	166, 133	$Hg[W(CO)_3Cp]_2$	Sn—Pt	210	$Pt(SnCl_3)_5^{3-}$
Hg—Mn	188, 161	$Hg[Mn(CO)_5]_2$	Pb—Mn	161	$Et_3PbMn(CO)_5$
Hg—Fe	200	$Hg[Fe(CO)_2Cp]_2$	In—Co	155	$In[Co(CO)_4]_3$
Hg—Ru	187, 170	$(ClHg)_2Ru(CO)_4$	Mn—Re	157	$MnRe(CO)_{10}$
Hg—Os	169, 156	$(ClHg)_2Os(CO)_4$	Hg—Co	196, 161	$Hg[Co(CO)_4]_2$

fundamentals consequently may be weak. However, strong vibrational mixing with polar groups can enhance the infrared intensity, and the antisymmetric normal modes, even in homonuclear M—M systems, may be expected to be quite intense.

In a recent review, Spiro[8] has discussed the use of Raman intensities as a criterion for metal–metal interaction. When the presence of a M—M bond is structurally unambiguous, as in the non-bridged binary metal carbonyls for instance, an intense Raman fundamental is found. In many structures, however, the metal cluster could be sustained purely by bridging ligands; the metal–metal distances are large compared with the sum of the covalent radii, but not so large as to preclude all metal–metal interaction. The Raman intensity of the cluster modes for intermediate cases such as this offers a means of assessing the extent of the M—M interaction. Conversely, the absence of intense cluster fundamentals in circumstances where other spectral features are clearly observed, may be taken as indicating the absence of M—M bonding. There are obvious dangers in such a 'proof by default', but when applied with care this criterion can be useful for making structural inferences.[9]

C. Interpreted spectra

In the following sections the salient features of the vibrational spectra of a number of M—M bonded compounds are described and their interpretation is discussed. It is convenient to classify the compounds according to the number of metal atoms in the molecule, thus emphasising the vibrational and structural relationships at the expense of the chemical similarities. Such a classification also provides a useful framework for the examination of trends among chemically related species, an approach adopted in the last section of this chapter. The following summary cannot be regarded as exhaustive, but it is hoped that most of the significant work published at the time of writing is represented.

(*a*) *Binuclear systems*

It was pointed out in the introduction that much vibrational data on simple diatomic M—M systems in the gas phase has been provided by ultraviolet emission spectroscopy (Table 16.2). At the sample temperatures presently accessible to vibrational spectroscopy only condensed phases can be examined, in which the type of binuclear M—M system is rather different, the M—M framework forming only part of a usually complicated polyatomic molecule.

Some of the simplest binuclear species are to be found in Group II. In addition to the well known result for the Hg_2^{2+} ion, and the subsequent observation of ν(Hg—Hg) in several other mercurous compounds[10] Raman fundamentals have recently been reported[11] for the ions Cd_2^{2+} and Zn_2^{2+}, the former at 183 cm^{-1} from the compound $Cd_2(AlCl_4)_2$ in the melt and solid phase, and the latter at 175 cm^{-1} from a zinc chloride melt containing dissolved zinc. Although no evidence for coordination has been found in the Raman spectra of the zinc and cadmium systems, X-ray crystallographic studies have shown[12] that linear digonal coordination, X—Hg—Hg—X, is a recurrent feature in the structural chemistry of Hg(I) compounds. Indeed, when the infrared and Raman spectra of a single crystal of mercurous nitrate dihydrate, $Hg_2(OH_2)_2(NO_3)_2$, are examined,[6] weak bands are found at ~380 cm^{-1} which may be assigned to ν(Hg—O) fundamentals. In the mercurous halides the corresponding features are more pronounced. There is a striking increase in the Hg—Hg bond length as the series of halides proceeds from Hg_2F_2 (2·43 Å) to Hg_2I_2 (2·69 Å), which provides a remarkable opportunity to attempt a correlation of metal–metal bond stretching force constant with bond length. It is therefore necessary to seek a complete assignment of the normal modes of these molecules. Treatment of the linear ($D_{\infty h}$) molecule as an isolated unit yields three bond-stretching and two bond angle-deformation modes.

Analysis based on the site group of the molecule in the crystal lattice, D_{4h}, yields the same selection rules for intramolecular motion, with the addition of a rotational lattice mode (e_g) and two translational lattice modes ($a_{2u} + e_u$):

ν_1	a_{1g}	ν(Hg—X)	Raman-active
ν_2	a_{1g}	ν(Hg—Hg)	
ν_3	e_g	δ(HgHgX)	
ν_4	e_g	lattice mode	
ν_5	a_{2u}	ν(Hg—X)	Infrared-active
ν_6	a_{2u}	lattice mode	
ν_7	e_u	δ(HgHgX)	
ν_8	e_u	lattice mode	

The descriptions of ν_1 and ν_2 are, of course, a formalism; the closer the two frequencies lie, the greater will be the mixing of the Hg—Hg and Hg—X motions in these normal modes.

The vibrational spectra of the mercurous halides as recorded by several workers[6, 10, 13, 14] appear to fit these selection rules quite satisfactorily; the very intense Raman fundamental lying in the range 100–180 cm^{-1} is assigned reliably as ν_2 (in-phase stretching of Hg—Hg and Hg—X) and a weaker Raman feature at higher frequency as ν_1 (stretching of Hg—Hg and Hg—X with opposite phase). The a_{1g} character of both these fundamentals is unambiguously demonstrated by the single-crystal Raman study by Beattie and Gilson.[14] ν_5 is observed in the infrared at a frequency comparable to ν_1, and the infrared-active deformation, ν_7, is found at lower frequency. Corresponding to the Raman-active deformation ν_3 and the rotational lattice mode ν_4, expected at low frequency, there are two features in the spectra—one, rather weak, ranging from 196 cm^{-1} in Hg_2F_2 to 65 cm^{-1} in Hg_2I_2, the other, very intense, ranging from 49 cm^{-1} in Hg_2F_2 to 31 cm^{-1} in Hg_2I_2. Both are shown to be of e_g symmetry by single crystal measurements; Beattie and Gilson assign the 40 cm^{-1} feature to the lattice mode ν_4 and the higher frequency line as the fundamental ν_3, whereas Durig *et al.*[13] prefer to assign the very intense 40 cm^{-1} feature as the fundamental ν_3, but then have no explanation for the weak line at higher frequency. The assignment of Beattie and Gilson seems preferable on several counts: it is supported by single crystal measurements; the deformation frequency shows a reasonable variation with changing halogen and parallels the behaviour of its infrared-active counterpart, and no inexplicable features of the spectrum remain. With the complete assignments as set out in Table 16.4, it might be thought that valid Hg—Hg

TABLE 16.4. *Assignment of fundamentals for the mercurous halides*

Mode	Hg_2F_2	Hg_2Cl_2	Hg_2Br_2	Hg_2I_2
ν_1 a_{1g} ν(HgX)	387	277	220	193
ν_2 a_{1g} ν(HgHg)	185	167	134	113
ν_3 e_g δ(HgHgX)	196	139	92	65
ν_4 e_g lattice	49	42	37	31
ν_5 a_{2u} ν(HgX)	—	252	180	138
ν_7 e_u δ(HgHgX)	—	109	71	49
Other weak Raman features	400, 230 138, 108	292	266	224
r(Hg—Hg) Å	2·43[a]	2·53[b]	2·58[b]	2·69[b]

The frequencies, in cm^{-1}, quoted here are average values of the data obtained by many workers; see references 6, 10, 13, 14, 15.

[a] Grdenić, D., and Djordjević, C., *J. Chem. Soc.*, 1956, 316.

[b] Havighurst, R. J., *J. Amer. Chem. Soc.*, 1926, **48**, 2113.

force constants could be calculated and a comparison made with the rapidly changing Hg—Hg bond length. The linear molecule would seem to be particularly favourable in this respect, because symmetry determines that the bending and stretching motions are completely orthogonal. The three stretching frequencies ν_1, ν_2 and ν_5 will suffice to determine three of the four parameters of a general valence force-field: k_1(Hg—Hg), k_2(Hg—X) and the interaction constants k_{12} and k_{22}. However, if one makes the apparently reasonable assumption that the long-range interaction constant k_{22} is zero, the calculated values of k_1, k_2 and k_{12} are physically quite unacceptable, and in some cases imaginary. The importance of using the most general potential field is shown quite clearly in the work of Durig *et al.*; physically reasonable solutions to the secular equation do exist, but only when both interaction constants are included. Unfortunately, it is then impossible to determine a unique set of values for the four force constants, and one must be content with exploring the regions of force constant space that provide admissible frequencies. A correlation of k(Hg—Hg) with r_0(Hg—Hg) is therefore not possible at present.

A further source of difficulty in such work arises from the fact that the spectra are necessarily obtained from the solid state, and may not accurately represent the unperturbed frequencies of free Hg_2X_2 molecules. There is little direct evidence for correlation field splitting in the spectra of the mercurous halides: the weak, supernumerary Raman features (Table 16.4) might be attributed to this cause, but can equally well be explained as overtones, possibly enhanced by Fermi resonance. However, it does appear that in other mercurous compounds of somewhat lower symmetry there is abundant evidence for large correlation field splittings in the solid state.

The vibrational spectra[15] (Table 16.5) of mercurous sulphate, selenate, bromate and iodate are surprisingly rich in bands. The sulphate and selenate are isomorphous,[16] belonging to the space group Pc(C_s^2) with two molecules per unit cell, each having C_1 site symmetry. The number of observed bands exceeds that predicted even under this very low symmetry; in particular two very strong bands are observed in the Raman spectra at frequencies that suggest they should be assigned to ν(Hg—Hg). It is therefore necessary to invoke the factor group, which is C_s, in order to obtain a satisfactory explanation of the multiplicity of bands. Mercurous bromate[16] belongs to the space group C2/c(C_{2h}^6) and the unit cell contains four molecules occupying sites of C_i point symmetry. Again the site group predicts insufficient bands and the factor group, C_{2h}, must be used. On the grounds of their very high intensity the Raman lines at 108 and 183 cm^{-1} can reasonably only be assigned as ν(Hg—Hg), showing a very large correlation field splitting. The crystal structure of mercurous iodate is unknown, but the fact that the vibrational spectra are much richer in bands than those of the bromate suggests that the two substances are not isomorphous. Moreover, unlike the bromate, the iodate shows many

TABLE 16.5. *Raman spectra in the low-frequency region for mercurous oxysalts*

Oxysalt	Frequency, cm^{-1}
Hg_2SO_4	234 w, **193 s**, 166 mw, **142 s**
Hg_2SeO_4	194 w, **183 m**, 128 s
$Hg_2(BrO_3)_2$	265 m, **183 s**, 159 vw, **108 s**, 71 wsh
$Hg_2(IO_3)_2$	265 m, **181 s**, 153 m, **120 s**, 102 w, **89 s**
$Hg_2(CO_2CH_3)_2$	194 w, **168 s**, **134 s**, 112 vw, 92 w
$Hg_2(CO_2CH_2Cl)_2$	182 vw, 164 w, **126 s**, **119 ms**?
$Hg_2(CO_2CHCl_2)_2$	196 mw, 183 vw, 171 w, **116 s**
$Hg_2(CO_2CCl_3)_2$	202 m, **168 s**, 132 mw, **113 s**
$[Hg_2N_2(COCH_3)_2]_n$	209 m, 184 w, **108 s**
$Hg_2(CO_2C_4H_9)_2$	**126 s**
$Hg_2(CO_2)_2C_6H_4$	192 mw, 168 m, 155 w, **122 s**

Tentative assignments of ν(Hg—Hg) are indicated by bold type.

coincidences in its spectra. Mercurous acetate also shows some evidence for correlation field splitting of ν(Hg—Hg) in the two strong Raman lines at 134 and 168 cm^{-1}. Vibrational spectra have also been recorded for several other mercurous carboxylates and the polymeric N,N′-diacetylhydrazide, in which the Hg—Hg bond achieves the remarkable length of 2·9 Å, and ν(Hg—Hg) at 108 cm^{-1} is correspondingly low. The Hg—Hg stretching frequencies are summarized in Table 16.5.

In conclusion, it must be admitted that a quantitative interpretation of the spectra of mercurous compounds has turned out to be much more difficult than the simple structure of the individual molecules might lead one to anticipate; in part, this may be a function of the unexpectedly large correlation field splittings in the solid state. Experience so far suggests that correlation field effects in the skeletal vibrations of other metal–metal compounds are not large enough to observe, and in no other case has it yet proved necessary to resort to factor group analysis. The marked difference between the behaviour of mercurous compounds and other M—M compounds may be due to the unique coordination geometry in Hg_2X_2, where the metal atoms are exposed to the influence of the neighbouring groups in directions perpendicular to the molecular axis. In other M—M compounds the metal atom skeleton is usually shielded from this influence by a 'sheath' of attached ligands.

Little is known about direct M—M bonds in Group III, but recently Evans and Taylor[17] have prepared and examined the interesting binuclear ions $Ga_2X_6^{2-}$ in which X = Cl, Br or I. In addition to the bands ascribable to Ga—X vibrations, the Raman spectra of these ions in solution show intense polarized lines at 233, 162 and 122 cm^{-1}, respectively, which are very reasonably assigned to ν(Ga—Ga), coupled with ν(Ga—X); the variation in frequency is even larger than that of ν(Hg—Hg) and provides an index of the coupling between ν(Ga—Ga) and ν(Ga—X).

In Group IV many of the binuclear analogues of ethane have been investigated. These molecules are of the type M_2X_6, where X may be a halogen, alkyl or aryl group. Complete assignment has been achieved only in the simplest cases, viz. Si_2H_6 and Ge_2H_6, but the a_1 ν(M—M) fundamental can usually be identified by its intensity and polarization in the Raman spectrum. It has been pointed out[8] that the presence of phenyl groups in the molecule may lead to erroneous assignments: a low-frequency deformation of the aromatic ring often appears with reasonable intensity at about 208 cm^{-1} in the Raman spectrum and has been mistakenly attributed[6] to the ν(M—M) fundamental. Since mixing with other low frequency vibrations of a_1 symmetry is unlikely to be negligible, Table 16.6 records the other chief frequencies of the a_1 block, ν(M—M), ν(M—X) and δ(XMX), in the cases where they have been identified.

TABLE 16.6. *Totally symmetric fundamentals in binuclear compounds of the non-transition metals*

Compound	Frequency, cm^{-1}			Reference
	ν(M—X)	ν(M—M)	δ(XMX)	
Ge_2H_6	2070	270	835	a
$Ge_2(CH_3)_6$	572	273	164	b
$Sn_2(CH_3)_6$	512	190	126	b
$Sn_2(C_2H_5)_3(C_4H_9)_3$	—	199	—	c
$Sn_2(C_6H_5)_6$	—	138	—	d
$Pb_2(CH_3)_6$	458	116	117	e
$Pb_2(C_6H_5)_6$	—	108	—	d
$Ga_2Cl_6^{2-}$	375	233	—	f
$Ga_2Br_6^{2-}$	311	162	—	f
$Ga_2I_6^{2-}$	286	122	—	f
$Te_2(CH_3)_2$	—	122	—	g
Te_2F_{10}	726	168	475	h

a. Crawford, V. A., Rhee, K. H., and Wilson, M. K., *J. Chem. Phys.*, 1962, **37**, 2377.
b. Brown, M. P., Cartmell, E., and Fowles, G. W. A., *J. Chem. Soc.*, 1960, 506.
c. Carey, N. A. D., and Clark, H. C., *Chem. Comm.* 1967, 292.
d. Bulliner, P. A., Quicksall, C. O., and Spiro, T. G., unpublished work; Clark, R. J. H., Davies, A. G., and Puddephatt, R. J., *Inorg. Chem.*, 1969, **8**, 457.
e. Clark, R. J. H., Davies, A. G., Puddephatt, R. J., and McFarlane, W., *J. Amer. Chem. Soc.*, 1969, **91**, 1334.
f. Ref. 17.
g. Chen, M. T., and George, J. W., *J. Organometallic Chem.*, 1968, **12**, 401.
h. Dodd, R. E., Woodward, L. A., and Roberts, H. L., *Trans. Faraday Soc.*, 1957, **53**, 1545.

Few binuclear compounds of the heavier elements of Groups V and VI are known and the description of some of these elements as metals is a matter of personal taste. Table 16.6 includes the data for Te_2F_{10} and $Te_2(CH_3)_2$.

Among the transition metals, the binuclear molecules that have received most attention are the dimetal dodecacarbonyls of Group VII: $M_2(CO)_{10}$ where M = Mn, Tc and Re. With the availability of superior instrumentation the original observations[6] of ν(Mn—Mn) at 157 cm^{-1} and ν(Re—Re) at 128 cm^{-1} have been confirmed;[8, 18] but it now seems that ν(Mn—Re) in the mixed carbonyl $MnRe(CO)_{10}$ should be reassigned to 157 cm^{-1}, in preference to the earlier figure of 182 cm^{-1}. In all cases the ν(M—M) fundamental is clearly distinguished in the Raman spectrum by its very high intensity and polarization. Reliable assignment of the normal modes most likely to mix with the ν(M—M) vibration—in particular the a_1 δ(CMC) vibration—is much more difficult; this point is discussed further in a later section.

In contrast to these simple species, binuclear metal carbonyls containing bridging CO groups yield much less clear-cut results; for instance, solid di-iron nonacarbonyl, $Fe_2(CO)_9$, shows[19] four strong low-frequency features in the Raman spectrum at 107, 124, 132 and 161 cm^{-1}, from which no direct inferences can be made concerning the Fe...Fe interaction. Interpretation will be possible only on the basis of single crystal Raman data, and then a very careful analysis of the kinematics of the complicated bridged system will be necessary. Similar considerations apply to the bridged carboxylato-complexes $Mo_2(COOCH_3)_4$, $Re_2(COOC_3H_7)_4Cl_2$ etc., for which Raman spectra are readily obtained[19] but prove to be uninterpretable. These species are of particular structural interest because of the very high formal bond order and remarkably short distance between the metal atoms. A more tractable problem vibrationally is presented by the ion $Re_2Cl_8^{2-}$ of D_{4h} symmetry, which may also be formally regarded[20] as containing a quadruple Re—Re bond. A preliminary examination of the Raman spectrum[21] has revealed features at 360, 335, 252 and 190 cm^{-1} of medium intensity and a very strong line at 274 cm^{-1}. Mixing is likely to be severe in this molecule, so it would be premature to do more than observe that these frequencies do indeed appear very high for an M—M compound.

A number of binuclear cyano-complexes of cobalt and nickel have recently been examined[22] and ν(M—M) assigned. The data for these, and the other binuclear compounds containing bonds between transition metals are summarized in Table 16.7.

The greatest number of binuclear compounds which have been synthesized contain a bond between a transition metal and a post-transition metal. Owing to the greater polarity of the M—M′ bond, some of the ν(M—M′)

fundamentals have also been identified in the infrared spectra. Table 16.8 lists the frequencies of the heavy atom skeletons.

TABLE 16.7. *ν(M—M) in binuclear compounds of the transition metals*

Molecule	Point group	ν(M—M), cm^{-1}	Reference
$Mn_2(CO)_{10}$	D_{4d}	157	a
$Re_2(CO)_{10}$	D_{4d}	128	a
$MnRe(CO)_{10}$	C_{4v}	157	a
$Co_2(CN)_{10}^{6-}$	D_{4d}	177	b
$Co_2(CN)_8^{8-}$	D_{3d}	168 or 184	b
$Co_2(CO)_6[P(OPh)_3]_2$	D_{3d} or D_{3h}	172	b
$Co_2(CNMe)_{10}^{4+}$	D_{4d}	190	b
$Ni_2(CN)_6(CO)_2^{4-}$	C_{2h}	180	b
$Ni_2(CN)_6^{4-}$	D_{2h}	190	b

a. Refs. 6, 8 and 18.
b. Ref. 22.

(b) Trinuclear systems

The heavy-atom skeleton may adopt three shapes: linear, bent or equilateral triangular. If the substituent ligands preserve the symmetry, the vibrational selection rules for the ν(M—M) fundamentals in the three cases will differ and vibrational spectroscopy may provide an indication of the geometry of the M_3 framework. This is well illustrated by the example of triosmium dodecacarbonyl, $Os_3(CO)_{12}$, and two classes of derivative: the dihalides $Os_3(CO)_{12}X_2$ (X = Cl, Br or I), and the alkoxides $Os_3(CO)_{10}Y_2$ (Y = O-alkyl). The parent carbonyl is known from X-ray crystallography[23] to contain an equilateral triangle of osmium atoms and belongs to point group D_{3h}. For the Os_3 skeleton the selection rules are:

$$\Gamma(Os_3)D_{3h} = a_1'(R) + e'(R, i.r.)$$

Two intense Raman lines[24] may be identified as the ν(M—M) fundamentals; if a simple valence force field, containing only one parameter k(M—M), is assumed, the a_1' mode is predicted to lie at higher frequency than the e′ mode (the ratio of the G matrix elements gives $\nu(a_1') = \sqrt{2}\nu(e')$. This assignment is confirmed by the observation[25] of the lower frequency fundamental only in the infrared spectrum, and is substantiated by exactly similar results for the isostructural ruthenium carbonyl.

In the dihalides, $Os_3(CO)_{12}X_2$, however, one ν(M—M) fundamental is found[25] in the Raman spectrum and a second, non-coincident ν(M—M)

TABLE 16.8. *ν(M—M) in binuclear compounds having a bond between a transition metal and a post-transition metal*

Molecule	ν(M—M), cm^{-1}	Ref.	Molecule	ν(M—M), cm^{-1}	Ref.
$Et_3PbMn(CO)_5$	155	a	$Ph_3SnMo(CO)_3Cp$	169	c
$Ph_3SnMn(CO)_5$	174	b	$Me_3SnMo(CO)_3Cp$	172	c
$BrPh_2SnMn(CO)_5$	176	a	$ClMe_2SnMo(CO)_3Cp$	186	c
$Br_2PhSnMn(CO)_5$	180	a	$Cl_2MeSnMo(CO)_3Cp$	195	c
$Me_3SnMn(CO)_5$	182	c	$Cl_3SnMo(CO)_3Cp$	190	c
$ClMe_2SnMn(CO)_5$	197	c	$Me_3SnW(CO)_3Cp$	168	c
$Cl_2MeSnMn(CO)_5$	201	c	$ClMe_2SnW(CO)_3Cp$	175	c
$Cl_3SnMn(CO)_5$	200	d	$Me_3SnFe(CO)_2Cp$	185	c
$BrMe_2SnMn(CO)_5$	191	c	$Ph_3SnFe(CO)_2Cp$	174	c
$Br_3SnMn(CO)_5$	184	d	$ClHgMn(CO)_5$	186	e
$IMe_2SnMn(CO)_5$	178	c	$BrHgMn(CO)_5$	156	e
$I_3SnMn(CO)_5$	153	d	$IHgMn(CO)_5$	120	e
$Me_3SnCo(CO)_4$	176	c	$Cl_3SnCo(CO)_4$	206	f
$Bu_3SnCo(CO)_4$	145	a	$Br_3SnCo(CO)_4$	188	a
$Ph_3SnCo(CO)_4$	175	a	$I_3SnCo(CO)_4$	159	a
$ClPh_2SnCo(CO)_4$	182	a	$ClBu_2SnCo(CO)_4$	182	a

a. Ref. 37.
b. Refs. 6, 30, 37.
c. Ref. 30.
d. Refs. 30, 37.
e. Refs. 31, 37.
f. Ref. 37 and Watters, K. L., Brittain, J. N., and Risen, W. M., Jr., *Inorg. Chem.*, 1969, **8**, 1347.

at higher frequency in the infrared, suggesting a linear Os_3 skeleton

$$\Gamma(Os_3)D_{\infty h} = \sigma_g^+(R) + \sigma_u^+(i.r.)$$

It is reasonable to suppose that the infrared-active deformation mode π_u δ(OsOsOs) is too low in frequency to be observed. The two halogen-sensitive Os—X stretching modes display the same selection rules; indeed the complete infrared and Raman spectra exhibit mutual exclusion, and may be satisfactorily assigned on the basis of a molecular symmetry point group D_{4h}. Despite the extensive mixing between ν(M—M) and ν(M—X), the ratio of the ν(M—M) fundamentals is in the order of that predicted $[\nu(a_{2u}) = \sqrt{3}\nu(a_{1g})]$ for a linear system by a simple valence force field.

The Raman spectrum of bis(methoxy)triosmium decacarbonyl, $Os_3(CO)_{10}(OCH_3)_2$, in contrast, shows[25] three lines that are probably identifiable as ν(M—M), two having counterparts in the infrared. The Os_3 skeleton clearly has lower symmetry in this type of derivative; a bent C_{2v}, structure with the metal atoms at the apices of an isosceles triangle implies three ν(M—M), active in both Raman and infrared:

$$\Gamma(\mathrm{Os_3})\mathrm{C_{2v}} = 2\mathrm{a_1}(\mathrm{R, i.r.}) + \mathrm{b_2}(\mathrm{R, i.r.})$$

It is reasonable to suppose that the highest frequency fundamental, which is the counterpart of the 'breathing' mode in the D_{3h} case, does not gain sufficient intensity to be observed in the infrared. While the selection rules point strongly to a degradation of molecular geometry from D_{3h} in the

TABLE 16.9. *ν(M—M) in compounds containing the Os_3 and Ru_3 skeletons*

Compound	Point group	ν(M—M), cm^{-1}, and assignment: Raman	ir.
$Ru_3(CO)_{12}$	D_{3h}	186 (a_1'), 152 (e′)	150 (e′)
$Os_3(CO)_{12}$	D_{3h}	161 (a_1'), 120 (e′)	121 (e′)
$Os_3(CO)_{12}Cl_2$	D_{4h}	116 (a_{1g})	163 (a_{2u})
$Os_3(CO)_{12}Br_2$	D_{4h}	100 (a_{1g})	157 (a_{2u})
$Os_3(CO)_{12}I_2$	D_{4h}	97 (a_{1g})	153 (a_{2u})
$Os_3(CO)_{10}(OMe)_2$	C_{2v}	172 (a_1), 136 (b_2), 119 (a_1)	132 (b_2), 121 (a_1)
$Os_3(CO)_{10}(OEt)_2$	C_{2v}	170 (a_1), 133 (b_2), 98 (a_1)	n.i.[a]
$Os_3(CO)_{10}H(OBu^n)$	C_s	178 (a′), 136 (a″), 90 (a′)	n.i.

[a] n.i. = not investigated.

parent carbonyl to C_{2v} in the bis-methoxide, they do not decide the direction of this distortion, i.e. whether the unique Os—Os bond is longer (as might be the case if the methoxide groups are bridging) or shorter (if the methoxide groups are terminal) than the two equivalent Os—Os bonds. A preliminary analysis suggested that the latter might be the case, but it has been pointed out that this solution is not unique. Recent X-ray crystallographic work has shown[26] that the Os_3 skeleton is indeed an isosceles triangle, but the apical angle is ~65° because of bridging methoxide groups which span the longer Os—Os bond. Thus we see that vibrational spectroscopy can give a simple and rapid, if rough, demonstration of the three basic shapes adopted by trinuclear metal compounds; some of the details are assembled in Table 16.9.

Many other linear systems have now been examined. Following the pioneer work of the late Professor Hans Stammreich,[27] spectra have now been obtained[28] for the complete series of compounds $M[M'(CO)_n]_2$, where M = Zn, Cd or Hg and M′ is Co ($n = 4$) or Mn ($n = 5$). All show the mutual exclusion demanded by a linear centrosymmetric skeleton, which is confirmed in $Zn[Co(CO)_4]_2$ and $Hg[Co(CO)_4]_2$ by X-ray structure determination.[29] Other, apparently linear systems have been investigated[30] in which mercury is bonded to two transition metal moieties containing carbonyl and cyclopentadienyl ligands, $Hg[M(CO)_n(C_5H_5)]_2$, M = Fe, Cr, Mo or W, or carbonyl and nitrosyl ligands, $Hg[Fe(CO)_3NO]_2$. The data for linear trinuclear molecules are summarized in Table 16.10.

TABLE 16.10. *ν(M—M) in compounds containing a linear tri-metal skeleton*

Compound	ν(M—M), cm⁻¹ Symmetric (Raman)	ν(M—M), cm⁻¹ Antisymmetric (i.r.)	Reference
$Zn[Mn(CO)_5]_2$	170	259	a
$Cd[Mn(CO)_5]_2$	160	209	a
$Hg[Mn(CO)_5]_2$	161	188	b
$Zn[Co(CO)_4]_2$	170	265	c
$Cd[Co(CO)_4]_2$	159	216	d
$Hg[Co(CO)_4]_2$	161	196	b
$Hg[Cr(CO)_3Cp]_2$	—	186	b
$Hg[Mo(CO)_3Cp]_2$	—	176	b
$Hg[W(CO)_3Cp]_2$	133	166	b
$Hg[Fe(CO)_2Cp]_2$	—	200	b
$Hg[Fe(CO)_3NO]_2$	152	186	e

a. Cram, A. G., and Ware, M. J., unpublished work.
b. Adams, D. M., Cornell, J. B., Dawes, J. L., and Kemmitt, R. D. W., *Inorg. Nuclear Chem. Letters*, 1967, **3**, 437; refs. 31, 37.
c. Ref. 28.
d. Refs. 27, 37.
e. Refs. 22, 30.

Many derivatives of the type $M(CO)_4(HgX)_2$ are readily prepared[31] for M = Fe, Ru or Os and X = Cl, Br or I. *Cis-* or *trans-*octahedral substitution at the transition metal is possible, giving rise to bent or linear trimetal skeletons, respectively. The vibrational spectra[32] in the ν(M—M) and ν(Hg—X) stretching regions strongly indicate a bent, or *cis-*, configuration for these species, since all four stretching modes appear to be observed in both Raman and infrared spectra. Similar bent configurations, where there is no likelihood of structural ambiguity, are encountered in the di-substituted tin halides $Cl_2Sn[Mn(CO)_5]_2$ and $Cl_2Sn[Co(CO)_4]_2$. The skeletal frequencies of bent trimetal species are summarized in Table 16.11.

TABLE 16.11. *$\nu(M—M)$ in compounds containing a bent tri-metal skeleton*

Compound	ν(M—M), cm^{-1}	ν(M—X), cm^{-1}	Reference
$(ClHg)_2Fe(CO)_4$	204, 186	291, 270	a
$(BrHg)_2Fe(CO)_4$	184, 169	226, 203	a
$(IHg)_2Fe(CO)_4$	154	210, 181	a
$(ClHg)_2Ru(CO)_4$	187, 170	280, 261	b
$(BrHg)_2Ru(CO)_4$	175, 157	220, 203	b
$(ClHg)_2Os(CO)_4$	169, 156	285, 270	b
$(BrHg)_2Os(CO)_4$	155, 140	216, 207	b
$(IHg)_2Os(CO)_4$	140, 135	192, 178	b
$Cl_2Sn[Mn(CO)_5]_2$	200, 170		c

a. Ref. 31.
b. Ref. 32.
c. Ref. 37.

(*c*) *Tetranuclear systems*

Of the tetrametal dodecacarbonyls of the cobalt group, the iridium compound, $Ir_4(CO)_{12}$, is the most attractive for vibrational study. Unlike its cobalt and rhodium analogues, the iridium compound has no bridging carbonyl groups; the metal cluster approximates closely to a regular tetrahedron in the crystal[33] and the molecular symmetry may be taken to be effectively T_d. On this basis, all three ν(M—M) fundamentals are predicted to be Raman active, but only one infrared active:

$$\Gamma(Ir_4)T_d = a_1(R) + e(R) + t_2(R, i.r.)$$

Assumption of a single-parameter S.V.F.F. for the cluster yields the predicted frequency ratios:

$$\nu(a_1) : \nu(t_2) : \nu(e) = 2 : \sqrt{2} : 1$$

Two of the ν(M—M) are readily identified[34] by their frequency and intensity in the Raman spectrum at 207 and 161 cm^{-1}. The former is assigned as the a_1 mode and the observation of the latter at 162 cm^{-1} in the infrared confirms its assignment as the t_2 mode. Identification of the remaining low frequency e mode is less straightforward, since it falls in the region of δ(CMC) fundamentals. Spiro has found at least six lines in the Raman spectrum below 160 cm^{-1}; the original assignment of a line at 105 cm^{-1} to the e mode—on the basis of the cluster frequency ratios—cannot be reconciled with a normal coordinate analysis, and a weaker line at 131 cm^{-1} is now preferred for this fundamental.

Recently, the vibrational spectra of the rather photosensitive $Rh_4(CO)_{12}$

have been recorded;[35] the crystal structure[36] shows a molecule of C_{3v} symmetry in which three edges of the Rh_4 tetrahedron each carry a bridging CO group. The degradation from T_d symmetry should lead to a splitting of the triply degenerate mode $t_2(T_d) \rightarrow (a_1 + e)(C_{3v})$ and a relaxation of the selection rules:

$$\Gamma(Rh_4)C_{3v} = 2a_1(R, i.r.) + 2e(R, i.r.)$$

As may be seen from Table 16.12, four fundamentals ascribable to ν(M—M) are indeed observed in the Raman spectrum, but only two of these (a_1 and e from the t_2 mode) show counterparts in the infrared. As in the case of

TABLE 16.12. *ν(M—M) of tetranuclear metal species*

Molecule	Point group	ν(M—M), cm^{-1}, and assignment	Reference
$Ir_4(CO)_{12}$	T_d	207 (a_1), 161 (t_2), 131 (e)	a
$Rh_4(CO)_{12}$	C_{3v}	225 (a_1), 200 (e), 176 (a_1), 128 (e)	b
$In[Co(CO)_4]_3$	C_{3h}	155 (e′), 93 (a′)	c
$Pb_4(OH)_4^{4+}$	T_d	130 (a_1), 87 (t_2), 60 (e)	d
$Tl_4(OEt)_4$	T_d	102 (a_1), 63 (t_2), 44 (e)	e
$(Me_3Pt)_4(OH)_4$	T_d	137 (a_1), 97 (t_2), 75 (e)	f
$(Me_3Pt)_4Cl_4$	T_d	99 (a_1), 79 (t_2), 57 (e)	f
$(Me_3Pt)_4I_4$	T_d	88 (a_1), 65 (t_2), 48 (e)	f

a. Ref. 34.
b. Refs. 22, 35.
c. Ref. 37.
d. Maroni, V. A., and Spiro, T. G., *Inorg. Chem.*, 1968, **7**, 188.
e. Maroni, V. A., and Spiro, T. G., *Inorg. Chem.*, 1968, **7**, 193.
f. Bulliner, P. A., Maroni, V. A., and Spiro, T. G., unpublished work.

$Os_3(CO)_{12}$ and $Os_3(CO)_{10}(OCH_3)_2$, it appears that although the lowered symmetry is sufficient to produce an observable frequency splitting in the triply degenerate mode, it does not lend sufficient intensity for the newly allowed infrared fundamentals to be detected.

A tetranuclear metal carbonyl derivative of rather different type may be obtained from the reaction of indium tribromide with sodium tetracarbonylcobaltate. The product, $In[Co(CO)_4]_3$, is a red, photosensitive material, decomposing rapidly in air. The Raman spectra obtained so far,[37] although of poor quality, are consistent with a planar trigonal metal skeleton, of probable symmetry C_{3h} if the orientation of the CO groups is included:

$$\Gamma(InCo_3)C_{3h} = a'(R) + e'(R, i.r.)$$

The e′ mode appears in the Raman and infrared spectra at 155 cm^{-1}, while the a′ fundamental may be tentatively identified with a Raman line at 93 cm^{-1}.

A number of tetrameric metal species containing bridging oxygen functions have been carefully examined by Maroni and Spiro[38] for evidence of metal–metal interaction. All have basically the same structure of twin interpenetrating tetrahedra (Kepler's 'Stella Octangula'), one composed of metal atoms the other of bridging groups lying symmetrically above each face, so that the molecular point group is T_d. Table 16.12 shows that the low-frequency Raman spectra of species such as $Pb_4(OH)_4^{4+}$, $Tl_4(OR)_4$, $[(CH_3)_3Pt]_4(OH)_4$ exhibit quite intense lines of the type expected for a T_d M_4 system. In a normal coordinate analysis it is found that the spectra can best be fitted by employing a metal–metal bond-stretching force constant in the potential function, but the authors are careful to point out that this analysis may not be unique; rather, it is argued, the relatively high Raman intensities must be taken as evidence for direct metal–metal interaction, especially in view of the absence of such intense low-frequency lines from the spectra of bridged polynuclear species where no metal–metal interaction can reasonably be expected to occur.

If other heavy atoms are present in the cluster, detection of metal–metal interaction may be more difficult. The tri-n-butylphosphine copper iodide tetramer $(Bu_3PCuI)_4$ has a structure analogous to the foregoing molecules, with a terminal phosphine ligand on each copper atom. Its Raman spectrum[19] shows three quite intense lines in the low frequency region at 113, 120 and 184 cm^{-1} which may be the three totally symmetric 'breathing' modes of the cluster, but much better spectra, including polarization measurements, will be needed before any firm interpretation can be made.

(*d*) *Pentanuclear systems*

Little is known of the vibrational spectra of compounds containing five metal atoms. The only structural type for which data have been obtained is the regular tetrahedral skeleton $M'M_4$, for which the selection rules are

$$\Gamma(M'M_4)T_d = a_1(R) + e(R) + 2t_2(R, \text{i.r.})$$

The two ν(M—M) stretching fundamentals, $a_1 + t_2$, have been identified[8] in the Raman spectrum of $Sn[SnPh_3]_4$ at 101 cm^{-1} (a_1) and 159 cm^{-1} (t_2). The bending fundamentals remain unknown. The dark red, photosensitive tin compound, $Sn[Co(CO)_4]_4$, has also been examined[37] but the very poor quality of the Raman spectrum allows only a very tentative assignment of ν(M—M) to lines at 111 cm^{-1} (a_1), and 159 cm^{-1} (t_2).

(*e*) *Hexanuclear systems*

The majority of hexanuclear systems studied to date contain a regular octahedral cluster of metal atoms. In compounds of the type[39] $(Mo_6X_8)X'^{2-}_6$ where X and X′ are halogens, the eight 'cage' halogens X occupy triply bridging positions above each face of the Mo_6 octahedron, thereby defining the corners of a cube; each of the six terminal halogens X′ is bonded to a molybdenum atom in a direction normal to a cube face. A highly symmetrical structure, point group O_h, results. In the Group V derivatives of type[40] $(Nb_6X_{12})X'_6$ the structure is similar, but the twelve cage halogen atoms now bridge the edges of the Nb_6 octahedron, thereby defining the vertices of a cuboctahedron. Although bands in the infrared spectra of the molybdenum compounds have been variously ascribed[41] to ν(M—M), a Raman investigation[42] in solution with approximate normal coordinate analysis of the vibrations of these unusually rigid species suggests that mixing may be very extensive and that no fundamental can be simply labelled as ν(M—M). It is therefore desirable to interpret the spectra as far as possible without preconceptions about group frequencies. The selection rules for the complete ion—$Mo_6Cl_8Br_6^{2-}$ serves as a convenient example—are

$$\Gamma = 3a_{1g} + 3e_g + 2t_{1g} + 4t_{2g} + a_{2u} + e_u + 5t_{1u} + 3t_{2u}$$

Ten of these normal modes $(3a_{1g} + 3e_g + 4t_{2g})$ are permitted as fundamentals in the Raman spectrum, five $(5t_{1u})$ in the infrared, and the remaining seven are totally inactive. The observed spectra fit these selection rules well.

The key to the analysis of the rather complicated spectra lies in identifying the three totally symmetric vibrations by their polarization in the Raman spectrum. In each case, two such features are easily found, but the third totally symmetric fundamental must be assigned to one of the very weak Raman lines whose states of polarization could not be determined. An approximate normal coordinate analysis for the a_{1g} block assists this assignment. With a diagonal $\mathscr{F}$ matrix containing k(Mo—Mo), k(Mo—X) and k(Mo—X′), it is found that one of the a_{1g} normal modes corresponds to an almost 'pure' breathing motion of the X_8 cube of halogen atoms; this is plausibly assigned to an intense, polarized Raman line, ν_1, whose frequency is almost independent of the nature of the terminal halogen, X′. The two remaining a_{1g} modes, ν_2 and ν_3, may be described as simultaneous breathing of the Mo_6 and X'_6 octahedra; in the lower frequency mode, ν_3, the breathing is in phase, in the other, ν_2, it is in antiphase. A rough estimate of the relative intensities of ν_2 and ν_3 may be formed on the assumption that Wolkenstein's bond polarizability theory is applicable to these species. Whereas the polarizability derivative is large in ν_3, there is a tendency for the polarizability derivatives of the two octahedra to cancel one another in the antiphase motion of ν_2. Thus the weakness of this fundamental receives an explanation, and an important exception is found to the general rule that totally symmetric stretching modes give rise to intense Raman fundamentals. There is evidence that similar effects occur in other molecules containing concentric polyhedra of atoms. The assigned a_{1g} modes of these cluster compounds are summarized in Table 16.13, together with the approximate force constants derived from them; the force constants seem physically reasonable and suggest a distinct Mo—Mo bonding interaction in the cluster that is consistent with the observed internuclear distances and with M.O. theories of bonding.

TABLE 16.13. *Fundamentals of* a_{1g} *block in octahedral* $Mo_6X_8X'_6$ *clusters*

Species	ν_1	ν_2 cm^{-1}	ν_3	k(Mo—Mo)	k(MoX) mdyn Å^{-1}	k(MoX′)
$Mo_6Cl_8Cl_6^{2-}$	366	318	236	1·6	1·1	1·8
$Mo_6Cl_8Br_6^{2-}$	356	318	159	1·6	1·1	1·6
$Mo_6Cl_8I_6^{2-}$	338	317	115	1·4	1·1	1·3
$Mo_6Br_8Cl_6^{2-}$	—	211	—	—	1·0	—
$Mo_6Br_8Br_6^{2-}$	—	207	158	—	1·0	—
$Mo_6Br_8I_6^{2-}$	—	212	111	—	1·0	—
$Mo_6I_8Cl_6^{2-}$	—	108	213	—	0·4?	—

Interatomic distances in $H_2[Mo_6Cl_8Cl_6].8H_2O$:

r(Mo—Mo) = 2·64; r(Mo—Cl) = 2·55; r(Mo—Cl′) = 2·48 Å (ref. 39).

Of the many related hexanuclear metal halide clusters, the tungsten analogues are currently under investigation;[43] the infrared spectra of the niobium compounds have been reported[44] and the hexameric dihalides of platinum and palladium have been examined in the infrared.[45]

The hexanuclear metal carbonyl $Rh_6(CO)_{16}$, which has an octahedral structure with four bridging CO groups located over alternate faces of the octahedron (point group T_d), has yielded[46] Raman lines at 221 and 173 cm^{-1} that have been assigned as ν(M—M) modes. Maroni and Spiro[47] have examined the Raman and infrared spectra of the hydrolyzed Bi(III) species, $Bi_6(OH)_{12}^{6+}$, which has a structure analogous to that of $Nb_6Cl_{12}^{2+}$, and find a high Raman intensity in the low-frequency fundamentals at 177 (a_{1g}), 107 (t_{2g}) and 88 (e_g) cm^{-1} that argues for some Bi—Bi interaction despite the large (3·71 Å) Bi—Bi distance.

Other hexanuclear M—M species are known in which several trihalogenotin groups are bonded to a transition metal. The ion $Pt(SnX_3)_5^{3-}$, which has a trigonal-bipyramidal structure, and the halogen-bridged complexes of type $Rh_2X_2(SnX_3)_4^{4-}$ have been investigated[48] in the infrared and bands assigned to ν(Pt—Sn) at 208 cm^{-1} and to ν(Rh—Sn) at 209 cm^{-1}.

D. Force constants and other bond parameters

A complete assignment of the fundamentals has been achieved for very few M—M compounds. It is therefore rare that force constants can be calculated from a complete set of normal frequencies, and even when this is possible—as in the case of the mercurous halides described earlier—a general valence force field requires the evaluation of more parameters than there are observed frequencies. The usual supplementary data, drawn, for example, from Coriolis coupling constants, isotopically substituted molecules or vibrational amplitudes, are rarely available for M—M compounds. In view of these serious limitations, it might well be argued that any attempt to calculate force constants for these molecules is bound to be profitless. However, even very approximate values for bond-stretching force constants, k(M—M), have more significance than the raw frequencies, ν(M—M), as an indication of trends in bond character within a series of related molecules, for the simple reason that the masses of the constituent metal atoms may vary so widely; some correction for this effect is better than none.

Much of the uncertainty in analyzing the vibrations of polynuclear metal carbonyls lies in the assignment of the δ(CMC) fundamentals. These fall in the same frequency region as ν(M—M) and therefore mixing between symmetry coordinates of the same species is likely to be severe. To take the simplest case, even in compounds of the type $XM(CO)_n$, it appears that the totally symmetric CMC deformation has not yet been unambiguously assigned: as pointed out by Long,[49] Wolkenstein's bond polarizability theory would suggest that such modes may have very low intensities in the Raman spectrum, and the polarization would therefore be correspondingly hard to detect. It may be hoped that single crystal spectra will, in the near future, remedy this deficiency.

To date, the best estimates of k(M—M) in polynuclear carbonyls have been obtained by Spiro[8] and co-workers. For the bi-, tri- and tetranuclear carbonyls, normal coordinate analysis of all the low-frequency motions has been performed, using a simple valence force field. Since the chosen assignment of the δ(CMC) modes is directly linked with the results of normal coordinate analysis and its inherent assumptions, there is some risk of circularity of argument, which is often present in such analyses of large molecules. Nonetheless, the resulting force constants and potential energy distribution appear to be quite informative. Table 16.14 summarizes the results of Spiro's calculations for several metal carbonyls. It is found that mixing between ν(M—M) and δ(CMC) is greater for the light metal atoms, and greater in the antisymmetric than in the symmetric ν(M—M). Particularly interesting is the indication that the bond-stretching force constant k(M—M) increases as we go down the Periodic Group, i.e. k(Os—Os) > k(Ru—Ru) and k(Re—Re) > k(Mn—Mn).

TABLE 16.14. *Metal–metal force constants obtained by Spiro*[c] *for polynuclear metal carbonyls*

Molecule	Point group	k(M—M)[a] mdyn Å^{-1}	k'(M—M)[b] mdyn Å^{-1}	r(M—M) Å
$Mn_2(CO)_{10}$	D_{4d}	0·59	0·41	2·92
$Re_2(CO)_{10}$	D_{4d}	0·82	0·82	3·02
$MnRe(CO)_{10}$	C_{4v}	0·81	0·62	
$Ru_3(CO)_{12}$	D_{3h}	0·82	0·68	2·85
$Os_3(CO)_{12}$	D_{3h}	0·91	0·93	2·89
$Ir_4(CO)_{12}$	T_d	1·69	1·22	2·68

[a] (M—M) bond-stretching force constant from normal coordinate analysis.
[b] Approximate force constant obtained by ignoring the CO groups ($M_e = m_M$) and using only the totally symmetric frequency.
[c] Ref. 8.

In seeking to obtain force constants for a wider range of compounds, yet more approximations must be made owing to the present inadequacy of the vibrational data. The most useful assumption, if indeed any is valid, would enable one to attach an effective mass, M_e, to an $M(CO)_n$ group considered as a single entity. The first comparisons were made[6] by simply taking the sum of the masses, a method that has proved quite successful for methyl compounds, i.e. $M_e = m_M + n.m_{co}$, but this gave force constants that were almost certainly too high. Spiro has suggested that when the attached ligands contain only first-row atoms the other extreme, namely that of taking the mass of the metal alone, i.e. $M_e = m_M$, is a better approximation. It may be seen from Table 16.14 that force constants obtained in this way are certainly closer to the values given by a fuller normal-coordinate analysis and, more important, reflect the same trends in k(M—M) with changing metal mass. In his critical assessment of the value of this approximation (the 'neglect of first-row ligands'), Spiro concludes that, *faute de mieux*, the method yields force constants that deviate by no more than 30 per cent from the correct values, and often by much less. There would thus seem to be a rough basis for comparing the nature of M—M bonds as judged by their force constants in related compounds.

Table 16.15 summarizes the results of such a survey, from which it may be observed that for the sequence of Group IIB metals, Zn, Cd, Hg, there appears to be an increase of k(M—M). This contrasts with the behaviour

TABLE 16.15. *Approximate metal–metal bond-stretching force constants*

Bond	k(M—M), mdyn $Å^{-1}$	Compound	Vibrational model
Zn—Zn	0·6	$Zn/ZnCl_2$	Diatomic Zn—Zn
Cd—Cd	1·1	$Cd_2(AlCl_4)_2$	Diatomic Cd—Cd
Hg—Hg	1·9	$Hg_2(OH_2)_2(NO_3)_2$	$D_{\infty h}$, 4-atomic; S.V.F.F.
Ge—Ge	1·3	$Ge_2(CH_3)_6$	D_{3d}, 8-atomic; CH_3 as point mass; S.V.F.F. a_{1g} block
Sn—Sn	1·0	$Sn_2(CH_3)_6$	
Pb—Pb	0·8	$Pb_2(CH_3)_6$	
Zn—Co	0·9	$Zn[Co(CO)_4]_2$	$D_{\infty h}$, triatomic; S.V.F.F. plus one interaction
Cd—Co	0·8	$Cd[Co(CO)_4]_2$	
Hg—Co	0·9	$Hg[Co(CO)_4]_2$	
Zn—Mn	0·9	$Zn[Mn(CO)_5]_2$	
Cd—Mn	0·8	$Cd[Mn(CO)_5]_2$	
Hg—Mn	0·8	$Hg[Mn(CO)_5]_2$	
Hg—Mn	0·9	$XHgMn(CO)_5$ (X = halogen)	$C_{\infty v}$, triatomic; S.V.F.F.
Sn—Co	0·7	$(CH_3)_3SnCo(CO)_4$	C_{3v}, 5-atomic; X_3SnM, $a_1 + e$ blocks S.V.F.F. + one interaction
Sn—Mn	0·7	$(CH_3)_3SnMn(CO)_5$	
Sn—Co	0·9	$X_3SnCo(CO)_4$	
Sn—Mn	0·9	$X_3SnMn(CO)_5$ (X = halogen)	

of the Group IVB elements, Ge, Sn, Pb, of the alkali metals (Table 16.2), and of typical non-metals, like the halogens. Although there are no grounds *a priori* for correlating the two quantities, a parallel may be noted (Table 16.2) between the force constant, k(M—M), and bond dissociation energy, D(M—M). An increased bond strength for the heavier transition metals such as rhenium is in keeping with the results of d-orbital overlap calculations, with the heats of atomization of the metals and with the general body of chemical knowledge of M—M compounds that has grown up in recent years. For instance, metal carbonyls containing bridging carbonyl groups are much commoner among the first- and second-row metals; in the third row the carbonyls usually contain simple metal–metal bonds unsustained by bridging carbonyl groups.

In conclusion, the evaluation of accurate M—M force constants demands more information than we have at present, although judicious approximations have revealed chemically significant trends; the most useful application of vibrational spectroscopy will continue to lie in the detection of M—M bonding and the inference of the shape of the M—M framework.

References

1. Woodward, L. A., *Phil. Mag.*, 1934, **18**, 823.
2. Baird, M. C., *Progr. Inorg. Chem.*, 1968, **9**, 1; Vyazankin, N. S., Razuvaev, G. A., and Kruglaya, O. A., *Organometallic Chem. Rev.*, 1968, **3A**, 323.
3. Herzberg, G., 'Molecular Spectra and Molecular Structure, Vol. I, Diatomic Molecules', Van Nostrand, Princeton, 1950.
4. Almenningen, A., Jacobsen, G. G., and Seip, H. M., *Acta Chem. Scand.*, 1969, **23**, 685.
5. Johnson, B. F. G., Lewis, J., Williams, I. G., and Wilson, J. M., *J. Chem. Soc.* (*A*), 1967, 341.
6. Gager, H. M., Lewis, J., and Ware, M. J., *Chem. Comm.*, 1966, 616.
7. Woodward, L. A., and Long, D. A., *Trans. Faraday Soc.*, 1949, **45**, 1131; Long, T. V., and Plane, R. A., *J. Chem. Phys.*, 1965, **43**, 457.
8. Spiro, T. G., *Progr. Inorg. Chem.*, 1970, **11**, 1.
9. Abel, E. W., Hendra, P. J., McLean, R. A. N., and Quarashi, M. M., *Inorg. Chim. Acta*, 1969, **3**, 77.
10. Cooney, R. P. J., Hall, J. R., and Hooper, M. A., *Austral. J. Chem.*, 1968, **21**, 2145; Goldstein, M., *Spectrochim. Acta*, 1966, **22**, 1389; Stammreich, H., and Teixeira Sans, T., *J. Mol. Structure*, 1968, **1**, 55.
11. Corbett, J. D., *Inorg. Chem.*, 1962, **1**, 700; Kerridge, D. H., and Tariq, S. A., *J. Chem. Soc.* (*A*), 1967, 1122.
12. Grdenić, D., *Quart. Rev.*, 1965, **19**, 303, and references cited therein.
13. Durig, J. R., Lau, K. K., Nagarajan, G., Walker, M., and Bragin, J., *J. Chem. Phys.*, 1969, **50**, 2130.
14. Beattie, I. R., and Gilson, T. R., *Proc. Roy. Soc.*, 1968, **A307**, 407.
15. Hartley, D., and Ware, M. J., unpublished work.
16. Dorm, E., *Acta Chem. Scand.*, 1969, **23**, 1607.
17. Evans, C. A., and Taylor, M. J., *Chem. Comm.*, 1969, 1201.
18. Adams, D. M., and Squire, A., *J. Chem. Soc.* (*A*), 1968, 2817; Cotton, F. A., and Wing, R. M., *Inorg. Chem.*, 1965, **4**, 1328; but see also Hyams, I. J., Jones, D., and Lippincott, E. R., *J. Chem. Soc.* (*A*), 1967, 1987; Quicksall, C. O., and Spiro, T. G., *Inorg. Chem.*, 1969, **8**, 2363.
19. Hartley, D., and Ware, M. J., unpublished observations.
20. Cotton, F. A., Curtis, N. F., Johnson, B. F. G., and Robinson, W. R., *Inorg. Chem.*, 1965, **4**, 326; Cotton, F. A., and Harris, C. B., *ibid.*, p. 330; Cotton, F. A., *ibid.*, p. 334.

21. Edwards, H. G. M., and Ware, M. J., unpublished work.
22. Griffith, W. P., and Wickham, A. J., *J. Chem. Soc.* (*A*), 1969, 834.
23. Corey, E. R., and Dahl, L. F., *Inorg. Chem.*, 1962, **1**, 521.
24. Quicksall, C. O., and Spiro, T. G., *Inorg. Chem.*, 1968, **7**, 2365.
25. Hartley, D., Kilty, P. A., and Ware, M. J., *Chem. Comm.*, 1968, 493.
26. Duckworth, V., and Mason, R., personal communication.
27. Stammreich, H., Kawai, K., Sala, O., and Krumholz, P., *J. Chem. Phys.*, 1961, **35**, 2175.
28. Adams, D. M., Crosby, J. N., and Kemmitt, R. D. W., *J. Chem. Soc.* (*A*), 1968, 3056; Burlitch, J. M., *J. Organometallic Chem.*, 1967, **9**, 11; Cram, A. G., and Ware, M. J., unpublished work.
29. Lee, B., Burlitch, J. M., and Hoard, J. L., *J. Amer. Chem. Soc.*, 1967, **89**, 6392; Sheldrick, G. M., and Simpson, R. N. F., *J. Chem. Soc.* (*A*), 1968, 1005.
30. Carey, N. A. D., and Clark, H. C., *Chem. Comm.*, 1967, 292; *Inorg. Chem.*, 1968, **7**, 94; Beck, W., and Noack, K., *J. Organometallic Chem.*, 1967, **10**, 307.
31. Brier, P. N., Chalmers, A. A., Lewis, J., and Wild, S. B., *J. Chem. Soc.* (*A*), 1967, 1889.
32. Adams, D. M., Cook, D. J., and Kemmitt, R. D. W., *Nature*, 1965, **205**, 589; Adams, D. M., and Chandler, P., *Chem. and Ind.*, 1965, 269; Radford, C. W., van Bronswyk, W., Clark, R. J. H., and Nyholm, R. S., *J. Chem. Soc.* (*A*), 1968, 2456.
33. Wilkes, G. R., *Diss. Abs.*, 1966, **26**, 5029.
34. Quicksall, C. O., and Spiro, T. G., *Chem. Comm.*, 1967, 839; *Inorg. Chem.*, 1969, **8**, 2011.
35. Abel, E. W., and McLean, R. A. N., *Inorg. Nuclear Chem. Letters.*, 1969, **5**, 381.
36. Wei, C. H., and Dahl, L. F., *J. Amer. Chem. Soc.*, 1966, **88**, 1821.
37. Cram, A. G., and Ware, M. J., unpublished observations.
38. Maroni, V. A., and Spiro, T. G., *J. Amer. Chem. Soc.*, 1967, **89**, 45; *Inorg. Chem.*, 1968, **7**, 188; *Inorg. Chem.*, 1968, **7**, 193.
39. Sheldon, J. C., *J. Chem. Soc.*, 1960, 1007; 1961, 750; 1962, 410; Brosset, C., *Arkiv Kemi, Min., Geol.*, 1945, **A20**, No. 7; **A22**, No. 4; *Arkiv Kemi*, 1950, **1**, 353.
40. See ref. 2.
41. Clark, R. J. H., Kepert, D. L., Nyholm, R. S., and Rodley, G. A., *Spectrochim. Acta*, 1966, **22**, 1697; Cotton, F. A., Wing, R. M., and Zimmerman, R. A., *Inorg. Chem.*, 1967, **6**, 11; Mattes, R., *Z. anorg. Chem.* 1968, **357**, 30.
42. Hartley, D., and Ware, M. J., *Chem. Comm.*, 1967, 912.
43. McCarley, R. E., Hartley, D., and Ware, M. J., unpublished work.
44. Boorman, P. M., and Straughan, B. P., *J. Chem. Soc.* (*A*), 1966, 1514; Mackay, R. A., and Schneider, R. F., *Inorg. Chem.*, 1968, **7**, 455.
45. Mattes, R., *Z. anorg. Chem.*, 1969, **364**, 290.

46. See ref. 22.
47. Maroni, V. A., and Spiro, T. G., *J. Amer. Chem. Soc.*, 1966, **88**, 1410; *Inorg. Chem.*, 1968, **7**, 183.
48. Adams, D. M., and Chandler, P. J., *Chem. and Ind.*, 1965, 296.
49. Long, D. A., *Proc. Roy. Soc.*, 1953, **A217**, 203.

CHAPTER 17

Infrared and Raman Spectra of Organometallic and Related Compounds

J. R. HALL

A. Introduction

It seems appropriate that in this volume a chapter be included dealing with methyl–metal and related silyl and germyl compounds which have been investigated in the laboratory of Dr. L. A. Woodward. The interest in the first group was initiated with the study of trimethylboron.[1] After this investigation was completed there was a natural extension to trimethylgallium and trimethylindium[2] and then subsequently to the dimethylthallium(III) cation[3] and to various methylmercury(II) systems.[4] The concern with silyl compounds began with the investigation of trisilylamine.[5] A number of silyl derivatives of Group VI elements[6] as well as trigermylphosphine[7] have since been examined in Woodward's laboratory.

Alkyl–metal compounds have been the subject of study by infrared and Raman spectroscopy for many years. Although the novelty of metal–carbon σ-bonding has long passed, the metal–carbon stretching frequency and associated force constant remain interesting aspects of the vibrational spectra. The stereochemistry of the metal in some compounds has been worked out from a consideration of the number of metal–carbon stretching modes observed in the infrared and Raman spectra. For example, the dimethylgallium(III)[8] and dimethylgermanium(IV)[9] species in aqueous solution adopt a bent C—M—C skeleton in contrast to the linear skeleton of the isoelectronic dimethylzinc.[10] Some trimethyltin compounds contain a planar SnC_3 skeleton[11] and exhibit one band in the infrared spectrum attributable to Sn—C stretching: others have a pyramidal skeleton[12] and display two infrared Sn—C stretching frequencies.

Considerable interest was aroused in silyl compounds after a paper had appeared on the electron diffraction investigation of trisilylamine, $(SiH_3)_3N$, which announced[13] that the NSi_3 skeleton was planar. This stereochemistry for nitrogen was supported by the vibrational spectra[5, 14]. Trisilylphosphine,[15] on the other hand, has a pyramidal skeleton, and recently the infrared spectrum of trigermylamine[16] has been interpreted in favour of a pyramidal NGe_3 skeleton. An explanation for these structural differences is desirable.

B. Linear dimethylmetal compounds

The Raman spectrum[17] of dimethylzinc was first obtained in 1930 and its infrared spectrum[18] in 1940. By 1949 the vibrational spectra of the cadmium and mercury compounds[19] had also been recorded. Linearity of the CMC skeleton was deduced from an investigation of the pure rotational Raman spectra.[20] In aqueous solution the linear species $(CH_3)_2Tl^+$,[3] $(CH_3)_2Sn^{2+}$ [21] and $(CH_3)_2Pb^{2+}$ [22] have been characterized by Raman spectroscopy. The infrared spectrum of the unsaturated vapour of $(CH_3)_2Be$ has been reported recently and interpreted in terms of a monomer with a linear skeleton.[23]

Since the establishment of essentially free rotation of the methyl groups in dimethylzinc and -mercury,[24] discussion has centred around the appropriate point group to be used for spectral analysis. For various 'rest' positions of the methyl groups, the model is described alternatively by D_{3h} (eclipsed), D_{3d} (staggered) and D_3 (skew). However, Longuet-Higgins[25] showed that conventional point groups cannot be used to classify the states of non-rigid ethane-type molecules and introduced the symmetry group G_{36}. Bunker[26] has demonstrated that to determine the selection rules on the rotational, torsional and vibrational quantum numbers, the rotational, torsional and vibrational parts of the wave functions must be classified separately in one group, viz., the double group $G_{36}^{\dagger}$. The group D_{3h}' introduced by Howard, and previously used for the assignment of observed frequencies to the various normal modes of vibration of many freely rotating, linear $(CH_3)_2M$ molecules, is a sub-group of $G_{36}^{\dagger}$.

The $3N-7$ normal coordinates (torsional coordinate ignored) are functions of the Cartesian displacement coordinates which have coefficients that are dependent on the torsional angle. The form of these coefficients depends on the force field in the molecule. Consequently their species and the species of the normal coordinates will also depend on the force field. The criterion for the choice of symmetry coordinates is to form block diagonals of the *G* and *F* matrices to the maximum practical extent, regardless of the force constants, and in such a manner that the *G* matrix is independent of the torsional angle.

The symmetry coordinates suitable for normal coordinate calculation are shown to be of the species $3a_{1s} + 3a_{4s} + 4e_{1d} + 3e_{2d}$. The details of the normal modes and their distribution among the symmetry species are set out in Table 17.1. The main point of interest lies in the derived activity

TABLE 17.1. *Distribution of $3N-7$ normal modes among the symmetry species for the linear molecule $(CH_3)_2M$ with freely rotating CH_3 groups*

Species	Activity	Mode
a_{1s}	Raman	C—H stretching CH_3 deformation M—C stretching
a_{4s}	Infrared	C—H stretching CH_3 deformation M—C stretching
e_{1d}	Infrared and Raman	C—H stretching CH_3 deformation CH_3 rocking C—M—C bending
e_{2d}	Infrared and Raman	C—H stretching CH_3 deformation CH_3 rocking

of the skeletal bending mode. From the early studies of the Raman spectra of dimethylzinc, -cadmium and -mercury concern has been expressed over the observation of lines at, respectively, 144, 150 and 155 cm^{-1} in the spectra of these compounds, ascribable to the skeletal bending mode, which, according to the point groups applied, was Raman-forbidden. Support for the assignment of these lines to skeletal bending was obtained by Woodward[27] who observed the C—Hg—C bending mode in the far-infrared spectrum of $(CH_3)_2Hg$ at 153 cm^{-1}, that is, in coincidence with the observed Raman line for the liquid. Selection rules under $G_{36}^{\dagger}$ predict the bending mode to be both infrared and Raman active.

The Raman spectrum of the deuterated compound has recently been reported[28] for the first time and a line attributable to the C—Hg—C bending mode was observed at 134 cm^{-1}. Table 17.2 sets out the fundamentals for $(CH_3)_2Hg$ and $(CD_3)_2Hg$ and their assignment. Infrared spectra refer to the gaseous compounds, Raman spectra to the liquid state.

TABLE 17.2. *Vibrational spectra and assignments for $(CH_3)_2Hg$ and $(CD_3)_2Hg$*

Frequency, cm^{-1}				
$(CH_3)_2Hg$		$(CD_3)_2Hg$		
Infrared	Raman	Infrared	Raman	Assignment
—	—		134 dp	δ(C—Hg—C), e_{1d}
153	155 dp	—	—	δ(C—Hg—C), e_{1d}
—	—	—	470 p	ν(Hg—C), a_{1s}
—	—	491	—	ν(Hg—C), a_{4s}
—	514 p	—	—	ν(Hg—C), a_{1s}
540	—	—	—	ν(Hg—C), a_{4s}
—	—	598		ρCD_3, e_{1d}
—	—	669	669 dp	ρCD_3, e_{2d}
700	700 dp	—	—	ρCH_3, e_{2d}
780		—	—	ρCH_3, e_{1d}
—	—	—	907 p	δCD_3, a_{1s}
—	—	931	—	δCD_3, a_{4s}
—	—	1030		δCD_3, e_{1d}
—	—		1033 dp	δCD_3, e_{2d}
—	1181 p	—	—	δCH_3, a_{1s}
1191	—	—	—	δCH_3, a_{4s}
1405		—	—	δCH_3, e_{1d}
—	1437 dp	—	—	δCH_3, e_{2d}
—	—	—	2111 p	ν(C—D), a_{1s}
—	—	2114	—	ν(C—D), a_{4s}
—	—		2223 dp	ν(C—D), e_{2d}
—	—	2224		ν(C—D), e_{1d}
2892	—	—	—	ν(C—H), a_{4s}
—	2911 p	—	—	ν(C—H), a_{1s}
2962		—	—	ν(C—H), e_{1d}
	2970 dp	—	—	ν(C—H), e_{2d}

The three polarized Raman lines in the spectrum of $(CH_3)_2Hg$ at 2911, 1181 and 514 cm^{-1} are assigned to the totally symmetric C—H stretching, methyl deformation and Hg—C stretching modes respectively. The infrared bands at 2892, 1191 and 540 cm^{-1} are ascribed to the antisymmetric C—H stretching, methyl deformation and Hg—C stretching modes of species a_{4s}. These bands do not appear to have counterparts in the Raman spectrum. On the assumption that e_{2d} modes will be more intense in the Raman effect than e_{1d}, the Raman lines at 2970, 1437 and 700 cm^{-1} are assigned to C—H stretching, methyl deformation and rocking vibrations of species e_{2d}. This leaves the low frequency feature of 155 cm^{-1} to be allocated to the skeletal bending mode (e_{1d}). Non-coincident infrared bands at 2962, 1405 and 780 cm^{-1} are assigned to the e_{1d} modes of C—H stretching, methyl deformation and rocking. The vibrational spectrum of the deuterated compound is assigned in a similar manner.

It is well known that in vibrational spectra of series of analogous compounds of zinc, cadmium and mercury the usual trend of decreasing frequencies with increasing atomic number is not followed. For example, the symmetrical stretching frequencies for ZnI_4^{2-}, CdI_4^{2-} and HgI_4^{2-} are 122, 117 and 126 cm^{-1}, respectively. The series of dimethyl compounds of these elements is no exception to this pattern. Figure 17.1 illustrates the

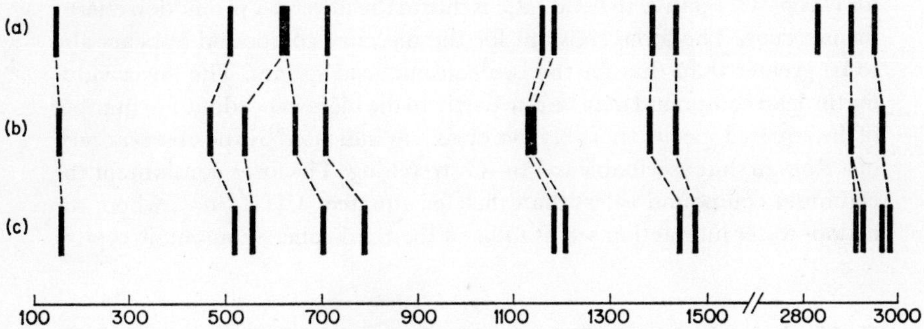

FIGURE 17.1. *Diagrammatic representation of the Raman spectra of (a) dimethylzinc, (b) dimethylcadmium and (c) dimethylmercury.*

trends in the frequencies of the various kinds of modes in the dimethyl derivatives. The assignment of features in the zinc and cadmium compounds may be obtained by correlation with the dimethylmercury spectrum which has been described above. The trend reversal is most marked in the cases of M—C stretching (about 500 cm^{-1}) and methyl deformation (about 1200 cm^{-1}). The decreasing separation between the symmetric and antisymmetric M—C stretching frequencies as the atomic number of M increases is normal.

Miles *et al.*[29] have recently carried out a normal coordinate analysis on a number of linear $(CH_3)_2M$ species including dimethylcadmium and dimethylmercury. Although calculations were based on the use of the D_{3d} point group, the relative values for the force constants are of significance. Table 17.3 contains M—C stretching force constants (U.B.F.F.) for the cadmium and mercury compounds, together with their isoelectronic species $(CH_3)_2Sn^{2+}$ and $(CH_3)_2Pb^{2+}$, respectively. The increase in force constant from $(CH_3)_2Cd$ to $(CH_3)_2Sn^{2+}$ is normal, as is the decrease from $(CH_3)_2Sn^{2+}$ to $(CH_3)_2Pb^{2+}$ (direction of increasing atomic number). However, there is a possible increase in force constant on passing from the

TABLE 17.3. *M—C stretching force constants for some linear $(CH_3)_2M$ species*

Species	k_{M-C} (mdyn Å^{-1})	Species	k_{M-C} (mdyn Å^{-1})
$(CH_3)_2Cd$	1·11 ± 0·23	$(CH_3)_2Sn^{2+}$	1·51 ± 0·02
$(CH_3)_2Hg$	1·25 ± 0·23	$(CH_3)_2Pb^{2+}$	0·96

cadmium to the mercury compound. Adams[30] claims that the higher value for ν(Hg—C) relative to ν(Cd—C) is due to the lower d-s promotion energy for mercury. The force constant for the mercury compound appears also to be greater than that for the isoelectronic lead species. The lower value for the lead compound may be due partly to the higher coordination number of the aquated species (possibly five or six), as indicated by the observation[22] of a Raman shift ascribable to Pb—O stretching. The force constant for the cadmium compound is less than that for aquated $(CH_3)_2Sn^{2+}$, where the cation–water interaction seems to be of the hard-sphere coulombic type.[21]

C. Methyl–platinum(IV) compounds

It is remarkable that the methyl–platinum(IV) system was ignored from a vibrational point of view for so long. The first compounds (trimethylplatinum compounds) were prepared by Pope and Peachey[31] in 1907, and forty years later the cubane structure of chlorotrimethylplatinum(IV) was deduced by X-ray analysis.[32] The arrangement of two interpenetrating tetrahedra, one of four $(CH_3)_3Pt$ groups, the other of four chloro atoms, in the case of $[PtCl(CH_3)_3]_4$ is not uncommon. $[ZnCH_3(OCH_3)]_4$ has a similar arrangement of four CH_3Zn groups and four methoxy groups.[33]

Gilman *et al.*[34] have reported the results of reaction between $PtCl_4$ and methylmagnesium iodide, and claim that the products include dimethyl- and monomethylplatinum compounds as well as the main product, iodotrimethylplatinum(IV). Ruddick and Shaw[35] have prepared complexes of the type $[Pt(CH_3)_xX_{4-x}L_2]$, where $x = 1$–3, X = Cl, Br and I, and L = tertiary phosphine or tertiary arsine, and report the first genuine cases of tetramethylplatinum(IV) compounds, *cis*-$[Pt(CH_3)_4\{As(CH_3)_2(C_6H_5)\}_2]$, the corresponding phosphine compound and $[Pt(CH_3)_4(PEt_3)_2]$. Recent attempts to prepare the fascinating compounds tetramethylplatinum(IV), $[Pt(CH_3)_4]_4$, with a structure analogous to $[PtCl(CH_3)_3]_4$ and long taken to illustrate six-valent carbon, and hexamethyldiplatinum, $Pt_2(CH_3)_6$, have failed. It has been concluded[36] that the original sample of the former compound was indeed hydroxotrimethylplatinum(IV), $[PtOH(CH_3)_3]_4$ and the 'hexamethyl compound' probably[37] iodotrimethylplatinum(IV), $[PtI(CH_3)_3]_4$.

Apart from a few desultory reports no record of the infrared spectra of methylplatinum(IV) compounds appeared until 1964. In a study of a number of trimethylplatinum(IV) compounds by Hoechstetter,[38] frequencies due to the most interesting vibrational modes of all, the Pt—C stretching modes were not observed in the spectra. Since 1965, a number of publications dealing with the infrared and Raman spectra of trimethylplatinum(IV) compounds have appeared.

(a) *Trimethylplatinum(IV) compounds*

(*i*) *Trimethyltriaquoplatinum(IV) and trimethyltriammineplatinum(IV) cations.* All trimethylplatinum(IV) compounds whose crystal structures have been determined have a *cis* arrangement of the methyl groups about an octahedrally coordinated platinum atom. The Raman spectra of aqueous solutions of trimethylplatinum(IV) sulphate, nitrate and perchlorate[39] contain a set of lines common to all and attributed to the aquated trimethylplatinum(IV) cation. There is no evidence for cation–anion association. Glass and Tobias[40] have found by ^{17}O n.m.r. that the hydration number of $(CH_3)_3Pt^+$ is three. Therefore the aquated ion may be formulated as $[Pt(CH_3)_3(H_2O)_3]^+$ having C_{3v} skeletal symmetry. The triammine,[41] $[Pt(CH_3)_3(NH_3)_3]^+$, is assumed to have the same skeletal symmetry, and this is supported by the ^1H n.m.r. spectrum which contains one signal corresponding to three equivalent CH_3 groups.

Table 17.4 lists the Raman frequencies of the triaquo and triammine complex ions of trimethylplatinum(IV), together with the N-deuterated species. The spectrum of the $(CH_3)_3Pt$ moiety may be divided clearly into regions of C—H stretching (~2900 cm^{-1}), CH_3 deformation (1250–1450 cm^{-1}), CH_3 rocking (~880 cm^{-1}), Pt—C stretching (~600 cm^{-1}) and PtC_3 deformation (~260 cm^{-1}). The remaining line in the triaquo species at 357 cm^{-1} is assigned to Pt—O stretching. It is shifted to ~345 cm^{-1} when the solvent is changed from normal to heavy water. It is noticed from Table 17.4 that the spectrum of $(CH_3)_3Pt$ is little affected by the nature of the groups *trans* to the CH_3 groups. In the spectra of the triammine are sets of lines due to the $(NH_3)_3Pt$ and $(ND_3)_3Pt$ moieties superimposed on the contributions from $(CH_3)_3Pt$. The $(NH_3)_3Pt$ spectrum consists of ν(N—H) at ~3200 cm^{-1}, δNH_3 at 1300 and 1636 cm^{-1}, ν(Pt—N) at 390 cm^{-1} and δPtN_3 at 201 cm^{-1}. Appropriately lower values are observed for the corresponding vibrations in the N-deuterated complex ion.

Of interest are the magnitudes of the Pt—O and Pt—N stretching frequencies. The values are quite low compared with those for the octahedral complexes $[Pt(OH)_6]^{2-}$ (515, 538 cm^{-1})[42] and $[Pt(NH_3)_6]^{4+}$ (545 and 569 cm^{-1})[41], respectively. The lowering by about 150 cm^{-1} is due to the weakening effect of the methyl groups on those bonds *trans* to

TABLE 17.4. *Raman frequencies of aqueous solutions of several complex ions of trimethylplatinum(IV)*

Frequency, cm^{-1}			
$[Pt(CH_3)_3(H_2O)_3]^+$	$[Pt(CH_3)_3(NH_3)_3]^+$	$[Pt(CH_3)_3(ND_3)_3]^+$	Mode
		180	δPtN_3
	201		δPtN_3
259	271	266	δPtC_3
357			ν(Pt—O)
		364	ν(Pt—N)
	390		ν(Pt—N)
600	584	585	ν(Pt—C)
882	887	870	ρCH_3
		970	δND_3
		1194	δND_3
1250	1245	1251	δCH_3
1290	1274	1281	δCH_3
	1300		δNH_3
1329			δCH_3
1427	1438	1434	δCH_3
	1636		δNH_3
		2327	ν(N—D)
		2381	ν(N—D)
2823	2825	2829	$2 \times \delta CH_3$
2909	2901	2904	ν(C—H)
2978	2958	2960	ν(C—H)
	3194		ν(N—H)
	3270		ν(N—H)

them. The effect is reflected in the different behaviour of the triammine and hexammine in aqueous solution. The former solution is alkaline but the latter is acidic.[42]

(ii) Compounds of the type $[PtX(CH_3)_3]_4$, X = OH, Cl, Br and I. The hydroxo-chloro- and iodo-trimethylplatinum(IV) compounds are known to exist as tetramers in the solid state. The bromo- compound is tetrameric in benzene solution and is presumed to be tetrameric in the solid state with a structure similar to that of the other members of the series. The basic structure for a tetramer is illustrated in Figure 17.2. Because of the bridges

that each X atom makes to three Pt atoms, one may regard the system as another example of a $\{Pt(CH_3)_3X_3\}$ species.

Raman spectra of these compounds in solution are not obtainable because of their low solubility; accordingly Table 17.5 contains infrared and Raman frequencies[44] of the solid compounds. In the case of the hydroxo-compound the point group for an isolated tetramer is T_d and therefore the distribution of modes among the symmetry species is as follows: $\Gamma = 10a_1(R) + 5a_2(ia) + 15e(R) + 19f_1(ia) + 24f_2(R, i.r.)$; that is, totals of 49 Raman lines and 24 infrared bands due to fundamentals are predicted.

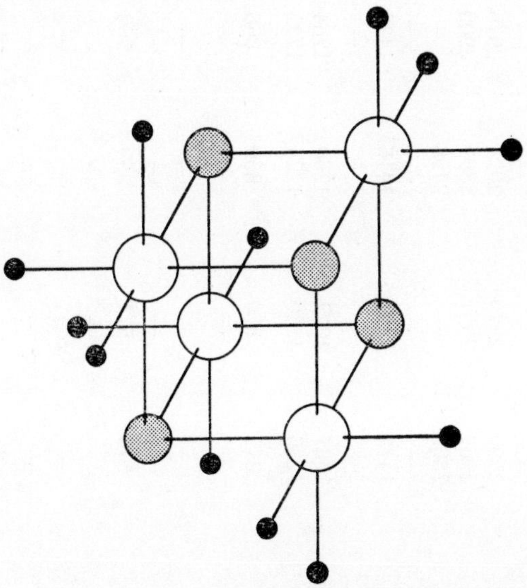

FIGURE 17.2. *Structure of the tetramer $[PtCl(CH_3)_3]_4$. Open circles = Pt; hatched circles = Cl; black circles = CH_3.*

15 Raman lines and 18 infrared bands are observed. Although some accidental degeneracy of modes of a similar type (e.g., C—H stretching and CH_3 deformation) is expected, the number of observable features differs from the predicted number of fundamentals for a single tetramer by a large margin. There is a similar gap between prediction and observation for the halogeno- compounds. In theory, the interpretation of these spectra should be attempted after a factor group analysis has been made, based on the space group of each compound. Since the unit cells of the hydroxo-, chloro- and iodo- compounds contain two, two, and eight tetramer molecules respectively, thus at least doubling the number of predicted fundamentals, such an analysis has not been attempted and only a qualitative assignment of modes has been made. This assignment is included in Table 17.5.

TABLE 17.5. *Vibrational frequencies of the compounds $[PtX(CH_3)_3]_4$, where $X = OH, Cl, Br$ and I, in the solid state*

Frequency, cm^{-1}								
$[PtOH(CH_3)_3]_4$		$[PtCl(CH_3)_3]_4$		$[PtBr(CH_3)_3]_4$		$[PtI(CH_3)_3]_4$		Approximate description of mode
Infrared	Raman	Infrared	Raman	Infrared	Raman	Infrared	Raman	
3598	3595	—	—	—	—	—	—	ν(O—H)
—	—	2985	2993	—	—	—	—	
2957	2965	2964	2976	2980	2979	2971	2978	ν(C—H)
2898	2895	2900	2901	2907	2901	2899	2900	
2801	—	2797	—	2802	—	2798	—	$2 \times \delta_{asym}CH_3$
1423	1427	—	—	—	—	—	—	
1408	1408	1411	1416	1418	1421	1416	1422	$\delta_{asym}CH_3$
1380	1376	—	—	—	—	—	—	
1279	1280	1271	1272	1267	1272	1260	1264	δCH_3
1243	1243	1234	1236	1230	1230	1224	1222	
877	876	877	878	875	870	859	868	ρCH_3
854	—	—	—	—	—	—	—	
722	716	—	—	—	—	—	—	δ(Pt—O—H)
—	700	—	—	—	—	—	—	
590	595	580	581	573	574	559	560	ν(Pt—C)
394	—	—	—	—	—	—	—	ν(Pt—O)
368	360	—	—	—	—	—	—	
—	—	331	—	—	—	—	—	
—	—	270	272	—	—	—	—	ν(Pt—Cl)
245	250	240	238	218	249	241	261	δPtC_3
—	—	216	—	176	—	164	—	ν(Pt—X)
—	—	—	—	152	—	122	—	
147	—	170	—	120	—	92	—	$\delta(Pt_4X_4)$
126	—	136	—	106	—	—	—	
—	—	108	—	—	—	—	—	
—	102	—	76	—	66	—	60	

The very sharp infrared band at 3598 cm^{-1} in the spectrum of the hydroxo- compound indicates that the OH bond does not participate in hydrogen-bonding to any significant degree. As for the triaquo and triammine compounds, the contribution of the $(CH_3)_3Pt$ moiety to the spectra of these compounds is reasonably constant. The occurrence in the infrared spectra of three main bands in the C—H stretching region at approximately 2970, 2900 and 2800 cm^{-1} is a characteristic commented upon by workers who have recorded the spectra of a variety of other trimethylplatinum(IV) compounds.[38, 45] Apart from the broad feature at about 1410 cm^{-1}, which shows structure in the hydroxo compound, the infrared spectra are characterized by two sharp bands in the region 1230–1280 cm^{-1} ascribed to methyl deformation modes. Pairs of Raman lines also occur in this region. The higher-frequency Raman line at about 1270 cm^{-1} in all of the compounds probably corresponds to the higher-frequency, polarized lines at 1290 and 1274 cm^{-1} in the spectra of the triaquo and triammine species respectively, while the lower Raman shift of about 1230 cm^{-1} may correspond to the depolarized Raman lines at 1250 and 1245 cm^{-1} in the triaquo and triammine complexes respectively. It follows that the higher-frequency line in the Raman spectra of $[PtX(CH_3)_3]_4$ may be due to a totally symmetric mode and the lower-frequency shift to a non-totally symmetric mode. The difference of 30–40 cm^{-1} between the frequencies is not expected, however, for in-phase and out-of-phase type modes as a result of coupling between molecules in the unit cell. The higher-frequency band in the infrared spectra grows in intensity relative to the other throughout the series and there is a noticeable decrease in the frequencies as X changes from OH to I. There is a similar drift in the Pt—C stretching frequencies. Both sets of frequencies decrease in the direction of decreasing electronegativity of X. Plots of δCH_3 vs. ν(Pt—C) are linear, as has been found for alkyl-Pt(II) compounds.[46] The methyl rocking frequencies decrease progressively from 877 cm^{-1} in the hydroxo- compound to 859 cm^{-1} in the iodo- compound.

The very strong infrared band at 722 cm^{-1} in the hydroxo- compound, not present in the spectra of the halogeno- compounds, is assigned to Pt—O—H bending (it shifts to 559 cm^{-1} on deuteration), and the bands at 394 and 368 cm^{-1} (which also shift to lower values on deuteration) are ascribed to Pt—O stretching.[47] The Raman spectrum contains a line at 360 cm^{-1} which is also attributed to this type of mode. The value compares with $\nu(Pt—OH_2) = 357$ cm^{-1}, for the triaquo species. It is interesting that the change from the PtC_3O_3 skeleton of the $[Pt(CH_3)_3(H_2O)_3]^+$ species to the 'box-like' Pt_4O_4 entity has so little effect on the stretching frequency. The expected decrease in frequency due to the bridging function of the OH group is presumably balanced by a gain achieved through formation of the rigid cage. The feature common to all spectra at about 250 cm^{-1} is ascribed to PtC_3 deformations.

Platinum–halogen stretching frequencies vary in magnitude depending on the oxidation state of platinum and the nature of the ligands other than halogen bound to it. Vibrations of the Pt_4X_4 skeleton (X = halogen) would be expected to occur around and below 250 cm^{-1}. For example, Pt—Cl stretching in Pt_4Cl_4 is expected to be lower in frequency than analogous modes in $[PtCl_6]^{2-}$ ($\nu_1 = 345$ cm^{-1}) on two grounds, firstly, because of the weakening effect of the methyl groups on the bonds *trans* to them and secondly, because of the bridging function of the chlorine atoms. The effect of the latter may be minor in view of the similar values for ν(Pt—O) in $[Pt(CH_3)_3(H_2O)_3]^+$ and $[PtOH(CH_3)_3]_4$. One might expect Pt—X symmetrical stretching modes to appear as strong features in the Raman spectra. However, in the spectrum of the chloro- compound only one line appears that may be attributed to Pt—Cl stretching, namely, the weak feature at 272 cm^{-1}. It is surprising that Raman lines due to Pt—Br and Pt—I stretching are not observed in the spectra of the corresponding compounds, since increasing bond polarizability is expected for the series Pt—Cl < Pt—Br < Pt—I. One concludes that the Pt—X bonds are markedly ionic in character. This is the conclusion reached in the case of the hydroxo-compound where the Pt—O—H bending frequency (722 cm^{-1}) is much lower than that in $[Pt(OH)_6]^{2-}$ (1058–1076 cm^{-1}),[42] and it is claimed[48] that the lower the value of δ(metal–O—H) the greater the ionic character of the metal–oxygen bond. The symmetric stretching of the Pt—X bonds in the Pt_4X_4 skeleton is clearly envisaged as a 'breathing' vibration, but the antisymmetric (f_2 type) modes must involve some degree of Pt—X bending. The analysis of the fairly rich far-infrared spectra is therefore not attempted in detail. There appear to be two well-defined groups of frequencies whose values change progressively from the hydroxo- to the iodo- compound. The higher-frequency group is assigned to modes which are predominantly Pt—X stretching, the lower-frequency group to modes which are essentially Pt—X bending. These are designated in Table 17.5.

It appears that the different packing arrangements of tetramers in the chloro- and iodo- compounds have no marked effects on the vibrational spectra. Indeed, the very simplicity of the spectra suggests that the tetramers are behaving as separate entities.

The Pt—C stretching frequencies of the compounds in Tables 17.4 and 17.5 indicate some dependence on the nature of the groups *trans* to the methyl groups. By taking an aqueous solution of trimethylplatinum(IV) sulphate it is possible to form in solution a complex by adding an excess of ligand. Complexes formed this way have been characterized by ^1H n.m.r. and

Raman spectroscopy.[49] Table 17.6 lists the Pt—C symmetrical stretching frequencies in descending order of magnitude for a number of $\{Pt(CH_3)_3X_3\}$ species in either the solid or solution state. Apart from the position of the nitro group, the list of ligand groups follows fairly closely the order of the *trans* effect series for platinum(II) and the trend in ν(Pt—C) for a series of *trans*-$[PtXCH_3(PR_3)_2]$ compounds,[46] where X = a univalent acid radical and R = methyl or ethyl.

TABLE 17.6. *Symmetric Pt—C stretching frequencies from Raman spectra of systems* $\{Pt(CH_3)_3X_3\}$

Frequency, cm^{-1}	Ligand X
600 ± 2	H_2O[a]
595	OH^-[b]
584	NH_3[a]
583	NO_2^-[a]
581	Cl^-[b]
578	C_5H_5N[a]
577	CH_3NH_2[a]
574	Br^-[b]
563	SCN^-[a]
560	I^-[b]
553	CN^-[a]

[a] Aqueous solution.
[b] Solid state.

(*iii*) *Dimethylplatinum(IV) compounds*. The preparation of some dimethylplatinum(IV) compounds has been reported.[35] These are of the type $[Pt(CH_3)_2X_2L_2]$, where X = halogen and L = a tertiary phosphine or arsine. Others of a simpler empirical formula have been prepared,[50] the composition of which corresponds to $Pt(CH_3)_2X_2$, where X = halogen. No structural studies have yet been made on the latter series of dihalogeno-compounds, but their chemical reactivity points to there being two kinds of halogen atom in the molecule. A tentative suggestion for the structure of these compounds is that they exist as tetramers made up of two interpenetrating tetrahedra, one of four $(CH_3)_2XPt$ groups and the other of four X groups. This arrangement would be represented by Figure 17.2 where one CH_3 group in each set of three is replaced by an X atom. Obviously a number of isomers is possible.

Infrared spectra (4000–250 cm^{-1}) of the compounds characterized at the present time are set out in Table 17.7. Of interest are the features attributable to the $(CH_3)_2Pt$ moiety. The three bands at about 2900 cm^{-1}, while reminiscent of the spectra of $(CH_3)_3Pt$ derivatives, are spread over a wider range. The low-frequency band, some 15–20 cm^{-1} lower in frequency than its counterpart in $(CH_3)_3Pt$ compounds, is ascribed to the overtone of the antisymmetric methyl deformation at 1408 cm^{-1}, which has in turn dropped by about 10 cm^{-1}. Pairs of sharp bands in the 1200–1250 cm^{-1} region are assigned to methyl deformation modes. The frequencies shift to lower values on passing from the bromo- to the iodo- compound. The bromo- compound shows splitting of features due to δCH_3 which may be the result of intermolecular coupling in the unit cell. Weak bands are observed

TABLE 17.7. *Infrared frequencies* (4000–250 cm^{-1}) *of* $[Pt(CH_3)_2X_2]$, $X = Br, I$

Frequency, cm^{-1}		
$Pt(CH_3)_2Br_2$	$Pt(CH_3)_2I_2$	Description of mode
2995 w	2990 w	C—H stretch
2911 m	2910 m	C—H stretch
2787 vw	2786 vw	$2 \times \delta_{asym}CH_3$
1407 m	1408 m	$\delta_{asym}CH_3$
1252 m	1241 s	δCH_3
1245 w		
1222 sh		
1220 s	1212 s	
557 w		Pt—C stretch
552 w	538 w	

in the 'platinum–carbon stretching' region. In the case of the bromo-compound a doublet is present, due either to a spectroscopic feature such as a lifting of degeneracy, coupling between molecules or discrimination between symmetric and antisymmetric stretching modes or to the presence of isomers. The Pt—C stretching frequencies are about 20 cm^{-1} lower than for the corresponding $(CH_3)_3Pt$ compounds.

D. Trisilyl compounds of Group V elements

Molecules are known of the type $(SiH_3)_3E$, where E is a group V element. Structural interest in this class arose when Hedberg[13] reported, as the result of an electron diffraction study, that the NSi_3 skeleton of trisilylamine is planar, in contrast to the pyramidal NC_3 skeleton of trimethylamine.

The planar stereochemistry was accounted for by means of sp^2 hybridization of nitrogen and involvement of the lone-pair electrons in $(p \rightarrow d)$ π-bonding with d orbitals of the silicon atoms. The vibrational studies of the molecule supported the planar arrangement.[5, 14]

Vibrational spectroscopy can, in principle, distinguish between the planar and pyramidal ESi_3 models, as shown by Table 17.8 where the

TABLE 17.8. *Selection rules for planar and pyramidal ESi_3*

Planar (D_{3h})			Pyramidal (C_{3v})		
	Activity			Activity	
Mode	Infrared	Raman	Mode	Infrared	Raman
E—Si antisym. stretch (e′)	Active	dp	E—Si antisym. stretch (e)	Active	dp
E—Si sym. stretch (a_1')	—	p	E—Si sym. stretch (a_1)	Active	p
ESi_3 sym. deformation (out-of-plane) (a_2'')	Active	—	ESi_3 sym. deformation (a_1)	Active	p
ESi_3 in-plane deformation (e′)	Active	dp	ESi_3 antisym. deformation (e)	Active	dp

modes and their activity for the two cases are set out. If there is negligible coupling between the skeletal vibrations of $(SiH_3)_3E$ and the modes of the SiH_3 moiety, then an examination of the skeletal region alone may enable a decision to be made concerning the stereochemistry about E. It is clear from Table 17.8 that for a planar skeleton one polarized Raman line is predicted together with two depolarized lines; three infrared bands are predicted, two of them coincident with the two depolarized Raman lines. For the C_{3v} model, all four modes are both infrared and Raman active, two of the Raman lines being polarized.

In the case of $(SiH_3)_3N$ the Raman spectrum fulfils the predictions for the planar model, viz., the observation of three lines in the 'skeletal' region, one of them polarized. The experimental results are summarized in Table 17.9. It is not possible to demonstrate that the observations conform completely to one or other of the two models since no absorption below 400 cm^{-1} has been observed in the infrared spectra. However, no convincing coincidence is observed in the infrared corresponding to the N—Si

TABLE 17.9. *Fundamental frequencies for NSi_3*

Frequency, cm^{-1}		
Infrared	Raman	Assignment
996 vs	987 m, br, dp	antisym. N—Si stretch
—	496 s, sharp, p	sym. N—Si stretch
—	—	sym. NSi_3 deformation
—	204 w, br, dp	antisym. NSi_3 deformation

symmetrical stretching mode observed in the Raman spectrum at 496 cm^{-1}, and no Raman line is observed that can be assigned to the out-of-plane NSi_3 bending mode. The overall symmetry for $(SiH_3)_3N$ depends on the positions of the H atoms in one silyl group relative to the others. The point group describing the molecule may be C_{3h}, C_{3v}, C_3, C_s or C_1, for different 'rest' positions of the silyl groups. For free rotation of the silyl groups the effective point group is D_{3h}. The rules of selection and polarization are different for each of these possible point groups and the vibrational spectra should, in principle, distinguish one from the others. In practice, however, the numbers of observed infrared and Raman lines, their coincidences and the number of polarized Raman lines provide insufficient information to allow a definite conclusion to be drawn. On the assumption that the partial double-bond character of the N—Si bonds due to $(p \rightarrow d)\pi$-bonding prevents free rotation of the silyl groups, the most favourable arrangement sterically is that where one H atom of each silyl group lies in the NSi_3 plane and points between two H atoms of an adjacent silyl group. This model is described by the point group C_{3h}. The fundamentals are distributed among the symmetry species as follows: $6a' + 5a'' + 7e' + 4e''$. Modes of species a' are active only in the Raman spectrum and should appear as polarized lines; those of species a'' are active only in the infrared; those of species e' are active in both spectra; those of e'' are active only in the Raman spectrum.

No difficulty is encountered in assigning the observed frequencies to modes of vibration. Table 17.10 sets out an assignment of the fundamentals. The features falling in the regions of 2100, 1000 and 700 cm^{-1} are assigned to Si—H stretching, SiH_3 deformation and SiH_3 rocking vibrations respectively.

Recently the X-ray crystal structure[51] of the ion $[N(SO_3)_3]^{3-}$ in $K_3[N(SO_3)_3]$, $2H_2O$ has been determined. The anion has C_{3h} symmetry with a 'head-to-tail' arrangement of oxygen atoms, that is, one O of each SO_3 group lies in the NS_3 plane and points between two O atoms of an adjacent SO_3 group. The selection of C_{3h} symmetry for trisilylamine therefore appears a reasonable one.

If partial double-bonding plays a significant role in this compound, it may be reflected in an enhanced value for the N—Si stretching force

TABLE 17.10. *Assignment of fundamentals of $(SiH_3)_3N$ on the basis of point group C_{3h}*

Symmetry species	Frequency, cm^{-1} Infrared (vapour)	Raman (liquid)	Approximate description of mode
a′		2168	Si—H stretch
		2168	Si—H stretch
		1011	SiH_3 deformation
		946	SiH_3 deformation
		697	SiH_3 rock
		496	N—Si stretch
a″	2165		Si—H stretch
	~945		SiH_3 deformation
	748		SiH_3 rock
	—		NSi_3 out-of-plane deformation
	—		SiH_3 torsion
e′	2165	2168	Si—H stretch
	2165	2168	Si—H stretch
	947	946	SiH_3 deformation
	941	946	SiH_3 deformation
	996	987	N—Si stretch
	695	697	SiH_3 rock
	—	204	NSi_3 in-plane deformation
e″		2168	Si—H stretch
		1011	SiH_3 deformation
		697	SiH_3 rock
		—	SiH_3 torsion

constant. The evidence is not easy to obtain since it is necessary to have for comparison examples where the N—Si bond is clearly single bond in type. The N—Si stretching force constant in $(SiH_3)_3N$ is 3·90 mdyn Å^{-1}, a value lying in the normal range for single-bond stretching force constants. It is larger, however, than values obtained for C—Si bonds where there is no possibility of partial double-bonding, e.g. k_{C-Si} for $(CH_3)_4Si$ is

2·9 mdyn $Å^{-1}$. This suggests that the degree of π-bonding in $(SiH_3)_3N$ is probably not very large.[5]

The first report[52] of the vibrational spectra of trisilylphosphine and -arsine came out in favour of a planar skeleton for each of these molecules. The arguments were based on the selection rules and on the apparent absence from the spectra of certain features after the fashion of those advanced in the trisilylamine case.[5] However, McKean[53] predicted from approximate force constant calculations that the molecules $(SiH_3)_3E$, where E = P, As and Sb, should have a pyramidal skeleton. McKean calculated k_{Si-E} and the stretch–stretch interaction force constant (k_{12}) for a range of angles SiESi and compared the results with k and k_{12} for $(SiH_3)_3N$, $(SiH_3)_2O$ and $(SiH_3)_2S$, whose shapes are known from electron diffraction studies. For these molecules k_{12} (expected to be positive on physical grounds) is either positive and small or zero within experimental error. On this basis McKean suggested a probable value of slightly less than 100° for SiPSi and SiAsSi angles and a value close to 90° for the SiSbSi angle. Electron diffraction studies have since confirmed the pyramidal arrangement and yield SiESi angles of 96·5°, 93·8° and 88·6° for the P, As and Sb compounds respectively.[15, 54]

It is invariably unwise to base structural conclusions on the absence of bands from vibrational spectra because of the possibility that the intensity of an allowed mode is inherently weak and below the threshold of detection. Indeed, it appears now that the vibrational spectrum for $(SiH_3)_3P$ first reported was not complete, since a re-examination[55] comes out in favour of a pyramidal PSi_3 skeleton. The skeletal frequencies for $(SiH_3)_3P$ are set down in Table 17.11. For C_{3v} symmetry, as outlined in Table 17.8, one expects all four modes to be infrared and Raman active, two of the Raman lines being polarized. Siebert and Eints[55] have observed a strongly polarized Raman line at 415 cm^{-1} assignable, as before, to the P—Si symmetrical stretching mode. A weakly polarized, broad feature at 115–150 cm^{-1} is attributed to both the symmetric and antisymmetric bending modes. The depolarized line at 458 cm^{-1} is assigned to the antisymmetric P—Si stretching mode. The main feature of the infrared spectrum is the observation of a weak band at 423 cm^{-1} attributable to the P—Si symmetric stretching mode.

These results differ from those of the $(SiH_3)_3N$ case mainly in respect of the state of polarization of the low-frequency Raman line. This line at 204 cm^{-1} in $(SiH_3)_3N$ is broad and no quantitative polarization measurement has been carried out. The infrared spectrum of $(SiH_3)_3N$ vapour includes a very weak band at 504 cm^{-1} and almost coincident with the Raman shift of 496 cm^{-1} for the liquid state. Although the 504 cm^{-1} band can be assigned as a difference frequency, it appears that further structural

TABLE 17.11. *Skeletal frequencies for $(SiH_3)_3P$*

Mode	Frequency, cm^{-1}	
	Infrared	Raman
P—Si antisym. stretch (e)	456, 463	458
P—Si sym. stretch (a_1)	423	415
PSi_3 sym. deform. (a_1)	—	143
PSi_3 antisym. deform. (e)	128	131

studies on $(SiH_3)_3N$ may be worthwhile. It is to be noted that the infrared spectrum[56] of $(SiH_3)_3N$ isolated in an argon matrix provides evidence for C_{3v} rather than C_{3h} symmetry.

The skeletal frequencies from the Raman spectra of $(SiH_3)_3E$, where E = P, As and Sb, are listed in Table 17.12. The separation between symmetric and antisymmetric stretching frequencies decreases as the atomic number of E increases. Failure to observe the a_1 symmetric deformation in the Raman spectra of the arsenic and antimony compounds may mean that the deformation modes are accidentally degenerate at about 100 cm^{-1}.

TABLE 17.12. *Skeletal frequencies from the Raman spectra of $(SiH_3)_3E$, E = P, As and Sb*

Mode	Frequency, cm^{-1}		
	P	As	Sb
E—Si antisym. stretch (e)	458	362	310
E—Si sym. stretch (a_1)	415	346	308
ESi_3 sym. deformation (a_1)	143	?	?
ESi_3 antisym. deformation (e)	131	115	99

E. Trigermyl compounds of Group V elements

The vibrational spectra[7] of trigermylphosphine indicate that the PGe_3 skeleton is pyramidal, and this arrangement is confirmed by an electron diffraction study.[54] The observed fundamentals in the spectra of $(GeH_3)_3P$ are listed in Table 17.13 and classified according to the point group C_{3v}. The 33 modes are distributed among the symmetry species according to the following scheme: $7a_1$(i.r. and R) + $4a_2$(ia) + 11e(i.r. and R). The

TABLE 17.13. *Assignment of fundamentals for $(GeH_3)_3P$ according to C_{3v} selection rules*

Symmetry species	Frequency, cm^{-1} Infrared (vapour)	Raman (liquid)	Approximate description of mode
a_1	2062	2044	Ge—H stretch
	2062	2044	Ge—H stretch
	872	864	GeH_3 deformation
	872	864	GeH_3 deformation
	516	—	GeH_3 rocking
	—	322	P—Ge stretch
	—	112	PGe_3 sym. deformation
a_2	inactive	inactive	Ge—H stretch
			GeH_3 deformation
			GeH_3 rocking
			GeH_3 torsion
e	2079	2061	Ge—H stretch
	2079	2061	
	2079	2061	
	839	826	GeH_3 deformation
	799	772	
	753	772	
	558	556	GeH_3 rocking
	558	556	GeH_3 rocking
	367	366	P—Ge stretch
	—	—	GeH_3 torsion
	—	88	PGe_3 antisym. deformation

Ge—H stretching, GeH_3 deformation and GeH_3 rocking modes are characteristically observed with frequencies of about 2100, 800–900 and 500 cm^{-1} respectively. The significant feature of the skeletal region in the Raman spectrum is the observation of four lines of which two are polarized. These are precisely the predictions for a C_{3v} skeletal arrangement. Complete agreement with prediction is not obtained since an absorption corresponding to the P—Ge symmetric stretching vibration is not detected in the infrared spectrum of the vapour or liquid.

The vibrational spectra (where available) of the analogous nitrogen,[16] arsenic[57] and antimony[57] compounds are all interpreted in terms of a pyramidal EGe_3 skeleton. The skeletal frequencies for all the germyl compounds are set out in Table 17.14. The skeletal deformation frequencies for trigermylamine are not known since the infrared study did not cover this region and no Raman spectrum has been recorded. However, the pyramidal NGe_3 arrangement is favoured by the observation in the infrared of a band ascribable to the symmetric N—Ge stretching mode. It is to be noted that the symmetric and antisymmetric frequencies draw closer together with increasing atomic number of E, and that the observation of one line in the skeletal deformation region of the Raman spectrum of the antimony compound probably means that both deformation modes are accidentally degenerate at about 68 cm^{-1}.

TABLE 17.14. *Skeletal frequencies for $(GeH_3)_3E$, E = N, P, As and Sb*

Mode	Frequency, cm^{-1}			
	N[a]	P[b]	As[b]	Sb[b]
E—Ge antisym. stretch (e)	848	366	263	216
E—Ge sym. stretch (a_1)	362	322	245	209
EGe_3 sym. deformation (a_1)	ns	112	90	(68)
EGe_3 antisym. deformation (e)	ns	88	78	68

[a] From infrared spectrum.
[b] From Raman spectrum.
ns. Not studied in this region.

Although the EGe_3 skeleton is non-planar, this does not rule out the possibility of $(p \rightarrow d)\pi$-bonding[58] and the apparent lack of activity of $(GeH_3)_3P$ as a Lewis base towards BF_3, for example, is of interest in this respect. This property is no criterion of $(p \rightarrow d)\pi$-bonding, however, since both $(SiH_3)_3N$ and $(GeH_3)_3N$ form adducts with BF_3 but not with $(CH_3)_3B$. A simple valence force field calculation[16] on $(GeH_3)_3N$ yields a GeNGe bond angle of 116° when the stretch–stretch interaction force constant is zero. The value of the N—Ge stretching force constant under these conditions is 3·67 mdyn $Å^{-1}$, suggesting possibly that $(p \rightarrow d)\pi$-bonding occurs to a lesser extent than in trisilylamine. The weaker π-acceptor property of germanium may be a result of the radial node in the 4d orbitals.[59]

Table 17.15 contains values of MEM angles for $(MH_3)_3E$ compounds. The usual decrease in angle as the atomic number of E increases is observed. The bond angles for $(MH_3)_3P$, where M = C, Si and Ge, decrease in that order, but for $(MH_3)_3N$ there is a maximum at M = Si. It may be significant that, where the electronegativity difference between E and M is

TABLE 17.15. *Experimental and calculated MEM angles (in °) for the molecules $(MH_3)_3E$*

M	E: N	P	As	Sb
C	109–111	98·6	96 ± 5	—
Si	119·6	96·5	93·8	88·6
Ge	(116)	95·4	(95)	(93)

Values in parentheses are calculated.

greatest, the MEM bond angle approaches 120°. Consequently the shape of the EM_3 skeleton may be influenced by the magnitude of the charge on the MH_3 groups, for the repulsion between these groups is then greatest for the most electronegative central atom E (i.e., nitrogen) and for the central atom of smallest covalent radius (again nitrogen).

References

1. Woodward, L. A., Hall, J. R., Dixon, R. N., and Sheppard, N., *Spectrochim. Acta*, 1959, **15**, 249.
2. Hall, J. R., Woodward, L. A., and Ebsworth, E. A. V., *Spectrochim. Acta*, 1964, **20**, 1249.
3. Goggin, P. L., and Woodward, L. A., *Trans. Faraday Soc.*, 1960, **56**, 1591.
4. Goggin, P. L., and Woodward, L. A., *Trans. Faraday Soc.*, 1962, **58**, 1495; 1966, **62**, 1423.
5. Ebsworth, E. A. V., Hall, J. R., Mackillop, M. J., McKean, D. C., Sheppard, N., and Woodward, L. A., *Spectrochim. Acta*, 1958, **13**, 202.
6. Ebsworth, E. A. V., Taylor, R., and Woodward, L. A., *Trans. Faraday Soc.*, 1959, **55**, 211.
7. Cradock, S., Ebsworth, E. A. V., Davidson, G., and Woodward, L. A., *J. Chem. Soc. (A)*, 1967, 1229.
8. Tobias, R. S., Sprague, M. J., and Glass, G. E., *Inorg. Chem.*, 1968, **7**, 1714.
9. Tobias, R. S., and Hutcheson, S., *J. Organometallic Chem.*, 1966, **6**, 535.
10. Feher, F., Kolb, W., and Leverenz, L., *Z. Naturforsch.*, 1947, **2a**, 454.
11. Clark, H. C., and O'Brien, R. J., *Inorg. Chem.*, 1963, **2**, 1020.
12. Kriegsmann, H., and Pischtschan, S., *Z. anorg. Chem.*, 1961, **308**, 212.
13. Hedberg, K., *J. Amer. Chem. Soc.*, 1955, **77**, 6491.
14. Robinson, D. W., *J. Amer. Chem. Soc.*, 1958, **80**, 5924; Kriegsmann, V. H., and Forster, W., *Z. anorg. Chem.*, 1959, **298**, 212.
15. Beagley, B., Robiette, A. G., and Sheldrick, G. M., *J. Chem. Soc. (A)*, 1968, 3002; 1968, 3006.

16. Rankin, D. W. H., *J. Chem. Soc. (A)*, 1969, 1926.
17. Venkateswaran, S., *Indian J. Phys.*, 1930, **5**, 145.
18. Thompson, H. W., Linnett, J. W., and Wagstaffe, F. J., *Trans. Faraday Soc.*, 1940, **36**, 797.
19. Gutowsky, H. S., *J. Chem. Phys.*, 1949, **17**, 128; *J. Amer. Chem. Soc.*, 1949, **71**, 3194.
20. Rao, K. S., Stoicheff, B. P., and Turner, R., *Canad. J. Phys.*, 1960, **38**, 1516.
21. McGrady, M. M., and Tobias, R. S., *Inorg. Chem.*, 1964, **3**, 1157.
22. Freidline, C. E., and Tobias, R. S., *Inorg. Chem.*, 1966, **5**, 354.
23. Kovar, R. A., and Morgan, G. L., *Inorg. Chem.*, 1969, **8**, 1099.
24. Boyd, D. R. J., Thompson, H. W., and Williams, R. L., *Discuss. Faraday Soc.*, 1950, **9**, 154.
25. Longuet-Higgins, H. C., *Mol. Phys.*, 1963, **6**, 445.
26. Bunker, P. R., *J. Chem. Phys.*, 1967, **47**, 718.
27. Woodward, L. A., *Spectrochim. Acta*, 1963, **19**, 1963.
28. Bribes, J. L., and Gaufres, R., *Compt. rend.*, 1968, **266C**, 584.
29. Miles, M. G., Patterson, J. H., Hobbs, C. W., Hopper, M. J., Overend, J., and Tobias, R. S., *Inorg. Chem.*, 1968, **7**, 1721.
30. Adams, D. M., 'Metal-Ligand and Related Vibrations', Edward Arnold, London, 1967, p. 206.
31. Pope, W. J., and Peachey, S. J., *Proc. Chem. Soc.*, 1907, **23**, 86.
32. Rundle, R. E., and Sturdivant, J. H., *J. Amer. Chem. Soc.*, 1947, **69**, 1561.
33. Shearer, H. M. M., and Spencer, C. B., *Chem. Comm.*, 1966, 194.
34. Gilman, H., Lichtenwalter, M., and Benkeser, R. A., *J. Amer. Chem., Soc.*, 1953, **75**, 2063.
35. Ruddick, J. D., and Shaw, B. L., *Chem. Comm.*, 1967, 1135.
36. Cowan, D. O., Krieghoff, N. G., and Donnay, G., *Acta Cryst.*, 1968, **B24**, 287.
37. Donnay, G., Coleman, L. B., Krieghoff, N. G., and Cowan, D. O., *Acta Cryst.*, 1968, **B24**, 157.
38. Hoechstetter, M. N., *J. Mol. Spectroscopy*, 1964, **13**, 407.
39. Clegg, D. E., and Hall, J. R., *Spectrochim. Acta*, 1965, **21**, 357.
40. Glass, G. E., and Tobias, R. S., *J. Amer. Chem. Soc.*, 1967, **89**, 6371.
41. Clegg, D. E., and Hall, J. R., *Spectrochim. Acta*, 1967, **23A**, 263.
42. Maltese, M., and Orville-Thomas, W. J., *J. Inorg. Nuclear Chem.*, 1967, **29**, 2533.
43. Emeléus, H. J., and Anderson, J. S., 'Modern Aspects of Inorganic Chemistry', 3rd edn, Routledge and Paul, 1960, p. 209.
44. Clegg, D. E., and Hall, J. R., *J. Organometallic Chem.*, 1970, **22**, 491.
45. Kite, K., and Truter, M. R., *J. Chem. Soc. (A)*, 1968, 934.
46. Adams, D. M., Chatt, J., and Shaw, B. L., *J. Chem. Soc.*, 1960, 2047.
47. Clegg, D. E., and Hall, J. R., *J. Organometallic Chem.*, 1969, **17**, 175; Bulliner, P. A., and Spiro, T. G., *Inorg. Chem.*, 1969, **8**, 1023.
48. Tarte, P., *Spectrochim. Acta*, 1958, **13**, 107.

49. Clegg, D. E., and Hall, J. R., *Austral. J. Chem.*, 1967, **20**, 2025.
50. Hall, J. R., and Swile, G. A., unpublished work.
51. Tillack, J. V., and Kennard, C. H. L., *J. Chem. Soc.* (*A*), 1970, 1637.
52. Davidson, G., Ebsworth, E. A. V., Sheldrick, G. M., and Woodward, L. A., *Spectrochim. Acta*, 1966, **22**, 67; 1967, **23A**, 2609.
53. McKean, D. C., *Spectrochim. Acta*, 1968, **24A**, 1253.
54. Rankin, D. W. H., Robiette, A. G., Sheldrick, G. M., Beagley, B., and Hewitt, T. G., *J. Inorg. Nuclear Chem.*, 1969, **31**, 2351.
55. Siebert, H., and Eints, J., *J. Mol. Structure*, 1969, **4**, 23.
56. Goldfarb, T. D., and Khare, B. N., *J. Chem. Phys.*, 1967, **46**, 3379.
57. Ebsworth, E. A. V., Rankin, D. W. H., and Sheldrick, G. M., *J. Chem. Soc.* (*A*), 1968, 2828.
58. Ebsworth, E. A. V., *Chem. Comm.*, 1966, 530; Randall, E. W., and Zuckerman, J. J., *Chem. Comm.*, 1966, 732; Perkins, P. G., *Chem. Comm.*, 1967, 268.
59. Urch, D. S., *J. Inorg. Nuclear Chem.*, 1963, **25**, 771.

CHAPTER 18

Structural Aspects of Sulphur–Fluorine Chemistry

H. L. ROBERTS

A. Introduction

Sulphur can assume the valency states 6, 4 and 2 in compounds with fluorine. The hexavalent compounds are easily handled in the laboratory as they are stable to air and moisture. The lower valency states, by contrast, are readily hydrolyzed and the divalent compounds easily disproportionate. As a consequence of this sulphur hexafluoride has been a well established compound since it was first prepared by Moissan in 1900. Disulphur decafluoride was obtained and correctly identified by Whytlaw-Gray and Denbigh in 1934.

Sulphur tetrafluoride was probably obtained by early workers in the field but such were the difficulties of adequate characterization that Sidgwick in 1950 doubted its existence. An unequivocal characterization was, however, obtained by Brown and Robinson in 1955. This was followed in 1960 by an elegant and simple preparation by Opegard, Smith, Muetterties and Engelhardt which opened the way to an exploitation of this compound as a synthetic reagent.

The preparation and characterization of the divalent disulphur difluoride proved even more difficult, in part because of its inherent instability and also because of the existence of an isomer, thiothionyl fluoride, in which the two sulphur atoms are in different valency states. While the methods of structural chemistry were important in establishing the chemistry of S^{IV} and S^{VI} fluorides, it is of interest that it was microwave spectroscopy which first really proved the existence of disulphur difluoride after years of confusion.

With the establishment of the basic binary fluorides the way was open for the exploration of the chemistry of the groups —SF_5, —SF_3 and —SF. This work proceeded rapidly after 1960 and many compounds are now known.[1]

This chapter sets out to illustrate how the use of spectroscopic techniques has established the molecular structures of some sulphur–fluorine compounds.

B. Hexavalent sulphur fluorides

(*a*) *Theoretical*

Simple considerations of orbital symmetry suggest that SX_6 compounds should be octahedral with sp^3d^2 hybridization. This simplicity, however, does not extend to quantitative calculations. An excellent review of the theoretical calculations carried out has been given by Cruickshank and Webster.[2] Two models have been used. One, due to Craig and his co-workers, starts from the fact that in the unperturbed sulphur atom the 3d orbitals are too diffuse to hybridize to any appreciable extent with the 3s and 3p orbitals. If, however, the ligand field of electronegative elements is considered, the effective radius of the 3d orbitals contracts while the 3s and 3p orbitals are relatively less affected. The alternative model, due to Rundle, postulates 3-centre orbitals using only sp^3 orbitals from sulphur. In effect this is $S^+(s^2p^3)\ (F_6)^-$ as opposed to $S(sp^3d^2)\ F_6$ in the Craig model. Both models involve substantial assumptions and neither predicts very accurately the total energy of atomization of SF_6. The Craig model, however, makes useful predictions concerning the type of compounds expected and qualitatively explains some structural features. The major prediction is that sulphur(VI) compounds in which sulphur is bonded to six other atoms will form only with small electronegative atoms. This is consistent with the known compounds where SF_5 is bonded to C, N, O or F in the first row, SF_5 and Cl only in the second row and Br in the third row of the Periodic Table. The only exception is a recently prepared compound in which SF_5 is bonded to platinum.

(*b*) *Bond distances in sulphur(VI) compounds*

Bond distances have been measured for six compounds containing the SF_5 group and are shown in Table 18.1. In addition, SOF_4, SO_2F_2 and NSF_3 have also been studied. Their dimensions are shown in Table 18.2.

The S—F bond distances for SF_6, SF_5SF_5 and SF_5OOSF_5 are identical, while in SF_5OF a lower figure is given. However, the upper range of the experimental error would allow even this value to be identical. The more

TABLE 18.1. *Bond distances in SF_5R compounds*

Compound	Bond length d(S—F), Å	Bond length d(S—R), Å	Method	Reference
SF_6	$1{\cdot}56 \pm 0{\cdot}02$	—	E.D.	3, 4
SF_5SF_5	$1{\cdot}56 \pm 0{\cdot}02$	$2{\cdot}21 \pm 0{\cdot}03$	E.D.	5
SF_5Cl	$1{\cdot}58 \pm 0{\cdot}01$	$2{\cdot}030 \pm 0{\cdot}002$	MW.	6
SF_5Br	$1{\cdot}5970 \pm 0{\cdot}0025$	$2{\cdot}1902 \pm 0{\cdot}0025$	MW.	7
SF_5OF	$1{\cdot}53 \pm 0{\cdot}04$	$1{\cdot}66 \pm 0{\cdot}05$	E.D.	8
SF_5OOSF_5	$1{\cdot}56 \pm 0{\cdot}02$	$1{\cdot}66 \pm 0{\cdot}05$	E.D.	9

E.D.: Electron diffraction.
MW.: Microwave spectroscopy

TABLE 18.2. *Bond distances in sulphur(VI) compounds*

Compound	Bond length d(S—F), Å	Bond length d(S—R), Å	Description	Reference
SO_2F_2	$1{\cdot}530 \pm 0{\cdot}003$	1·405 (S═O)	Distorted tetrahedron	10
SOF_4	1·58 (polar) 1·55 (eq)	1·41 (S═O)	Distorted trigonal bipyramid	11
NSF_3	1·552	1·416 (S≡N)	Distorted tetrahedron	12

precise microwave results for SF_5Cl and SF_5Br do indicate an increase in the S—F distance from SF_5Cl to SF_5Br, as predicted by the 'd orbital' model.

The S—F bonds in SOF_4 are similar in length to those in SF_6 and R—SF_5. The S—F distance in SO_2F_2 is significantly shorter than in SF_6; similarly the S—O distance is shorter than in SO_3. This may be compared with the fact that the reaction

$$SF_6 + 2SO_3 \rightarrow 3SO_2F_2$$

proceeds in the direction shown, which implies that the bond energies in SO_2F_2 are higher than the corresponding energies in SF_6 and SO_3.

(*c*) *Vibrational spectra*

The vibrational spectra of SF_6 and some simple SF_5X derivatives have been studied in detail while infrared spectra have been recorded for numerous $SF_5.R$ compounds, and some simple correlations have been made.

The infrared[13] and Raman[14] spectra of SF_6 were the first to be studied on the grounds of both simplicity and availability. The SF_6 molecule has symmetry O_h and the assignment of the observed frequencies to the normal modes of vibration is shown in Table 18.3.

TABLE 18.3. *Assignment of the vibrational spectra of SF_6, point group* O_h

Number	Symmetry	Activity	Frequency, cm^{-1}	Approximate description
ν_1	a_{1g}	R (pol)	775	S—F stretching
ν_2	e_g	R	644	S—F stretching
ν_3	f_{1u}	i.r.	932	S—F stretching
ν_4	f_{1u}	i.r.	613	SF_4-square out-of-plane deformation
ν_5	f_{2g}	R	524	SF_4-square in-plane deformation
ν_6	f_{2u}	Inactive	344	F–S–F wagging

R: Raman-active.
i.r.: infrared-active.

The only other SF_5 compounds for which both infrared and Raman spectra have been observed are SF_5Cl[15] and S_2F_{10},[16] for which assignments have been made on the basis of the point group C_{4v} for SF_5Cl and D_{4d} for S_2F_{10}. These assignments are shown in tables 18.4 and 18.5. For SF_5Cl it has been possible to make a complete assignment of observed frequencies to the fundamental modes of vibration of a monosubstituted octahedron. The assignment given for S_2F_{10} is that of Dodd, Woodward and Roberts and differs in detail from that of Bernstein and Wilmshurst. It has the merit of being consistent with the assignment of Te_2F_{10}, which was not available to Bernstein and Wilmshurst, and with that of SF_5Cl which was observed only at a later date.

The infrared spectra of a number of R.SF_5 compounds, where R = alkyl or perfluoroalkyl, all show strong bonds at 580–600 cm^{-1} and very strong bands at 860–910 cm^{-1}, which are diagnostic for this structural type.[17] For Ar.SF_5 (Ar = aryl) characteristic strong absorptions are found in the range 820–880 cm^{-1}.[18]

The infrared and Raman spectra of SOF_4 have been observed[19] and tentatively assigned on the basis of the point group C_{2v} as the more symmetric possibilities C_{3v} and C_{4v} were not consistent with the observed spectra. This symmetry has subsequently been shown to be correct both by electron diffraction and n.m.r. studies.

(d) N.m.r. spectra

Nuclear magnetic resonance spectroscopy has proved a powerful structural tool in the characterization of SF_5R and $R'R''SF_4$ derivatives. The SF_5 group contains two types of fluorine atom. A single fluorine atom is *trans* to the group to which the SF_5 group is bonded and the other four are in a plane at 90° to the F—S—R axis. This leads to an AB_4 pattern which is observed in the ^{19}F n.m.r. spectra of all SF_5 compounds. For

TABLE 18.4. *Assignment of the vibrational spectra of SF_5Cl, point group* C_{4v}

Number	Symmetry	Activity		Frequency, cm^{-1} i.r.[a]	R	Approximate description
ν_1	a_1	i.r.	R (p)	854	834 (p)	Axial S—F stretching
ν_2	a_1	i.r.	R (p)	706	703 (p)	SF_4-square stretching
ν_3	a_1	i.r.	R (p)	602	599 (p)	SF_4 out-of-plane deformation
ν_4	a_1	i.r.	R (p)		404 (p)	S—Cl stretching
ν_5	b_1	–	R (dp)		624	SF_4-square stretching
ν_6	b_1	–	R (dp)		396	SF_4 out-of-plane deformation
ν_7	b_2	–	R (dp)		504	SF_4 in-plane deformation
ν_8	e	i.r.	R (dp)	908	916	SF_4-square stretching
ν_9	e	i.r.	R (dp)	578	579	SF wag
ν_{10}	e	i.r.	R (dp)		422	SF_4 in-plane deformation
ν_{11}	e	i.r.	R (dp)		270	S—Cl wag

[a] i.r. frequencies below 550 cm^{-1} not observed for instrumental reasons.
R: Raman-active.
i.r.: infrared-active.
p: polarized.
dp: depolarized.

$R'R''SF_4$ there are two possibilities. Firstly the R′—S—R″ unit is linear, in which case there are four identical fluorine atoms. Secondly R′—S and R″—S are at 90°, in which case there are two similar fluorine atoms at 90° to both R′ and R″ and two non-equivalent fluorine atoms *trans* to R′ and R″ respectively. No examples of R_3SF_3 or more highly substituted compounds are known. For compounds where the SF_5 group is bonded to carbon n.m.r. spectra have been studied by several workers. Their results on the AB coupling constant moduli and chemical shifts for the SF_5 group, summarized by Barlow, Dean and Lee,[20] are shown in Table 18.6.

TABLE 18.5. *Assignment of the vibrational spectra of S_2F_{10}, point group D_{4d}*[16]

Number	Symmetry	Activity	Frequency, cm^{-1}	Approximate description
ν_1	a_1	R (p)	913 (p)	SF axial stretching
ν_2	a_1	R (p)	690 (p)	SF_4-square stretching
ν_3	a_1	R (p)	Not observed	
ν_4	a_1	R (p)	247 (p)	S—S stretching
ν_5	b_1	Inactive	—	
ν_6	b_2	i.r.	938	SF axial stretching
ν_7	b_2	i.r.	684	SF_4-square stretching
ν_8	b_2	i.r.	571	SF_4 out-of-plane deformation
ν_9	e_1	i.r.	826	SF_4-square stretching
ν_{10}	e_1	i.r.	544	SF axial wag
ν_{11}	e_1	i.r.	410	SF_4 out-of-plane deformation
ν_{12}	e_1	i.r.	Not observed	
ν_{13}	e_2	R (dp)	728	SF_4-square stretching
ν_{14}	e_2	R (dp)	624	SF_4 out-of-plane deformation
ν_{15}	e_2	R (dp)	509	SF_4 in-plane deformation
ν_{16}	e_3	R (dp)	860	SF_4-square stretching
ν_{17}	e_3	R (dp)	634	SF axial wag
ν_{18}	e_3	R (dp)	425	SF_4 out-of-plane deformation
ν_{19}	e_3	R (dp)	188	SF_5 rocking

R: Raman-active.
i.r.: infrared-active.
p: polarized.
dp: depolarized.

Muller, Lauterbur and Svatos[21] have studied several $R'R''SF_4$ compounds. In $(C_2F_5)_2SF_4$ the resonance due to the $—SF_4—$ group is a single line with a chemical shift of −104 ppm relative to CF_3COOH implying a *trans* arrangement of the C_2F_5 groups. In the molecule

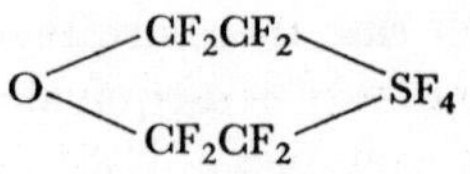

where the ring forces a *cis* arrangement, resonances are observed with chemical shifts of −123 and −98 p.p.m., consistent with *cis*-R_2SF_4. No examples of simple *cis*-$R'R''SF_4$ molecules are known.

A study of the n.m.r. spectra of SF_5 compounds where the group is bonded to elements other than carbon has been undertaken by Cady and his co-workers,[22] who were also responsible for many of the preparative reactions in this field. The data obtained are presented in Table 18.7. In compiling this table some data from reference (20) are also used and, for the sake of

TABLE 18.6. *Chemical shifts and* AB *coupling constant moduli for the SF_5 group in alkyl- and alkoxysulphur pentafluorides*[20]

Compounds	J_{AB} Hz	$\delta(SF_A)$ p.p.m.[a]	$\delta(SF_B)$ p.p.m.[a]
$CF_2BrCHFSF_5$	145·2	−148·0	−132·2
$CF_2ClCHFSF_5$	145·3	−146·6	−130·3
$CF_2BrCFBrSF_5$	145·9	−145·6	−131·0
$CF_2ClCFClSF_5$	145·6	−144·3	−128·2
$CFCl_2CF_2SF_5$	145·4	−142·6	−126·1
$MeOCF_2CHFCHFSF_5$	145·5	−149·2	−129·6
$CHFClCF_2SF_5$	143·4	−143·0	−119·8
$CF_3CF_2CFISF_5$	146·2	−147·1	−131·4
$CF_2ClCFBrSF_5$	146·5	−145·8	−130·4
$CF_2BrCFClSF_5$	145·6	−145·2	−130·0
$CF_3CF_2SF_5$	145·0	−137·0	−118·0
	152·2	−138·0	−118·5
$CF_3CF_2CF_2SF_5$	?	−137·5	−118·7
$CF_3CF_2CF_2CF_2SF_5$	146·0	−136·9	−119·5
CF_3SF_5	145·4	−139·0	−115·2
$CF_2ClCH_2SF_5$	144·4	−157·7	−140·9
$CH_3CHClCH_2SF_5$	144·2	−158·8	−141·5
$CH_2ClCH_2CH_2CH_2CH_2SF_5$	143·6	−160·6	−139·3

[a] p.p.m. to high field of CF_3CO_2H (ext.); 76·6 and 133·6 p.p.m. subtracted when referenced with $CFCl_3$ (int.) and SF_6 (int.), respectively.

comparability, the chemical shifts are adjusted to CF_3COOH as the reference by subtraction of 133·6 p.p.m. from the values referred to SF_6 as standard. As in the case of the SF_5—C compounds, all the J_{AB} values are close to 150 Hz.

One of the most interesting results of n.m.r. spectroscopy has been to establish that in the products of the reactions[23–26]

$$SF_4 + FSO_2.O.O.SO_2F \rightarrow (FSO_2O)SF_4$$
$$SF_4 + SF_5OOSF_5 \rightarrow (SF_5O)_2SF_4$$
$$SF_4 + CF_3OOCF_3 \rightarrow (CF_3O)_2SF_4$$
$$2SF + CF_3OOCF_3 + N_2F_4 \rightarrow 2CF_3O.SF_4.NF_2$$

the —SF_4 group is an A_2B_2 spin system. This suggests that addition of the bulky groups has proceeded so as to give a *cis*-disubstituted octahedron rather than the expected *trans* form. It would have been difficult to demonstrate this by other structural methods.

TABLE 18.7. *Chemical shifts and* AB *coupling constant moduli for SF_5 bonded to elements other than carbon*

Compound	J_{AB}, Hz	$\delta(SF_A)$[a]	$\delta(SF_B)$[a]
SF_5Br	143·1	−138·6	−221·8
SF_5Cl	148·5	−137·4	−202·0
SF_5OSO_2F	153·5	−132·0	−148·4
$(SF_5O)_2SO_2$	153·4	−133·2	−148·7
$(SF_5O)_2SF_4$	156	—	—
SF_5OSF_5	150	−138·1	−147·2
SF_5OOSF_5	—	−135·6	−134·4
CF_3OSF_5	153·0	−137·8	−146·1
$C_2F_5OSF_5$	152·8	−136·9	−148·8
$C_4F_9OSF_5$	154·9	−137·6	−148·3
$CFH_2CH_2OSF_5$	153·8	−151·6	−136·1
$FC_2H_4OSF_5$	153·8	−151·6	−136·1
SF_5OCl	161·5	—	—

[a] δ values relative to CF_3CO_2H.

C. Tetravalent sulphur fluorides

(*a*) *Theoretical*

In compounds where sulphur is bonded to four monovalent atoms or groups, there are 10 valence electrons to be accommodated on the sulphur atom. As in the case of hexavalent sulphur compounds, it is possible to postulate either d-orbital participation (sp^3d) or three-centre bonds using only 3p orbitals from sulphur in the bonding. The model involving d-orbital participation is used here.

As the tetravalent sulphur is co-ordinatively unsaturated, it is vulnerable to nucleophilic attack. This makes tetravalent sulphur fluorides susceptible to hydrolysis and much more difficult to handle. Also it is probable that SF_3X compounds, where X is a halogen or pseudo-halogen, disproportionate, and this restricts the range of compounds available. As a consequence there have been far fewer structural determinations in the sulphur(IV) than in the sulphur(VI) series.

(*b*) *Bond distances*

Bond distances for SF_4, SOF_2 and NSF have been obtained from the microwave spectra of these compounds. The bond distances are shown in Table 18.8. The structure of sulphur tetrafluoride is depicted in Figure 18.1.

The mean S—F bond length in SF_4 is approximately equal to that in SOF_2 and SSF_2. Tolles and Gwinn[27] describe the structure of SF_4 as being

TABLE 18.8. *Bond distances in sulphur(IV) fluorides*

Compound	d(S—F), Å	d(S—X), Å	Description	References
SF_4	1·646 ± 0·003		Distorted trigonal bipyramid[a]	27
	1·545 ± 0·003			
SOF_2	1·585 ± 0·001	1·412 ± 0·001	Distorted tetrahedron[a]	28
SSF_2	1·598 ± 0·012	1·860 ± 0·015	Distorted tetrahedron[a]	29
NSF	1·646	1·446	Bent	30

[a] One vertex of polyhedron occupied by lone-pair of electrons.

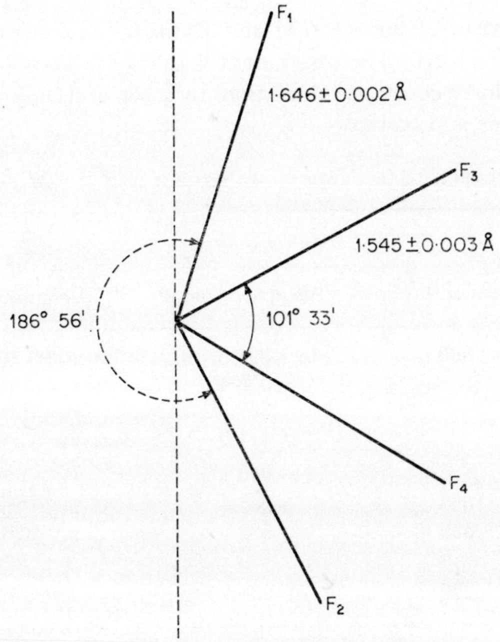

FIGURE 18.1. *Structure of sulphur tetrafluoride.*

intermediate between the sp^3d structure which predicts a trigonal bipyramid and the form using only 3p orbitals to form four bonds. The short equatorial bonds and the lone pair in the same plane can be visualized as involving sp^2 hybrid orbitals while the two longer polar bonds involve pd hybrid orbitals, the weaker bonding being due to the participation of the high-energy d orbital. Despite these deviations from a simple trigonal bipyramid, the overall symmetry remains C_{2v}, as indicated by studies of the infrared, Raman and n.m.r. spectra.

(c) Vibrational spectra

The infrared and Raman spectra of SF_4[31] provided the first structural information about the molecule. These studies indicated the symmetry C_{2v} as the number of bands observed was too great for the alternative, more symmetric forms. More recently a re-examination of the infrared spectrum of gaseous SF_4 has led to a revision of the original assignment of the fundamental vibrations; the details of the revised assignment are given in Table 18.9. Spectroscopic studies of SF_4 trapped in an argon matrix at ~4 K indicate that in the condensed phase the molecules suffer association, a conclusion which may account for some features of the Raman spectrum of the liquid which appear not to be characteristic of monomeric SF_4 molecules.[31]

The assignment of the infrared and Raman frequencies for SOF_2 are shown in Table 18.10. The assignment is that of O'Loane and Wilson,[32] supported by force constant calculations by Long and Bailey[33] rather than that of Gillespie and Robinson.[34]

TABLE 18.9. *Raman and infrared spectra of SF_4*[31]

Frequencies, cm^{-1}		Assignment	
Raman, liquid	Infrared, vapour		
898 p	891·5	ν_1	a_1
	558·4	ν_2	a_1
	353	ν_3	a_1
	171·0	ν_4	a_1
	645 (?)	ν_5	a_2
858 dp	867·0	ν_6	b_1
536 dp	532·2	ν_7	b_1
	728	ν_8	b_2
~239	226	ν_9	b_2

p: polarized.
dp: depolarized.

TABLE 18.10. *Vibrational spectra of SOF_2*

Frequencies, cm^{-1}			
Infrared, vapour	Raman, liquid	Assignment	
1333	1312	ν_1	a′
808	795	ν_2	a′
530	529	ν_3	a′
(410)	326	ν_4	a′
748	720	ν_5	a″
390	395	ν_6	a″

(*d*) *N.m.r. spectra*

Studies of the n.m.r. spectra of sulphur tetrafluoride have been made by Cotton, George and Waugh[35] and by Muetterties and Phillips.[36] At 23°C the ^{19}F spectrum consists of a single broad line but at −85°C this is resolved into two triplets thus indicating C_{2v} symmetry for the SF_4 molecule. By studying the temperature dependence of the ^{19}F spectrum it was possible to give a value of $4{\cdot}5 \pm 0{\cdot}8$ kcal for the activation energy of fluorine exchange between the two types of fluorine atom. It has been suggested that this takes place through fluorine-bridged dimers, a view consistent with the behaviour of matrix-isolated SF_4 molecules.[31]

Rosenberg and Muetterties[40] have also studied the ^{19}F spectra of $(CF_3)_2CF.SF_3$ and $[(CF_3)_2CF]_2SF_2$ and the results are consistent with substitution in the SF_4 structure.

Chemical shifts relative to $CFCl_2CFCl_2$ are:

	$>SF_2$, p.p.m.	$>\!\!-SF$, p.p.m.
$[(CF_3)_2CF]_2SF_2$	−56·9	
$(CF_3)_2CFSF_2CF_3$	−53·8	
$(CF_3)_2CFSF_3$	−129	−13·5

D. Divalent sulphur fluorides

Only two compounds of divalent sulphur with S—F bonds have been adequately characterized. These are FSSF and i-C_3F_7SF. The highly reactive nature of these compounds makes them very difficult to study. The very confused earlier literature contained many conflicting reports of the nature and reactivity of the fluorides of divalent sulphur. The position was first clarified by an elegant use of microwave spectroscopy. Kuczkowski[37] caused sulphur to react with silver fluoride and studied the microwave spectrum of the gaseous product. He did not obtain a single pure compound but was able to use the high resolution of his spectra to pick out features due to FSSF, SSF_2 and SOF_2. These he was able to assign completely and the results were used to obtain the molecular dimensions of FSSF and SSF_2. With this firm evidence of the existence of the isomeric pair FSSF and SSF_2 other workers[38, 39] were able to isolate and characterize the two compounds individually.

Disulphur difluoride has a hydrogen peroxide-type structure. The S—S—F bond angle is $108 \pm 0{\cdot}5°$ and the dihedral angle defined by the F—S—S and F′—S—S planes is $87{\cdot}9 \pm 1{\cdot}5°$. The bond lengths in S_2F_2 are (i) S—S, $1{\cdot}88 \pm 0{\cdot}01$ Å and (ii) S—F, $1{\cdot}635 \pm 0{\cdot}01$ Å. These may be compared with (i) S—S in S_2 of 1·88 Å and (ii) S—F in SF_6 of 1·56 Å. The S—S bond is therefore close to a double bond while the S—F bond is comparable to the weaker bonds in SF_4 (d(S—F) = 1·646 Å). The position is similar to that of the structure of FOOF where the O—O distance is close to that of the O_2 molecule; in marked contrast, the O—O distance in SF_5OOSF_5 is comparable with that in hydrogen peroxide and has the characteristics of a single bond.

The only other definitely established sulphur(II) fluorides are $(CF_3)_2CF.SF$, the structure of which was proved by ^{19}F n.m.r. spectroscopy,[40] and $CF_nCl_{3-n}.SF$ (n = 0–3).[41] It is of note that the SF fluorine resonance of $(CF_3)_2CF.SF$ is moved upfield by 422 p.p.m. relative to the SF_2 resonance of $(CF_3)_2CF.SF_3$;[40] this remarkably high chemical shift appears to be characteristic of sulphenyl fluorides.[41]

References

1. Roberts, H. L., 'Inorganic Sulphur Chemistry', ed. G. Nickless, Elsevier, Amsterdam, 1968, p. 419.
2. Cruickshank, D. W. J., and Webster, B. C., 'Inorganic Sulphur Chemistry', ed. G. Nickless, Elsevier, Amsterdam, 1968, p. 7.
3. Braune, H., and Knoke, S., *Z. phys. Chem. (Frankfurt)*, 1933, **B21**, 297.
4. Brockway, L. O., and Pauling, L., *Proc. Nat. Acad. Sci. U.S.A.*, 1933, **19**, 68.
5. Harvey, R. B., and Bauer, S. H., *J. Amer. Chem. Soc.*, 1953, **75**, 2840.
6. Kewley, R., Murty, K. S. R., and Sugden, T. M., *Trans. Faraday Soc.*, 1960, **56**, 1732.
7. Neuvar, E. W., and Jache, A. W., *J. Chem. Phys.*, 1963, **39**, 596.
8. Crawford, R. A., Dudley, F. B., and Hedberg, K. W., *J. Amer. Chem. Soc.*, 1959 **81**, 5287.
9. Harvey, R. B., and Bauer, S. H., *J. Amer. Chem. Soc.*, 1954, **76**, 859.

10. Fristrom, R. M., *J. Chem. Phys.*, 1952, **20**, 1.
11. Hencher, J. L., Cruickshank, D. W. J., and Bauer, S. H., *J. Chem. Phys.*, 1968, **48**, 518.
12. Kirchoff, W. H., and Wilson, E. B., *J. Amer. Chem. Soc.*, 1962, **84**, 334.
13. Lagemann, R. T., and Jones, E. A., *J. Chem. Phys.*, 1951, **19**, 543.
14. Gaunt, J., *Trans. Faraday Soc.*, 1953, **49**, 1122.
15. Cross, L. H., Roberts, H. L., Goggin, P. L., and Woodward, L. A., *Trans. Faraday Soc.*, 1960, **56**, 945.
16. Dodd, R. E., Woodward, L. A., and Roberts, H. L., *Trans. Faraday Soc.*, 1957, **53**, 1545; Wilmshurst, J. K., and Bernstein, H. J., *Canad. J. Chem.*, 1957, **35**, 191.
17. Cross, L. H., Cushing, G. H., and Roberts, H. L., *Spectrochim. Acta*, 1961, **17**, 344.
18. Sheppard, W. A., *J. Amer. Chem. Soc.*, 1962, **84**, 3064.
19. Goggin, P. L., Roberts, H. L., and Woodward, L. A., *Trans. Faraday Soc.*, 1961, **57**, 1877.
20. Barlow, M. G., Dean, R. R., and Lee, J., *Trans. Faraday Soc.*, 1969, **65**, 321.
21. Muller, N., Lauterbur, P. C., and Svatos, G. F., *J. Amer. Chem. Soc.*, 1957, **79**, 1043.
22. Merrill, C. I., Williamson, S. M., Cady, G. H., and Eggers, D. F., Jr., *Inorg. Chem.*, 1962, **1**, 215.
23. Merrill, C. I., and Cady, G. H., *J. Amer. Chem. Soc.*, 1963, **85**, 909.
24. Duncan, L. C., and Cady, G. H., *Inorg. Chem.*, 1964, **3**, 850.
25. Duncan, L. H., and Cady, G. H., *Inorg. Chem.*, 1964, **3**, 1045.
26. Shreeve, J. M., and Cady, G. H., *J. Amer. Chem. Soc.*, 1961, **83**, 4524.
27. Tolles, W. M., and Gwinn, W. D., *J. Chem. Phys.*, 1962, **36**, 1118.
28. Ferguson, R. C., *J. Amer. Chem. Soc.*, 1954, **76**, 850.
29. Kuczkowski, R. L., *J. Amer. Chem. Soc.*, 1964, **86**, 3617.
30. Kirchoff, W. H., and Wilson, E. B., *J. Amer. Chem. Soc.*, 1963, **85**, 1726.
31. Dodd, R. E., Roberts, H. L., and Woodward, L. A., *Trans. Faraday Soc.*, 1956, **52**, 1052; Levin, I. W., and Berney, C. V., *J. Chem. Phys.*, 1966, **44**, 2557; Redington, R. L., and Berney, C. V., *ibid.*, 1965, **43**, 2020.
32. Bender, P., and Wood, J. M., *J. Chem. Phys.*, 1955, **23**, 1316; O'Loane, J. K., and Wilson, M. K., *ibid.*, p. 1313.
33. Long, D. A., and Bailey, R. T., *Trans. Faraday Soc.*, 1963, **59**, 792.
34. Gillespie, R. J., and Robinson, E. A., *Canad. J. Chem.*, 1961, **39**, 2171.
35. Cotton, F. A., George, J. W., and Waugh, J. S., *J. Chem. Phys.*, 1958, **28**, 994.
36. Muetterties, E. L., and Phillips, W. D., *J. Amer. Chem. Soc.*, 1959, **81**, 1084.
37. Kuczkowski, R. L., *J. Amer. Chem. Soc.*, 1964, **86**, 3617.

38. Brown, R. D., Pez, G. P., and O'Dwyer, M. F., *Austral. J. Chem.*, 1965, **18**, 27.
39. Seel, F., Budenz, R., and Werner, D., *Chem. Ber.*, 1964, **97**, 1369.
40. Rosenberg, R. M., and Muetterties, E. L., *Inorg. Chem.*, 1962, **1**, 756.
41. Seel, F., Gombler, W., and Budenz, R., *Angew. Chem.*, *Internat. Edn.*, 1967, **6**, 706; Seel, F., and Gombler, W., *ibid.*, 1969, **8**, 773.

Index